Summary of FORTRAN Statements

KEYWORD	EXAMPLE	EXPLANATION OF THE EXAMPLES
BLOCK DATA	BLOCK DATA	Start of a subroutine which initializes the variables in a labeled COMMON statement.
CALL	CALL SUB(A,B)	Transfers control to the subroutine named SUB.
CHARACTER	CHARACTER A*12	Defines A to be character type and a maximum of 12 characters long.
CLOSE	CLOSE(5,ERR=50)	Closes files associated with unit 5.
COMMON	COMMON A,B	Causes the variables A and B to be accessible to both the main program and the subprograms.
COMPLEX	COMPLEX A,B	Defines A and B to be complex variables.
CONTINUE	10 CONTINUE	The last statement of a DO loop.
DATA	DATA PIE/3.14/	Initializes the value of the variable PIE.
DIMENSION	DIMENSION A(10)	Defines A to be an array containing 10 elements.
DO	DO 10 I=1,30	All statements from here to statement 10 are to be repeated thirty times.
DOUBLE PRECISION	DOUBLE PRECISION A	Defines A to be double precision.
END	END	The final statement of a program.
EQUIVALENCE	EQUIVALENCE(A,B)	Allows A and B to share the same storage unit.
EXTERNAL	EXTERNAL FUN1	Allows the function name FUN1 to be an argument of a function or subroutine.
FORMAT	FORMAT(1X,I7,5X,F5.2) FORMAT(A10) FORMAT(E12.5) FORMAT(D10.4) FORMAT(L4)	Indicates the type of the input or output: integer (I), real (F), character (A), exponential (E), double precision (D), or logical (L) data.
FUNCTION	FUNCTION SUM(X,Y,Z)	The first statement of the function SUM.
GO TO	GO TO 10	Jump to statement 10.
	GO TO (10,20,30),IA	Jump to statement 10, 20, or 30 if IA = 1, 2, or 3, respectively.
IF	IF(A .EQ. B)GOTO 100	If the condition inside the parentheses is satisfied, execute the imperative statement; otherwise ignore it.

(continued on page 550)

PROGRAMMING BYTE BY BYTE

Little, Brown Computer Systems Series

Gerald M. Weinberg, *Editor*

Barnett, Michael P., and Graham K. Barnett
Personal Graphics for Profit and Pleasure on the Apple II Plus Computer

Basso, David T., and Ronald D. Schwartz
Programming with FORTRAN/WATFOR/WATFIV

Chattergy, Rahul, and Udo W. Pooch
Top-down, Modular Programming in FORTRAN with WATFIV

Coats, R. B., and A. Parkin
Computer Models in the Social Sciences

Conway, Richard, and David Gries
An Introduction to Programming: A Structured Approach Using PL/I and PL/C, Third Edition

Conway, Richard, and David Gries
Primer on Structured Programming: Using PL/I, PL/C, and PL/CT

Conway, Richard, David Gries, and E. Carl Zimmerman
A Primer on Pascal, Second Edition

Cripps, Martin
An Introduction to Computer Hardware

Easley, Grady M.
Primer for Small Systems Management

Finkenaur, Robert G.
COBOL for Students: A Programming Primer

Freedman, Daniel P., and Gerald M. Weinberg
Handbook of Walkthroughs, Inspections, and Technical Reviews: Evaluating Programs, Projects, and Products, Third Edition

Graybeal, Wayne, and Udo W. Pooch
Simulation: Principles and Methods

Greenfield, S. E.
The Architecture of Microcomputers

Greenwood, Frank
Profitable Small Business Computing

Healy, Martin, and David Hebditch
The Microcomputer in On-Line Systems: Small Computers in Terminal-Based Systems and Distributed Processing Networks

Lemone, Karen A., and Martin E. Kaliski
Assembly Language Programming for the VAX-11

Lias, Edward J.
Future Mind: The Microcomputer—New Medium, New Mental Environment

Lines, M. Vardell, and Boeing Computer Services Company
Minicomputer Systems

Mashaw, Bijan
Programming Byte by Byte: Structured FORTRAN 77

Mills, Harlan D.
Software Productivity

Monro, Donald M.
Basic BASIC: An Introduction to Programming

Morrill, Harriet
Mini and Micro BASIC: Introducing Applesoft, Microsoft, and BASIC Plus

Mosteller, William S.
Systems Programmer's Problem Solver

Nahigian, J. Victor, and William S. Hodges
Computer Games for Businesses, Schools, and Homes

Nahigian, J. Victor, and William S. Hodges
Computer Games for Business, School, and Home for TRS-80 Level II BASIC

Orwig, Gary W., and William S. Hodges
The Computer Tutor: Learning Activities for Homes and Schools

Parikh, Girish
Techniques of Program and System Maintenance

Parkin, Andrew
Data Processing Management

Parkin, Andrew
Systems Analysis

Pizer, Stephen M., with Victor L. Wallace
To Compute Numerically: Concepts and Strategies

Pooch, Udo W., William H. Greene, and Gary G. Moss
Telecommunications and Networking

Reingold, Edward M., and Wilfred J. Hansen
Data Structures

Savitch, Walter J.
Abstract Machines and Grammars

Schneiderman, Ben
Software Psychology: Human Factors in Computer and Information Systems

Simpson, Tom, and Shaffer & Shaffer Applied Research & Development, Inc.
Visicalc Programming: No Experience Necessary

Walker, Henry M.
Problems for Computer Solutions Using FORTRAN

Walker, Henry M.
Problems for Computer Solutions Using BASIC

Weinberg, Gerald M.
Rethinking Systems Analysis and Design

Weinberg, Gerald M.
Understanding the Professional Programmer

Weinberg, Gerald M., Stephen E. Wright, Richard Kauffman, and Martin A. Goetz
HIgh Level COBOL Programming

Windeknecht, Thomas G.
6502 Systems Programming

PROGRAMMING BYTE BY BYTE
STRUCTURED FORTRAN 77

Bijan Mashaw

Little, Brown and Company
Boston   Toronto

TO MY
FAMILY

Library of Congress Cataloging in Publication Data

Mashaw, Bijan.
   Programming byte by byte.

   (Little, Brown computer systems series)
   1. FORTRAN (Computer program language)
2. Structured programming. I. Title. II. Series.
QA76.73.F25M39   1982   001.64′24   82-24980
ISBN 0-316-54908-8

Copyright © 1983 by Bijan Mashaw

All rights reserved. No part of this book may be reproduced in any form or by any electronic or mechanical means including information storage and retrieval systems without permission in writing from the publisher, except by a reviewer who may quote brief passages in a review.

Library of Congress Catalog Card No. 82-24980

ISBN 0-316-54908-8

9 8 7 6 5 4 3 2 1

MV

Published simultaneously in Canada by
Little, Brown & Company (Canada) Limited

Printed in the United States of America

Disclaimer of Liabilities: Due care has been exercised in the preparation of this book to insure its effectiveness. The author and publisher make no warranty, expressed or implied, with respect to the programs or other contents of this book. In no event will the author or publisher be liable for direct, indirect, incidental, or consequential damages in connection with or arising from the furnishing, performance, or use of this book.

Foreword

I wrote a book about FORTRAN in 1966. The reviewers were not shy about mentioning flaws in the work, but I can't recall that any reviewer questioned the need for another FORTRAN text. After two decades, however, *every* review of a FORTRAN text starts with the same question: "Why do we need another one?"

It's a fair question. And since I'm not the author of this book, I believe I can give a fair answer. Put simply, *Programming Byte by Byte: Structured FORTRAN 77* is a beautifully crafted, carefully planned, step-by-step teaching tool that has obviously been tested in real classes with real students. That may sound like heavy praise, but it's praise that any experienced teacher reading two or three pages can immediately verify.

Those who read more deeply will be pleased to discover something even more praiseworthy about *Programming Byte by Byte*—the aptness and effectiveness of its many examples. I was so struck by this feature that I queried Professor Mashaw, who replied by sending a copy of his paper, "An Inductive Method for Teaching Computer Programming." The paper explains why programming is best taught by induction—by moving the student up a ladder of carefully chosen examples to a high level of both practical ability and cognitive understanding.

To some people, induction means learning by random trial and error. That sort of induction not only wastes time, but locks many bad practices into the student's repertoire. On the other hand, some texts have overreacted to such unstructured induction by taking a rarified and theoretical deductive approach. Although such an approach can be effective for advanced students, I believe that a well-structured inductive method is the proper way to *introduce* programming, whether in a classroom or for individual study. I know of no other book that implements this inductive approach with the consistent workmanship of Mashaw's *Programming Byte by Byte: Structured FORTRAN 77*.

Gerald M. Weinberg

Preface

About FORTRAN

FORTRAN was one of the first higher-level computer languages. Subsequent to its successful introduction in 1957, over 200 additional languages have been developed. Yet, FORTRAN remains as powerful as ever and is still one of the most widely used computer languages. FORTRAN compilers can be found on almost any computer system. FORTRAN is also used as a vehicle for teaching computer programming because it is powerful, flexible, and easy to learn. Mastery of its programming patterns facilitates learning other computer languages.

The Unique Approaches in This Text

The purpose of this text is to teach FORTRAN in the most efficient manner possible. The approach has been designed specifically for readers who have had no previous instruction in programming. The method is inductive; the reader learns a computer language by examples. Each example has been carefully designed to introduce the reader to the most important underlying concepts. After spending a few minutes on an example, the reader should be able to understand and apply the concepts illustrated. From the beginning of the text, the reader is able to write a program.

The Organization of the Text

Each section starts with an example, followed by several important points which illustrate the learning objective for that example. The examples are designed to help the reader understand the progression of a program's logic, structure, and techniques. Next, the reader is presented with solved problems to help reinforce the learning process. The solved problems illustrate some of the common mistakes a novice programmer might make. Finally, exercises provide practice to help master the fundamentals presented within each section. For most of the text, no knowledge of mathematics beyond basic algebra is required. Only Chapter 10 discusses the application of programs in mathematics.

Language syntax, techniques of problem solving, structured programming, modular approach, planning, and good programming practices are introduced gradually for ease of learning. Abundant exercises at the end of each chapter with a variety of applications challenge the student to practice what has been learned. They are arranged from simple to complex.

Topics Covered

Topics in the text are arranged to accommodate the induction method of delivery. Chapter 1 introduces the fundamentals of programming, program logic, and program writing. Format is discussed in Chapter 2, and is emphasized throughout the text because it is the most important feature of FORTRAN, distinguishing it from other programming languages such as BASIC. The format is presented in a clear enough fashion so that the programmer will find it easy to use. After learning (in chapters 1 and 2) what a program is and how to write a simple program in Chapter 3 the reader learns about how to run a program, program planning, and tools. In chapters 4 and 5 IF statements, structured-programming concepts, and loops are covered. DO loops are emphasized throughout the text. In Chapter 6 data types are introduced, and character-type data are discussed in detail. Chapters 7 and 8 present arrays in a manner that eliminates the usual fear many novice programmers experience about arrays. Chapter 9 explains the importance of subprograms and the techniques of subprogramming. Chapter 10 discusses the program efficiency and sample algorithms which are not straightforward. Further applications of programs in data processing, mathematics, statistics, graphing, and simulation are also presented in this chapter. Chapter 11 further discusses the data type (double precision, logical, and complex) and additional features of FORTRAN statements. Finally, in Chapter 12, structured programming and writing better programs are discussed.

Why FORTRAN 77?

Because standard FORTRAN 77 is becoming widely used, the most important feature of this compiler (or any compiler which supports the CHARACTER type data, IF block, and format-free I/O) is covered in the text. However, whenever it is not possible to use a feature with other compilers, this is noted.

Alternative Ways of Using This Text

Although this text covers some rather broad topics, it remains flexible enough to be utilized in a variety of ways. It can be readily used as a text for a three-hour course on FORTRAN. Chapters 1 through 7 can be used for a one-hour introduction to programming. (The arithmetic IF, the computed GO TO in Chapter 3, and part of Chapter 6 can be eliminated for this purpose.) It could also serve as a supplementary text for a course in data processing. Finally, it will be an excellent self-teaching guide for individuals interested in learning FORTRAN.

Acknowledgments

I would like to express my thanks and appreciation to the many individuals who assisted me in undertaking this project. First, I would like to thank the editor of the series, Gerald Weinberg, for his support, suggestions, and encouragement. I gratefully acknowledge those who reviewed the manuscript and made numerous useful suggestions, including Syed Shahabuddin, Freeman Moore of Central Michigan University, Ron Schwartz of University of Akron, and especially Henry Walker of Grinnell College, who very carefully reviewed the entire text. However, as customary, I am responsible for any error. A special thanks to James Pahz who got me started and made innumerable editorial suggestions.

I express thanks to many other individuals who contributed to this project. In particular, I thank Annette Cook, who started the typing and craftsmanship of the manuscript and never gave up until the end, Sheila Dailey, Chris Christopherson, Darrel Martine, Edward Fisher, Carol Billingham, Bea Windgoston, Jeanine Sharland, Joy Pashee, Rick Eickler, Jim Mayes, and Larry Mateo. For their assistance, I am grateful.

Bijan Mashaw

Brief Contents

Chapter 1 **BASIC PROGRAMMING CONCEPTS** 1

Chapter 2 **INPUT–OUTPUT** 33

Chapter 3 **PROGRAMMING DEVELOPMENT AND EXECUTION** 81

Chapter 4 **DECISION MAKING, COMPARING, AND BRANCHING** 123

Chapter 5 **LOOPING** 177

Chapter 6 **DATA TYPES AND MORE ABOUT INPUT AND OUTPUT** 211

Chapter 7 **ONE-DIMENSIONAL ARRAYS** 263

Chapter 8 **MULTIDIMENSIONAL ARRAYS** 311

Chapter 9 **SUBPROGRAMS** 349

Chapter 10 **PROBLEM SOLVING AND PROGRAMMMING TECHNIQUES** 413

Chapter 11 **ADDITIONAL FEATURES OF FORTRAN** 463

Chapter 12 **STRUCTURED PROGRAMMING** 517

Contents

Chapter 1 BASIC PROGRAMMING CONCEPTS 1

INTRODUCTORY CONCEPTS   2

   Introduction   2
   Organization of the Text   2
   Computer Programs and Languages   3

WRITING PROGRAMS   4

   A Simple Program   4
   The PROGRAM Statement   5
   The PRINT Statement   5
   Arithmetic Operators   6
   Use of Parentheses in Arithmetic Operations   9
   Priorities of Arithmetic Operators   10
   Variables   12
      Variables and Constants   13
   The READ Statement   18
   Statement Labels   22
   Looping   22
   The GO TO Statement   23

xvii

xviii Contents

SOME PROGRAMMING TECHNIQUES   25

    Generating Sequence Numbers   25
    Calculating the Sum of Integer Numbers   26

FINAL NOTE   26

EXERCISES   28

SUMMARY OF CHAPTER 1   29

SELF-TEST REVIEW   29

ANSWERS TO SELECTED EXERCISES   32

Chapter 2 INPUT–OUTPUT   33

REAL AND INTEGER NUMBERS   34

    Numbers in FORTRAN   34
    Mixed Modes of Operation   35
    Field Descriptions   40

THE FORMAT STATEMENT: USING FIELD DESCRIPTORS   41

    Formatted READ Statement   41
    Formatted WRITE Statement   53
    Selecting a FIELD Descriptor   54
        The Use of Spaces   47
        Omitting the Decimal Point   53
        Blank Spaces in Data Fields   54
    Explanatory Comments in the Output   58
        Printing a Heading   61
    More Notes about FORMAT   63
    Formatted PRINT Statement   64

SOME PROGRAMMING TECHNIQUES   65

    The STOP Statement   65
    Controlling an Infinite Loop   65
    Generating Sequence Numbers   67
    Calculating a Sum   68

EXERCISES   72

PROGRAMMING EXERCISES   74

SUMMARY OF CHAPTER 2   77

SELF-TEST REVIEW   78

Chapter 3 PROGRAMMING DEVELOPMENT AND EXECUTION   81

RUNNING A PROGRAM   82

    Card-Oriented Batch Processing   82
        The Card Deck   83

On-Line Processing 83
Position of Instructions 84
   FORTRAN Coding Form 85
   The Print Chart 85
The Printout 86
Field, Record, File 86
Carriage Control 88
   The Slash 89
Documentation within the Program 91

THE COMPILERS 95

FORTRAN Compilers 97
FORTRAN Statements 98

ERROR DETECTION 98

Syntax Errors 98
Execution Errors 101
Logical Errors 101

PROGRAM PLANNING 102

Programming Tools 102
   Flowcharts 102
   Pseudocode 104
   Logic-Charts 105
Algorithms 105
Programming Cycle 107

A SAMPLE PROBLEM 109

EXERCISES 117

SUMMARY OF CHAPTER 3 120

SELF-TEST REVIEW 121

Chapter 4 DECISION MAKING, COMPARING, AND BRANCHING   123

THE LOGICAL IF STATEMENT 124

Important Notes about the Logical IF Statement 127

PROBLEMS REQUIRING THE IF STATEMENT 129

Finding the Largest Number 129
Terminating a Loop 132
   Header Record 132
   Trailer Record 133

PROGRAMMING STYLE AND STRUCTURED IF BLOCK 136

GO TO Style—Style One 137
GO TO-less Style—Style Two 138
Structured IF—Style Three 139
   Rules of Structured IF 140
   Advantages of Using Structured IF 142

CONDITIONAL BRANCHING   146

    Arithmetic IF Statement   146
    Computed GO TO Statement   148

A SAMPLE PROBLEM   153

EXERCISES   163

SUMMARY OF CHAPTER 4   170

SELF-TEST REVIEW   171

ANSWERS TO SELECTED EXCERCISES   175

Chapter 5 LOOPING   177

THE DO LOOP   178

    DO-Loop Flowcharting   181
    Rules of a DO Loop   181
      The Control Variable   181
      The CONTINUE Statement   185
      Normal and Abnormal Termination of a DO Loop   186
      Transfer within and from a DO Loop   188
      DO Loop and IF Block   189
      Summary of the Rules about DO Loops   189
    Nested DO Loops   195

EXERCISES   201

PROGRAMMING EXERCISES   203

SUMMARY OF CHAPTER 5   206

SELF-TEST REVIEW   206

ANSWERS TO SELECTED EXERCISES   209

Chapter 6 DATA TYPES AND MORE ABOUT INPUT AND OUTPUT   211

DATA TYPES   212

    REAL and INTEGER Statements   212
    Character Data   213
      CHARACTER Statement   213
      Character Variables   214
      Character Input and Output; A-field   215
      Character Constants   216
      Substring Reference   216
      Character Expressions   216
      Important Notes about Alphanumerics   217

DATA STATEMENT   225

OTHER FEATURES OF FORMAT   225

    Repeating Field Descriptors   226
        FORMAT Scanning   227
        T Descriptor   230
        H Descriptor   233
    Printing an Apostrophe   233
    Reading and Writing Literals   234
    The Slash   235
        Summary of Rules about the Slash   236

LIST-DIRECTED INPUT AND OUTPUT   240

    List-Directed Input   240
    List-Directed Output   241
    Important Notes about List-directed READ and WRITE   241

FORMATTED PRINT STATEMENT   242

EXERCISES   246

PROGRAMMING EXERCISES   253

SUMMARY OF CHAPTER 6   257

SELF-TEST REVIEW   258

Chapter 7 ONE-DIMENSIONAL ARRAYS   263

INTRODUCTION   264

SUBSCRIPTED VARIABLES   266

DIMENSION STATEMENT   267

REASON FOR ARRAYS: AN EXAMPLE   270

ARITHMETIC EXPRESSIONS WITH ARRAYS   272

ARRAY INPUT AND OUTPUT   276

    Array I/O with a DO Loop   277
    Implied DO Loop   278
    The Entire-List Method   281
    Mirror Printing   282

INITIALIZING THE ARRAY AND THE DATA STATEMENT   283

IMPORTANT NOTES ABOUT ARRAYS   289

SOME EXAMPLES OF ARRAY USE AND PROGRAMMING TECHNIQUES   290

PROGRAMMING EXERCISES   301

SUMMARY OF CHAPTER 7   306

SELF-TEST REVIEW   307

Chapter 8 MULTIDIMENSIONAL ARRAYS 311

INTRODUCTION   312

DIMENSION STATEMENT   313

ARITHMETIC EXPRESSIONS WITH MULTIDIMENSIONAL ARRAYS   314

ARRAY STORAGE   317

MULTIDIMENSIONAL ARRAY INPUT AND OUTPUT   317
   The Entire-List Method   320

THE DATA STATEMENT   321

SUMMARY OF THE IMPORTANT RULES   327

EXAMPLES OF TWO-DIMENSIONAL ARRAYS AND PROGRAMMING TECHNIQUES   327

EXERCISES   336

PROGRAMMING EXERCISES   338

SUMMARY OF CHAPTER 8   343

SELF-TEST REVIEW   343

Chapter 9 SUBPROGRAMS 349

INTRODUCTION   350

SUBROUTINES   350

   Structure of a Subroutine   350
     The Name of the Subroutine   352
     Actual Arguments and Dummy Arguments   352
     Completeness and Independence of Subroutines   354
     The RETURN Statement   356
     Subroutines and Arrays   356
     Adjustable Array Size   360
   Summary of the Important Points about Subroutines   361

FUNCTIONS   366

   Introduction   366
   Library Functions   367
   Function Statement   371
   Function Subprograms   375
     Function Name   377
     Arrays and Functions   378
       The Difference Between an Array and a Function   381
       Summary of the Important Points about Function Subprograms   381
   Comparing Functions with Subroutines   384

COMMON BLOCK 386
    Unlabeled COMMON Block 386
    Labeled COMMON Block 390

ORDER OF STATEMENTS 396

EXERCISES 396

PROGRAMMING EXERCISES 404

SUMMARY OF CHAPTER 9 408

SELF-TEST REVIEW 409

Chapter 10 PROBLEM SOLVING AND PROGRAMMING TECHNIQUES 413

INTRODUCTION 414

    Algorithms 414
    Program Efficiency 414
        Language Dependence 415
        Array Use 415
        Algorithm Design 416
        Statement Design 417
        Human Factors 418

DATA PROCESSING 419

    Sorting 419
    Searching 422
        Binary Search 422
    Merging 424

STATISTICS 425

    Median 426
    Summation Notation 426
    Regression Analysis 427

MATHEMATICS 428

    Binary Numbers 428
    Finding Roots of an Equation—Bisection Method 429
    Simultaneous Linear Equations 430

GRAPHING TECHNIQUES 434

    Graphing 434
    Plotting a Histogram 435

SIMULATION 437

    Random Numbers 437
    Generating Random Numbers 438
    Monte Carlo Simulation 439

MANAGEMENT INFORMATION AND DECISION SUPPORT SYSTEMS   444

EXERCISES   446

SUMMARY OF CHAPTER 10   458

SELF-TEST REVIEW   459

ANSWERS TO SELECTED EXERCISES   460

Chapter 11 ADDITIONAL FEATURES OF FORTRAN   463

MORE ABOUT DATA TYPE   464

    Exponent Form of Real Data   464
        Input and Output of Data in E Form   465
    Precision of Data in a Program   466
    Double Precision   469
        Double-Precision Constants   469
        DOUBLE PRECISION Statement   469
        Double-Precision Input and Output   470
    Logical Data   472
        Logical Constants   472
        LOGICAL Statement   472
        Logical Expressions   472
        Logical Assignment   473
        Logical Operators   473
        IF Statement and Logical Expressions   475
        Input and Output of Logical Data   477
        Hierarchy of Operations   479
    Complex Data   482

MORE ABOUT FORMAT   486

    G Descriptor   486
    P Descriptor—Scale Factor   489
    BN and BZ Descriptors—Blank Interpretation   493

ADDITIONAL FEATURES AND STATEMENTS   493

    EQUIVALENCE Statement   493
    IMPLICIT Statement   496
    BLOCK DATA Subprograms   497
    General Form of READ and WRITE   498
    Variable FORMAT   499
    OPEN and CLOSE Statements   501
    EXTERNAL and INTRINSIC Statements   502
    PAUSE Statement   503

EXERCISES   504

PROGRAMMING EXERCISES   510

SUMMARY OF CHAPTER 11   512

SELF-TEST REVIEW   514

Chapter 12 STRUCTURED PROGRAMMING — 517

WRITING BETTER PROGRAMS — 518

Introduction   518
Structured Design   518
   Top-Down Design   519
   Modularity   520
Readability   521
Documentation   525
Reliability   525

A SAMPLE PROBLEM — 526

Step 1: The Problem   526
Step 2: Input-Output Formulation   526
Step 3: Layout of the Input and Output   527
Step 4: Process Design   527
Step 5: Pseudocode   531
Step 6: The Program   532

Appendix A AN OVERVIEW OF COMPUTER ORGANIZATION — 537

COMPUTER SYSTEMS — 537

Input Devices   538
Ouput Devices   539
Central Processing Unit   540
Auxiliary Devices   540

INTERNAL DATA STORAGE — 540

Binary Numbers   541
Storing Character Data   542
Storing Integer Data   542
Storing Real Data   542

Appendix B INTERACTIVE FORTRAN — 545

INDEX — 547

Chapter 1
Basic Programming Concepts

INTRODUCTORY CONCEPTS

   Introduction
   Organization of the Text
   Computer Programs and Languages

WRITING PROGRAMS

   A Simple Program
   The PROGRAM Statement
   The PRINT Statement
   Arithmetic Operators
   Use of Parentheses in Arithmetic Operations
   Priorities of Arithmetic Operators
   Variables
      Variables and Constants
   The READ Statement
   Statement Labels
   Looping
   The GO TO Statement

SOME PROGRAMMING TECHNIQUES

   Generating Sequence Numbers
   Calculating the Sum of Integer Numbers

INTRODUCTORY CONCEPTS

☐ Introduction

This text introduces you to the world of computer programming. You may not realize it, but you are affected by computers frequently in everyday life. When you use your credit card, pay bills, or make use of your checking account, you indirectly utilize the services of computers although you are not actually involved in the programming itself. You can even use computers directly without knowing programming by employing what are called "ready to use" programs. Nevertheless, there are several reasons why you might like to learn computer programming. Some of these are: (1) computers being an ever-increasing part of our lives, one needs to have some knowledge about the way they operate; (2) the increasing use of computers has prompted an increasing demand for computer education; (3) it is desirable to learn to get computers to do your work; and (4) programming is fun—writing a computer program and getting the computer to do what you want it to do is enjoyable, challenging, and rewarding.

With this background, perhaps you are ready to learn a programming language. It is easy to learn programming. You don't need to be mathematically minded. In fact, the knowledge of mathematics beyond basic algebra is not required in most parts of this text. But you should be able to think logically, as programming is a step-by-step, logical process. When you can follow the logic and the pattern of a program, you can say, "I am beginning to learn programming." When you can create the logic and pattern of your own program, you have accomplished the most difficult phase of the work.

☐ Organization of the Text

Throughout this text you will learn by examples, using the programming language called FORTRAN. Each section includes several examples, and in each example the learning objectives will be explained as "notes." It is important that you understand the notes. The most important note in each example is explained as a "rule." The solved problems in each section are also a vital part of the learning process. Thus, it is very important to fully understand each of the solved problems. Don't look at the answers to the problem until you have given enough thought to the problem and have an answer ready.

Understanding the material in the text is not enough. To really learn programming skills you need to practice the concepts discussed in the text. The exercises in each chapter are designed to help you do this. You should complete these before proceeding to the next chapter. Furthermore, you should run several programs with the computer, because programming, like any other skill, is not learned unless it is practiced.

Each chapter has the following divisions:

1. *Examples*, with several "notes" as the learning objective of the examples.
2. *Rules*, the most important notes following examples.
3. *Solved problems*, also an important part of the learning process.
4. *Exercises*.
5. *The summary* of each chapter.
6. *A self-test*, which makes sure that you understand the materials covered in the chapter. The answers are given at the end of the test.

Since it is assumed that FORTRAN is the first programming language that you are learning, you should read each section carefully in order to understand the material thoroughly. You will benefit most from the text by taking time with each chapter rather than just scanning it.

The following section will introduce you to general programming concepts.

☐ Computer Programs and Languages

After reviewing this section, you should be able to answer the following questions.

1. What is a program?
2. What are the basic functions that a computer can perform?
3. What is a computer language?

People are impressed by the computer's problem solving ability, yet computers do not really solve problems—people do! A programmer expresses the solution procedure to a computer in terms of a series of instructions called a program.

> *A computer program is a series of instructions which enables the computer to perform a designated task.*

These instructions must be:

- Specific
- Detailed
- Logical
- Sequential
- Clearly phrased

As you will come to understand, you must tell a computer what to do in complete and precise detail. It is not enough merely to tell a computer to "solve this problem." A computer can, however, be instructed through a program to perform a specific function.

1. Accept data
2. Do arithmetic operations such as addition, subtraction, multiplication, and division
3. Store the data in its memory
4. Print stored data

All of these functions can be accomplished by a program.

The specific instructions in a program are conveyed to the computer through a computer language. Computer language, like any language, is a combination of predetermined symbols, terms, words, rules, and functions used as a medium of communication. A computer language is a way of conveying the instructions to the computer and translating the results. A part of learning a language is to learn the words, rules, and phrases that are used in that language; but more importantly, you must learn the logic, techniques, and the appropriate style of solving a problem. You will have the opportunity to learn and practice all of these in this text.

There are several computer languages available—each with a different name such as BASIC, FORTRAN, COBOL, or PL/I. FORTRAN, which is an acronym for "FORmula TRANslation," is one of the most widely used languages. Throughout this text you will learn FORTRAN.

4 Basic Programming Concepts

WRITING PROGRAMS

The basic concepts of programming—the foundation—are explained in the following sections. These concepts, although simple, are very important.

A Simple Program

EXAMPLE 1.1

Problem: Find the floor area of a given room which has a length of 15.5 feet and a width of 9.5 feet.

Solution Procedure: The length of the room will be called A, the width will be called B, and the area will be called C. The area can be found through five steps:

1. Start
2. Assign the data to A (length) and to B (width)
3. Calculate the area as: C = A × B
4. Print the area
5. End

Program:

```
PROGRAM AREA
A = 15.5
B = 9.5
C = A*B
PRINT * , C
END
```

Printout: The printout will show:

```
147.25
```

Notes:

1. Each instruction is on one line. Each line must have at least a six-space margin.
2. The program starts with the word PROGRAM followed by the name of the program.
3. Names for the variables are made up, such as A for width, B for length, and C for the area. The program itself is called AREA.
4. The * in C = A*B is a multiplication sign.
5. Computation is done on the right side of the "=" and the result is assigned to the variable C on the left side.
6. A and B are defined before calculating C. That is, the value of A and B must be known before calculating C.
7. The PRINT statement prints the result.
8. The PROGRAM statement is the first and the END is the last statement in a program.

☐ The PROGRAM Statement

The program statement is used to "start" a program and define the name of the program. The word PROGRAM must be spelled correctly, but the name of the program is chosen by the programmer. The name of the program must not be the same as other names in the program. It must be a unique name. This statement is optional with standard FORTRAN.

☐ The PRINT Statement

The PRINT statement causes the desired variables to be printed. This statement has the following form, with four main components:

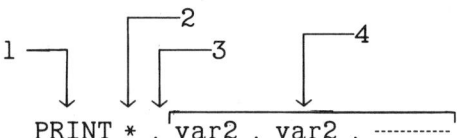

1. The word PRINT (must be spelled correctly).
2. An asterisk (*).
3. A comma.
4. The list of the desired variables to be printed.

For example:

```
PRINT * , A , B , C , D
```

prints the contents of four variables: A, B, C, and D. One or more spaces before or after the asterisk and before or after a comma are optional.

EXAMPLE 1.2

■ **Problem:** Calculate the sum and the average of numbers 72.50, 96.10, and 84.90. Print the data on one line, the average and the sum on the next line.

Program:

```
PROGRAM TWOPRT
A = 72.50
B = 96.10
C = 84.90
S = A + B + C
G = S/3.0
PRINT * , A , B , C
PRINT * , S , G
END
```

Printout: The printout will show

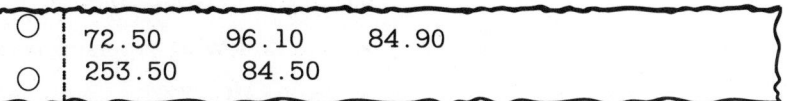

```
72.50    96.10    84.90
253.50   84.50
```

Notes:

1. More than one operation (calculation) can be performed in a program, namely, the sum and the average.

> 2. The slash is a division sign.
> 3. More than one variable is printed by each PRINT statement: A, B, C, by the first one, and S, G by the second one.
> 4. Each time that a PRINT statement is carried out, it starts printing the data at the beginning of a new line.

☐ Arithmetic Operators

The arithmetic operators are used to do the arithmetic operations. They are:

- ****** for exponentiation; example: Z = X**2 (for X^2).
- ***** for multiplication, as mentioned before; example: P = R*Q.
- **/** for division; example: A = B/C $\left(\text{for } \dfrac{B}{C} \text{ or } B \div C\right)$.
- **+** for addition; example: X = Y + 2.0.
- **−** for subtraction; example: W = V − X.

The following rules also apply to the arithmetic operators:

1. Parentheses can be used freely for grouping the variables.
2. A space before or after an operator is allowed.
3. Two successive operators may not appear next to each other. For example, X + − 3 is not legal. However, it can be written legally as X + (−3).

A few proper expressions are listed below in both ordinary arithmetical language and FORTRAN.

Ordinary	FORTRAN
X = A + B + C − D − H	X = A + B + C − D − H
W = 2X + A · B	W = 2 * X + A * B
$A = \dfrac{A1 + A2 + A3}{3X + B}$	A = (A1 + A2 + A3)/(3.0 * X + B)
$Z = (A - B)^2 + \dfrac{3A}{B}$	Z = (A − B)**2 + 3.0 * A/B
Y = X^3	Y = X * X * X or Y = X**3
X = $\sqrt{A}$ (is the same as X = $A^{1/2}$ or X = $A^{.5}$)	X = A**.5

■ SOLVED PROBLEMS

■ 1. Are the following programs correct?

    a. PROGRAM XYZ
        X = 9.6
        Y = X**2
        END
        PRINT * , Y

b. PROGRAM ABC
```
      A = 3.5
      C = A + B
      B = 92.1
      PRINT * , C
      END
```
c. PROGRAM XXX
```
      A = 3.5
      B = 92.1
      A + B = C
      PRINT * , C
      END
```
d. PROGRAM YOU
```
      A = 3.5
      B = 92.1
      C = A + B
      PRINT C
      END
```
e. PROGRAM AYE
```
      A = 6.956
      PRNT * , A
      END
```
f.
```
      X = 5.0
      Y = 6.0
      Z = X + Y
      PRINT * , Z
      END
```

☐ **Answers**

a. No, the END statement should be the last statement.
b. No, A and B should be defined before calculating C.
c. No, computation should be done on the right side of the "=".
d. No, the logical sequence is right, but an asterisk (*) and a comma are needed after the word PRINT.
e. No, the word PRINT must be spelled correctly.
f. The program should start with the PROGRAM statement. (However, this is optional with standard FORTRAN.)

■ **1.2** Write the following expressions in FORTRAN.

a. $d = 2(a + b)(a - b)$

b. $X = \dfrac{c + d + a}{a + d}$

c. $Y = \dfrac{a}{g + h + b}$

d. $Y = 2(5b + 3c)$

e. $Y = \dfrac{a \cdot b \cdot c}{a + b + c}$

f. $Y = X^4$

g. $Z = (x + y + a)^{c+d}$

8 Basic Programming Concepts

> ☐ **Answers**
> a. D = 2 * (A + B) * (A − B)
> b. X = (C + D + A)/(A + D)
> c. Y = A/(G + H + B)
> d. Y = 2*(5*B + 3*C)
> e. Y = (A*B*C)/(A + B + C) or Y = A*B*C/(A + B + C)
> f. Y = X**4 or Y = X*X*X*X
> g. Z = (X + Y + A)**(C + D)

EXERCISES

1.1 Correct the following programs:

*a. ```
PROGRAM QUE
C = 56.1
D = C*A ÷ 2.5
A = 1000.5
PRINT A
END
```

*b. ```
PROGRAM MY
PRINT B
B = A*A
A = 1296.5
END
```

*c. ```
X = 3296.50
Y = 56.20
X + Y = Z
PRINT Z
END
```

d. ```
P = 32.9
Q = 1932.0
R = (P + Q) ÷ P
PRINT R
```

e. ```
E = 5.0
F = E * G
G = 9.1
```

f. ```
S = (A * B + C) ÷ D
PRINT S
```

*1.2 Write the following expressions in FORTRAN:

a. $A = \dfrac{1}{X} + \dfrac{1}{Y}$

b. $B = \dfrac{1}{X + Y}$

c. $X = (A \cdot B) + (C \cdot D)$
d. $P = 2.5R + 3.14R^2 + 6.28$
e. $Q = A^5 + (5A)^3$

* Answers to starred exercises are provided at the end of the chapter.

*1.3 Translate each of these FORTRAN expressions into its equivalent mathematical context:

a. X = B**3 + A
b. R = A + B + C/D
c. Q = X**3/(A + B + C + 4.5*X)

☐ Use of Parentheses In Arithmetic Operations

You have learned how a program can be started. You also should have an idea how mathematical expressions can be explained in a program. Now let's proceed with some more examples.

EXAMPLE 1.3

Problem: Find the average of four numbers: 89.5, 95.8, 81.2, and 98.9. Print the data and the average.

Solution Plan:

1. Start.
2. Assign the values to the variable names.
3. Calculate the average by adding all the variables and then dividing by 4.
4. Print the data and the average.

Program:

```
PROGRAM AVRG
A = 89.5
B = 95.8
C = 81.2
D = 98.9
X = (A + B + C + D)/4.0
PRINT * , A , B , C , D , X
END
```

Printout: The printout will show:

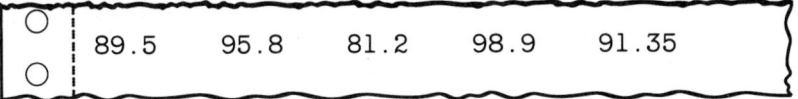

Notes:

1. One must know how to calculate the average. We could not ask the computer to find "the average." In other words, the method of arriving at the solution of a problem must be known before programming.
2. We use parentheses for a correct calculation. If parentheses are not used in this example, only D will be divided by 4.
3. The PRINT statement prints variables A, B, C, D, and X on one line.

10 Basic Programming Concepts

☐ Priorities of Arithmetic Operators

The computer scans an expression from left to right and performs all the exponentiation first,[1] then all the multiplications and/or divisions, and finally the addition or subtraction. Thus, a mathematical expression is performed according to the following priorities:

1. ** exponentiation.
2. *,/ multiplication and/or division.
3. +,− addition and/or subtraction.

If two operators have the same priorities, the one farthest left is performed first. For example, the expression

A + X/Y * 2 + 3 * X**2

will be performed in the following order:

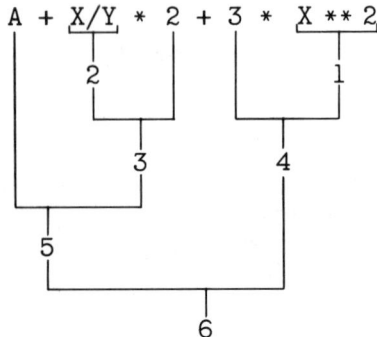

However, a pair of parentheses will overrule these priorities. Mathematical expressions between parentheses have the first priority. If there are more than one pair of parentheses, the innermost pair of parentheses will be treated first. For example, the expression

(5 * X + 3 * (5 + 3 * X))/B**2

is evaluated as follows:

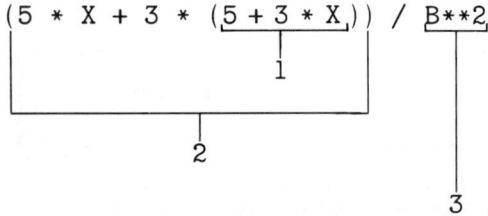

Parentheses must often be used to force the computer to perform the operations in the desired order. It is advisable to use extra parentheses for correctness, or, more importantly, extra spaces and parentheses should be used to make the expression more readable. Spacing is also helpful for readability. Examples are given in the following solved problems.

---

[1] The exponentiations are performed from right to left in the case of "double exponentiations."

SOLVED PROBLEMS

1.3 Can any of the parentheses in the following expressions be deleted without changing the order of operations? If so, rewrite the expression.

  a. P = (A + B − C) + D
  b. Y = ((A * B)/C) + D
  c. X = (A * B) * (C) ** D
  d. Y = (8 * X) + 13 − (5 * (Y ** 3))
  e. Z = ((A + B) + (X − Y) + A/(X + Y))
  f. X = ((A/B) * (C * B)/A) + B + (−C)
  g. Q = (((X * Y) * Z)/(B * A)) + 2
  h. R = ((A + B + C)/(3 * X))
  i. Z = ((A/B) * C) + ((A * C)/B) + A/(B*C)

☐ **Answers**

  a. P = A + B − C + D
  b. Y = A * B/C + D
  c. X = A * B * C ** D
  d. Y = 8 * X + 13 − 5 * Y ** 3
  e. Z = A + B + X − Y + A/(X + Y)
  f. X = A/B * C * B/A + B + (−C)
  g. Q = X * Y * Z/(B * A) + 2
  h. R = (A + B + C)/(3 * X)
  i. Z = A/B * C + A * C/B + A/(B * C)

1.4 Write the following expressions in FORTRAN.

  a. $Y = \dfrac{(3X^5 + 2)^3 + 15(Z + 2)(3P - 7)}{(5X^2 - 3)^3}$

  b. $A = 6Q^2 + \dfrac{9X^2 + 3}{2X + 1} - \dfrac{10P - 3X}{12X + 1}$

  c. $B = \dfrac{\dfrac{A + 3}{B} + Y}{A - 2 - \dfrac{Y}{A}}$

☐ **Answers**

  a. Y = ((3 * X**5 + 2)**3 + 15 * (Z + 2) * (3 * P − 7))/(5 * X**2 − 3)**3
  b. A = 6 * Q**2 + (9 * X**2 + 3)/(2 * X + 1) − (10 * P − 3 * X)/(12 * X + 1)
  c. B = ((A + 3)/B + Y)/(A − 2 − Y/A)

☐ Variables

> **EXAMPLE 1.4**
>
> ■ **Problem:** Write a program which finds the floor area of a room. The length is 14 and the width is 9.5.
>
> **Program:**
>
> ```
> PROGRAM CALCUL
> LENGTH = 14
> WIDTH = 9.5
> AREA = LENGTH * WIDTH
> PRINT * , LENGTH , WIDTH , AREA
> END
> ```
>
> **Printout:** The printout will show:
>
> ```
>    14    9.5    133.0
> ```
>
> **Note:** The symbolic name of a variable can be more than one letter in length.

> **RULE 1.1**
> A FORTRAN symbolic name can consist of one to six letters or digits, the first of which must be a letter.

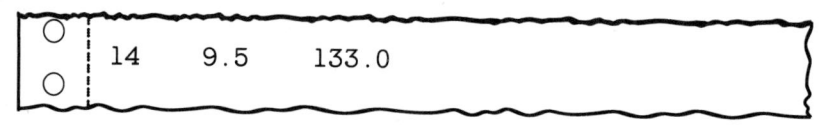

For example, the following are proper names for a variable.

```
MASS      VOLT      COUNT     XL    TAX     D87943
ALPHA1    FORCE     PRICE1    A1    AVG     X5489A
AREA      JOHN      MARY      X9    X3P2    P32X1
```

The rule implies that special characters such as "." or "$" *cannot* be used for a symbolic name. For example, some improper symbols are:

VOLTAGE	Too many characters, more than six.
A.12	Illegal character, cannot use "."
X$Y	Illegal character, cannot use $.
1X	The first character is not a letter.
3PX	The first character is not a letter
AVERAGE	Too many characters.

> **EXAMPLE 1.5**
>
> ■ **Problem:** What number will the following program print?
>
> **Program:**
>
> ```
> PROGRAM CUMULT
> ```

```
SUM = 2.1
SUM = SUM + 5.2
PRINT * , SUM
END
```

Printout: The printout will show:

```
7.3
```

Note: In the statement SUM = SUM + 5.2, the value of SUM on the right side of the equation is 2.1, and the *new* value of SUM (on the left side of the equation) is calculated to be 7.3.

This method of assigning a new value to a variable is a technique frequently used in programming. The usefulness of the technique will become apparent when you learn more about programming in the following chapters.

Variables and Constants

You may have been wondering why we often use names for data. Is this necessary? There are, in fact, several reasons for doing this. The most important one is that data are stored in the memory of a computer. In order to use a memory location, we need a name for that memory cell. Once we choose a name for a memory cell, we can store the data in it, as well as update, change, or erase its content. The name of the memory cell is also called a variable, because the content of that cell can vary. Thus, a variable refers to a storage location. Whenever the variable is used, it refers to the value currently stored in that location. Variables are essential in writing a program because all data manipulations are done through variables. You may also use actual numbers in a program. They are called *constants*.

> **Constants** *are actual numbers in a program that do not change.*

For example, in

```
Y = 3.0 * X + 2.0
```

Y and X are variables, but 3.0, and 2.0 are constants.

You will learn several ways to store a value into a memory cell; one way, though, is with an *assignment statement*. For example, X = 3.5 places 3.5 in the memory cell called X. Thus:

> *An* **assignment statement** *is a statement that assigns a value to a variable.*

Notice that an = sign is used by an assignment statement for this purpose. That is:

14 Basic Programming Concepts

> **RULE 1.2: EQUAL SIGN**
> *An equal sign places the value on the right in the memory cell named on the left.*

As mentioned before, the content of a memory cell (a variable) can be assigned a value, copied, renewed, changed, or erased at any time in a program. The following program is an example.

EXAMPLE 1.6

■ **Problem:** What values will be printed by the following program?

Program:
```
PROGRAM COPY
X = 2.1
Y = X
X = 6.5
X = X + 5.2
PRINT * , X , Y
END
```

Printout: The printout will show:

```
11.7    2.1
```

Notes:

1. In the first assignment statement, 2.1 is assigned to X.
2. In the second assignment statement, X is copied into Y; thus the value of Y is also 2.1.
3. In the third assignment statement, the value of X is changed to 6.5.
4. In the fourth assignment statement, the new value of X (on the left) is calculated as the previous one (6.5) plus 5.2.
5. By placing a new value in the memory cell, the previous one is erased.

As was mentioned before, the variables are an essential part of a program. They allow us to write a flexible program to solve a problem in a general form rather than by doing specific calculations. For example,

```
AREA = LENGTH * WIDTH
```

can be used for calculating any area by using different data.

Choosing a symbolic name for a variable is also important. Since FORTRAN allows us to choose a name up to six characters, it is always advisable to pick a descriptive name for a variable. For example, it is better to choose AREA for the area of a room rather than X. The value of this becomes apparent later when your program becomes complicated and long.

SOLVED PROBLEMS

1.5 Which of the following are proper FORTRAN symbolic names?

    a. SUM   b. KOUNT   c. INIGER
    d. J   e. E.WEST   f. BANKING
    g. 1XYZ   h. X+Y   i. SUM$

☐ **Answer**

The first four symbols are correct, but the last five are not.

1.6 Which of the following statements are not correct in FORTRAN?

    a. X = A + B + C + D + E + F
    b. Y9 = (X1 + X2)/(X3 + X4)
    c. AREA = WIDTH ** WIDTH/2.5
    d. 2X + 3Y = 3(X + Y)
    e. X = Y ** .5
    f. A + B = C
    g. GEE = 3.5 + (A + B + C) ** 3
    h. SUM = SUM + NEW
    i. A ** (I + 2) = OLD
    j. JAY = 2 * ICONT * JCONT ** N
    k. FOR = ((X + Y) + Z) ** 2
    l. PROGRAM AVERAGE
    m. PRINT * , PACK , JACK , MACK

☐ **Answer**

d, f, i, and l are not correct because more than one variable cannot be used on the left side of an assignment statement, and AVERAGE in the PROGRAM statement is more than six characters.

1.7 Determine what values are printed by using the following programs. Also, correct any errors you may find.

```
a. PROGRAM SUM
   A = 2.5
   B = 11.5
   PRINT * , A , B
   C = A + B
   PRINT * , C
   END
b. PROGRAM QUE
   W = 2.1
   PRINT * , W
   END
c. PROGRAM AXE
   X = 5.5
   X = X + 3.4
   Y = 2.0*X
   PRINT * , Y
   END
```

     d. PROGRAM DOUBLE
```
   A = 1.2
   B = A + 2.1
   B = 2.0*B
   PRINT * , B
   END
```
     e. PROGRAM SUM
```
   ISUM = 0
   I = 1
   ISUM = ISUM + 1
   J = I + 2
   ISUM = ISUM + J
   PRINT * , I , J , ISUM
   END
```
     f. PROGRAM AVG
```
   X = 5.93
   Y = 3.07
   AVG = (X + Y)/2.0
   PRINT * , AVG
   END
```

☐ **Answers**

    a. 2.5   11.5 in the first line, and 14.0 in the second line.
    b. 2.1
    c. 17.8
    d. 6.6
    e. 1 3 4
    f. The program will not print anything but error messages because the name of the program is the same as one of the variables.

EXERCISES

1.4 Convert the following into FORTRAN expressions.

    a. $\dfrac{(A + B)^2 + C}{2} = X$

    b. $(2X + 3Y)^2 + \dfrac{(5X + Y)^2 + X^2}{2} = Z$

    c. $X = \dfrac{A + A^2 + (-4AB)}{2A}$

    d. $(1/2)(5X \cdot Y \cdot Z^2 + X) = W$

*1.5 Which of the following are *not* proper names in FORTRAN?

    a. BOM   **b.** CODE   **c.** SOLVE   **d.** SOLUSION
    e. AAA   **f.** AA.AA   **g.** NORMAL **h.** MEAN
    i. MEAN1   **j.** 1X   **k.** A5   **l.** 5A

1.6 Correct the following programs.

a.
```
END
PRINT X
X = A + B
A = 53.1
B = 23.1
```

b.
```
PRINT Y
Y = 5.1
END
```

c.
```
V = 2.1
W = V + X
X = 1.3
PRINT X, V, X
END
```

d.
```
AVRG = C + D/2.0
C = 5.1
D = 2.3
PRINT, C, D, AVRG
END
```

e.
```
PROGRAM CUMULATIVE
FIRST NUMBER = 1.0
SECOND NUMBER = 2.3
AVERAGE = (FIRST NUMBER + SECOND NUMBER)/2.0
PRINT * , AVERAGE
END
```

f.
```
PROGRAM WAGE
HRS = 40.0
RATE = 8.75
WAGE = HRS * RATE
PRINT * HRS , RATE , WAGE
END
```

1.7 What do the following programs print after execution?

a.
```
PROGRAM AB
A = 2.5
B = 2*A
PRINT * , B
END
```

b.
```
PROGRAM ABC
Y = 2.1
X = 5.5
Y = Y + X
PRINT * , Y
END
```

*c.
```
PROGRAM MIXED
J = 2
I = 3
ISUM = I + J
MULT = I * J
IN = MULT * ISUM
IN = IN + 1
PRINT * , J , I , IN , MULT , ISUM
END
```

18 Basic Programming Concepts

1.8 Suppose the width of a room is 19.5, the length is 25.6, and the height is 12.5. Write a program which prints the width, length, height, and volume of the room. (Hint: the volume is equal to width × length × height.)

1.9 If the temperature in Fahrenheit is equal to 75, write a program to calculate and print the centigrade temperature. Fahrenheit can be changed to centigrade by the following formula:

$$C = \frac{100(F - 32)}{180}$$

☐ The READ Statement

In the examples in the previous section, you saw that the data were given in the program by an assignment statement. In many cases, however, the data are given to the computer in a section called the "data section." In this case, a READ statement must be used to read the data. Look at the following example to see how this technique is utilized. In Chapter 3, it will be shown how the main program is separated from the data section. At this point, you may assume that the data section is after the END statement.

EXAMPLE 1.7

■ Problem: Find the average of three exam marks. The data in one occasion are 85.6, 78.5, and 90.3.

Solution Plan:

1. Start.
2. Read the values of the variables from the data section.
3. Calculate the average.
4. Print the average.

Program:

```
PROGRAM NEW
READ * , A , B , C
AVERAG = (A + B + C)/3.0
PRINT * , AVERAG
END
```

Data: Then in the data section, we place the numbers as:

```
85.6    78.5    90.3
```

Printout: The printout shows:

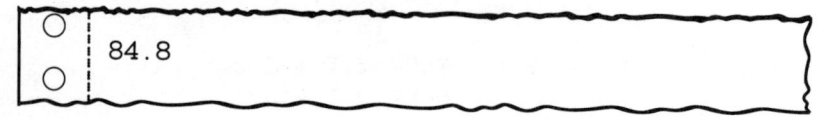

```
84.8
```

Notes:

1. The READ statement, which must be spelled correctly, is followed by an asterisk (*) and a comma, followed by the name of the variables to be read.

> 2. Because of the READ statement, the computer automatically reads the data from the data section. It reads the first number for A, the second number for B, and the third number for C. Then the average is calculated and printed.
> 3. "READ * , A, B, C" does the same job as saying A = 85.6, B = 78.5, and C = 90.3 in this program.
> 4. The data can start in the first column.

The READ statement places values in the memory cell listed. The values will be read from the data section. The main program is separated from the data section; the order is:

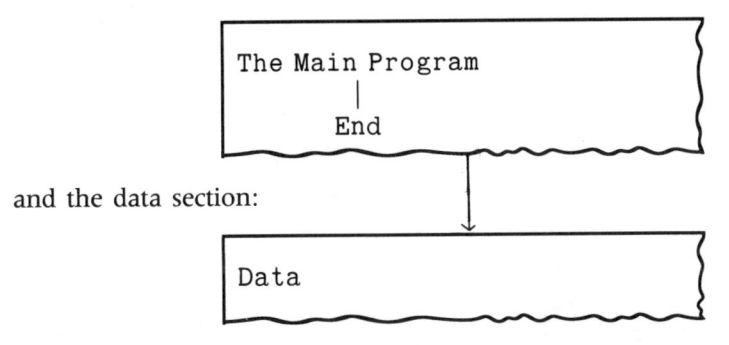

```
The Main Program
       |
      End
```

and the data section:

```
Data
```

The data given to the computer in the data section are called the *input data*. There is a difference between the input data and the data provided by the assignment statements, as the input data are always furnished by the READ statement. A READ statement is preferable to an assignment statement because we can write a single program and use it over and over with different sets of data.

Hereafter we refer to a line of data in the data section as a *record* or *data line*. A *data card* is also a record which contains the values of the variables.

The following solved problems will help you to better understand the READ statement.

SOLVED PROBLEMS

1.8 Correct the following programs:

a.
```
PROGRAM ABC
AA = 201.50
BB = 6321.1
READ * , AA, BB
CC = (AA + BB)/2.0
PRINT * , CC
END
```
In the data section:
```
201.50    6321.1
```

b. PROGRAM NAMES
READ * , JOHN , KAY
MASS = JOHN + KAY
PRINT * , MASS
END
In the data section:

```
JOHN = 96    KAY = 36
```

c. PROGRAM WWW
READ * , WIDTH , WIDE , AREA
AREA = WIDTH * WIDE
PRINT * , AREA
END
In the data section:

```
12.0    18.0
```

d. PROGRAM ABCD
PRINT * , AB , CD
READ * , AB, CD
X = AB + CD
PRINT * , X
END
In the data section:

```
6328.0    93285.65
```

e. 3.9 6.2 1.2
W = V + U + Y
READ * , V , U , Y
PRINT , V , U , Y , W
END

☐ **Answers**

a. The data should be given by *either* AA = 201.50, BB = 6321.1, or the READ statement, but not by both. As explained before, READ * , AA, BB, performs the same function as saying AA = 201.50 and BB = 6321.1. Thus, the program should look like this:

PROGRAM ABC
READ * , AA , BB
CC = (AA + BB)/2.0
PRINT * , CC
END

b. The data in the data section should look like:

```
96    36
```

without the names of the variables. Look at Note 2 of Example 1.7.

c. The variable AREA which is to be calculated should not be read by the READ statement. The READ statement should look like:

```
READ * , WIDTH , WIDE
```

d. PRINT * , AB , CD should be either omitted or put after the READ statement. Remember, a variable should be defined before any calculation or printing upon that variable.

e. The READ * , V , U , Y should be placed before calculating W, and the data should go to the data section instead of being placed at the beginning of the program. The program also needs the PROGRAM statement. The correct program is:

```
PROGRAM PALE
READ * , V , U , Y
W = V + U + Y
PRINT * , V , U , Y , W
END
```

and in the data section:

```
3.9    6.2    1.2
```

EXAMPLE 1.8

Problem: Read the values of A, B, and C from the first data line and the values of D, E, and F from a second line. Calculate the sum of the variables. Print the data and the sum.

Program:

```
PROGRAM XXX
READ * , A , B , C
READ * , D , E , F
SUM = A + B + C + D + E + F
PRINT * , A , B , C , D , E , F , SUM
END
```

Sample data:

Line 1 → | 15.0   22.0   18.0
Line 2 → | 59.0   98.0   75.0

Printout: The printout will show:

```
15.0  22.0  18.0  59.0  98.0  75.0  167.0
```

Statement Labels

A statement in FORTRAN can start with a number. The following is an example:

```
        PROGRAM EXMPLE
10      READ * , A , B , C
        SUM = A + B + C
20      PRINT * , A , B , C , SUM
        END
```

Each number before each statement is called *the statement label* or *the statement number*. The statement labels must be in the margin, in columns one to five. Rule 1.3 explains this.

RULE 1.3: STATEMENT LABEL
1. *A statement in a FORTRAN program may have a number as a label.*
2. *The labels can be any positive number smaller than (or equal to) 99999 (5 digits).*
3. *The labels do not have to be in sequence.*
4. *Two labels cannot have the same number (each one must be unique).*
5. *The labels appear at the beginning of the statement. They must be between columns 1 and 5.*

We may assign a label to any statement in a program. As an example, the following program with or without labels is correct:

```
        PROGRAM XYZ
100     READ * , A , B , C
        SUM = A + B + C
200     AVG = SUM/3.0
        PRINT * , A , B , C
300     PRINT * , SUM
400     PRINT * , AVG
        END
```

As mentioned before, the labels do not have to be in sequence. However, choosing labels in sequence helps to locate a statement in a large program.

Looping

One advantage of using a computer is its ability to process a mass of data quickly and accurately without getting tired, as people easily do in manual computation. Therefore, computers are a very useful tool if we have a lot of data to be processed. Furthermore, if there are several sets of data, where the same kind of calculation is applied to all of them, we do not need to write individual programs for each set. Instead, we can instruct the computer to repeat the program for the different sets of data. The term *loop* is used when a segment of a program is repeated automatically over and over, a certain number of times.

> *Automatic repetition of a segment of a program is called* **looping.**

Looping is a very common practice in programming. It is simple yet very important. Looping avoids the necessity of having to rewrite certain statements in a program. There are several techniques for looping in FORTRAN, and they will be discussed throughout the text. The simplest method is using a GO TO statement.

☐ The GO TO Statement

Normally a program is executed in the order in which the statements appear, one after another. The programmer, however, can change the normal sequence of execution by using a GO TO statement. This statement is called an *unconditional jump*, or *branching*, because it transfers the control from one fixed point in the program to another. For example, a GO TO statement can be used to transfer the control to the beginning of a program in order to execute the same program for a new set of data, thereby making a loop. The following example demonstrates this technique.

EXAMPLE 1.9

■ **Problem:** There are several sets of data (say one thousand sets). On each set, there are three numbers. Find and print the average of the numbers of each set.

Program:

```
        PROGRAM LOOP
10      READ * , X , Y , Z
        AVERAG = (X + Y + Z)/3.0
        PRINT * , AVERAG
        GO TO 10
        END
```

Data: We put the data in the data section as:

86.5	78.5	90.5
99.5	62.8	70.1
71.9	82.3	73.7
⋮	⋮	⋮
65.0	71.5	83.2

24 Basic Programming Concepts

> **Printout:** The output will show:
>
> ```
>  85.16666
>  77.46666
>  75.96666
>     ⋮
>  73.23333
> ```
>
> **Notes:**
>
> 1. At the beginning, the READ statement causes the computer to read the first set of data; thus the values will be X = 86.5, Y = 78.5, Z = 90.5. The average of the first set of data will then be calculated and printed. After printing the first average, the "GO TO 10" statement transfers the control to statement number 10, the READ statement, for reading a *new* set of data. Each time the READ statement is executed, the next set of data will be read automatically. Thus, for the second time the values are X = 99.5, Y = 62.8, Z = 70.1, and so on.
> 2. The loop created for reading the data is an infinite loop. The methods for terminating this kind of loop will be discussed in Chapter 2.
> 3. When a new set of data is read, the previous values are erased—just like on audio tape (when a new recording is made, the previous one is erased).

It is very important to remember that:

> **RULE 1.4**
> *Each time that the READ statement is executed, it automatically starts at the beginning of a new line of data; thus a data line will not be read twice.*

GO TO and GOTO are both permissible.

Another use of the GO TO statement is to transfer the control from one point in the program to another point in order to skip one or more instructions. This is illustrated by the following example.

> **EXAMPLE 1.10**
>
> ■ **Problem:** What number will be printed if the following program is executed?
>
> **Program:**
>
> ```
>        PROGRAM SKIP
>    50  READ * , APE , CAT , DOG
>        SUM = APE + CAT + DOG
>        GO TO 115
> ```

```
60      PRINT * , SUM
70      PRINT * , APE
115     PRINT * , DOG
        END
```

Data:

```
32.1    5.1    2.3
```

Answer: Only the value of the DOG variable (2.3) will be printed.

Note: This program shows only that a GO TO statement can be used for unconditional branching.

SOME PROGRAMMING TECHNIQUES

☐ Generating Sequence Numbers

EXAMPLE 1.11

■ **Problem:** Print the sequence of numbers 1, 2, 3, ...

Program:

```
        PROGRAM LOOP2
        I = 0
30      I = I + 1
        PRINT * , I
        GO TO 30
        END
```

Notes:

1. The program does not need any input data.
2. A value of zero is assigned to variable I at the beginning of the program.
3. Statement number 30 (I = I + 1) generates numbers 1, 2, 3, ... It works as follows: The first time that the program is executed, I is equal to 1 (I = 0 + 1), so the number 1 is printed. Then the control is transferred to statement number 30. The second time, I becomes equal to 2 (I = 1 + 1) and is printed; the third time, I becomes equal to 3 (I = 2 + 1); and the program continues in this manner. I = 0 before the loop is necessary to start generating the sequence numbers.

The technique shown in Example 1.11 is a very common technique of generating sequence numbers in a program. The statement I = I + 1 is also called a *counter block* because it counts the number of times that the program goes

26 Basic Programming Concepts

through the loop. Setting the initial value of a variable to zero (or any initial value) is called initializing the variable.

☐ Calculating the Sum of Integers

EXAMPLE 1.12

■ **Problem:** Calculate the successive sums of the integers 1, 2, 3, ... in a program.

Program:

```
      PROGRAM SUM
      J = 0
      JSUM = 0
190   J = J + 1
      JSUM = JSUM + J
      GO TO 190
      END
```

Notes:

1. The second and third statements initialize J and JSUM. Setting the initial values of the variables to zero is a necessary step before the loop starts.
2. The fourth statement is the beginning of the loop. Each time that this statement is carried out, J will be increased by 1 to generate the sequence numbers 1, 2, 3, ...
3. Each time that the fifth statement is carried out, J will be added to JSUM. Thus, JSUM cumulates the numbers 1, 2, 3, . . . in the loop.
4. The GO TO statement returns the action to the beginning of the loop, and the loop repeats itself infinitely.
5. This example shows how the numbers can be cumulated.
6. Two major flaws in the program are (1) the infinite loop, and (2) that no value is printed.

You certainly have noticed the flaw in the loop of the previous examples. How many times will the loop be repeated? The answer is "an infinite number of times" because there is no statement for terminating the loop. This flaw can be removed by using certain conditional statements, which will be explained in the following chapters. For the time being, this is a deliberate error, and has been tolerated in order to show certain techniques. It should not be of concern at this time.

■
FINAL NOTE

How to run a program will be discussed in Chapter 3. However, if you have access to a computer and you would like to try to run any of your programs at this time, do the following:

1. Do not use a program which has an infinite loop. Punch each statement on an 80-column card, starting the statement at column 7 of the card.

Punch statement numbers (labels) in columns 1–5. Each line of instruction must be punched on a separate card.
2. Consult with your computer service staff about your account number and control statements for running a FORTRAN program.
3. Submit the deck of cards (including the control cards) to the operator.

SOLVED PROBLEMS

1.9 There are several sets of data. In each set there are two numbers. Write a program to calculate and print the sum and the product of the numbers in each set. Use an infinite loop.

1.10 Write a program which calculates the multiplication of the numbers 1, 2, 3, ... through infinity. Print the numbers and the multiplication at each stage.

1.11 What does the following program do? Is there any flaw in the program?

```
      PROGRAM SUM
      N = 0
      IS = 0
5     N = N + 2
      IS = IS + N
      GO TO 5
      END
```

Answers

1.9

```
      PROGRAM LOOP
100   READ * , FIRST , SECOND
      SUM = FIRST + SECOND
      PRODUC = FIRST * SECOND
      PRINT * , SUM , PRODUC
      GO TO 100
      END
```

1.10

```
      PROGRAM CUTE
      N = 0
      MULT = 1
100   N = N + 1
      MULT = MULT * N
      PRINT * N , MULT
      GO TO 100
      END
```

1.11 The program adds up the even numbers 2, 4, 8, ... The flaws include the infinite loop as well as the fact that no values are printed.

EXERCISES

1.10 There are several sets of data. On each set there are three numbers. Write a program to calculate and print the sum, the product, and the average of each set of numbers. Use an infinite loop.

1.11 Write a program which prints the odd numbers 1, 3, 5, 7, . . . through infinity.

1.12 Write a program which adds up the odd numbers 1, 3, 5, 7, . . . through infinity.

1.13 Write a program which calculates the product of the odd numbers 1, 3, 5, 7, . . . through infinity.

1.14 What does the following program *print*?

```
62      READ * , A , B
        SUM = A + B
        X = SUM/2.0
        GO TO 10
        X = 0.0
10      PRINT * , X
        END
```

Data:

```
82.5    87.5
```

1.15 If A dollars are invested at an annual interest rate of R (expressed as a decimal), after N years the total amount of money compounded (TOT) can be calculated by

$$TOT = A(1 + R)^N$$

Write a program which calculates and prints the total amount of each of the following data lines after 20 years:

Amount	Interest Rate
55.5	.05
565.5	.10
6960.0	.12
25.0	.04
100.0	.08

Use a READ statement and an infinite loop to read the data. However, if you choose to run this program, use only one data line at a time and no loop.

1.16 It is said that in 1627 Manhattan Island was bought from Indians for $24. Write a simple program to calculate how much the $24 would be worth today if it had been placed in a savings account at 5.25% annual interest, and also at 14.25%. Use the formula in the previous problem.

1.17 Write a complete program which calculates T from the following formula:

$$T = (U + V + W)(U - V)(U - W)(V - W).$$

SUMMARY OF CHAPTER 1

You have learned:

1. A computer program is the detailed set of instructions which directs the computer to do a specific task.
2. A computer language uses some predetermined words, terms, and rules to convey the instructions to the computer.
3. Basic operations that a computer can perform are: accepting data, doing arithmetic calculations, storing the data, and printing the desired data.
4. FORTRAN is an acronym for FORmula TRANslation.
5. A variable represents a quantity which can be changed in a program.
6. By creating a name for a variable the computer assigns a memory space to that name.
7. A variable's symbolic name cannot exceed six letters or digits, and must begin with a letter.
8. A variable must be defined before being used.
9. Mathematical expression can be written in FORTRAN by using the mathematical operators **, *, /, +, −, and parentheses.
10. By using a READ statement, the computer automatically reads the data from the data section.
11. Looping is used when a segment of a program must be repeated several times.
12. A GO TO statement can be used for unconditional branching; it may be used for looping in a program.
13. A statement may begin with a number, which is the statement label.
14. A statement such as I = I + 1 can be used for generating a sequence of numbers.
15. Some of the words that the computer can understand in the FORTRAN language are:

    PROGRAM   for starting a program and defining its name
    PRINT   for printing values of the variables
    END   for ending a program (this is the last statement in a program)
    READ   for reading the data from the data section
    GO TO   for transferring control to another statement

SELF-TEST REVIEW

1.1 Explain briefly.
   a. What is a computer program?
   b. What is a computer language?
   c. Where does the term FORTRAN come from?
   d. What is a loop in a program?
   e. What is the purpose of a READ statement?

1.2 Write FORTRAN expressions for the following mathematical expressions:
   a. $B = X^a - P(3X + Y)$
   b. $B = \dfrac{X + Y + Z}{X + 2Y} + 3Z$
   c. $B = X \cdot Y \cdot Z + A - 3X^4$

1.3 Which of the following are *not* proper symbolic names in FORTRAN?
    a. XMAS   **b.** CHRIS   **c.** CHRISTMAS
    d. A   **e.** AVG   **f.** AVERAGE
    g. AAA3   **h.** A*1   **i.** 1XAF

1.4 Which of the following expressions are *not* correct expressions in FORTRAN?
    a. (A + B)/2.0 = AVG
    b. XYZ + 2 = X
    c. XER = (3X + 5Y)(2X − 1)

1.5 Correct the following programs.

a.
```
ABC = 89.5
AVG = (ABC + S)/2.0
S = 99.3
PRINT ABC, S, AVG
END
```

b.
```
PROGRAM ION
JOHN = 269
KAY = 596
ION = JOHN + KAY
END
PRINT * ION
```

c.
```
SCORE1 = 92.5
SCORE2 = 78.2
CALCULATE AVERAGE
PRINT AVERAGE
END
```

d.
```
PROGRAM ABC
A = 5.5
B = 6.2
READ * , A , B
S = A + B
PRINT * , S
END
```

e.
```
PROGRAM XYZ
READ * , X , Y
P = X * Y
PRINT * , X , Y , P
END
```

The data:

```
X = 5.1    Y = 6.2
```

1.6 What will the following programs print? Are there any flaws?

a.
```
        PROGRAM INFINT
        N = 0
58      READ * , U , V
        S = U + V
        N = N + 1
        PRINT * , N , U , V , S
        GO TO 58
        END
```

b.
```
        PROGRAM SUM
        N = 0
        JSUM = 0
   88   N = N + 1
        JSUM = JSUM + N
        PRINT * , N , JSUM
        GO TO 88
        END
```

1.7 Write a complete program which calculates SUM and PROD from the following formulas for several sets of data:

SUM = A + B + C
PROD = A * B * C

Print the data on one line, and SUM and PROD on the next line.

☐ **Answers**

1.2
  a. B = X**A − P*(3*X + Y)
  b. B = (X + Y + Z)/(X + 2*Y) + 3*Z
  c. B = X*Y*Z + A − 3*X**4

1.3 c., f., h., i.

1.4 Neither a., b., nor c.

1.5
  a.
```
  PROGRAM FIRST
  ABC = 89.5
  S = 99.3
  AVG = (ABC + S)/2.0
  PRINT * , ABC , S , AVG
  END
```
  b. First, the name of the program cannot be the same as one of the variables; second, END must be the last statement; and finally, the PRINT statement needs a comma.
  c. The procedure for calculating the average must be written out. The program should be similar to **1.5a**.
  d. READ *, A, B does the same thing as A = 5.5 and B = 6.2. Therefore, A = 5.5 and B = 6.2 should be omitted.
  e. The data should be inserted without X= and Y=, thus:

```
5.1    6.2
```

1.6
  a. It prints the values of N, U, V, S respectively, where U, V are the input data, S is the sum of each set of data, and N represents the sequence numbers 1, 2, 3, ... Notice also the infinite loop.
  b. Prints N and JSUM, where N is the sequence number and JSUM is the sum of the sequence numbers. The printout will look like this:

32 Basic Programming Concepts

```
              1      1
              2      3
              3      6
              4     10
              ⋮      ⋮
```

There is again an infinite loop.

1.7
```
         PROGRAM TEST
    15   READ * , A , B , C
         SUM = A + B + C
         PROD = A * B * C
         PRINT * , A , B , C
         PRINT * , SUM , PROD
         GO TO 15
         END
```

■

ANSWERS TO SELECTED EXERCISES

1.1

a.
```
PROGRAM QUE
A = 1000.5
C = 56.1
D = C*A/2.5
PRINT * , A
END
```
b.
```
PROGRAM MY
A = 1296.5
B = A*A
PRINT * , B
END
```
c.
```
PROGRAM SUM
X = 3296.50
Y = 56.20
Z = X + Y
PRINT * , Z
END
```

1.2

a. A = 1/X + 1/Y
b. B = 1/(X + Y)
c. X = A * B + C * D
d. P = 2.5 * R + 3.14 * R ** 2 + 6.28
e. Q = A ** 5 + (5 * A) ** 3

1.3

a. $X = B^3 + A$

b. $R = A + B + \dfrac{C}{D}$

c. $Q = \dfrac{X^3}{A + B + C + 4.5X}$

1.5 d., f., j., l.

1.7 c. 2 3 31 6 5

Chapter 2
Input–Output

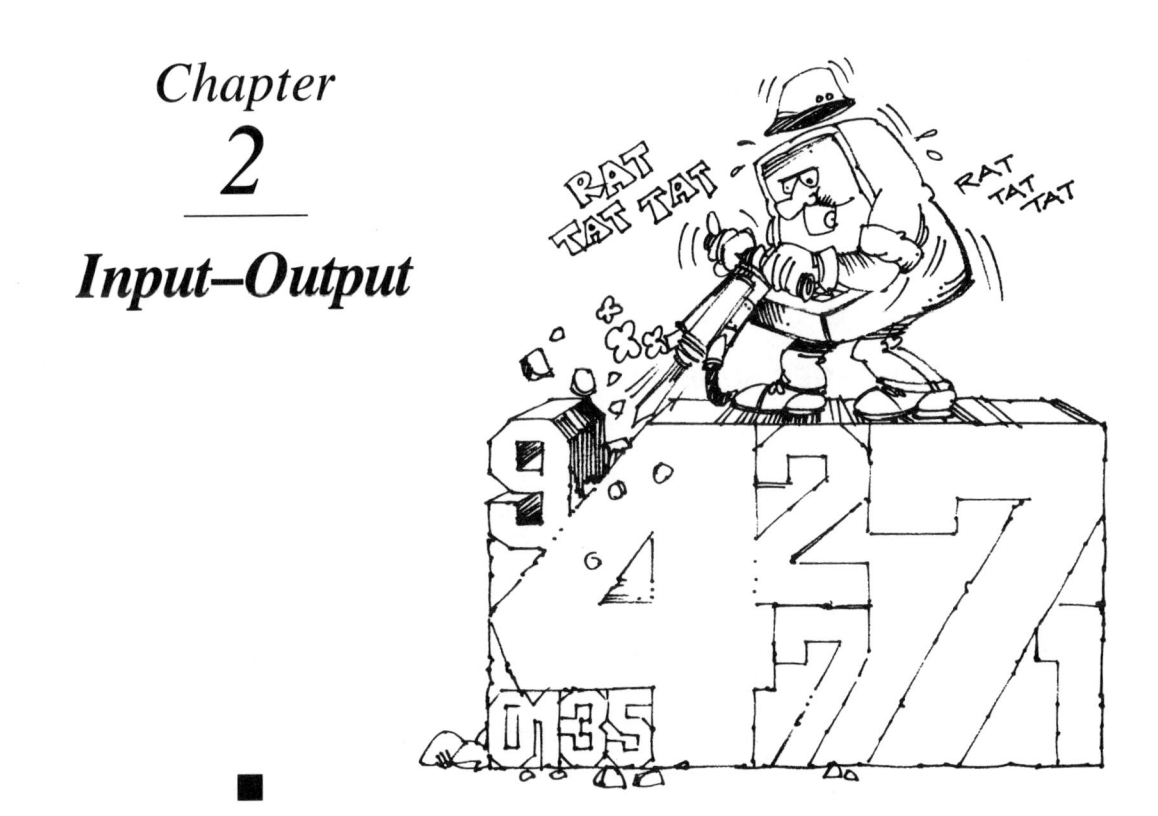

REAL AND INTEGER NUMBERS

    Numbers in FORTRAN
    Mixed Modes of Operation
    Field Descriptions

THE FORMAT STATEMENT: USING FIELD DESCRIPTORS

    Formatted READ Statement
    Formatted WRITE Statement
    Selecting a Field Descriptor
        The Use of Spaces
        Omitting the Decimal Point
        Blank Spaces in Data Field
    Explanatory Comments on the Output
        Printing a Heading
    More Notes about FORMAT
    Formatted PRINT statement

SOME PROGRAMMING TECHNIQUES

    The STOP Statement
    Controlling an Infinite Loop
    Generating Sequence Numbers
    Calculating a Sum

Typically, data are recorded on a piece of paper or punched on a card. Before any processing can occur, the data must be transferred to the computer. FORTRAN allows us to place the data in a desired form and to direct the computer to read them in whatever way we wish. Furthermore, we can also direct the computer to print the information in a desired form, length, and location. This can be accomplished with a FORMAT statement in FORTRAN, as explained in this chapter. Before discussing the FORMAT statement itself, a few important points should be explained.

REAL AND INTEGER NUMBERS

Numbers in FORTRAN

There are several types of data in FORTRAN. Two important types of numerical data are *real* and *integer*.

1. A *real* number is a numeral written with a decimal point, such as 93.56, 8.1, 9321.12, −56.20.
2. An *integer* number is a numeral written without a decimal point, such as 53, 193216, 131, or −92.

Let's look at some more examples of real numbers:

```
562.1234    32.19
   -1.0     60.
+8000.0      5.0
12000.       0.0
```

More examples of integer numbers are:

```
-621      +123     10001    1
600196    10000      -52    0
```

An integer-type variable is a variable which can accept integer data, and a real-type variable is a variable which can accept real data. FORTRAN distinguishes between the two, so it is very important to determine whether the variable is real or integer. The following rules apply to a variable's symbolic name.

> **RULE 2.1**
> *The first character of an integer variable's symbolic name must be one of the following letters:* I, J, K, L, M, N.

Some proper names for integer variables are I, ICOUNT, IOWA, JEAN, KOUNT, MOUNT, MULT, LARRY, NOON, I1, IRK, K5.

> **RULE 2.2**
> *The first character of a real variable's symbolic name must be a letter* other *than* I, J, K, L, M, and N.

Some proper names for real variables are AVERAG, COUNT, EKON, ENEGER, X, A1, AAA, WIDE, BOB, B1, X5.

You must follow these rules throughout your program. The reason they exist is that FORTRAN handles real and integer data differently. The programmer, therefore, must keep them separate so that FORTRAN can generate the proper code for each. In chapter six we learn how to change these rules explicitly in a program if desired.

Remembering the rule is easy: you need only recall the first two letters of the word *integer* as a mnemonic device, since the letters I through N are used to begin every integer variable's symbolic name.

SOLVED PROBLEMS

2.1 **a.** Which of the following may *not* be used as names for *real* variables?

    a. FORT   **b.** GEES   **c.** JEE   **d.** A152B
    e. COURT   **f.** KEEN   **g.** OPZ   **h.** 2CDE
    i. X$P   **j.** M*N   **k.** X.Y   **l.** X+Y

b. Which of the following may *not* be used as names for *integer* variables?

    a. I   **b.** ME   **c.** LIST   **d.** IN
    e. COPY   **f.** JOHN   **g.** TERRY   **h.** L1
    i. KAY   **j.** MEAN   **k.** MN+Y   **l.** K56
    m. X   **n.** M.N   **o.** M   **p.** ENTEGER

☐ **Answers**

    a. c, f, h (starts with a number); and i, j, k, l (include special characters).
    b. e, g, k, m, n, and p.

☐ **Mixed Modes of Operation**

The term *mixed mode of operation* applies to cases when both integers and reals are found in the same statement. For example, all of the following equations contain mode mixing:

```
A = 52 + 285.0
X = 52 + A
K = 5.5 + 3
B = X + 1
R = K * X
MULT = A*B + 2
```

Mixed modes should be avoided when possible, for two reasons:

  1. Since FORTRAN handles integers and reals differently, the processing of mixed modes is less efficient than the processing of either real data or integer data.
  2. When reals and integers are mixed in an equation and the programmer is not aware of it, the resulting values can be other than expected. This

36 Input–Output

Table 2.1 Examples of Mixed Modes

STATEMENT	WILL YIELD
PROD = 5.6 * 3.29	PROD = 18.424
MULT = 5.6 * 3.29	MULT = 18
P = 63/4[a]	P = 15.0
X = 5/8 + 3	X = 0 + 3 = 3.0
XX = 5.0/8.0	XX = .625
Z = 5.0 * (5/8)	Z = 0.0
ZZ = 5.0 * 5/8	ZZ = 3.125
A = 35/3 + 8.4	A = 19.4
AA = 35.0/3.0 + 8.4	AA = 20.066

[a] 63/4 yields 15; conversion to real yields 15.0.

may happen because the integer operations do not carry the fractional part. In particular, in an integer division such as 10/3 the result will be simply 3. Also, whenever a real number is assigned to an integer variable, the result is an integer. That is, the value of the integer variable will not have the fractional part. For example, the statement

I = 5.5 + 3.2

will yield

I = 8

where the fractional part (.7) has been cut off (truncated).

If you use mixed modes, make sure you are aware of the outcome of the operations. Table 2.1 illustrates the outcome of several equations when mixed modes are used.

The outcome of an operation depends on the mode of the values in that operation. Table 2.2 specifies the outcome of simple operations.

Sometimes we may purposely assign a real value to an integer variable in order to truncate the decimal portion. But unexpected results may occur when the integers and reals are mixed unintentionally. To minimize the possible pitfalls of mode mixing and to facilitate more efficient processing:

1. Always assign an expression to an appropriate variable name, integer or real. For example, you might use Z = X + Y rather than I = X + Y if an integer result is not desired.
2. Write the constant numbers in an appropriate mode. For instance, you might use 3.0 instead of 3 and vice versa.

Table 2.3 provides further examples.

Table 2.2 The Outcome of Simple Operations

MODES IN THE OPERATION			THE OUTCOME	EXAMPLE
Real * Real	or	Real/Real	Real	21.6/2.0 = 10.8
Real * Integer	or	Real/Integer	Real	21.6/2 = 10.8
Integer * Integer	or	Integer/Integer	Integer	9/3 = 3; 9/12 = 0
Real + Integer	or	Real − Integer	Real	6.3 + 2 = 8.3

Table 2.3 Examples of Mixed Modes Which Can Be Avoided

MIXED MODE (AVOID IF POSSIBLE)	SAME MODE
X = 1093	I = 1093 or X = 1093.0
AVG = SUM/3	AVG = SUM/3.0
K = 5.5 + 3	R = 5.5 + 3.0
MULT = A * B * C	PROD = A * B * C
W = 3 * Y + C	W = 3.0 * Y + C
IV = 3.0 * D	ZV = 3.0 * D
I = K + 3.0 * L + 3.0	I = K + 3 * L + 3

SOLVED PROBLEMS

2.2 When writing a FORTRAN program, which of the following expressions should be avoided if mixed modes are not desired?

  a. I = 629.19
  b. BETH = 5962
  c. COOL = I + J
  d. JIM = 10001
  e. CAL = −6296.51
  f. AVG = (A + B)/2
  g. M = N + I
  h. REED = GEE/FALL
  i. MULT = A * B * C
  j. JAVG = SUM/30
  k. XY = VW
  l. IVY = J + I
  m. N = M/L
  n. K = 3 * J + I
  o. I = I + 1

2.3 Suppose X = 5.1, Y = 1.6, and L = 2; what is the value of W or I in each of the following statements?

  a. I = X + L
  b. W = X + Y
  c. I = X + Y
  d. I = (X + Y)/L
  e. W = X * Y
  f. I = X * Y
  g. W = 2.0 * Y + X/2.0 + L
  h. I = 2.0 * Y + X/2 + L
  i. W = X * X * Y * Y/10
  j. I = X * X * Y * Y/10.0

2.4 What would be printed if the following programs were executed?

  a. PROGRAM MODE
     I = 629.19

```
             PRINT * , I
             END
         b.  PROGRAM CUTE
             J = 98.5/10.0
             PRINT * , J
             END
         c.  PROGRAM MIXED
             A = 10.5
             N = 4
             B = A + N
             K = A + N
             PRINT * , B, K
             END
         d.  PROGRAM WRONG
             R = 9/10
             PRINT * , R
             END
         e.  PROGRAM VOID
             X = 25/2 + 5.7
             Y = 25.0/2.0 + 5.7
             PRINT * , X, Y
             END
         f.  PROGRAM ZIP
             X = 59.600 * (7/8)
             PRINT * , X
             END
```

☐ **Answers**

2.2 Although all the expressions are permitted, the variables name could be chosen to reflect the appropriate mode in **a., b., c.,** and **i.** The equation in **f.** can be written as AVG = (A+B)/2.0: One is unlikely to want the result of a division operation to be an integer; therefore, the variable on the left in **j.** and **m.** should be real variables (unless it is chosen intentionally).

2.3
  a. I = 7
  b. W = 6.7
  c. I = 6
  d. I = 3
  e. W = 8.16
  f. I = 8
  g. W = 7.75
  h. I = 7
  i. W = 6.65856
  j. I = 6

2.4
  a. 629
  b. 9
  c. 14.5   14
  d. 0.0
  e. 17.7   18.2
  f. 0.0

EXERCISES

2.1 In FORTRAN would the following numbers be integer or real?

    a. 5629.5   **b.** 621   **c.** +921.500
    d. 29.0   **e.** 62.   **f.** −521.1
    g. 32   **h.** −3291   **i.** −29.50000
    j. .0069   **k.** −5.328917   **e.** 5632897

2.2 Which of the following normally specify integer variables?

    a. LENGTH   **b.** COOL   **c.** JOOL
    d. XYZ   **e.** ENEGER   **f.** ISUM
    g. MULTI   **h.** NORM   **i.** J1

2.3 Which of the following normally specify real variables?

    a. MUST   **b.** MODE   **c.** MEAN1
    d. WAGE   **e.** RATE   **f.** NORMAL
    g. SUM   **h.** PLUS   **i.** III

2.4 Which of the following might be avoided because of mixed modes?

    a. WAGE = 11.50 + 40 * R
    b. RATE = SUM/HOURS + TOT/N
    c. MULT = A * B * C
    d. PROD = V * W + 3 * X
    e. TOTAL = 3 * B/2
    f. TOP = MULT * J
    g. IRON = 3.0 * R + 5.0 * X
    h. SUM = A1 + B1 + C1 + MAY
    i. MEAN = SUM/5.0
    j. PROD = Y * Z * I
    k. INCOME = WAGE * RATE
    l. MARK = JAHN/N

2.5 Suppose C = 3.2, D = 4.5, and M = 3; what would be the value for Z and N in the following expressions?

    a. Z = C + D
    b. N = C + D
    c. Z = (C + D)/2.0
    d. N = (C + D)/2
    e. N = 2 * C + D/2 + M
    f. A = 2.0 * C + D/2.0 + M
    g. Z = M/4
    h. Z = 5.0 * (M/5)
    i. Z = (5.0 * M)/5
    j. Z = C * D
    k. N = C * D
         Z = N/10
    l. N = C * D * M/10
    m. N = M/5
    n. N = 97.9/10.0

Field Descriptions

It was previously mentioned that we can place data in any desired form and then direct the computer to *read* them. To do this, we should specify (1) whether the data item is real or integer, (2) how many columns it occupies, and (3) if the data item is real, the number of digits that appear after the decimal point. As we have learned, the symbolic name of the variable specifies its type. For example,

```
READ*, X, K
```

implies that the first datum (X) is a real number and the second one (K) is an integer.

The specific area that is assigned to datum is called a *field*. The following important rules are applied to describe the length of a field. These rules are called *field description rules*.

RULE 2.3 FIELD DESCRIPTIONS

1. *I field (for integer numbers):*
   In is used for showing the length of an integer type of data, where n shows the number of columns which the number occupies. For example, I3 might be used to display the number 197. The following are additional examples of using an I specification:

Number:	1979	123456789	−2319610	3	+193	−5230
Field descriptor:	I4	I9	I8	I1	I4	I5

2. *F field (for real numbers):*
   F$n.m$ is used for showing the length of real data, where n shows the number of columns which the number occupies (including all the digits, the decimal point, and the minus or plus sign if any), and m shows the number of columns which appear to the right of the decimal point. For example, F7.2 might be used to display the number 1973.45 (i.e., a total of 7 columns, including 2 decimal places and 1 decimal point). The following are some examples of F specifications:

Number:	621.2	+62.0	162821.621	62.	−93.5	1.00
Field descriptor:	F5.1	F5.1	F10.3	F3.0	F5.1	F4.2

3. *X field (for spaces):*
   When recording data, we use spaces to separate the fields. nX is used for showing blanks (or spaces) between the fields on the data line, where **n** represents the number of blanks. Examples:

1 space	5 spaces	51 spaces
1X	5X	51X

A few more field descriptors will be discussed in chapters 6 and 11.

Note that the letters I, F, and X are always the same for field descriptors; that is, whatever the name of the variable, I, F, and X will represent integer, real, and spaces. The next section demonstrates how these notations are used to describe the data. But before going further, look at the following solved problems.

SOLVED PROBLEMS

2.5 Use the appropriate field descriptors for showing the following data. Separate the descriptors by a comma. Note: ƀ is used for showing a blank space on the paper (X is for specifying a blank between the fields).

a.

```
6293.15ƀƀƀ6212ƀƀƀƀ321.0ƀ62ƀ-3.5
```

b.

```
ƀƀƀƀƀƀƀ629ƀƀƀ5ƀƀ123456ƀ6231.912ƀ-532
```

c.

```
ƀ6219ƀƀƀƀƀƀƀƀƀƀƀ1962.ƀƀ62.ƀƀ5ƀ-32.5
```

☐ **Answers**
  a. F7.2, 3X, I4, 4X, F5.1, 1X, I2, 1X, F4.1
  b. 7X, I3, 3X, I1, 2X, I6, 1X, F8.3, 1X, I4
  c. 1X, I4, 11X, F5.0, 2X, F3.0, 2X, I1, 1X, F5.1

THE FORMAT STATEMENT: USING FIELD DESCRIPTORS

A FORMAT statement, accompanied by a READ or WRITE statement, describes the exact position of each datum through the use of field descriptors. This is explained in detail in the following sections.

☐ Formatted READ Statement

To read the data with a desired specification, a combination of READ and FORMAT statements is used. Example 2.1 demonstrates the method.

EXAMPLE 2.1

Problem: Write a program to calculate the floor area of a room. The length of the room is 12.25 and the width is 9.5. The data is to be placed on a line as follows:

```
ƀƀ12.25ƀƀƀƀƀ9.5
```

Program:

```
        PROGRAM FIRST
        READ(5,20) A, B
20      FORMAT (2X , F5.2 , 5X , F3.1)
        C = A * B
        PRINT *, C
        END
```

Data: In the data section:

```
ƀƀ12.25ƀƀƀƀƀ9.5
```

Printout: The printout will show:

```
116.375
```

Notes:

1. The READ statement contains two numbers in parentheses. The first, 5, is a fixed number and refers to the unit from which the data are transferred; this number depends on the computer installation. The second, 20, is the label of the FORMAT for the READ statement, and is chosen by the programmer.
2. The FORMAT is a statement in FORTRAN which shows the data specifications.

Here is another example: The following program reads and calculates the average of two numbers. The program has a GO TO statement to allow the program to reexecute a new set of data. The loop is an infinite loop.

```
        PROGRAM FORM
220     READ (5,99) FIRST, SECOND
99      FORMAT (F6.2 , 1X , F6.2)
        SUM = FIRST + SECOND
        AVRG = SUM/2.0
        PRINT * , AVRG
        GO TO 220
        END
```

There should be more than one data line and each should look like this:

```
329.65ƀ512.35
```

Note that there is no space at the beginning of the data record.

Thus, a *formatted read* statement generally has the following form with nine important components:

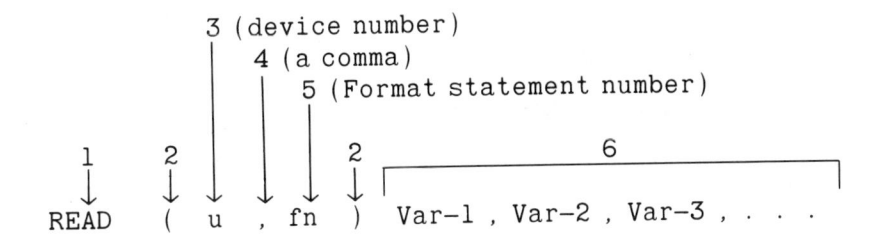

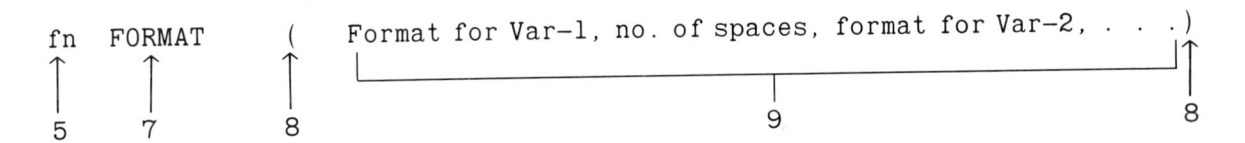

```
                    3 (device number)
                     │ 4 (a comma)
                     │ │ 5 (Format statement number)
       1   2         │ │ 2            6
       ↓   ↓         ↓ ↓ ↓ ┌──────────────────────┐
       READ  (   u   ,  fn )  Var-1 , Var-2 , Var-3 , . . .


  fn   FORMAT   (   Format for Var-1, no. of spaces, format for Var-2, . . . )
  ↑      ↑      ↑   └──────────────────────┬──────────────────────┘          ↑
  │      │      │                          │                                 │
  5      7      8                          9                                 8
```

1. The word READ.
2. A pair of parentheses enclosing items 3–5.
3. An integer number u: this specifies the unit from which the data is transmitted. Many installations use 5 to refer to the INPUT unit.
4. A comma for separating u and fn.
5. The integer fn, the label for the FORMAT.
6. The variable list (variables are separated by commas).
7. The word FORMAT.
8. A pair of parentheses enclosing item 9.
9. The format list, composed of the field descriptors for the variables and for the spaces between the fields. Field descriptors are separated by commas.

Pay attention to the following points when writing a FORMAT statement:

1. The variables must agree with their field descriptors from the standpoint of both order and type. For example,

         READ (5,10) AAA, KODE
    10   FORMAT (I2 , 1X , F5.2)

   is not correct, because the variables do not match their field descriptors.
2. If there is any space between the fields, it must be specified precisely. However, spaces are not always used. For example: if A = 69.5, B = 295.7, you may place them as

   ┌─────────────┐
   │ 69.5295.7   │
   └─────────────┘

   Then the READ and FORMAT for reading A and B are

         READ (5,20) A, B
    20   FORMAT (F4.1 , F5.2)

3. If you have *k* identical field descriptors, you can summarize them by writing *kFn.m*, or *kIn*. For example (2F3.2 , 3I2) is equivalent to

   (F3.2, F3.2, I2, I2, I2)

SOLVED PROBLEMS

2.6 Find any errors in the following statements:

a.
```
          READ (5, 110) A, B, L
110       FORMAT (F6.2 1X F5.2 IX I2)
```

b.
```
          READ (5, 66) X, Y, Z
 66       FORMAT (F10.2 , I5 , F5.0)
```

c.
```
          READ (99, 5) JAY, FAY
 99       FORMAT (I10 , F6.1)
```

d.
```
          READ (5, 20) V, W
 20       (F10.2 , F8.2)
```

e.
```
          READ (5, 121) A, B, C, D
121       FORMAT (F10.2 , F9.1 , F8.2)
```

f.
```
          READ (5,511) II, JJ, X, Y
511       FORMAT (I5 , F8.2 , I3 , F5.1)
```

g.
```
          READ (5, 11) K, L, M
 11       FORMAT (K3 , L5 , M2)
```

h.
```
          READ (5, 39) A, B
 39       FORMAT (2F5.1)
```

i.
```
          READ (5, 100) X, Y, Z, KODE, MODE
100       FORMAT (3F6.2 , 2I5)
```

☐ **Answers**

a. The commas for separating field descriptors are missing.
b. Y is a real variable, and is given an integer specification (I5). The type of the variables and their field descriptors must match.
c. (99, 5) should be (5, 99).
d. The word FORMAT is missing.
e. There are four variables (A, B, C, D), but there are only three field descriptors. (However, you will see in Chapter 6 that this kind of FORMAT will work under certain conditions.)

f. The variables JJ and X do not match their field descriptors.
g. The field descriptors cannot be K, L, or M (they should be I3, I5, and I2).
h. This is correct. Note: (2F5.1) is equivalent to: (F5.1, F5.1).
i. This is correct.

☐ Formatted WRITE Statement

To see the results of a program, we formerly used a PRINT statement, which did not allow control over the location of the printed data. However, we *can* use a FORMAT statement with a PRINT or WRITE statement to control the location and the sizes of the data to be printed. Because the general arrangement of the formatted WRITE is quite similar to that of the formatted READ, it is explained first in this chapter.

EXAMPLE 2.2

■ **Problem:** The floor area of the room in Example 2.1 is calculated and printed in the following program, using a WRITE statement.

Program:

```
        PROGRAM SECOND
        READ(5,20)A,B
20      FORMAT(2X, F5.2, 5X, F3.1)
        C = A * B
        WRITE(6,30)C
30      FORMAT(1X, F8.3)
        END
```

Sample data:

```
bb12.25bbbb9.5
```

Output:

```
 116.375
```

Notes:

1. The number 6 is used in the parentheses for the WRITE statement (where the number 5 was used for the READ statement). This represents the device number which records the output (such as the printer) and depends on the computer installation.
2. The numbers 30 and 20 are arbitrarily chosen for the labels of the formats.
3. The value of the variable C is to be printed in 8 columns; three of them are digits after the decimal point. This is specified by the programmer.

> 4. The 1X before F8.3 in the FORMAT of the WRITE statement is used as a carriage control character. It allows the printer to start a line, and will be explained in Chapter 3. At this time, however, keep the following rule in mind:

RULE 2.4

A carriage control character is always used at the beginning of the format of a WRITE statement, before any description. Use the code for a blank (1X) to start a new line.

Thus, the general form of a *formatted WRITE* statement is composed of ten components and appears as follows:

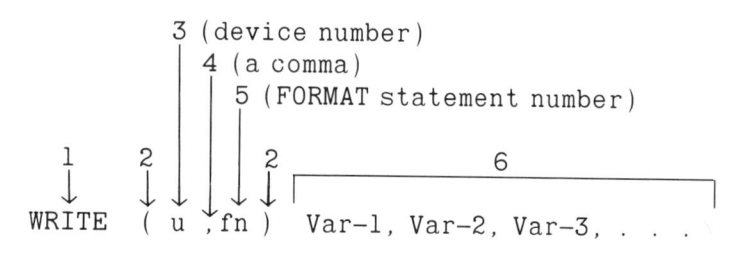

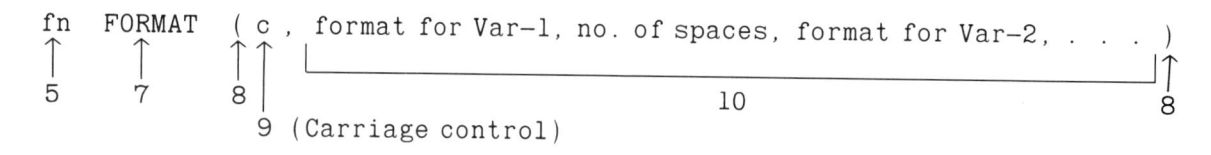

```
              3 (device number)
              4 (a comma)
              5 (FORMAT statement number)
    1    2       2         6
    ↓    ↓       ↓   ┌───────────────┐
  WRITE ( u , fn )  Var-1, Var-2, Var-3, . . .

  fn  FORMAT  ( c , format for Var-1, no. of spaces, format for Var-2, . . . )
  ↑     ↑     ↑ │   └─────────────────────────────────────────────────────┘ ↑
  5     7     8 │                          10                               8
                9 (Carriage control)
```

1. The word WRITE.
2. A pair of parentheses enclosing items 3–5.
3. An integer u; this specifies the unit which the data are transmitted to. Many installations use 6 to refer to the OUTPUT unit (the printer).
4. A comma for separating u and fn.
5. The integer fn which is the label for the FORMAT.
6. The variable list; each variable is separated by a comma.
7. The word FORMAT.
8. A pair of parentheses.
9. A carriage control character (1X is such a character).
10. The format list, composed of the field descriptors for the variables and the spaces between them, separated by commas.

☐ Selecting a Field Descriptor

Choosing the proper field descriptor for a variable is a critical matter. The collection of field descriptors that constitute a single format statement become a road map to the exact location and length of each datum on a record.

You will see in Chapter 3 that the layout for the output and input must be designed before writing the FORTRAN statements. For input, after the exact location of each data item has been designed, the field descriptors and spaces must then be precisely specified in the FORMAT. You should pay particular attention to the format of the READ statement. This is of great importance and is particularly troublesome for beginning programmers.

For output design, choosing a field descriptor for the WRITE is an easier matter. It is not necessary to specify the *exact* length of a field when writing. Therefore, when the length of an output *data* field varies from one data item to another, or when the size of a field cannot be estimated, you should select a large enough field descriptor, within a reasonable range, to allow ample space for any number. For instance, if you wish to print the result of

```
A = 25.5 * 41.1
```

you may simply choose F10.2 to allow enough space for a very large number to fit. Thus:

RULE 2.5
For printing the data, choose a larger field descriptor than appears necessary.

Even if you know the exact length of a number, you do not have to use it in the WRITE field descriptor. For example, it is quite possible to choose F10.2 for a number which was read with a field descriptor of F3.1.

If you make a field length too small for a number to be printed, asterisks (*) will be printed instead. For example, if you choose F4.1 for the number 653.2, then

```
****
```

will be printed. If the field descriptor is larger than necessary, the value will be printed right justified; that is, the value will be printed on the right side of the field with extra spaces on the left.

The number of digits after the decimal point can always be selected according to the desired precision. For example, if you would like to print an average of exam marks with only two decimal places, you may use F5.2 as the field descriptor. In this case, even if the average has more than two decimal places, it will be rounded to two decimal places.

The Use of Spaces

Still another point which should be remembered is the use of X for specifying the spaces between the fields. Spaces do not have to be used in the input fields. In fact, some programmers prefer not to use blanks to save spaces. Look at the following example:

Suppose A = 56.21, B = 8.1, C = 95.231, and D = 5.6, and the data are placed as:

```
56.218.1b95.231bbbb5.6
```

The correct format to read these numbers is:

```
        READ(5,21)A,B,C,D
    21  FORMAT(F5.2 , F3.1 , 1X , F6.3 , 4X , F3.1)
```

It is apparent that you may use one or more spaces between two fields to separate the data as long as you specify it in the FORMAT. You may also use a comma to separate the fields, simply by treating it as a space in the FORMAT. For example, assume that commas are used to separate the numbers on a data card, such as:

48 Input–Output

> 562.90,32.1,56,932.563

The correct format to read these numbers is:

```
        READ(5,100)A,B,I,C
100     FORMAT(F6.2 , 1X , F4.1 , 1X , I2 , 1X , F7.3)
```

> **RULE 2.6**
> When a comma is used to separate the fields on the data line, enter it as 1X in the FORMAT.

In the WRITE statement, make sure to use spaces generously between the fields in order to make the printout more readable.

As an example, the following WRITE and FORMAT print the values of A, B, C, and D with five spaces in between:

```
        WRITE (6,25) A, B, C, D
25      FORMAT (11X , F10.2 , 5X , F10.2 , 5X , F10.2 , 5X , F10.2)
```

Note that there are *ten* spaces at the beginning. (The first X field is for the carriage control.)

SOLVED PROBLEMS

2.7 Find all the errors in the following statements.

a.
```
        WRITE (6,25) A, B
25      FORMAT (F6.1 , F5.1)
```

b.
```
        WRITE (6, 55), N, X
55      FORMAT (1X, I5, I4)
```

c.
```
        READ(5,100)A
100     FORMAT (F6.1)
        WRITE (5, 62) A
62      FORMAT (1X, F10.1)
```

d.
```
        WRITE (6, 17) A, B, C
17      FORMAT (1X, 3F10.2)
```

e.
```
        WRITE (6,77) I, P, Q
77      FORMAT (21X, I5, 5X, F10.2, 5X, F10.2)
```

f.
```
        WRITE (99, 6) ART, JOY
99      FORMAT (1X, F4.9, I5)
```
g.
```
        WRITE (6, 21) P
21      (1X, F10.2)
```
h.
```
        WRITE (6, 35) X, Y, Z,
35      FORMAT (1X, F10.2F8.1F5.1)
```
i.
```
        WRITE (6, 55) AAA, III
        FORMAT (1X, F10.2, I5)
```
j.
```
        WRITE (3, 30) A
30      FORMAT (21X, F10.2)
```
k.
```
        WRITE (6, 80)
80      FORMAT (21X , F10.2 , 5X , F8.2)
```
l.
```
        WRITE (6, 90) A, B, C
90      FORMAT (21X, 10X, 5X, 10X, 5X, 10X)
```
m.
```
        WRITE (6, 100) W, X, Y, Z
100     FORMAT (11X, F10.2, 15X, F8.2, 10X)
```
n.
```
        WRITE (6, 110) A, I, B, K, C, L
110     FORMAT (1X, F6.1, I3, F7.1, K4, F8.1, L5)
```
o.
```
        WRITE (6, 120) X, N, P
120     FORMAT (11X, F6.1, F8.0, I5)
```
p.
```
        WRITE (6, 130) A, B, C,
130     FORMAT (11X, F6.1, 5X, F7.1, 5X, F8.1)
```

☐ **Answers**

    a. The carriage control, 1X, at the beginning of the format is missing.

    b. The comma after the parenthesis is not necessary, and the field descriptor for X does not agree with the type of the variable.

    c. The device number for the WRITE is the same as the device number for the READ. This could be correct if the input file is to be rewritten. But that is unlikely for this example.

50 Input–Output

d. This is correct. Note that (1X, 3F10.2) is equivalent to (1X,F10.2,F10.2,F10.2), and there is no space between the fields.
e. This is correct. Note that there are twenty spaces before printing I, and five spaces between I and P and between P and Q.
f. The device number and FORMAT label should be reordered. Also, the field descriptor for ART is wrong. Do you know of any number which would consist of a total of *four* digits, where *nine* of them are after the decimal point?
g. The word FORMAT is missing.
h. The commas separating the field descriptors are missing. Also, there is no need for the comma after Z. This comma will in fact cause an error in your program.
i. The label for the FORMAT is missing.
j. This is correct; note that the device number is 3 in this case.
k. The WRITE statement does not print any value, because there are no variables listed.
l. The field descriptors of A, B, and C are missing.
m. The number of field descriptors is not equal to the number of variables. (You will learn in Chapter 6 that this kind of FORMAT will work with some precautions.)
n. Only I may be used for field descriptions for integer variables. K4 and L5 must be changed to I4 and I5.
o. The field descriptors do not agree with the type of variables.
p. The comma after C is not necessary.

2.8 What is the value of each variable read by the following READ statements?

a.
```
        READ (5, 999) PAUL, WIDTH, LENGTH
999     FORMAT (2X , F5.2 , 1X , F8.2 , 7X , I3)
```
Data:

```
8953.21,39823.9600,bbb6952
```

b.
```
        READ (5, 29) FIRST, SECOND, THIRD
29      FORMAT (F2.0 , 1X , F2.0 , 2X , F3.0)
```
Data:

```
56,67,8915
```

c.
```
        READ (5,30) X , Y
30      FORMAT (F5.0 , F6.0)
```
Data:

```
           -562.+1389.123
```

d.

```
        READ (5, 22) EEE, Y, I
22      FORMAT (5X, F5.2, 5X, F3.1, 3X, I2)
```

Data:

```
2222222.22222222.2123456789
```

e.

```
        READ (5, 33) JAY, KAY
33      FORMAT (10X, I5, 10X, I5)
```

Data:

```
5621932108000000000000000000
```

f.

```
        READ (5, 15) FAR, ITEM
15      FORMAT (F7.2, I3)
```

Data:

```
5621.50,562
```

☐ **Answers**

a.

PAUL = 53.21
WIDTH = 39823.96
LENGTH = 952

The data are read as follows:

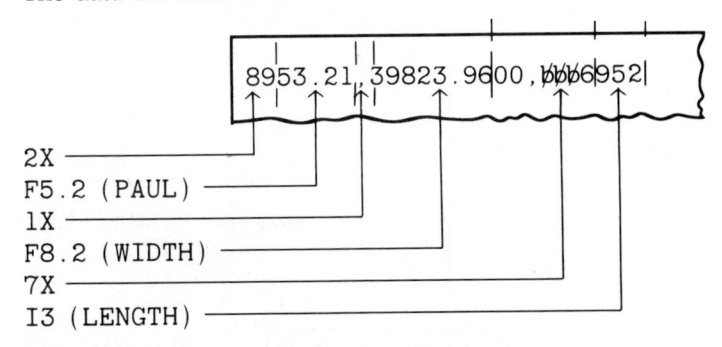

```
8953.21, 39823.9600,  bbb952
```

2X
F5.2 (PAUL)
1X
F8.2 (WIDTH)
7X
I3 (LENGTH)

Note: A comma is considered a blank space.

b.

FIRST = 56   (real, the value is 56.0),
SECOND = 67   (real, the value is 67.0),

THIRD = 915   (real, the value is 915.0).

The data are read as follows:

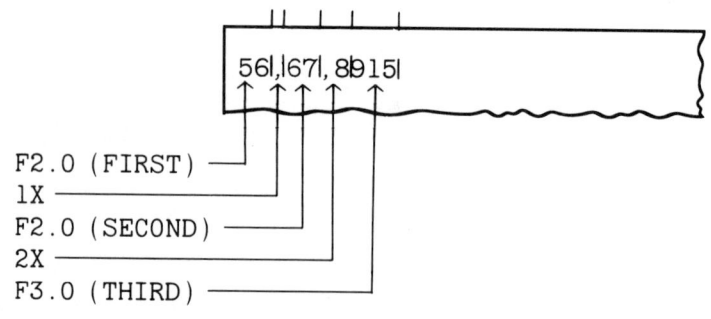

```
          |56|,|67|,|89|15|
           ↑   ↑   ↑  ↑
F2.0 (FIRST) ──┘   │   │  │
1X ────────────┘   │  │
F2.0 (SECOND) ─────┘  │
2X ───────────────────┘
F3.0 (THIRD) ─────────┘
```

c.
   X = −562.   (same as −562.0),
   Y = +1389. (same as +1389.0).

d.
   EEE = 22.22
   Y   = 2.2
   I   = 45

e.
   JAY = 00000
   KAY = 00000

f.
   FAR = 5621.50
   ITEM will not be read, because of the comma before 562. To be correct, the comma must be specified as a space. Thus, the correct format would be:

```
15    FORMAT (F7.2 , 1X , I3)
```

2.9 Which of the following programs are correct? Pay attention to the field descriptors of the variables in the WRITE statements.

a.
```
         PROGRAM AAA
         A = -123.45
         WRITE (6, 20) A
20       FORMAT (F5.1)
         END
```

b.
```
         PROGRAM ABE
         READ (5, 20) A, B
20       FORMAT (F7.1 , 1X , F7.1)
         C = A*B
         WRITE (6, 30) A, B, C
30       FORMAT(1X, F5.1, 2X, F5.1, 2X, F5.1)
         END
```

c.
```
         PROGRAM XXX
         READ (5, 50) NUMBR, X, W
50       FORMAT (I5 , 1X , F5.2 , 1X , F6.2)
         Z = X + W
         WRITE (6, 60) NUMBR, X, W, Z
```

```
              60      FORMAT(1X,I4,5X,F5.3,5X,F6.3,5X,F7.3)
                      END
```

☐ **Answers**

a. The carriage control character is missing; also, the field descriptor F7.1 should be chosen instead of F5.1. The FORMAT should then read:

```
20 FORMAT (1X,F7.1)
```

b. The field descriptors of A, B, and C in FORMAT 30 must be larger. A better FORMAT would be:

```
30 FORMAT (1X,F8.1,2X,F8.1,2X,F10.2)
```

c. Again, the field descriptors in the FORMAT for the WRITE should be larger. A better FORMAT would be:

```
60 FORMAT(1X,I6,5X,F8.3,5X,F8.3,5X,F10.3)
```

Omitting the Decimal Point

When recording the data, it is sometimes convenient to record a real number without a decimal point. This may save space and time when dealing with a large number of data. FORTRAN allows us to do this, and inserts the decimal point automatically in the correct place when it reads the data by using an appropriate field descriptor. For example, suppose a test score of 94.50 is to be placed on a line; if we place it as 9450, then by using field descriptor F4.2, the decimal point will be inserted in front of the last two digits (because F4.2 indicates that the number occupies four columns, and that two of them are assigned for decimals). In fact, in the absence of a decimal point in the field, the field descriptors determine the value of a number. Table 2.4 illustrates the value of several data when read by the field descriptors shown.

If the decimal point is *present* in the field, however, the following rule applies:

RULE 2.7
When the decimal point is present in the field, its existence takes precedence over the specified decimal places described by its field descriptor in the FORMAT statement.

Table 2.4 Values of Numbers Given Field Descriptors

RECORDED DATA	FIELD DESCRIPTION	VALUE
63293	F5.3	63.293
−9999	F5.2	−99.99
563921	F6.6	.563921
329315	F6.5	3.29315
639	F3.0	639.
8930000	F7.3	8930.000

54 Input–Output

Table 2.5 Values of the Numbers When the Decimal Point Is Present

PUNCHED IN CARD COLUMNS 1–5	FIELD DESCRIPTOR	VALUE OF A
02391	F5.2	023.91
23.91	F5.2	23.91
2.391	F5.2	2.391
.2391	F5.2	.2391
−2391	F5.2	−23.91
23910	F5.2	239.10

In other words, the *m* in an F*n*·*m* field descriptor will be ignored if the decimal point is entered directly in the data field (where *n* represents the total number of columns which are occupied by the number including minus or plus signs and the decimal point). As an example, look at the following READ statement:

```
      READ (5,10) A
10    FORMAT (F5.2)
```

Table 2.5 shows what the value of A will be if the data are placed as shown (remember that the field descriptor is F5.2).

Blank Spaces in Data Fields

Sometimes we need a large field to accommodate different sizes of data. For example, if the data are

```
12
49325
58
635
```

and they must be read with the same field descriptor, then we must assign five columns to the field. Of course, if a datum takes less than five columns, some of the columns will be left blank. In normal processing, a blank space in a data field is ignored, except that a field of all blanks has a value of zero.[1] Therefore, a number which is less than the field size can be placed anywhere in the field. Look at the following solved problems for some examples.

SOLVED PROBLEMS

2.10 What is the value of X read by the following READ statement, if any of the following data lines are used?

```
      READ (5, 20) X
20    FORMAT (F8.2)
```

---

[1] Conventional FORTRAN processes all blank spaces included in a field as though they were zeros. Therefore, it is very important to place the data right justified with these systems. In FORTRAN 77, however, blank spaces in a data field are ignored (under a default value) unless otherwise specified by the OPEN statement. The OPEN statement is discussed in Chapter 11.

Data:

|←—8—→|
a.

```
56.78
```

b.

```
56.78
```

c.

```
56.78
```

d.

```
5678
```

e.

```

```

f.

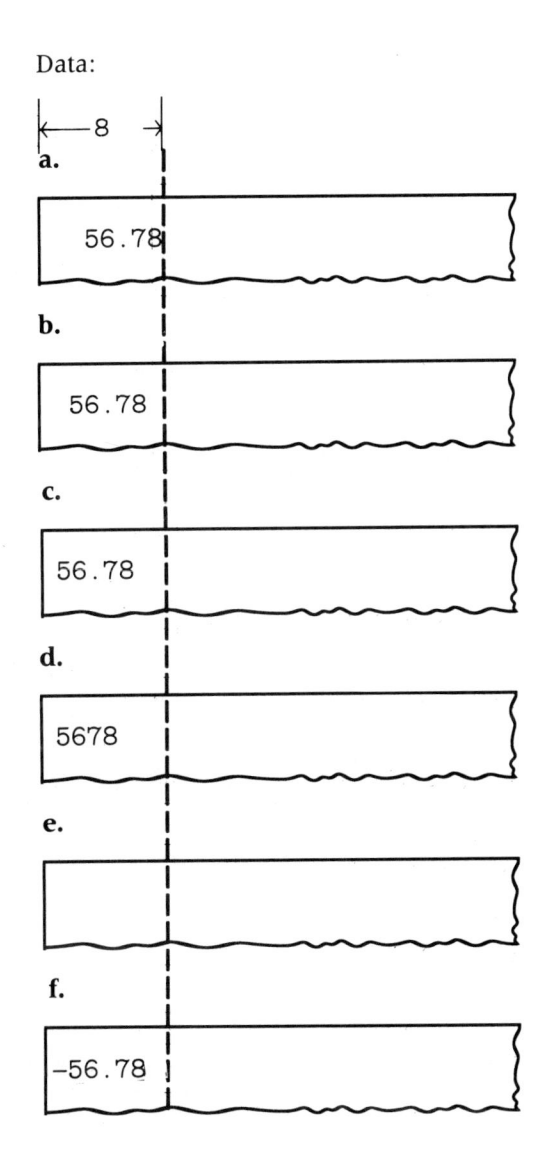

```
-56.78
```

☐ **Answers**

    a. X = 56.78
    b. X = 56.78
    c. X = 56.78
    d. X = 56.78
    e. X = 0.0
    f. X = −56.78

■ **2.11** What is the value of K read by the following READ statement if any of the following data lines are used?

```
        READ (5,70) K
70      FORMAT (I5)
```

Data:

|←—5—→|
a.

```
45
```

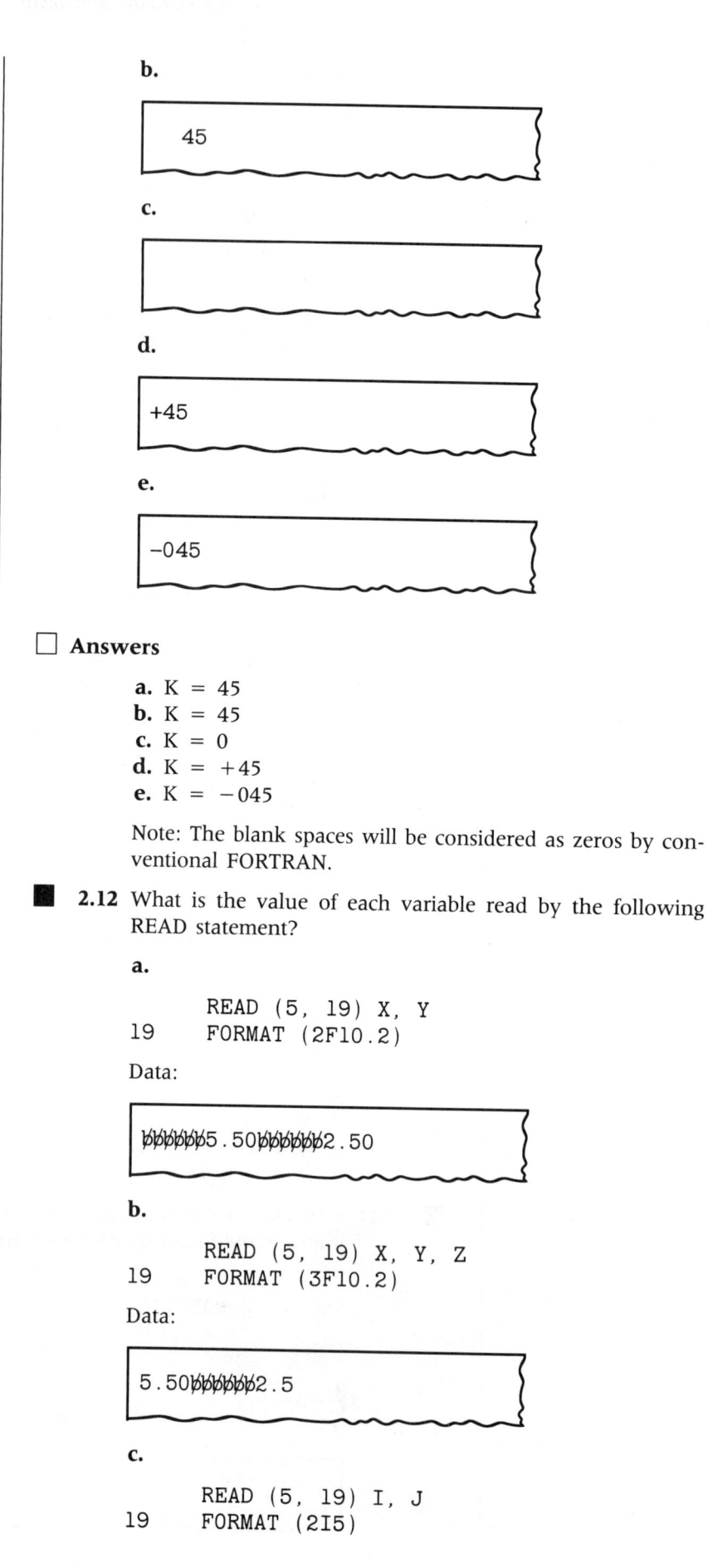

b.

```
45
```

c.

```
```

d.

```
+45
```

e.

```
-045
```

☐ **Answers**

a. K = 45
b. K = 45
c. K = 0
d. K = +45
e. K = −045

Note: The blank spaces will be considered as zeros by conventional FORTRAN.

2.12 What is the value of each variable read by the following READ statement?

a.
```
        READ (5, 19) X, Y
19      FORMAT (2F10.2)
```
Data:

```
bbbbbb5.50bbbbbbb2.50
```

b.
```
        READ (5, 19) X, Y, Z
19      FORMAT (3F10.2)
```
Data:

```
5.50bbbbbb2.5
```

c.
```
        READ (5, 19) I, J
19      FORMAT (2I5)
```

Data:

```
ØØØ5623
```

d.

```
      READ (5, 19) I, J, K
19    FORMAT (3I2)
```

Data:

```
56ØØØ23
```

e.

```
      READ (5, 110) A, I, B
110   FORMAT (F5.2, I5, F5.2)
```

Data:

```
562132161956342
```

☐ **Answers**

a. X = 5.50
    Y = 2.50
b. X = 5.50
    Y = 2.5
    Z = 0.0
c. I = 56
    J = 23
d. I = 56
    J = 0
    K = 2
e. A = 562.13
    I = 21619
    B = 563.42

■ **2.13** Write a READ and a FORMAT statement to read the value of each variable shown.

a. If

AA = 56.25
BB = 19.5
IC = 5

and the data are placed as:

```
56.2519.55
```

b. If

ASACK = 956.25
PALACE = 756.2

58 Input–Output

KAY = 92

and the data are placed as:

```
95625756292
```

c. If

POOL = 89.5
COOL = 99.9
DOOL = 163.1
KOOL = 56

and the data are placed as:

```
89.5,ϕ99.9,163156
```

☐ **Answers**

a.
```
        READ (5,20) AA,BB,IC
20      FORMAT (F5.2,F4.2,I1)
```
b.
```
        READ (5,25) ASACK, PALACE, KAY
25      FORMAT (F5.2,F4.1,I2)
```
c.
```
        READ (5,59) POOL, COOL, DOOL, KOOL
59      FORMAT (F4.1,2X,F4.1,1X,F4.1,I2)
```

☐ **Explanatory Comments in the Output**

So far in our examples, the value of the desired variable has been printed without any explanation. If there is more than one variable, however, the reader of the printout would like to have some description of the numbers. The explanation of the variables can be printed on the printout by the FORMAT of the WRITE statement. The following example shows how:

EXAMPLE 2.3

■ **Problem:** Calculate the average of two numbers. Print the numbers and the average with their descriptions.

Program:

```
        PROGRAM AVRAGE
        READ (5,15) X, Y
15      FORMAT (F3.1, 1X, F4.1)
        Z = (X + Y)/2.0
        WRITE (6,35) X, Y, Z
```

```
    35    FORMAT(1X,'X,Y:',F3.1,3X,F4.1,5X,'AVERAGE:',F5.1)
          END
```

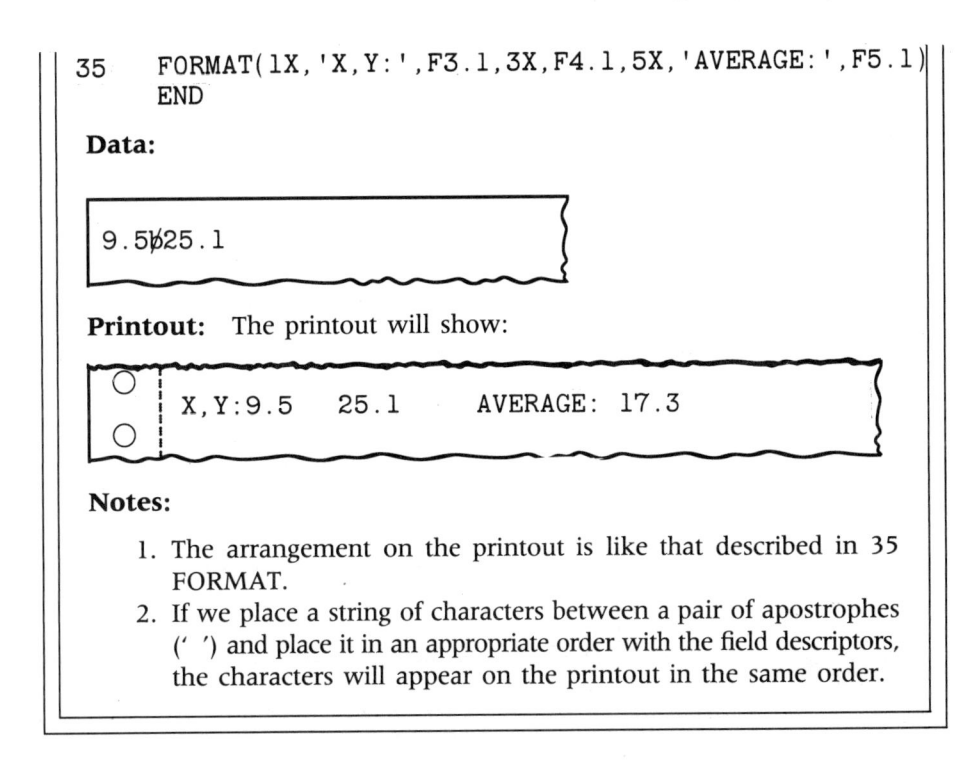

Data:

```
9.5ϕ25.1
```

Printout: The printout will show:

```
X,Y:9.5    25.1     AVERAGE: 17.3
```

Notes:

1. The arrangement on the printout is like that described in 35 FORMAT.
2. If we place a string of characters between a pair of apostrophes (' ') and place it in an appropriate order with the field descriptors, the characters will appear on the printout in the same order.

RULE 2.8
A string of characters in a pair of apostrophes in a FORMAT statement will appear on the printout exactly as they appear in the FORMAT.

Following are more examples of using the WRITE and FORMAT for remarks. It is assumed that A = 56.1 and B = 2.1 in the following examples.

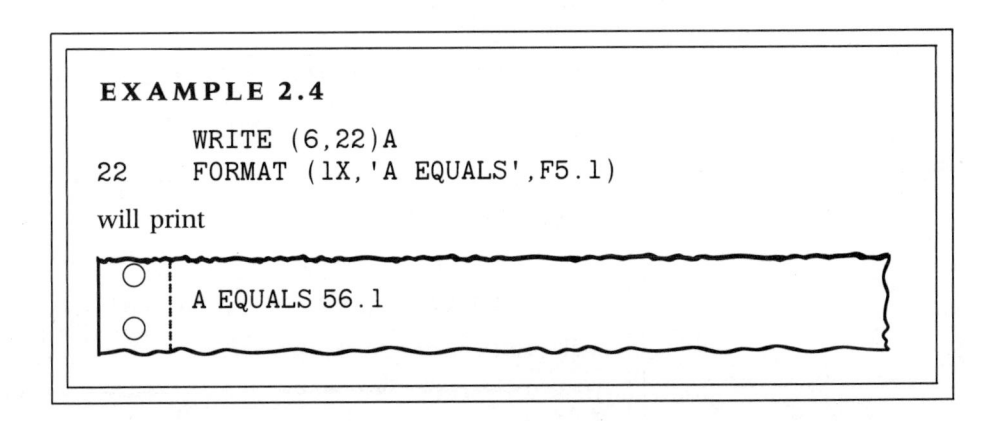

EXAMPLE 2.4

```
          WRITE (6,22)A
    22    FORMAT (1X,'A EQUALS',F5.1)
```

will print

```
A EQUALS 56.1
```

EXAMPLE 2.5

```
          WRITE (6,25)A
    25    FORMAT (1X,F5.1,1X,'IS EQUAL TO A')
```

will print

```
56.1 IS EQUAL TO A
```

EXAMPLE 2.6

```
      WRITE (6,28) A, B
28    FORMAT (1X,'A =', F5.1,5X, ' B=', F5.1)
```
will print

```
A = 56.1     B =  2.1
```

EXAMPLE 2.7

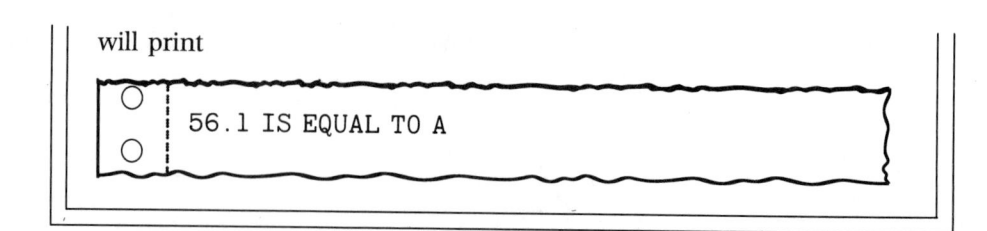

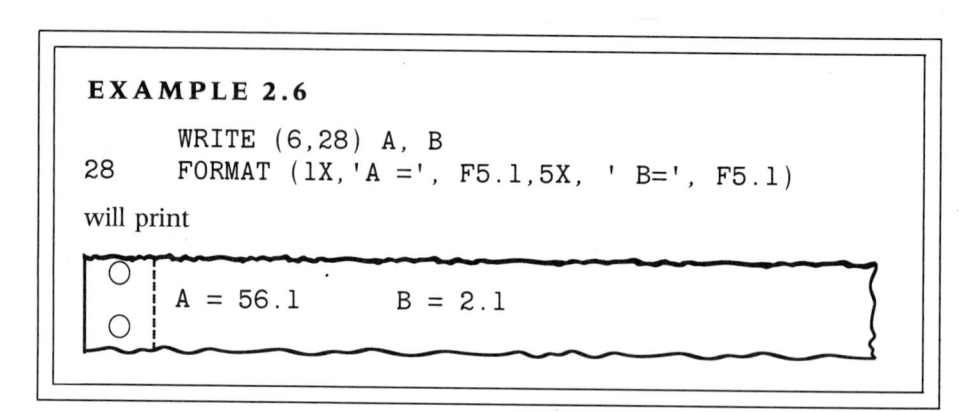

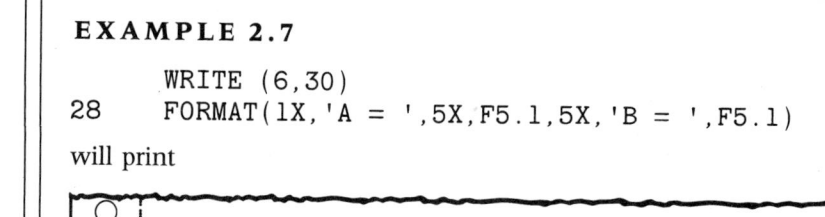

```
      WRITE (6,30)
28    FORMAT(1X,'A = ',5X,F5.1,5X,'B = ',F5.1)
```
will print

```
A =        B =
```

Notice that there is no value for A or B because they are not listed with the WRITE statement.

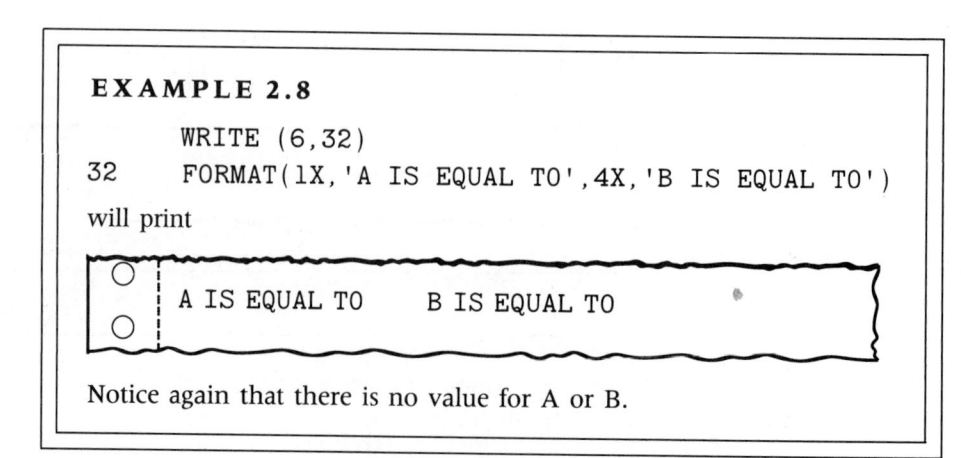

EXAMPLE 2.8

```
      WRITE (6,32)
32    FORMAT(1X,'A IS EQUAL TO',4X,'B IS EQUAL TO')
```
will print

```
A IS EQUAL TO     B IS EQUAL TO
```

Notice again that there is no value for A or B.

Printing a Heading

Sometimes we wish to print words or sentences at the top of a printout as a "heading." This can be done by using a WRITE and FORMAT, as explained in the following example.

EXAMPLE 2.9

Problem: There are several sets of data. In each set, there are two numbers. Write a program which calculates the average of the two numbers, prints a heading, and then prints the data and averages below the heading.

Program:

```
          PROGRAM LOOP
          WRITE (6,10)
10        FORMAT (2X,'THE 1ST NO.',4X,'THE 2ND NO.',
          *4X,'THE AVERAGE')
50        READ (5,20) A,B
20        FORMAT (2F4.1)
          AVG = (A + B)/2.0
          WRITE (6, 30) A, B, AVG
30        FORMAT (5X,F5.1,10X,F5.1,9X,F6.2)
          GO TO 50
          END
```

Notes:

1. To print the heading, we use a WRITE statement with no variable listed and the FORMAT.
2. Writing the heading is the first step in the program and should come before any loop.
3. The spaces in 30 FORMAT are designed to align the values with the heading. The printout will look like this:

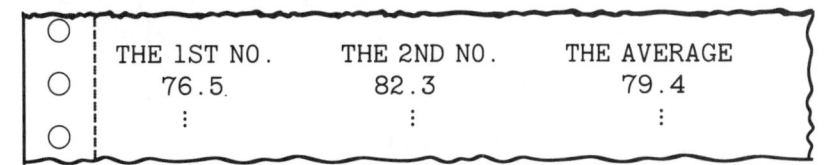

```
THE 1ST NO.     THE 2ND NO.     THE AVERAGE
   76.5            82.3             79.4
    :               :                :
```

4. Notice the infinite loop in the program. This problem is solved in the next section.
5. The * in the second line of the FORMAT means that line is the continuation of the previous line. This is explained in Chapter 3.

By using a WRITE-FORMAT pair, any headings, endings, descriptions, and explanations can be printed in an appropriate place and order. For more examples, look at the solved problems below.

SOLVED PROBLEMS

2.14 What do the following WRITE and FORMAT pairs print? Also find any errors.

a.
```
      WRITE (5,33) A
33    FORMAT (1X,F10.1,'THIS IS THE ANSWER
     *A')
```

b.
```
      WRITE (5,250)
250   FORMAT (51X,'THIS IS A HEADING')
```

c.
```
      WRITE (5,33) Y
33    FORMAT (1X,'THIS IS THE ANSWER')
```

d.
```
      WRITE (6,133) HEDING
133   FORMAT (31X,'THE BEGINNING OF PROGRAM')
```

e.
```
      WRITE (6,150) A, I, J
150   FORMAT (21X,I5,'IS A',F5.2,'IS I',
     *F8.1,'IS J')
```

f.
```
      A = 5.5
      B = 3.2
      C = 6.1
      WRITE (6,125) A, B, C
125   FORMAT (11X,'C = ',F10.1,5X,
     *'B = ',F8.1,4X,'A =',F8.1)
```

☐ **Answers**

a. The value of variable A will be printed, followed by: THIS IS THE ANSWER A.
b. THIS IS A HEADING will be printed after 50 spaces.
c. Only the phrase THIS IS THE ANSWER will be printed. However, the computer will print error messages indicating the missing field descriptor for Y.
d. THE BEGINNING OF PROGRAM will be printed. Note again that the word HEDING causes an error message because it will be considered a variable with no field descriptor.
e. This is not a correct WRITE-FORMAT pair, because the field descriptors do not match with the types of variables (real, integer).

> **f.** The printout will be as follows:
> C =   5.5   B =   3.2   A =   6.1
> Note the order of the values of A, B, C (how can you correct them?)

☐ More Notes about FORMAT

1. The FORMAT statement always has a label which is referred to by the READ or WRITE statements. Furthermore, the FORMAT statements do not do anything by themselves. They only provide information about the layout of the data. Consequently, placing a FORMAT statement anywhere in the program does not change the logical sequence of a program. As a matter of fact, it is good programming practice to separate FORMAT statements from the body of the program in order to make the logic clearer and more readable. FORMAT statements are normally placed near the beginning or near the end of a program; however, remember that the END statement must be the last statement in a program.

> **RULE 2.8**
> *The format statement can be placed anywhere before the END statement in a program.*

For example, the following program calculates the average of two numbers and prints the data as well as the results:

```
      PROGRAM NICE
      READ (5,15) X, Y
      Z = (X + Y)/2.0
      WRITE (6, 35) X, Y
      WRITE (6, 40) Z
15    FORMAT (F3.1, 1X, 5X, F5.1)
35    FORMAT (11X, F5.1, 5X, F5.1)
40    FORMAT (21X, F8.2)
      END
```

2. Any formatted READ or WRITE statement must have a FORMAT statement, but a format statement can be used by several READ or WRITE statements. For example, the following statements are correct in a program.

```
      READ (5,20) A, X, Y
      WRITE (6,20) A, X, Y
      WRITE (6,20) U, V, W
20    FORMAT (1X,F5.0,5X,F6.1,5X,F10.2)
```

3. The programmer should always be aware of the record length, i.e., the maximum number of positions available in a record. The record length on input files is normally 80 characters. Thus the statement

```
      READ (5,30) A, B, C, D, E, F, G, H, X
30    FORMAT (9F10.2)
```

is not valid because the record length (9*10 = 90) exceeds 80. The record length on print files depends on the device, and hereafter we assume it is 132 characters. This is discussed further in Chapter 3. The first character is always used as carriage control and is never printed. The carriage control characters are also discussed in Chapter 3.

4. The FORMAT for the READ statement does not include a carriage control character, whereas the FORMAT for the WRITE statement does.

☐ Formatted PRINT Statement

The PRINT statement can have a format statement if desired. The general form of a formatted PRINT statement is

PRINT fn, list

where fn is the format identifier, and list is the list of items to be printed. The following are examples:

```
        PRINT 10, A, B, C
10      FORMAT (1X,F6.2,5X,F5.1,3X,F8.2)
        PRINT 20, X, Y
20      FORMAT (1X,'X = ',F5.1,'Y = ',F6.1)
```

Note that the first 1X is for the carriage control.

SOLVED PROBLEMS

2.15 Find the errors, if any, in the following statements. Assume the input medium is an 80-column card.

a.
```
        READ (5,100)X,Y,Z
100     FORMAT (F10.2,35X,F10.2,35X,F10.2)
```
b.
```
        WRITE (6,50) A
50      FORMAT (120X,'A IS EQUAL TO',F10.2)
```
c.
```
        READ (5, 55) A, B, C
        WRITE (6, 55) A, B, C
55      FORMAT (1X, 3F5.2)
```
d.
```
        READ (5,15) U, P, X
        WRITE (6, 15) U, P, X
          ⋮
        END
15      FORMAT (1X,3F6.2)
```

> ☐ **Answers**
> a. The record length, including blank spaces, is more than 80 characters.
> b. The record length is more than 132 characters.
> c. This is correct; note, however, that there must be one space at the beginning of the data card.
> d. This is not correct, because END must be the last statement.

SOME PROGRAMMING TECHNIQUES

☐ The STOP Statement

A STOP statement may be used in a program to terminate the execution of the program. The following is an example:

```
          READ (5,10) A, B, C
          SUM = A + B + C
          PROD = A * B * C
          WRITE (6,20) A, B, C, SUM, PROD
    5     STOP
    10    FORMAT (3F5.2)
    20    FORMAT (11X, 4F9.2, 5X, F12.2)
          END
```

A program can contain more than one STOP statement. Normally, a STOP statement is placed at a logical end of a program. If there is only one logical end to the program, it may be used prior to the END statement if desired. A STOP statement may have a statement label.

A STOP statement can also contain a string of characters. The characters will be printed when the STOP statement is encountered. For example:

```
STOP 'THIS IS THE END OF THE PROGRAM'
```

prints the message when it is encountered.

☐ Controlling an Infinite Loop

As you remember, a loop allows us to repeat a part of a program over and over. You also learned that a GO TO statement can be used for "looping." But once you have started a loop, how do you stop it?

The infinite loop created by a GO TO statement to read the data must be terminated by some conditional statement. Basically, there are two methods in FORTRAN to terminate this kind of loop:

1. An implied END in the READ statement.
2. A conditional IF statement.

The first technique is only for controlling an infinite loop created for reading data. It is a built-in END in the READ statement. The following example demonstrates this technique. Other methods for terminating loops are discussed in detail in chapters 4 and 5.

66 *Input–Output*

> ### EXAMPLE 2.10
>
> ■ **Problem:** Write a program which:
> a. reads two numbers as a set of data,
> b. prints the data and the sum of the numbers for each set of data,
> c. has a loop for reading and executing several sets of data,
> d. stops execution and prints FINISHED when there are no more data to be read.
>
> **Program:**
>
> ```
>         PROGRAM LOOP
> 55      READ (5,50,END=100) X, Y
>         S = X + Y
>         WRITE (6, 60) X, Y, S
>         GO TO 55
> 100     STOP 'FINISHED'
> 50      FORMAT (2F5.2)
> 60      FORMAT (1X,F10.2,5X,F10.2,5X,F10.2)
>         END
> ```
>
> **Note:** END = 100 in the parentheses of the READ statement means: "at the end of the data go to statement number 100." Consequently, when there are no more data to be read, the control will be transferred to statement number 100, which is a STOP statement.

Thus the built-in END in the READ statement has the following form:

```
50      READ(u,fn,END=n) Var. 1, Var. 2, ...
        ⋮
        The loop
        ⋮
        GO TO 50
n       a statement (other than FORMAT)
```

where n is the label for the statement to which the control will be transferred at the end of the data. The statement cannot be a FORMAT.

> **RULE 2.9**
> END = n in the READ statement means: at the end of the data, GO TO statement number n.

Statement n, identified by the END = n in the READ statement, need not always be a STOP statement. For example:

> ### EXAMPLE 2.11
>
> ■ **Problem:** Repeat the previous problem, but print THIS IS THE END OF THE REPORT at the end of the program with a WRITE statement.

Some Programming Techniques

Program:

```
        PROGRAM LOOP2
85      READ (5,50,END=100) X, Y
        S = X + Y
        WRITE (6, 60) X, Y, S
        GO TO 85
100     WRITE (6, 65)
50      FORMAT (2F5.2)
60      FORMAT (1X,F10.2,5X,F10.2,5X,F10.2)
65      FORMAT (16X,'THIS IS THE END OF THE REPORT')
        END
```

Note: After there are no more data to be read (an end-of-record is detected), the control will be transferred to statement number 100, and the message THIS IS THE END OF THE REPORT will be printed.

☐ Generating Sequence Numbers

As mentioned previously, a statement such as

  N = N + 1

in a loop can generate the sequence numbers 1, 2, 3, ... Look at the following example.

EXAMPLE 2.12

■ **Problem:** Write a program which reads two numbers as a set of data, then prints the sequence number, the data, and the sum of the numbers in each set. There are several sets of data. Print the number of sets at the end of the program.

Program:

```
        PROGRAM SEQUEN
        N = 0
33      READ (5, 40, END=99) FIRST, SECOND
        SUM = FIRST + SECOND
        N = N + 1
        WRITE (6, 60) N, FIRST, SECOND, SUM
        GO TO 33
99      WRITE (6, 20) N
20      FORMAT (5X,'THE NUMBER OF DATA SET =',I3)
40      FORMAT (2F5.2)
60      FORMAT (1X,I2,5X,F6.2,5X,F6.2,5X,F7.2)
        END
```

Data:

```
5.5    396.8
6.7     59.8
 ⋮       ⋮
```

68 Input–Output

Notes:

1. The statement N = N + 1 counts the number of times that the loop is executed. N = 0 before the loop is necessary to set the initial value of N. Each time the program goes through the loop, 1 is added to N. N also indexes the sets of data. For the first set of data, N is equal to 1, and for the second set of data N is equal to 2, etc.
2. N is an integer variable, and must be initialized before the loop starts.
3. The WRITE statement prints the N, the data, and the sum for each set of data on one line. The printout will look like this:

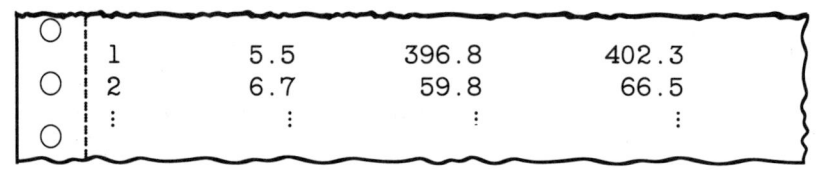

```
1        5.5        396.8        402.3
2        6.7         59.8         66.5
⋮         ⋮           ⋮            ⋮
```

4. After the end of the data, the last N, which shows the number of data, will be printed.

☐ Calculating a Sum

It is quite simple to calculate the sum of a series of numbers. The following program demonstrates this technique.

EXAMPLE 2.13

■ **Problem:** There are several hundred records, each containing one number. Write a program which reads the data, finds the sum and the average of the numbers, and prints the sum and the average of the data at the end.

Program:

```
        PROGRAM TECH
        N = 0
        SUM = 0.0
5       READ (5, 20, END=40) A
        N = N + 1
        SUM = SUM + A
        GO TO 5
40      AVG = SUM/N
        WRITE (6, 50) SUM, AVG
50      FORMAT (11X,'SUM:',F9.2,'AVERAGE:',F9.2)
20      FORMAT (F8.2)
        END
```

Notes:

1. SUM = 0.0 and N = 0 must be placed before the loop.
2. The statement: SUM = SUM + A, inside the loop accumulates the values of *A*. This always works because:

   the new SUM = the old SUM + a new datum (A)

For example, when the first datum, A1, is read, we have

SUM1 = 0 + A1 = A1

After reading the second datum,

SUM2 = SUM1 + A2 = A1 + A2

and after reading the third datum

SUM3 = SUM2 + A3 = A1 + A2 + A3

The program goes on until all of the data are read and the total is calculated.

3. N = N + 1 counts the number of data. We need this to keep track of the number of data in order to calculate the average at the end.
4. After reading all the data, control goes to statement 40, where the average is calculated.

SOLVED PROBLEMS

2.16 Find the errors, if any, in the following statements.

a. READ (5, 20, END = STOP) X, Y
b. READ (20, END = 5) A, B
c. READ (5, 10, END) K, L
d. READ (5, 30 END = 50) A, B
e. READ (5, 40END =) X
f. READ (5, 70, END = GO TO 70) U, P
g.
```
50      READ(5,10,END=100) A
        ⋮
        GO TO 50
100     FORMAT (1X,F8.2)
        STOP
        END
```

☐ **Answers**

a. One cannot use the word STOP in the READ statement.
b. The device number is missing.
c. There must be an equal sign and a statement number following the word END.
d. The comma after 30 is missing.
e. A comma after 40 and a number after END = are missing.
f. One cannot use GO TO in the READ.
g. The control cannot be transferred to the FORMAT statement.

2.17 What do the following programs do? Also, explain what will be printed when the end-of-record is detected.

a.

```
              PROGRAM P1
              SUM = 0.0
       33     READ (5,20,END=53) A, B
                  S = A + B
                  SUM = SUM + S
              GO TO 33
       53     WRITE (6,10) SUM
       10     FORMAT (1X,'SUM OF THE NUMBERS:',F8.1)
       20     FORMAT (2F3.1)
              END
```

b.

```
              PROGRAM P2
              S = 0.0
       55     READ (5,30,END=99) X
                  S = S + X
              GO TO 55
       9      WRITE (6,50) X
              WRITE (6,60) S
       99     WRITE (6,70)
              STOP 'END OF OUTPUT'
       30     FORMAT (F5.1)
       50     FORMAT (1X,F6.1)
       60     FORMAT (1X,'SUM=',F6.1)
       70     FORMAT (1X,'END OF RECORD')
              END
```

c.

```
              PROGRAM P3
       35     READ (5,30,END=5) K
              GO TO 35
       5      WRITE (6,30) K
       30     FORMAT, (1X,I3)
              STOP
              END
```

d.

```
              PROGRAM P4
              N = 0
              X = 0.0
       150    READ (5,100,END=200) A
              N = N + 1
              X = X + A
              WRITE (6,300) N, A
              GO TO 150
       200    WRITE (6,50) N, X
       50     FORMAT (1X,I3,'SETS OF DATA',
             *'SUM=',F9.2)
       100    FORMAT (F5.1)
       300    FORMAT (1X,I3,17X,F6.1)
              END
```

e.
```
        PROGRAM P5
        WRITE (6,10)
10      FORMAT (1X,'SET NO.',15X,'FIRST NO.',
       *15X,'SECOND NO.',15X,'SUM')
        N = 0
        TOTSUM = 0.0
33      READ (5,40,END=99) FIRST, SECOND
        SUM = FIRST + SECOND
        TOTSUM = TOTSUM + SUM
        N = N + 1
        WRITE (6,60) N, FIRST, SECOND, SUM
        GO TO 33
40      FORMAT (2F5.2)
60      FORMAT (1X,I2,15X,F6.2,15X,F6.2,F7.2)
100     FORMAT (11X,'THERE WERE',1X,I3,5X,
       *'SETS OF DATA',5X,'THE TOTAL IS',
       *F10.2)
        END
```

☐ **Answers**

a. The program reads two numbers (A and B) as a data record. The sum of A and B is calculated as S. The total of S's is calculated as SUM. When the end of the data record is detected, SUM will be printed.

b. It appears that the program's intention, in addition to printing the data, is to calculate and print the sum of a series of numbers. However, after the end of the data, the control will be transferred to statement 99, and the END OF RECORD will be printed without printing X or S. "END OF OUTPUT" will be printed last.

c. The program reads a series of integer numbers. At the end of the data, control will be transferred to statement 5, where only the value of the last K will be printed. (This is because when a new datum is read, the previous one will be erased, and only the value of the last datum remains in the memory.)

d. The program calculates the sum of several numbers and prints the sequence numbers as well as the data. At the end of data, it prints the last N, indicating the number of data, and the sum of numbers.

e. This program is similar to program in item a. It firsts adds two numbers on a data set (SUM), and then prints the sequence number (N), the data, and SUM. It also calculates the totals of the numbers (sum of the sums). At the end of the data, it prints the number of data and the totals. Notice the use of appropriate headings in this program.

EXERCISES

2.6 Find all errors in the following statements:

a.
```
         READ(5,12) P, Q, R
12       FORMAT (I5, X7, F2.10, X2, F9.3)
```

b.
```
         READ(2,5) I, J, K, L, M
5        FORMAT (I4, J3, K6, L5, M3)
```

c.
```
         READ(5,22) Q, V, L
22       FORMAT (I5, 7F5.2, I7)
```

d.
```
         READ(2,5) RST, BEST
2        FORMAT (I3, F1.10)
```

e.
```
         READ(220,5) JJ, A, G, KLM
220      FORMAT (I5, X12, 2F8.3, I5)
```

f.
```
         WRITE (5,12) J,
12       FORMAT (1X, I10)
```

g.
```
         WRITE (5, 150) J, JJ, B, WW, L
150      FORMAT (1X, 2I8, 2I5, I3)
```

h.
```
         WRITE (6, 64) X, Y, Z,
64       FORMAT (1X, F10.2, 2F8.1)
```

i.
```
         WRITE (6, 250) A, B, C
250      (15X, F10.1, 5X, F10.2, 5X, F10.2)
```

2.7 What values are read by the following READ statements for each variable?

a.
```
         READ (5, 250) A, B, I
250      FORMAT (10X, 2F5.1, I5)
```
Data:

```
ƀƀƀƀƀƀƀƀƀƀ39.5015.5ƀ23
```

b.
```
         READ (5, 100) IS, FI, JJ, XI
100      FORMAT (I5, F5.2, I5, F5.2)
```

Data:

```
9312521346628386238
```

c.
```
        READ (5, 1000) COST, RATE, LENGTH
1000    FORMAT (F4.1, F7.1, I3)
```
Data:

```
56.1b639.2b569
```

d.
```
        READ (5, 190) POOL, M, JAR
190     FORMAT (F5.1, 2X, I2, 2X, I2)
```
Data:

```
639.5,329,56
```

e.
```
        READ (5, 150) X, Y, I, J
150     FORMAT (F5.0, F3.0, 4X, I3, I8)
```
Data:

```
56.21bbb62.1bbb562394
```

f.
```
        READ (5, 100) X, Y, I, Z
100     FORMAT (F5.1, F3.1, I2, 5X, F6.2)
```
Data:

```
56.21bbb62.1bbb5626298256
```

2.8 What will be printed by the following statements? Also, correct any errors you may find.

a.
```
        WRITE (6,101) SUM
101     FORMAT (11X, 'THE SUM PROGRAM')
```

b.
```
        WRITE (6,102) POOR
102     FORMAT(11X,'THE POOR VARIABLE IS EQUAL',F10.2)
```

c.
```
        WRITE (6,105) MORK, RORK
105     FORMAT(6X,'THE 1ST AND 2ND VARIABLES:',2F9.2)
```

d.
```
        WRITE (6,110) DSTANC, TIME, SPEED
110     FORMAT (11X,'WHEN X=',F10.2,'T=',
       *F10.2,'SPEED IS=',F10.2)
```

e.   `WRITE (6,120) HOURS, RATE, PAY`
`120`   `FORMAT (11X,'HOURS=',F8.2,'RATE=',`
       `*F8.2,'PAY=',F9.2)`

f.   `WRITE (6,130)`
`130`   `FORMAT (11X,'HOURS',10X,'RATE',10X,'PAY')`

g.   `WRITE (6,140)`
`140`   `FORMAT (11X)`

h.   `WRITE (6,150) X, Y`
`150`   `FORMAT (21X,'THE VARIABLES ARE')`

2.9 What does the following program do? Do you see a flaw in the program?

```
        PROGRAM XXX
5       READ (5,160) FST, SND
160     FORMAT (2F5.2)
        SUM = FST + SND
        WRITE (6, 30) FST, SND, SUM
30      FORMAT (1X, F10.2, 5X, F10.2, 5X, F10.2)
        GO TO 5
        END
```

2.10

a. What is the difference between the previous program (2.9) and the following program?

```
        PROGRAM XXX
        WRITE (6,10)
10      FORMAT (31X, 'SUM CALCULATION')
        WRITE (6,20)
20      FORMAT (8X, 'FIRST NUMBER', 5X, 'SECOND NUMBER',
       *10X, 'SUM')
5       READ (5,160) FST, SND
160     FORMAT (2F5.2)
        SUM = FST + SND
        WRITE (6,30) FST, SND, SUM
30      FORMAT (8X, F10.2, 7X, F10.2, 10X, F10.2)
        GO TO 5
        END
```

b. Rewrite this program:

1. Pick better (more descriptive) variable names.
2. Place all the FORMAT statements at the end of the program.
3. Calculate the total of the numbers and print it at the end.
4. Correct the flaw.

2.11 Rewrite the program in Example 2.13 in the text so that it prints the sequence numbers and the data in the loop. Use also appropriate headings for this example.

PROGRAMMING EXERCISES

Write a FORTRAN program for the following problems. In all of them, include the FORMAT for READ and WRITE, appropriate headings, spaces, and loops. If data are not given, make up your own data. Assume there is more than one set of data for each problem. Terminate the loop with the END = n in the READ statement.

2.12 Write a program which calculates and prints the average of four numbers placed similar to the following line:

```
 bbb6253.129,bbb32.19b,562.1b,bb62.1
```

2.13 Calculate and print the weekly salary of the employees of a company. Each employee's wage rate and number of hours worked during a week are placed as follows:

```
12.59,39.50
```

2.14 Calculate and print the sum, multiplication, and average of three numbers placed on a data line similar to the following:

```
bbbbbbbb89.2138b,b2.83b,b23
```

Print the data as well as the results. (Hint: Use F2.0 for the last number to prevent mode mixing.)

2.15 The Celsius temperature C can be calculated by the following formula:

$$C = \tfrac{5}{9}(F - 32)$$

where F is the temperature in degrees Fahrenheit. Write a program to read the temperature in Fahrenheit and to calculate and print the equivalent temperature in Celsius.

*__2.16__ A deck of data consists of several cards. Each card contains (1) the customer number, (2) the number of units of a product ordered, and (3) the price of the product. The sales tax is 4% of the gross sales. Write a program which calculates gross sales for the item, and the total charge (sales + tax) to each customer. Print the customer number, number of units ordered, price, gross sales, tax, and total charge. Print also the total sales at the end of the report.

2.17 Given is a deck of cards. Each card contains the following information for a student:

1. The ID number (columns 1–6) and
2. The scores of two tests (columns 8–12 and 14–18).

Write a program to calculate the average of two scores. Print (1) the sequence number, (2) the ID, (3) the two scores, and (4) the average of the two scores for each student in a report form. Calculate also the grand average (the average of all the scores). Print the number of students and the grand average at the end of the report. (Hint: Examples 2.12 and 2.13 will be of great help for this problem.)

2.18 A set of data has been prepared as:

1. An employee's ID number, 6 digits	Columns 1 to 6
2. Hourly wage rate, such as 11.5	Columns 8 to 11

* Answer provided at the end of the chapter.

3. Number of hours worked during a week,
   such as 38.5   Columns 13 to 16

The gross pay can be calculated by the number of hours worked multiplied by the wage rate. Deductions are (1) federal tax, 12%, and (2) state tax, 6% of the gross pay. Write a program which calculates the gross pay, deductions, and net pay for the employee. Print ID number, gross pay, federal tax, state tax, total deductions (federal and state), and net pay. Assuming there is more than one employee, print the total pay, total deductions, and the total nets at the end of the report.

2.19 The velocity (V) and the distance (X) traveled are given by the following formulas:

$$X = \tfrac{1}{2}at^2,$$
$$V = at,$$

where a is the constant acceleration and t is the time in seconds. Write a program which reads two numbers as input for a and t, and prints a, t, X, and V.

2.20 Write a FORTRAN program for the following problem. Given is a file of several customers' records for a gas company. Each line contains information in the following fields:

Customer's number	5 columns	Columns 10–14
Last meter reading	7 columns	Columns 15–21
Current meter reading	7 columns	Columns 23–29

a. Calculate the amount of gas used.
b. Calculate the cost at 23.5¢ per unit.
c. Print the customer's number, last meter reading, current meter reading, amount of gas used, and cost for each customer in a report form.
d. Calculate and print the total gas used for all customers and the total cost to all users at the end of report.

2.21 Write a program to calculate kinetic energy T as

$$T = \tfrac{1}{2}PV^2,$$

where P (the mass) and V (the velocity) are given as input.

2.22 An annuity in which the yearly payment at the end of each of N years is P has a present value of

$$T = P\left(\frac{1 - (1 + R)^{-N}}{R}\right),$$

where R is the interest rate (as a decimal fraction) compounded annually. Suppose several P, R, and N are given as input. Write a program to calculate T for each set of data. Design the input and the output first.

2.23 The area of a triangle can be calculated by

$$A = \sqrt{S(S - a)(S - b)(S - c)}$$

where $S = (a + b + c)/2$ and a, b, c represent the three sides of the triangle. Write a program which accepts the three sides, calculates the area, and prints the data and area. Assume there is more than one data record.

SUMMARY OF CHAPTER 2

You have learned:

1. Real and integer are two types of numerical data in FORTRAN.
2. The first character of an integer variable's symbolic name must be I, J, K, L, M, or N; the first character of a real variable's symbolic name can be any other letter.
3. The variables on both sides of an equation should be of the same type, either real or integer.
4. A FORMAT statement in FORTRAN directs the computer to read or write the data in a desired way (order and width). The following field descriptors are used to show the type and the length of the field of the numeric data in FORMAT statement:

    In For integer values, n shows the number of columns the value occupies.

    F$n.m$ For real values, n shows the number of columns the value occupies and m shows the decimal places.

    nX   For showing n spaces on the data line.

5. An example of formatted READ is:

    ```
          READ (5, 100) A, B, K
    100   FORMAT (F5.2, F5.1, I3)
    ```

6. An example of the formatted WRITE is:

    ```
          WRITE (6, 200) A, B, K
    200   FORMAT (1X, F6.2, F6.1, I4)
    ```

7. An example of the formatted PRINT is:

    ```
          PRINT 100, X, Y, I
    100   FORMAT (1X, F5.1, 2X, F6.1, 2X, I3)
    ```

8. The FORMAT statement does not have to immediately follow the READ or WRITE; it may be placed anywhere in the program. It is recommended that all the formats be placed near the end of the program.
9. At the beginning of a FORMAT for a WRITE or PRINT statement, we need a carriage control character.
10. If we don't enter a number with the decimal point, an appropriate field descriptor can automatically place the decimal point in the appropriate position. For example, if we record 38.5 as 385, then F3.1 will put the decimal point in the appropriate place.
11. In normal processing, a space or blank which is included in a data field is ignored unless specified otherwise.
12. A string of characters—an explanation—can be caused to appear on a printout by placing the characters in a pair of apostrophes in a FORMAT statement.
13. One method for terminating an infinite loop for reading data is by placing an END=n in the READ statement.
14. We can count the number of loops or generate sequence numbers in a program by using a statement like

    N = N + 1

    in a loop.

78 Input–Output

15. We can add up the numbers in a loop by using a statement like

    SUM = SUM + A

    in a loop.

■
SELF-TEST REVIEW

2.1 Which of the following is an appropriate symbolic name for an integer variable?
  a. ENIGER
  b. MERY
  c. COOL
  d. AREA

2.2 Which of the following might be avoided because of mixed modes?
  a. KAR = 2 * N + 5
  b. AVG = (M + N)/3
  c. WIDTH = AREA/SIDE
  d. LENGTH = 2 * LONG

2.3 Write a formatted READ statement to read four numbers placed as:

```
6253.51bb,bb325bb,bb262.395b,b98
```

2.4 What are the values of A, B, and N read by the following READ statement?

```
        READ (5,20) A, B, N
20      FORMAT (2X, F5.1, 1X, F7.1, 3X, I4)
```

Data:

```
5629315823.32196,256
```

2.5 Write a FORTRAN program for printing the invoice for a customer who has ordered several items.

Input: A set of records for the customer; each record contains:

a. The item number, 5 columns, columns 1–5.
b. The number of units ordered, 2 columns, columns 7–8.
c. The unit price, 5 columns, columns 10–14.

Calculate the amount for each order and the grand total.
  Output:

a. The item number, number of units ordered, unit price, and amount for each item.
b. The total amount at the end of the invoice.
c. The word THANKS at the end.

Use appropriate headings, spaces, and loops.

Answers

2.1 b.

2.2 b.

2.3
```
        READ (5,10) A, K, B, L
10      FORMAT (F7.2, 5X, I3, 5X, F7.3, 3X, I2)
```

2.4 A = 2931.5
B = 23.3219
N = 56

2.5
```
        PROGRAM ORDER
        GRDTOT = 0.0
        WRITE (6, 10)
10      FORMAT (44X, 'YOUR BILL')
        WRITE (6, 20)
20      FORMAT (18X, 'ITEM NO.', 10X 'U-ORDERED',
       *10X, 'PRICE', 10X, 'AMOUNT')
50      READ (5, 30, END=99) ITN, UNORD, UPRICE
            AMT = UNORD * UPRICE
            WRITE (6, 40) ITN, UNORD, UPRICE, AMT
            GRDTOT = GRDTOT + AMT
        GO TO 50
99      WRITE (6, 55) GRDTOT
        WRITE (6, 170)
30      FORMAT (I5, 1X, F2.0, 1X, F5.2)
40      FORMAT (19X, I6, 14X, F3.0, 12X, F7.2, 10X,
       *F8.2)
55      FORMAT (53X, 'GRAND TOTAL =', F10.2)
170     FORMAT (45X, 'THANKS')
        STOP
        END
```

■

ANSWERS TO SELECTED EXERCISES

2.16
```
        PROGRAM SALES
        GTOT = 0.0
        WRITE (6,20)
20      FORMAT (31X,'PRICE CALCULATION')
        WRITE (6,30)
30      FORMAT (9X,'ID NO.',5X,'UNIT',5X,'PRICE',
       5X,'SALES',5X,'TAX',5X,'TOTAL')
95      READ (5,250,END=45) IDNUM, UNIT, PRICE
250     FORMAT (I5,1X,F5.0,F6.2)
        SALES = UNIT * PRICE
        TAX = SALES * .04
        TOT = SALES + TAX
        WRITE (6,300) IDNUM,UNIT,PRICE,SALES,TAX,TOT
```

```
300   FORMAT (9X,I6,4X,F6.0,3X,F7.2,3X,F7.2,
     *3X,F5.2,3X,F8.2)
      GTOT = GTOT + TOT
      GO TO 95
45    WRITE (6,400) GTOT
400   FORMAT (42X,'GRAND TOTAL =',F10.2)
      END
```

Chapter 3

Programming Development and Execution

RUNNING A PROGRAM

    Card-oriented Batch Processing
        The Card Deck
    On-Line Processing
    Position of Instructions
        FORTRAN Coding Form
        The Print Chart
    The Printout
    Field, Record, and File
    Carriage Control
        The Slash
    Documentation within the Program

THE COMPILERS

    FORTRAN Compilers
    FORTRAN Statements

ERROR DETECTION

    Syntax Errors
    Execution Errors
    Logical Errors

PROGRAM PLANNING

    Programming Tools
        Flowcharts
        Pseudocode
        Logic-Charts
    Algorithms
    Programming Cycle

A SAMPLE PROBLEM

In the previous chapters we discussed basic programming concepts and FORTRAN. By now you should be able to write a simple FORTRAN program. In this chapter you will learn:

- How to run a program
- How English words in your program are translated for the computer
- What kinds of errors a program may have
- How a program should be planned before writing the FORTRAN statements

RUNNING A PROGRAM

There are basically two ways you can run your program. The first way is by *card-oriented batch processing*, when punched cards are used. The second way is by *on-line processing*, when the information is keyed directly into the computer through a terminal.

☐ Card-oriented Batch Processing

In card-oriented batch processing, the information is punched onto cards and read by a "card reader," which is a device for entering the information into the computer. With punched cards, each line of your program is punched into a separate card. A typical card is shown in Figure 3.1. Each horizontal position of the card is referred to as a *column* of the card. In Figure 3.1, the number 5 is in column (or position) 1, and the R of READ is in column (or position) 7. A typical card has 80 columns, and one or more holes punched in each column represents a code for numbers, letters, or symbols. When punching the cards, you do not have to worry about the holes because the *keypunch* machine will punch them for you. The operation of a keypunch is similar to the operation of a typewriter. By depressing a key—for a letter, a number, or any other character—a hole or a combination of holes will be punched into the card.

Figure 3.1 A typical punch card.

The character punched in a column will be printed at the top of the card so that you are able to read it. The card reader, however, reads the holes and not the printing.

Putting several cards together, along with the control information and input data, creates what we call a *card deck*, or a *job*, or simply a *deck*. The deck has three main parts: (1) control cards, called *JCL* cards (which stands for job control language); (2) the main program, called the *source* program; and (3) the data. The control cards inform the computer about the account number of the user, the language used, the data section, and the end of the deck.

The Card Deck

Figure 3.2 shows the arrangement of a typical job:

1. One or several control cards, which contain information such as the name of the user, account number, and name of the language being used (ours is FORTRAN).
2. The source program.
3. A control card indicating the end of a source program.
4. The data.
5. A control card indicating the end of a job.

The *JCL* cards, or control cards (numbered 1, 3, and 5), vary from computer to computer and are not part of the FORTRAN language. Their purpose is to run a FORTRAN program with a particular computer model. Whenever you need to run a program, your instructor or the computer center staff can help you to prepare the appropriate control cards.

Punched cards are becoming less popular. Most of the installations use an on-line processing system, which is explained next.

☐ On-Line Processing

In *on-line processing*, the information is keyed into the computer, one line at a time, directly through a terminal. A *terminal* is a device for both transmitting the information to the computer and receiving the information from the computer. Terminals may be classified (among other ways) into *CRT* and *hard-copy* terminals. The *CRT* (cathode-ray tube) looks much like a TV monitor attached to a keyboard, and all the results appear on the screen. A hard-copy terminal resembles a typewriter and is used in much the same way as the *CRT*, except that you get

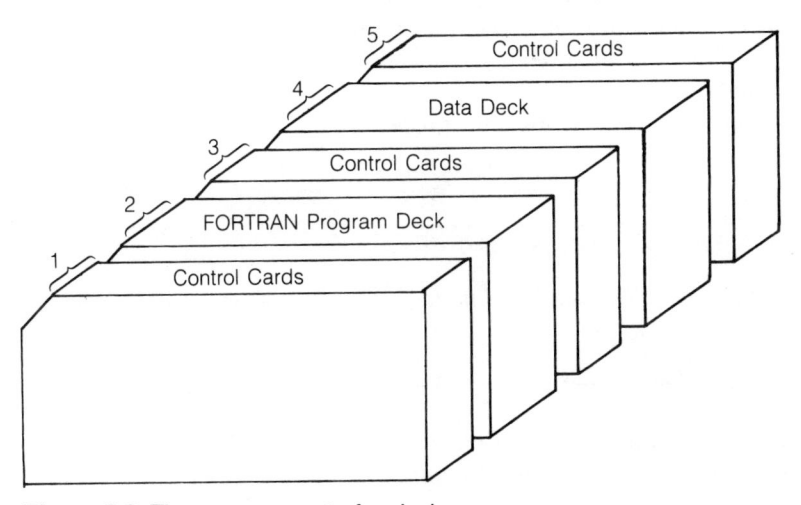

Figure 3.2 The arrangement of a deck.

a printout. There are many types of hard-copy terminals, each with a different name given by the manufacturer (TI, DTC, etc.). You must connect a terminal to the computer, log in, and key the program through a series of "system commands." The program then is run with *JCL* on line. When working with a terminal each line represents a punched card, and the position of each character on the line is identical to the position of that character on the card. Hereafter, when the word "card" is used, it is equivalent to "line" on a terminal, and vice versa.

Interactive processing is a mode of on-line processing in which an interactive program causes a continuous dialogue between the user and the computer. An interactive program is written so that the user has the illusion that the computer is talking back and forth with him. Some of the computer systems support using interactive FORTRAN programs. A sample of an interactive FORTRAN program is presented in Appendix C.

☐ Position of Instructions

When punching statements onto a card or keying them into a terminal, certain columns are reserved for a particular item of each statement. In FORTRAN, columns 1 to 5 are assigned to the statement number. Columns 7 to 72 are assigned to the statement. This is very important to remember:

> **RULE 3.1**
> *Columns 1–5:*   *Statement numbers*
> *Columns 7–72:*   *FORTRAN statements*

Blank spaces are allowed anywhere as long as they do not disturb the meaning of the word or the term. For example: P RINT *, A is not allowed whereas PRINT *, A is valid. *To make FORTRAN statements more readable, it is good practice to use blank spaces freely before or after each item;* i.e., before or after a word, a variable symbolic name, or a field descriptor. Furthermore, a statement may be indented at (or after) column 7 (as long as it does not go beyond column 72). Later, in chapter four, you will see how useful indentation is in making a program more readable.

If a statement is too long to fit on one line (card), the statement can be continued onto the next line (card). But in order to do this, a character must be placed in column 6 of the continuation line. You may have as many as 19 continuation lines, depending on the computer model used. Statements on the continuation line must also begin at (or after) column 7, and not go beyond column 72.

> **RULE 3.2**
> *A character in column 6 of a line means the following: This line is the continuation of the previous one.*

The continuation character may be any nonblank symbol other than zero. Usually, a programmer will punch an asterisk (*), or the number 1 in the first continuation line, the number 2 in the second, and so on.

Columns 73–80 can be used for identification information. Any information in these columns is ignored, but when the program itself is printed out, it will be listed as placed. Some programmers like to use this space for comments; others use it for sequence numbers to make the job easier to recover in case the deck is dropped.

A data line is not a FORTRAN statement, and the data section is not part of the source program. Consequently, the rules regarding the placement of FORTRAN statements on a line do not apply to data lines. The data must be placed according to the format specifications. For example, when using punched cards, all 80 columns of a card may be used for data.

FORTRAN Coding Form

> *Translating the solution of a problem into FORTRAN statements is called* coding.

You can use a preprinted coding form for coding your program. Figure 3.3 shows an example of such a form with a coded program.

The FORTRAN coding form is designed to indicate the place of the statement label (columns 1 through 5), the continuation column (column 6), and the FORTRAN statement (7 through 72). You can obtain a pad of coding paper in almost any supply store.

The Print Chart

Before coding a program, the output must be designed. Normally the output is designed on a *print chart*. A print chart is also called a printout layout form or printer spacing chart. It is a preprinted form used for determining the exact location of each item of information, the centering of headings, the required spaces between columns, and other necessary information about a desired printout. This information is necessary then for preparation of the output FORMAT statements. Figure 3.4 shows an example of a print chart.

```
      READ(5,6)A,B
6     FORMAT(2F4.1)
      WRITE(6,20)A,B
20    FORMAT(9X,'A=',F5.1,' B=',F5.1)
      END
```

Figure 3.3 A coding form.

86 *Programming Development and Execution*

Figure 3.4 A typical print chart.

☐ The Printout

In card-oriented batch processing, or when using a hard-copy terminal, the program and the results are printed by a printer. Figure 3.5 shows an example of a printout. If you use a *CRT* terminal, this information appears on the screen, and then it can be sent to the printer if desired.

Normally when your program is executed, the program itself will be listed first, followed by some information and statistics about it, and finally the results of the program. Of course, this assumes that you use an error-free program and control statements. Note also in Figure 3.5 that the computer assigns a sequential line number to each statement. The printout normally is about 15 inches wide with 132 print positions.

☐ Field, Record, and File

As discussed in chapter two, a *field* is the specific area assigned to a piece of information—i.e., the number of consecutive columns or positions reserved for a datum.

A *record* is a combination of one or more fields:

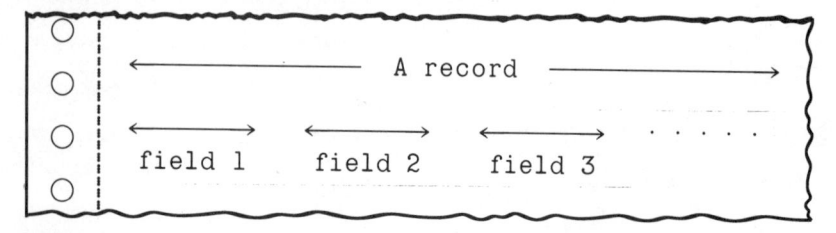

For example, the data on a single punched card, the line of data ready to be keyed into the terminal, or a line of information ready to be printed can all be considered as a record. A record does not have to be limited to one line,

```
PROGRAM ORDER       73/172   OPT=0                              FTN 5.0+508

    1                    PROGRAM ORDER
    2              C  INITIALIZE THE VARIABLE   GRANDTOTAL (GRDTOT)
    3                    GRDTOT=0.0
    4              C  WRITE THE HEADING
    5                    WRITE(6, 10)
    6                 10 FORMAT('1',44X,'YOUR BILL')
    7                    WRITE(6,20)
    8                 20 FORMAT(18X,'ITEM NO.',10X,'U-ORDERED',10X,'PRICE',10X,'AMOUNT')
    9                 50 READ (5, 30, END=89) ITN, UNORD, UPRICE
   10                    AMT=UNORD*UPRICE
   11                    WRITE(6,40) ITN, UNORD, UPRICE, AMT
   12                    GRDTOT = GRDTOT + AMT
   13                    GO TO 50
   14              C  END OF THE LOOP
   15              C  PRINT THE GRAND TOTAL
   16                 89 WRITE(6,55) GRDTOT
   17                    WRITE(6,170)
   18                 30 FORMAT(I5, 1X, F2.0, 1X, F5.2)
   19                 40 FORMAT(19X, I6, 14X, F3.0, 12X, F7.2, 10X, F8.2)
   20                 55 FORMAT(  // 53X, 'GRAND TOTAL =', F10.2)
   21                170 FORMAT( 45X, 'THANKS')
   22                    STOP
   23                    END

--VARIABLE MAP--(LD=A)
-NAME---ADDRESS--BLOCK-----PROPERTIES-------TYPE---------SIZE

 AMT       1368                              REAL
 GRDTOT    1328                              REAL
 ITN       1338                              INTEGER
 UNORD     1348                              REAL
 UPRICE    1358                              REAL

--STATEMENT LABELS--(LC=A)
-LABEL-ADDRESS-----PROPERTIES----DEF      -LABEL-ADDRESS-----PROPERTIES----DEF

   10     428      FORMAT         6    -     50     148                      9
   20     468      FORMAT         3          55     658     FORMAT          20
   30     568      FORMAT        18          89     278                     16
   40     608      FORMAT        19         170     718     FORMAT          21

--ENTRY POINTS--(LD=A)
 -NAME---ADDRESS--ARGS---
   --STATISTICS--
     PROGRAM-UNIT LENGTH       1378  =     95
     CM STORAGE USED         575008  =  24384
     COMPILE TIME            0.213 SECONDS

                                   YOUR BILL
           ITEM NO.        U-ORDERED         PRICE           AMOUNT
           63284              12.            18.99           227.88
           63327               5.            17.99            89.95
           63633              10.            10.25           102.50
           64115              17.             3.79            64.43
           64206               4.           115.25           461.00
           65909              20.              .99            19.80

                                        GRAND TOTAL =      965.56
                              THANKS
```

Figure 3.5 A printout.

but it is convenient to consider a line as a record. In this case, you should be aware of the number of *allowable spaces* on a line. For example, if one data card is used as a record, the maximum number of spaces available is 80 columns. The maximum number of spaces on a print line is between 72 and 200, depending on the printer, but usually it is 80, 120, 132, 136, or 160. We assume in this text that the printer can print up to 132 characters per line.

A *file* is a combination of several items, normally several records, to be considered as a unit. For example, data records about students in a class can

88 Programming Development and Execution

constitute the "students data file." There are different kinds of files; for example, data file, program file, input file, output file, master file, transaction file, and working file.

☐ Carriage Control

When the printer is printing, it needs to know when to go to the next line, to skip a line, or to skip a page. The command which gives such information to the printer is called the *carriage control*, and it is the first character of the record directed to the printer. Some useful carriage control characters are:

Character	Meaning
a blank: 1X or 'b'	Start a new line, then print
'1'	Eject to the beginning of the next page, then print
'+'	Stay on the same line (suppress), no advance before printing; allows over printing
'0'	Skip a line, then start printing on a new line; used for double spacing

For better understanding of how these carriage control characters can be used, consider the following examples:

EXAMPLE A

```
        WRITE( 6 , 10 ) A , B
10      FORMAT( ' ' , F10.2 , 5X , F8.2 )
```

will print the variables starting at the first position of the paper. The format is the same as

```
10      FORMAT( 1X , F10.2 , 5X , F8.2 )
```

EXAMPLE B

```
        WRITE( 6 , 20 ) X , Y
20      FORMAT( 11X , F5.2 , 5X , F5.2 )
```

will print the variables starting at the *tenth* position of the paper. The format is the same as

```
20      FORMAT( 1X , 10X , F5.2 , 5X , F5.2 )
```

EXAMPLE C

```
        WRITE( 6 , 10 ) A
10      FORMAT( 1X , F5.2 )
        WRITE( 6 , 30 ) U , W
30      FORMAT( '0' , F5.2 , 5X , F5.2 )
```

prints A, skips a line, and then prints U and W starting at the first position of the line. (One line is skipped before printing U and W.)

EXAMPLE D

```
        WRITE( 6 , 50 )
50      FORMAT( '1' , 20X , 'GMP COMPANY' )
```

will print the heading GMP COMPANY at the top of a new page starting at position 20.

EXAMPLE E

```
            WRITE( 6 , 60 )
    60      FORMAT( 1X , 'XXXXXXXXXX' )
            WRITE( 6 , 70 )
    70      FORMAT( '+' , '==========' )
```

will print XXXXXXXXXX, then print ========== over them, causing the following line to be printed:

X̄X̄X̄X̄X̄X̄X̄X̄X̄X̄

Note that the printer never returns to the *previous* line, but '+' causes it to stay on the same line, rather than advancing to the next line.

The carriage control character is never printed, but is only used for identifying the position of the paper. In most printers, if the programmer does not identify a carriage control character, then the first character of the record will be taken as the carriage control. Consequently, if you forget to identify any carriage control character at the beginning of a write FORMAT, the first character of the first record will disappear. For example, if you run the program

```
            X = 319.5
            WRITE( 6 , 59 ) X
    59      FORMAT( F5.1 )
            END
```

The printout will show

```
    19.5
```

where the first character (3) has been taken as carriage control character. If the value of X had been 139.5, the value 39.5 would have been printed on the next page (1 being the code for ejecting to the next page). If the character in the carriage control position is not one of the carriage control characters, it will generally be ignored and 1X will be assumed.

The carriage control characters are required only for output directed to the printer.

The Slash

We may use the slash (the same character that we used for the division operator) anywhere in the FORMAT to introduce *a new record*. For example, a slash anywhere in the FORMAT for a WRITE statement causes the printer to start a new line, and in the FORMAT for a READ statement causes the card reader to start a new card if cards are used.

EXAMPLE A

```
            WRITE( 6 , 70 ) A , B
    70      FORMAT( 1X , F6.2 , / , 1X , F6.2 )
```

will print the value of A at the beginning of the line, and the value of B on the next line, under the value of A.

EXAMPLE B

```
            WRITE( 6 , 40 ) M , N
    40      FORMAT( ////1X , I5 , / , 1X , I5 )
```

will cause four lines to be skipped before printing the value of M. Then the value of N will be printed under M, on the next line.

EXAMPLE C

```
         READ( 5 , 50 ) X , Y , Z
  50     FORMAT( F6.2 , / , F4.1 , / , F5.2 )
```

will read the value of X from the first line, the value of Y from the second line, and the value of Z from the third line.

There is a difference between a slash and carriage control characters: the slash causes the current record to be terminated and a new one to be started, but the carriage characters are used at the beginning of a record for controlling the position of the printer. Note also the following:

1. A slash can be placed anywhere in a format for either the READ or the WRITE statement.
2. After a slash, a carriage control character such as 1X is needed, because a carriage control character is required for any record to be printed, including new records introduced by a slash.
3. Normally a comma is used to separate two field descriptors in the FORMAT. However, a slash also can be a separator. A slash can be used with or without a comma. As an example, the statements

```
         WRITE( 6 , 20 ) A , B , C
  20     FORMAT( ////1X , F5.1/1X , F6.2/1X , F8.2 )
```

print the values of A, B, and C under each other after four blank lines.

We will be discussing more about the slash in Chapter 6.

SOLVED PROBLEMS

3.1 Describe the print position of each variable if the following statements are used:

a.
```
         WRITE( 6 , 40 ) A , B , AREA
  40     FORMAT('0',F6.2/1X,F8.2/1X,F10.2 )
```
b.
```
         WRITE( 6 , 65 ) AA , II , CC
  65     FORMAT('1',F6.2,2X,I5,2X,F8.2 )
```
c.
```
         I = 15329
         WRITE( 6 , 100 ) I
  100    FORMAT( I5 )
```
d.
```
         WRITE( 6 , 30 ) X
         WRITE( 6 , 40 ) Y
  30     FORMAT( 1X , F6.2 )
  40     FORMAT( '+' , 15X , F8.2 )
```
e.
```
  150    WRITE( 6 , 50 ) A
  50     FORMAT( '1' , F5.2 )
         GO TO 150
```

☐ Answers

a. The value of variable A will be printed in the second available line (one line will be skipped), the value of B will be printed under A, and the value of AREA under B.
b. The values of AA, II, and CC will be printed on the line at the top of a new page.
c. The number 5329 will be printed at the top of a new page. The 1 will be taken as a carriage control character.
d. The values of X and Y will be printed on the same line.
e. This is an infinite loop. The loop causes the printer repeatedly to skip to a new page after printing the value of A. Of course, an infinity of pages are not going to be printed, because of the limited computer time that a user has available.

☐ Documentation within the Program

One of the characteristics of a good program is that it is easy to follow and understand. The following tips should be helpful.

First, choose a descriptive name for a variable to help you easily remember what each variable refers to. Second, as you learn more about programming, you will see there may be several ways to write a program; always choose the simplest and most straightforward method. This will be apparent as the programs become longer and more complicated. Third, you will find that describing the important points of a program is a great help. One way to do so is to write comments about them. Comments are very useful to describe the sequence, logic, variables, and the critical point of a program. FORTRAN allows us to write a line of comment by placing the letter C in the first column of a line (card).

RULE 3.3
The letter C in the first column stands for "comment." The comment will be printed in the program text exactly.

A comment line may be placed anywhere in the program (as long as it is before the END statement, of course). After the letter C in the first column, the programmer can write anything as a comment. FORTRAN does not execute comment lines. They are treated as "comments" only, not instructions. They will appear in the listing of the program, but not on the output of the program.

It is a good programming practice to document a program with comment lines. Normally, several comment lines are used at the beginning of a program and include the name of the programmer, the date, a summary of the purpose of the program, and the solution procedures, as well as throughout the program to explain variables, loops, or other critical points. Comment lines are also useful for dividing a program into logical pieces. The comments should not be mere repetitions of FORTRAN statements; rather they should explain the points which are not self-explanatory.

FORTRAN 77 permits also an asterisk (*) instead of the letter C, and a completely blank line in the program (anywhere before the END statement) as a comment "spacer."

The following example demonstrates the use of the comments.

```
      PROGRAM EXAMPLE
C ****************************************************************
C * THIS PROGRAM CALCULATES THE TAX ON AN AMOUNT BY READING THE AMOUNT *
C * AND MULTIPLYING IT BY .04                                          *
C *                                                                    *
C * PROGRAMMED BY LYNN A. MADISON, APRIL 16, 19--                      *
C ****************************************************************
C THIS IS THE INPUT OF THE PROGRAM;
C AMT IS THE AMOUNT, IT IS PUNCHED ON THE 1ST FOUR COLUMNS OF A CARD
C
      READ( 5 , 33 ) AMT
33    FORMAT( F4.1 )
C TAX IS 4% OF THE AMOUNT
C THIS IS THE TAX CALCULATION
C
      TAX = AMT * .04
C
C THIS IS THE OUTPUT
C
      WRITE( 6 , 100 ) AMT , TAX
100   FORMAT( 11X , 'THE AMOUNT IS' , F5.1 , 'THE TAX IS' , F10.4 )
      END
```

Documentation of a program is important for the programmer as well as for anyone else who may review the program at a later time.

SOLVED PROBLEMS

3.2 Do you find any errors in the following statements?

a.
```
COMMENT: THIS IS A READ STATEMENT
      READ( 5 , 22 ) A , B
22    FORMAT( 2F5.2 )
```
b.
```
CTHIS IS VARIABLE X
      READ( 5 , 22 ) X
22    FORMAT( F5.1 )
```
c.
```
THIS IS A COMMENT
C THIS IS THE INPUT
      READ( 5 , 10 ) A
10    FORMAT( F5.1 )
```
d.
```
THIS IS THE INPUT
      READ( 5 , 5 ) X
5     FORMAT( F6.2 )
```
e.
```
C READ( 5 , 50 ) A , B , C
50    FORMAT( 3F5.2 )
```

```
            f.
            C OUTPUT
            WRITE( 6 , 20 ) A , B
            20    FORMAT( 1X , F10.2 , 5X , F10.2 )
            g.
            COPY X TO Y
                  Y = X
                  READ( 5 , 100 ) A
            h.
            C THIS IS TO CALCULATE C
            C = X
            i.
            C THIS PROGRAM CALCULATES THE DISCOUNT
            OF AN AMOUNT BY KNOWING THE AMOUNT AND
            THE DISCOUNT RATE.
            j.
            C THIS IS THE START OF THE PROGRAM
                  PROGRAM CUTE
```

☐ **Answers**

a. The statements are correct. Notice that any character may follow C in the first column.
b. The statements are correct; however, after C a space might be inserted for the sake of readability.
c. The first statement as it stands is neither a FORTRAN statement nor a comment. It needs a C in the first column.
d. If the first statement is a comment, it needs a C in the first column.
e. The READ statement will not read A, B, and C, because it will be considered as a comment.
f. The first line is correct, but the WRITE statement is incorrect because it starts in column 1 instead of column 7.
g. Correct. Note that the C in column 1 causes the word COPY to be considered as a comment.
h. The first line is a comment. The second will also be read as a comment, although it may be intended as an assignment statement. (It doesn't start from column 7.)
i. The letter C should be placed in the first column of the second and third lines. Remember that a multiple-line comment must have a C in column 1 of each line.
j. This is correct; note that the comments can be placed before the PROGRAM statement.

The following example illustrates the use of headings, carriage control characters, comment statements, and the separation of FORMATs from the program. It sums up what we have learned so far.

EXAMPLE 3.1

```
      PROGRAM SOFAR
C AUTHOR G. H. DOE, NOV. 19--
C ******************************************
```

```
C THIS PROGRAM READS TWO EXAM MARKS FOR EACH      *
C STUDENT, FINDS THE AVERAGE FOR EACH STUDENT,    *
C AND PRINTS THE MARKS AS WELL AS THE AVERAGE.    *
C THE PROGRAM ALSO FINDS THE GRAND AVERAGE FOR    *
C THE CLASS                                       *
C ********************************************
C ********************************************
C VARLABLE DICTIONARY:                            *
C SCORE1 :   THE TEST SCORE NO. 1                 *
C SCORE2 :   THE TEST SCORE NO. 2                 *
C AVRG   :   THE AVERAGE OF SCORE1 AND SCORE2     *
C TOTAVG :   THE SUM OF AVERAGES                  *
C GAVRG  :   THE GRAND AVERAGE                    *
C   N    :   THE NO. OF RECORDS                   *
C ********************************************
C HEADING NUMBER 1, STARTING A NEW PAGE
C
      WRITE( 6 , 10 )
10    FORMAT( '1' , 40X , 'CLASS AVERAGE' )
C
C HEADING NUMBER 2, AFTER SKIPPING 3 LINES
C
      WRITE( 6 , 20 )
20    FORMAT( ///11X , 'SEQUENCE NUMBER' , 5X ,
     *        'FIRST SCORE' , 5X , 'SECOND SCORE' ,
     *        5X , 'THE AVERAGE' )
C SETTING THE COUNTER BLOCK AND TOTAL OF AVERAGES
C (TOTAVG) TO ZERO
C
      N = 0
C     TOTAVG = 0.0
C READING THE DATA
C SCORE1 AND SCORE2 ARE THE VARIABLES. AVRG IS
C THE AVERAGE OF SCORE1 AND SCORE2. THIS IS THE
C START OF THE LOOP.
C
C READING THE DATA, WHEN FINISHED READING,
C TRANSFER TO STATEMENT 100
C
30    READ( 5 , 40 , END=100 ) SCORE1, SCORE2
C
C CALCULATING THE AVERAGE
C
      AVRG = (SCORE1 + SCORE2) / 2.0
C
C COUNTING THE NUMBER OF DATA AND GENERATING
C THE SEQUENCE NUMBERS
C
      N = N + 1
C
C WRITING THE DATA
C
      WRITE( 6 , 50 ) N , SCORE1 , SCORE2 , AVRG
C
C FINDING THE TOTAL AVERAGES FOR CALCULATING
```

```
C  THE GRAND AVERAGE
C
       TOTAVG = TOTAVG + AVRG
C
C  MAKING A LOOP TO READ ADDITIONAL DATA
C
       GO TO 30
C
C  CALCULATING AND WRITING THE GRAND AVERAGE
C  AT THE END OF DATA.
C  GAVRG IS THE GRAND AVERAGE
C
100    GAVRG = TOTAVG / N
       WRITE( 6 , 60 ) GAVRG
C
C  THE FORMAT STATEMENTS. THE FIRST SCORE IS PLACED
C  IN COLUMN 1 THROUGH 5 LIKE 99.99. THE SECOND
C  SCORE IS PLACED IN COLUMN 7 THROUGH 11.
C
40     FORMAT( F5.2 , 1X , F5.2 )
C  WRITING THE DATA IN EVERY OTHER LINE
50     FORMAT( '0' , 15X , I2 , 15X , F6.2 , 10X,
      *         F6.2 , 10X , F6.2 )
60     FORMAT( ///// ,41X,'THE GRAND AVERAGE IS' ,
      *         F6.2 )
C  A LAST WORD INDICATING THE END OF THE PROJECT
C  AFTER SKIPPING 5 LINES.
       WRITE( 6 , 70 )
70     FORMAT( ///// , 41X , 'END OF REPORT' )
       STOP
       END
```

THE COMPILERS

In this section we discuss some terminology with which you should be familiar. After reviewing, you should be able to define the following:

machine language   machine-independent language
higher-level language   source program
compiler   object program
compilation   execution

Each computer is designed by manufacturers to recognize special instruction codes. These codes can be executed only if they are in terms of some predetermined numerical codes. The instructions written in such a way are said to be in *machine language* form.

> **Machine language** *is a series of instructions in terms of numeric codes which can be interpreted by a certain computer model.*

96 *Programming Development and Execution*

Machine languages are *machine dependent*. That is, a program written for one model cannot be processed by another model. Writing a program in machine language is extremely difficult and time consuming.

However, *higher-level* programming languages have been developed to be used by a programmer without the need of knowing machine language, to be independent of the machine being used, and to be easier to learn. There are a large number of higher-level languages currently in use. FORTRAN is one of them. It has been designed to be somewhat like English, easy to learn, general enough to handle a variety of problems, and capable of being run on almost any computer.

Now you may be wondering how a higher-level language, such as FORTRAN, can be understood by almost any computer. The *source program* (the original program) is translated into machine language by a process called *compilation*. Computers can perform compilation without human intervention by using a special program called the *compiler*.

> **Compilation** *is the process of translating a source program into a machine-language program.*

> *A* **compiler** *is a special program which translates a source program into a particular machine language.*

The translated source program, which now is stated in terms of a series of numeric codes and is ready to be interpreted by the computer, is called an *object program*. (See Figure 3.6.)

Each compiler is made for a special model of computer; often it is provided by the manufacturer. FORTRAN—or any higher-level language—is then a machine-independent language because it does not have to be written for a specific computer model.

A successful program may be done in one batch run, but normally a program goes through a computer twice (see Figure 3.7):

1. The program is first translated into machine language by the compiler, which also checks the program for grammatical errors at this stage.
2. If the program does not have serious grammatical errors, the object

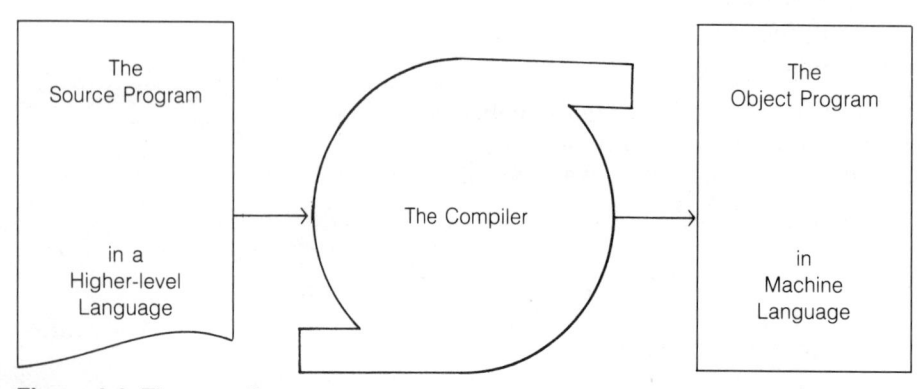

Figure 3.6 The compiler.

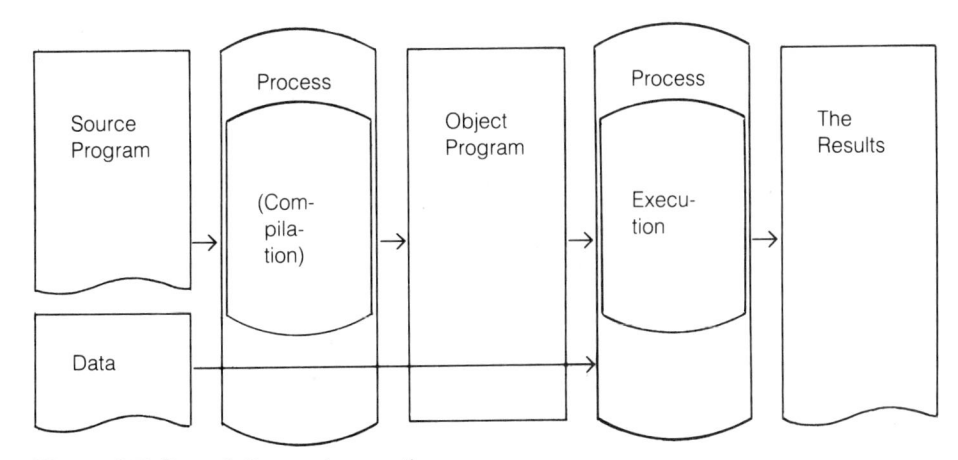

Figure 3.7 Compilation and execution.

program goes through the computer to be *executed*, that is, for the instructions to be carried out. The object program can also be saved for later use in order to save time in subsequent runs.

EXERCISES

3.1 Define the following terms:

  a. Machine language.
  b. Higher-level language.
  c. Machine-dependent language.
  d. Machine-independent language.
  e. Compiler.
  f. Compilation.
  g. Source program.
  h. Object program.
  i. Execution.

☐ **FORTRAN Compilers**

FORTRAN was the first higher-level language to be widely used. The first version of FORTRAN was developed in 1956 with the advent of commercial computers; since then several versions of FORTRAN compilers have been introduced. These versions have many common characteristics; they may be rather different because of add-on features and other modifications. Each such version has its own name. The following is a brief list of subsequent versions of the original FORTRAN:

- FORTRAN IV (pronounced "Fortran four"): Developed in 1962.
- ANS FORTRAN: Developed in 1966 by the *American National Standards Institute*, as an attempt to standardize FORTRAN.
- WATFOR (pronounced "Wat four," for Waterloo FORTRAN), and
- WATFIV (pronounced "Wat five"), and
- WATFIV-S: Developed by the University of Waterloo in Canada. Both WATFIV and WATFIV-S are modifications of WATFOR and very close to ANS FORTRAN.
- FORTRAN 77: The most recent revised version of ANS FORTRAN, developed in 1977 and published in 1978 by the American National Standards Institute.

98 Programming Development and Execution

Since this is an introductory text, we will include only the most common and useful features of FORTRAN 77.

☐ FORTRAN Statements

Some FORTRAN statements instruct the computer to carry out a specific task. For example, a READ statement causes the computer to read the data from the data section. Such statements are called *executable* statements.

Some statements, however, provide only information for execution of the statements, and do not cause a task to be performed. For example, a FORMAT statement furnishes supportive information about the layout of the data. These statements are *nonexecutable*. Such nonexecutable statements will not result in any machine-language instructions by the compiler, whereas the executable statements will result in some kind of machine-language instruction.

Executable statements build the logical sequence of a program. By changing the order of an executable statement, you may change the entire program. But the nonexecutable statement does not influence the logic of a program. Some of them—including FORMAT—may be placed anywhere in the program without affecting the program's logic. However, some nonexecutable statements must be placed in specific order. There are many kinds of nonexecutable statements, which we will be discussing as we progress.

■
ERROR DETECTION

After punching the FORTRAN statements onto cards, or keying them into a terminal, the entire deck, including the control cards, must be submitted to the computer for execution. If you have used the proper control statements, and if the program is correct, the source program will be listed followed by program information and the desired output as it was illustrated previously in Figure 3.5.

If there has been any error, you need to correct and resubmit the entire program. After each run, you must study the printout carefully for errors. A program may have to be rerun several times before it is completely correct. Removing the errors ("bugs") from a program is called *debugging*. Debugging is an important part of the programming process.

> **Debugging** *is the process of detecting and correcting programming errors.*

A program may have three types of errors:

1. Syntax errors
2. Execution errors
3. Logical errors

☐ Syntax Errors

A *syntax error* is a grammatical error. Remember that a language follows a series of predetermined grammatical rules. Therefore, when you make a grammatical error in a statement, the compiler cannot figure out what that statement is, only that there has been an error.

Some typical syntax errors are:

- Misspelling key words (typographical errors)
- Making variable names too long (more than six characters)
- Starting statements from column 1 instead of 7
- Having unbalanced parentheses (that is, the left or right parenthesis may be missing)
- Forgetting to place a character in the continuation column (column 6)
- Forgetting to punch a C in the first column of a comment line
- Omitting a necessary comma or incorrectly using commas and spaces
- Typing zero instead of the letter O, or one instead of I; and vice versa

Errors of syntax are detected by the compiler during compilation. Furthermore, a syntactical error can be either a *serious error* or a *warning*. In the case of a serious error, the program will not go through execution and must be corrected and resubmitted. A warning, however, does not stop execution of the program. As an example, misspelling a key word such as READ or WRITE is a serious error. But a variable's symbolic name with more than six characters could be a warning. Most compilers automatically truncate the variable to six characters. In any case, the computer will not correct any violation of rules; instead it prints some error messages when it detects the error. These messages, although not always straightforward in their meaning, will help you identify the errors and their location in your program. The message itself depends on the compiler being used. Some compilers print clearer messages than others.

Look at the following solved problems for examples of syntax errors.

SOLVED PROBLEMS

3.3 Find the syntax errors in the following statements. Can you guess what kind of messages will be printed for each error?

a.
```
      WRITE( 6 , 20 ) A , B
      FORMAT( 1X , 2F5.2 )
```
b.
```
      FORMAT( F7.2F8.2 )
```
c.
```
      FORMAT( 6X,5F5.2
```
d.
```
      AVERAGE= (A + B) /2
```
e.
```
100   COMMENT
```
f.
```
      REED( 5 , 10 ) U , V
```
g.
```
      THIS IS A COMMENT
```
h.
```
      NET-PAY = PAY - DED
```
i.
```
      AMT$=A*B
```

100 *Programming Development and Execution*

```
    j.
            READ( 5 , 20 ) A , N , C
    20      FORMAT( F6.1,F6.2,
            F3.1
    k.
            W RITE ( 6 , 10 ) A , B , C
    l.
            WRITE( 6 , 55 ) X , Y
         55FORMAT( 1X , F5.1 , 5X , F6.1 )
    m.
            AMOUNT = 369,000
    n.
              ⋮
            GO TO 30
            WRITE(6,10)TOTAL
    o.
            READ(5,10, END = STOP)A,B
```

☐ **Answers**

The error messages depend on the system being used. Examples of error messages are as shown below for each statement. Notice that sometimes several error messages can be printed for one error.

    a. 'STATEMENT LABEL 20 REFERENCED BUT NOT DEFINED'
'STATEMENT LABEL EXPECTED BUT NOT FOUND'
'FORMAT MUST HAVE STATEMENT LABEL'
    b. 'SEPARATOR MISSING AT -----'
    c. 'TERMINAL RIGHT PARAN MISSING'
    d. 'NAME EXCEEDS 6 CHARACTERS'
    e. 'THIS IS NOT A FORTRAN STATEMENT'
    f. 'THIS IS NOT A FORTRAN STATEMENT'
    g. 'THIS IS NOT A FORTRAN STATEMENT'
    h. 'LEFT SIDE OF EQUAL SIGN IS ILLEGAL'
    i. 'THIS IS NOT A FORTRAN STATEMENT'
    j. 'NON-NULL LABEL FIELD ON CONTINUATION LINE'
'TERMINAL RIGHT PARANTHESES MISSING'
'THIS IS NOT A FORTRAN STATEMENT'
'STATEMENT LABEL 20 REFERENCED BUT NOT FOUND'
    k. 'THIS IS NOT A FORTRAN STATEMENT'
    l. 'THIS IS NOT A FORTRAN STATEMENT'
'STATEMENT 55 REFERENCED BUT NOT DEFINED'
'NON-NULL LABEL FIELD ON CONTINUATION LINE'
    m. 'ILLEGAL FORM INVOLVING THE USE OF COMMA'
    n. 'NO PASS TO THIS STATEMENT'
(A statement label is expected after any GOTO statement.)
    o. 'E= IN A READ STATEMENT MUST BE FOLLOWED BY A LABEL'

☐ Execution Errors

Unlike syntactical errors, execution errors are not detected by the compiler. Generally they occur during execution because of:

1. Timing
2. Programming error

A timing error occurs when the program needs more than "normal time" to be completely executed. For example, an infinite loop in a program which runs in batch will cause the execution to stop because it needs "infinite time" (and an infinite number of data if it is reading the data).

A programming error is caused also by some type of mistake in the actual programming. For instance, dividing a number by zero is mathematically undefined. Thus, the program segment

```
A = 0.0
B = 5.0
C = B/A
```

has an execution error and causes the computer to stop executing.

If the program has to be terminated due to an execution error, the computer will print a numerical code at the end of a printout to indicate the reason for termination. In every computer center you should be able to find reference manuals which will show you the meaning of these error codes. Nevertheless, it is more difficult to locate and detect an execution error than a syntax error. One of the reasons is that messages regarding the execution errors frequently are ambiguous. Another reason is that the execution error can be caused by logical errors.

☐ Logical Errors

Logical errors have occurred if the program runs, but the output is incorrect. Such an error may develop because of omitting or misplacing a statement, or specifying an action incorrectly. For example, if a program does not print a desired output, it is possible that the WRITE statement is missing, or that the appropriate variable is not included in the WRITE statement. The logical error can also be the result of incorrect specification; for example, if the average of three variables is calculated by

```
AVG = A + B + C / 3.0
```

the result will not be correct because of the incorrect expression. The formula should be corrected to read

```
AVG = (A + B + C) / 3.0
```

Some logical errors are easy to find, but others may not be. If the printout, for instance, shows that the floor area of a room is a negative number, you know right away either something is wrong with calculations or the input data are not correct. If the listing of a program is correct and there is no execution error, but there is no output, it is possible that the WRITE statements are not correct. Other logical errors are more difficult to diagnose and correct. The computer may detect syntax and time execution errors, but it cannot detect logical errors, because a computer cannot understand the logical sequence of your program.

It is your responsibility then, as a programmer, to test the output for the correct results. A wise approach is to run all programs with a set of test data

102 *Programming Development and Execution*

for which the results are already known. If any discrepancies are found between the result on the printout and your expectations, look for a logical error.

Logical errors can sometimes be found by a method called "desk checking" or "role playing." The programmer tries to act like a computer and carries out the instructions, statement by statement, until the error is found.

It is not always obvious how to find and correct some logical errors. It requires a lot of checking, testing, and careful review of the program. The following hints should be helpful in reducing the chance of errors:

- Careful planning and designing of a program (explained in the next section)
- Using descriptive variable names
- Careful documentation
- Careful review of the program after coding
- Careful checking of the statements after punching or keying

PROGRAM PLANNING

☐ Programming Tools

As your program becomes longer and more complicated, you will find it necessary to plan ahead. The FORTRAN coding form and the print chart, discussed in previous sections, are two useful tools for planning the input and output of a program. There are also several tools available for expressing the solution procedures of your program, among them *flowcharts, pseudocode* and Logic-charts. Prepared as the initial step in writing a program, they serve two useful purposes:

1. as a guide to actual programming,
2. as a document of the logic of the program.

Flowcharts

A flowchart is an easy-to-read diagram of a program. It is a pictorial presentation of the necessary steps for solving a problem. Each step, represented by a symbol, indicates a necessary action. The sequence of the steps is shown by arrows. The most important symbols used in flowcharting are shown in Figure 3.8.

EXAMPLE A

The flowchart in Figure 3.9 shows how to find the floor area of a room. The inputs are two variables: length and width. The area is calculated by multiplying them together. The output is length, width, and area.

EXAMPLE B

The flowchart in Figure 3.10 shows how to find the floor area of a room if there is more than one set of data, and how to print the heading.

These two examples are simplified versions of flowcharting. Throughout this text you will be seeing more examples of flowcharts.

Flowcharting seems to be a time-consuming process; nevertheless, it is a very useful problem-solving tool. Some of the advantages of flowcharts are as follows:

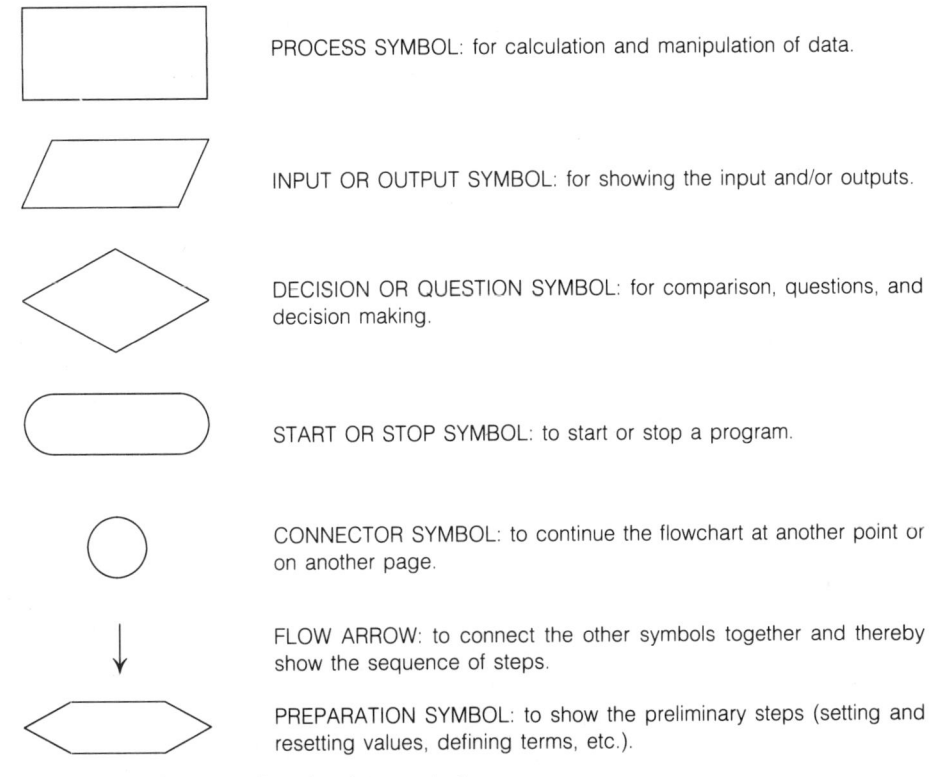

	PROCESS SYMBOL: for calculation and manipulation of data.
	INPUT OR OUTPUT SYMBOL: for showing the input and/or outputs.
	DECISION OR QUESTION SYMBOL: for comparison, questions, and decision making.
	START OR STOP SYMBOL: to start or stop a program.
	CONNECTOR SYMBOL: to continue the flowchart at another point or on another page.
	FLOW ARROW: to connect the other symbols together and thereby show the sequence of steps.
	PREPARATION SYMBOL: to show the preliminary steps (setting and resetting values, defining terms, etc.).

Figure 3.8 Common flowcharting symbols.

1. It is used as a plan for solving a problem.
2. It makes it easier to code a program.
3. It is useful for finding logical errors.
4. It is used as a document which shows just how the program should work.
5. It is used as a medium of communication between the problem solver and the programmer (if there's any difference between the two).

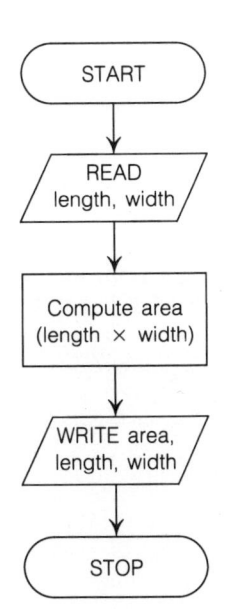

Figure 3.9 Flowchart for finding floor area.

104 Programming Development and Execution

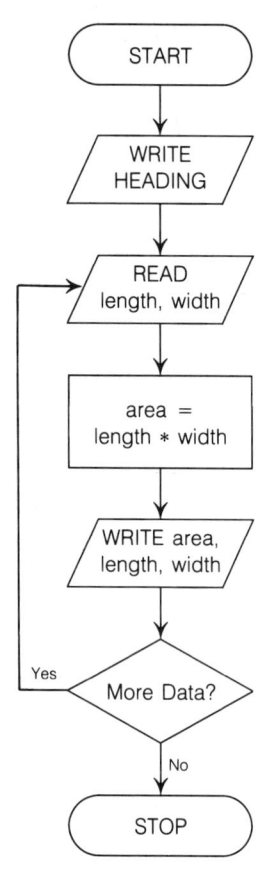

Figure 3.10 Flowchart for finding floor area with additional data.

Flowchart symbols can be hand drawn, but it is better to use a template. Flowcharting templates are available in most supply stores.

Pseudocode

Pseudocode is another tool for program design which has recently gained popularity among programmers. Pseudocode, just like flowcharts, represents the solution of a problem. However, it does not use graphic symbols and is an informal way of expressing the plan. The programmer writes, in a narrative form, the necessary steps for solving a problem.

The following is a simple example of a pseudocode:

Start
Read Length, Width
AREA = Length * Width
Print Length, Width, Area
END

The following is a more complete form of this example:

Start
Write the headings
Read Length, Width
AREA = Length * Width
PRINT Length, Width, Area
IF more data, THEN
        GO TO READ
ELSE
End of program

Pseudocode has not been standardized yet. There are several variations. One of the advantages of pseudocode is that it can be written as modules and submodules (like the outline for a report—chapter, paragraphs, and sentences). The following is an example:

1. Writing headings
    1.1 heading #1, Company's Title
    1.2 heading #2, Employee #, pay, deduction, net
2. Initializing variables
    2.1 Total pay = 0.0
    2.2 Total deductions = 0.0
    2.3 Total net = 0.0
3. Beginning of the loop
4. Reading data, at the end go to step 8
    4.1 Read employee #, wage rate, hours worked
5. Calculating
    5.1 Pay = wage × hours
    5.2 Deduction = pay × .12
    5.3 Net = pay − deduction
    5.4 Calculating totals
        5.4.1 Total pay = total pay + pay
        5.4.2 Total deduction = total deduction + deduction
        5.4.3 Total nets = total nets + nets
6. Writing the details
    6.1 Write employee #, pay, deduction, net
7. End of the loop, GO TO Step 3
8. Writing totals
    8.1 Write total pay, total deductions, total nets
9. End of the program

Logic-Charts

A logic-chart (also called Nassi-Shneiderman diagram) is a chart for showing the logical structure of a program. It is a table of the steps, logic, and flow of the solution methods to a problem. Figure 3.11 shows two examples of this kind of chart. The logic-chart uses a combination of the following:

1. A table for showing the structure of the solution procedure
2. A rectangle diagram to show one or several statements
3. An L-shaped box for showing a loop
4. Two triangles for showing the IF-THEN-ELSE conditions

☐ Algorithms

> *A set of step-by-step instructions, in a simple language, for solving a problem is called an* **algorithm.**

The algorithm, or solution procedure, must be defined in detail before coding. Most of the time, developing an algorithm needs ample thought, research, and time. An algorithm will lend itself very easily to a flowchart or pseudocode. Each step of an algorithm will eventually be performed by FORTRAN statements.

106 *Programming Development and Execution*

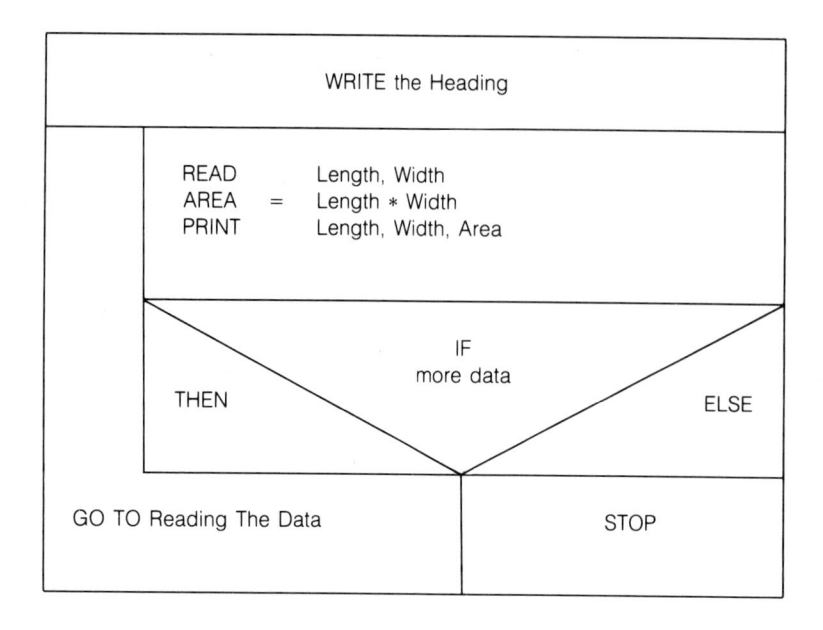

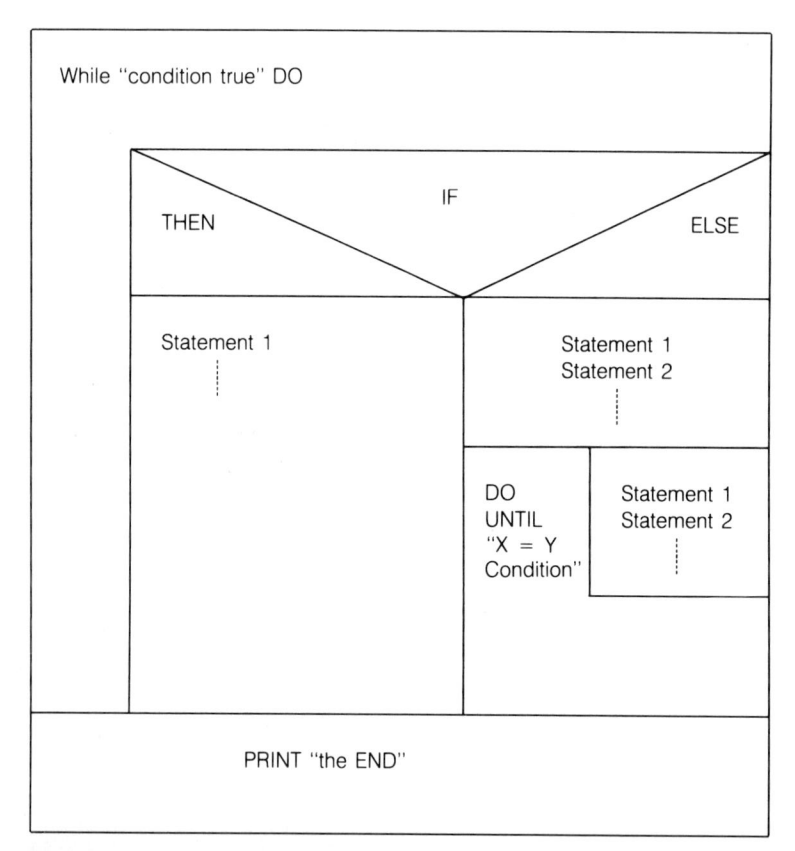

```
┌─────────────────────────────────────────────────────────┐
│                    WRITE the Heading                    │
├─────────────────────────────────────────────────────────┤
│   ┌─────────────────────────────────────────────────┐   │
│   │   READ       Length, Width                      │   │
│   │   AREA    =  Length * Width                     │   │
│   │   PRINT      Length, Width, Area                │   │
│   ├─────────────────────────────────────────────────┤   │
│   │ \                   IF                        / │   │
│   │   \              more data                  /   │   │
│   │     \                                     /     │   │
│   │ THEN  \                                 /  ELSE │   │
│   │         \                             /         │   │
│   ├─────────────────────────────────────────────────┤   │
│ GO TO Reading The Data         │        STOP            │
└─────────────────────────────────────────────────────────┘
```

```
┌─────────────────────────────────────────────────────────┐
│  While "condition true" DO                              │
│   ┌─────────────────────────────────────────────────┐   │
│   │  \                 IF                        /  │   │
│   │    \                                       /    │   │
│   │ THEN \                                   / ELSE │   │
│   ├──────────────────────┬──────────────────────────┤   │
│   │                      │                          │   │
│   │  Statement 1         │   Statement 1            │   │
│   │     ⋮                │   Statement 2            │   │
│   │                      │      ⋮                   │   │
│   │                      │   ┌──────────┬─────────┐ │   │
│   │                      │   │ DO       │Statement1│ │   │
│   │                      │   │ UNTIL    │Statement2│ │   │
│   │                      │   │ "X = Y   │   ⋮      │ │   │
│   │                      │   │ Condition"│         │ │   │
│   │                      │   └──────────┴─────────┘ │   │
│   └─────────────────────────────────────────────────┘   │
│                  PRINT "the END"                        │
└─────────────────────────────────────────────────────────┘
```

Figure 3.11 Examples of logic-charts.

The following are two examples of algorithms.

EXAMPLE A

An algorithm for finding the sum of a series of numbers.

    1. Start, set SUM = 0.0
    2. Read a number A
    3. Accumulate the numbers by SUM = SUM + A

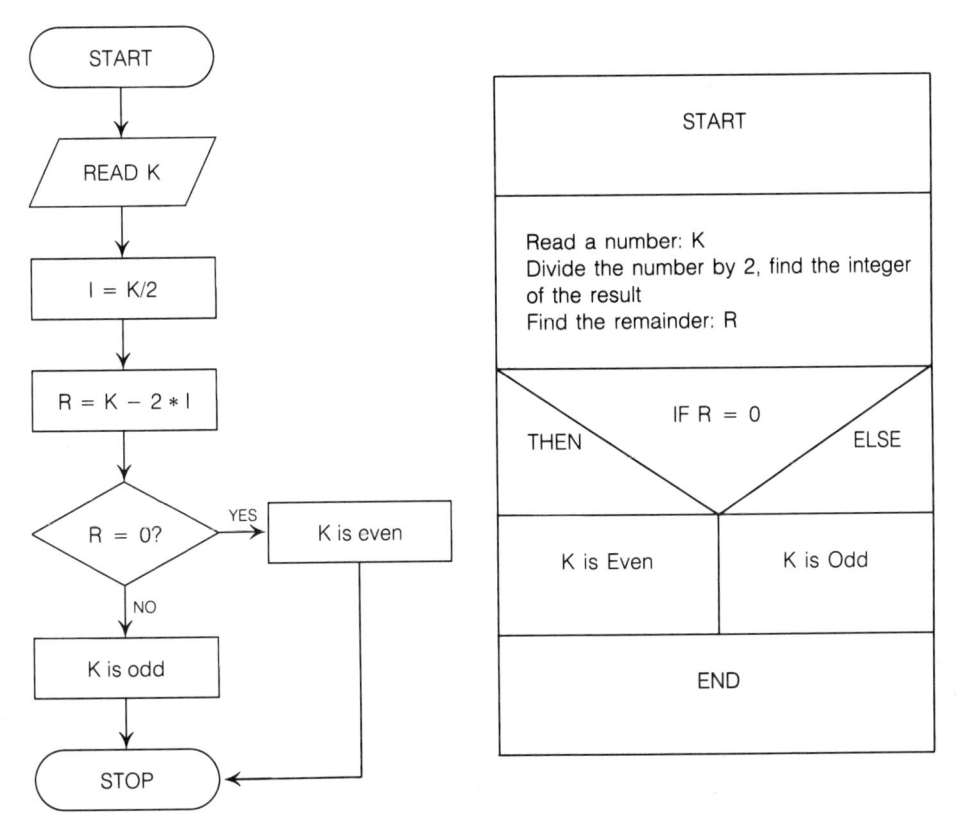

Figure 3.12 A flowchart and corresponding logic-chart.

4. If there are more data then
      go to step 2
   Else
      print the sum
   End

EXAMPLE B

An algorithm for finding whether an integer number is even or odd.

1. Start
2. Read a number K
3. Divide the number by 2, find the integer of the result
4. Find the remainder R
5. If R = 0 then
      K is even
   Otherwise
      K is odd
6. End

Notice how easily this algorithm is charted in Figure 3.12.

Algorithms are discussed further in Chapter 10, with examples of more complicated ones.

☐ Programming Cycle

As mentioned previously, writing a program in FORTRAN (or any other language) is called coding. Coding is not the only task of a programmer, and FORTRAN (or any computer language) is only a tool for solving problems. We must first understand the problem to be solved, analyze it, break it down into modules or components, develop an algorithm, and plan the solution methods before coding.

> **RULE 3.3**
> *A clear understanding of the problem, objectives, requirements, constraints, and solution procedure are necessary steps prior to coding.*

It is advisable to devote plenty of time and thought to the planning process. The following general steps will help you in the analysis of the problem and the programming process.

STEP 1: PROBLEM ANALYSIS

Analyze the problem carefully. Try to state its nature and list its objectives in writing. This phase clarifies what the problem is and how it can be solved by a program.

STEP 2: INPUT-OUTPUT FORMULATION

List the output and input requirements. First, list the items which must appear on the output as well as their length and type (real, integer). Then, list the input items as well as their length and type. Avoid unnecessary or redundant information. The input should all be information which is needed to produce the appropriate output. Remember that the items which will be calculated or generated by the program must not appear in the input list. For example, you can always calculate the average of the test scores, or generate the sequence numbers in the program. It is important to emphasize that the output should be designed first because that is the actual objective.

STEP 3: LAYOUT DESIGN

Describe the necessary data location of the information for input and output. Design the layout for the output and input. The output layout on a print-chart shows what the output should look like, along with the exact location of each item. The input layout shows the form and location of all input items.

STEP 4: PROCESS DESIGN

Formulate the solution procedures. This step defines the procedures or calculations necessary to arrive at the outputs from the inputs. This requires breaking the entire problem into modules. Each module is a small, workable component of the whole problem. For example, reading the data, a calculation, and printing the information can each be a module. Naturally, it is easier to define the required procedures and calculations for a small module than the problem as a whole.

STEP 5: SCHEME PREPARATION, FLOWCHARTING

Prepare a flowchart, a pseudocode, or a logic-chart for each module and the program. This represents the necessary steps in solving the problem: inputs, outputs, and process requirements. Either a chart or pseudocode or both are blueprints of the actual program.

STEP 6: CODING

Write the program in FORTRAN statements. Make sure to review the program carefully after coding to detect any logical or syntactical errors. Pay particular attention to the format of the read statements, verifying their correspondence with the input data layout. This is commonly overlooked by beginners.

STEP 7: RUN PREPARATION

Punch the statements and data onto cards or key them into a terminal. The following points are very important at this stage:

1. Review each line carefully to detect any typing errors.
2. Make sure the data are keyed (punched) according to the data layout, which should in turn correspond to the READ format. As was mentioned, this check is very important for beginners.
3. Check the control cards or system commands which vary from system to system, to insure they are correct.

STEP 8: EXECUTION, TESTING, AND DEBUGGING

Submit the program for execution. After running the program, correct any errors which may exist and resubmit the program for execution. After correcting all the syntax errors, the first run of the program should be with test data, for which you already know the correct output.

STEP 9: DOCUMENTATION

Describe the program, the solution procedure, and its limitations. This essential part of the programming process is necessary for future reference. After a time, even the original programmer may need this documentation to understand the details of a program. Documentation should start from the beginning of the programming cycle. The documentation includes flowcharts, pseudocode, format specifications, and comments within the source program. A final report, explaining the nature of the problem, the solution methods, how the program works, its limitations, and the other important points about the program completes the documentation phase.

Sometimes the programmer is different from the problem solver (the analyst). In this case, steps 1 through 5 are done by the problem solver, and steps 6 to 9 are done by the programmer. In any case, the programmer should fully understand the problem and the solution procedures. The following sample problem illustrates the process.

A SAMPLE PROBLEM

The following example is designed to demonstrate the general procedures for translating a problem into a program.

STEP 1: THE PROBLEM

XYZ Gas Company's managers would like a monthly report showing the dollar amount and total gas consumption by their customers for each month. Analysis shows that:

1. The billing rate is different for different customers (commercial, residential, etc.).

110 Programming Development and Execution

2. The last meter reading and the current meter reading are available.
3. The managers would like information about each customer, and the totals at the end of the report.

The objective is, therefore, to write a program which can generate such a report.

STEP 2: INPUT-OUTPUT FORMULATION

A detailed analysis shows that the variables listed in Table 3.1 are necessary for the output. The amount and the totals can be calculated if the rate is known. Therefore, we need the information shown in Table 3.2 for input.

STEP 3: LAYOUT OF THE INPUT AND OUTPUT

The OUTPUT: The output is designed as shown in the print-chart in Figure 3.13.

The INPUT: It is determined to place the input data as follows:

1. Customer's identification number   Columns 1–5
2. Previous meter reading for each customer Columns 7–12
3. Current meter reading for each customer   Columns 14–19
4. Rate   Columns 21–25

Notice that there is a space between each data item and the next. This facilitates reading the input data visually. The following sample data are used for this program:

```
54985 562.26 721.60 .35
 4982 205.50 296.75 .38
36872 911.50 983.25 .33
99995  59.75 295.90 .39
45632 199.50 358.30 .33
33333 0.0    123.50 .39
45678 250.50 423.50 .38
```

STEP 4: PROCESS DESIGN

The program is broken down into six modules:

1. Housekeeping: writing the headings, initializing the variables
2. Reading the input data
3. Calculating
   3.1 the gas used and
   3.2 the dollar amount for each customer

Table 3.1 Output Analysis Form for the Sample Program

ITEM	VARIABLE'S SYMBOLIC NAME	TYPE	FIELD LENGTH (COLUMNS)
1. Customer's identification number	ID	Integer	5
2. Quantity of gas used by each customer	GU	Real	6
3. Billing rate	RATE	Real	6
4. Total dollar amount charged to each customer	AMT	Real	7
5. Total quantity of gas used by all the customers	TGU	Real	7
6. Total dollar amount for all customers	TAMT	Real	7

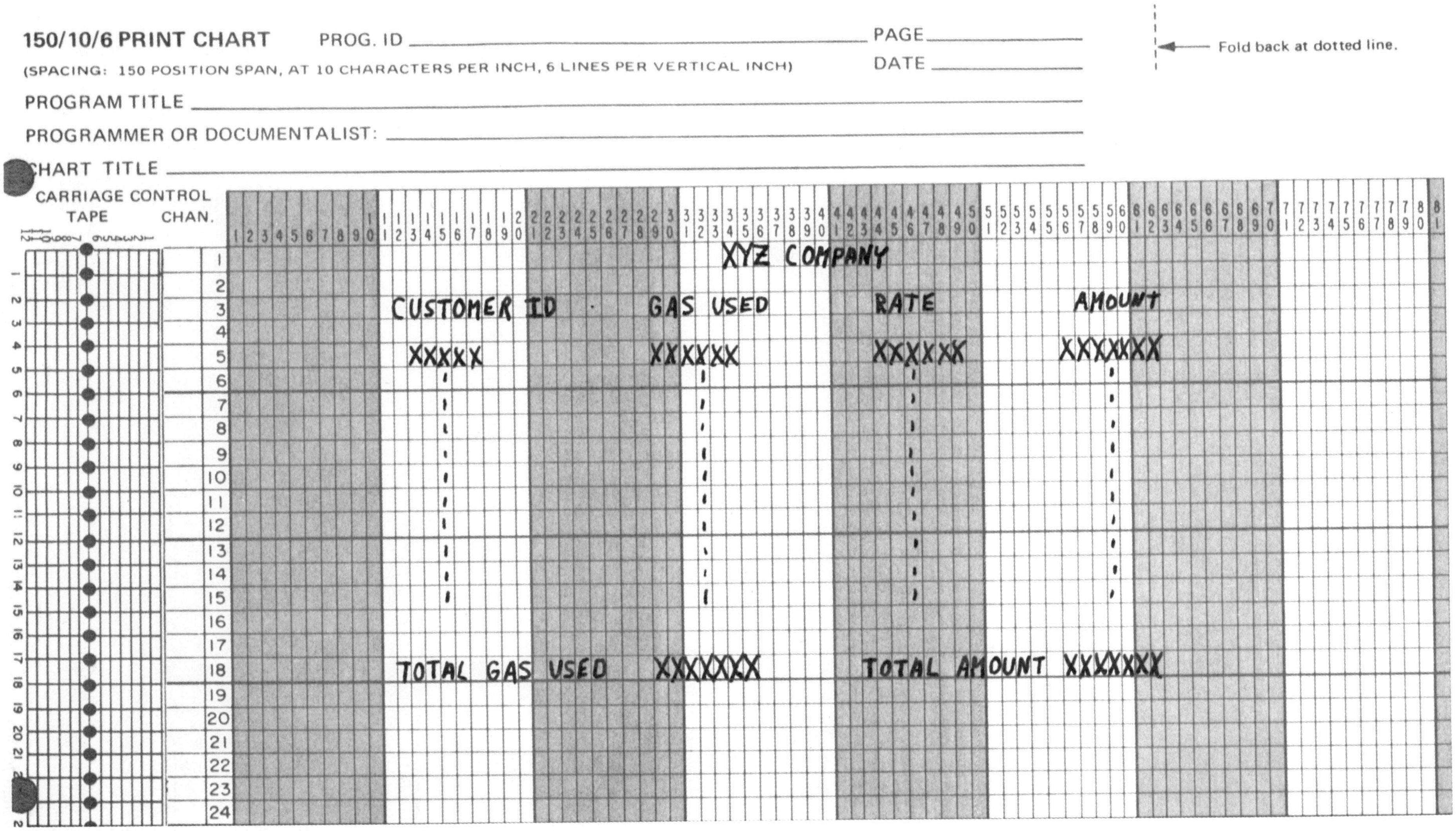

Figure 3.13 The design of the output for the sample problem.

111

Table 3.2 Input Analysis Form for the Sample Program

ITEM	VARIABLE'S SYMBOLIC NAME	TYPE	FIELD LENGTH (COLUMNS)
1. Customer's identification number	ID	Integer	5
2. Previous meter reading	PMR	Real	6
3. Current meter reading	CMR	Real	6
4. Rate per unit	RATE	Real	5

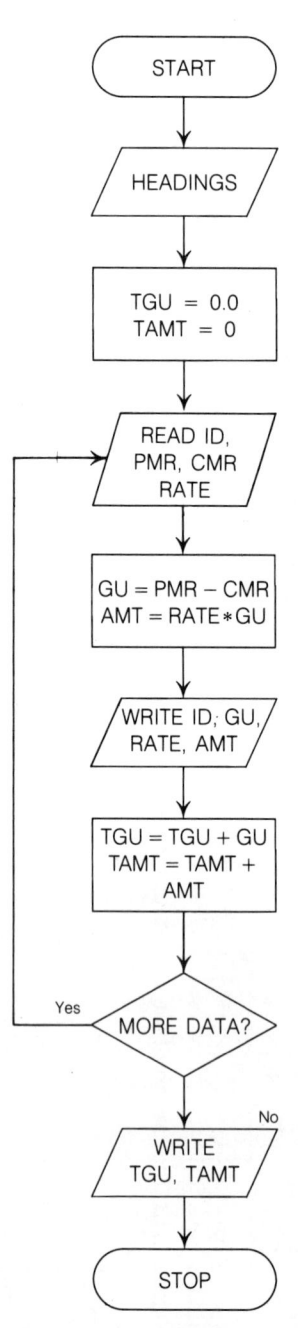

Figure 3.14 Flowchart for the sample program.

4. Writing the information about each customer
5. Calculating the totals
6. Writing the totals after the end of the data

The expanded outline is as follows:

1. Housekeeping
   1.1 Write headings
   1.2 Set:
       TGU = 0.0
       TAMT = 0.0
2. Read variables for each customer; at the end of data go to step 7.
3. Calculate:
   3.1 Gas used:
       GU = CMR − PMR
   3.2 Amount of charge:
       AMT = RATE∗GU
4. Write the information
5. Cumulate the gas used and amount:
       TGU = TGU + GU
       TAMT = TAMT + AMT
6. Go to Step 2.
7. Write the total gas used and total amounts
8. END

STEP 5: FLOWCHARTING

The flowchart is prepared as in Figure 3.14, on page 112.

STEP 6: CODING

The coded program is shown in Figure 3.15. After writing the program, the flowchart and the program are carefully reviewed again for any logical or syntactical errors.

```
      PROGRAM METER
C     THIS PROGRAM GENERATES A REPORT FOR THE GAS USERS OF THE COMPANY.
C     THE INPUTS ARE CUSTOMER'S ID, PREVIOUS METER READING (PMR), CURRENT
C     METER READING (CMR), AND THE RATE FOR EACH CUSTOMER (RATE).
C     AUTHOR T. H. SMALL, AUGUST 19, 19--
C
C     HEADINGS
      WRITE( 6 , 10 )
C     TGU = TOTAL GAS USAGE              TAMT = TOTAL AMOUNT
      TGU = 0.0
      TAMT = 0.0
C     READING THE DATA, THE START OF THE LOOP
15    READ( 5 , 20 , END=99 ) ID , PMR , CMR , RATE
C
      GU = CMR − PMR
      AMT = RATE ∗ GU
      WRITE( 6 , 30 ) ID, GU , RATE , AMT
C     CALCULATE TOTALS FOR ALL CUSTOMERS
      TGU = TGU + GU
      TAMT = TAMT + AMT
C     END OF THE LOOP
      GO TO 15
C
C     WRITE THE FOOTINGS
99    WRITE( 6 , 40 ) TGU , TAMT
C
C     THE FORMATS USED IN THE PROGRAM
10    FORMAT('1' , 35X , 'XYZ COMPANY' //13X , 'CUSTOMER ID' , 6X , 'GAS USED' ,
     *7X , 'RATE' , 7X , 'AMOUNT'//)
20    FORMAT( I5 , 1X , F6.2 , 1X , F6.2 , 1X , F5.3 )
30    FORMAT( 14X , I5 , 11X , F6.2 , 7X , F6.3 , 7X , F7.2 )
40    FORMAT( ///12X , 'TOTAL GAS USED' , 3X , F7.2 , 7X , 'TOTAL AMOUNT' ,
     *2X , F7.2)
      END
```

Figure 3.15 Coding of the sample problem.

114 *Programming Development and Execution*

STEP 7: RUN PREPARATION

The program is keyed into a terminal. After both the program and control statements are reviewed, the program is submitted for execution.

```
          PROGRAM METER       73/172   OPT=0                         FTN 5.1+528
       1                 PROGRAM METER
       2          C THIS PROGRAM GENERATES A REPORT FOR THE GAS USERS OF THE COMPANY.
       3          C THE INPUTS ARE CUSTOMER (ID), PREVIOUS METER READING (PMR),
       4          C CURRENT METER READING (CMR), AND THE RATE FOR EACH CUSTOMER (RATE).
       5          C AUTHOR:  T. H. SMALL, AUGUST 19, 19--
       6          C
       7          C HEADINGS
       8                 WRITE(6,10
       9          C
      10          C TGU = TOTAL GAS USAGE              TAMT = TOTAL BILLINGS
FATAL    *        PREMATURE E.O.S. IN I/O CONTROL LIST
      11                 TGU + 0.0
FATAL    *        THIS IS NOT A FORTRAN STATEMENT
      12                 TAMT = 0.0
      13          C
      14          C READING THE DATA, THE START OF THE LOOP
      15            15   READ(5,20,END=99) ID, PMR, CMR, RATE
      16          C
      17                 GU = PMR - CMR
      18                 AMT = RATE * GU
      19                 WRITE(6,30) ID, GU, RATE, AMT
      20          C
      21          C CALCULATE TOTALS FOR ALL CUSTOMERS
      22                 TGU = TGU + GU
      23                 TAMT = TAMT + AMT
      24          C
      25          C END OF THE LOOP
      26                 GO TO 15
      27          C
      28          C WRITING THE FOOTINGS
      29            99   WRITE(6,40) TGU, TAMT
      30          C
      31          C THE FORMATS USED IN THE PROGRAM
      32            10   FORMAT('1',35X,'XYZ COMPANY'//13X,'CUSTOMER ID',6X,'GAS USED',7X,
      33               *'RATE',7X,'AMOUNT'//)
      34            20   FORMAT(I5,1X,F6.2,1X,F6.2,1X,F5.3)
      35            30   FORMAT(14X,I5,11X,F6.2,7X,F6.3,7X,F7.2)
      36            40   FORMAT(///12X,'TOTAL GAS USED',3X,F7.2,7X,'TOTAL AMOUNT',2X,F7.2)
      37                 END

--VARIABLE MAP--(LO=A)
 -NAME---ADDRESS--BLOCK-----PROPERTIES-------TYPE---------SIZE      -NAME---ADDRESS--BLOCK-----PROPERTIES

  AMT        0B                              REAL                    PMR        0B
  CMR        0B                              REAL                    RATE       0B
  GU         0B                              REAL                    TAMT       0B
  ID         0B                              INTEGER                 TGU        0B

--STATEMENT LABELS--(LO=A)
 -LABEL-ADDRESS------PROPERTIES----DEF         -LABEL-ADDRESS------PROPERTIES----DEF

   10        0B      FORMAT          32          30        0B       FORMAT          35
   15        0B                      15          40        0B       FORMAT          36
   20        0B      FORMAT          34          99        0B                       29

--ENTRY POINTS--(LO=A)
 -NAME---ADDRESS--ARGS---

  METER      0B       0

--I/O UNITS--(LO=A)
 -NAME---PROPERTIES-------------

  TAPE5   FMT/SEQ
  TAPE6   FMT/SEQ

--STATISTICS--

 PROGRAM-UNIT LENGTH              0B  =     0
 CM STORAGE USED              60600B  = 24960
 COMPILE TIME                 0.167 SECONDS

      2  FATAL      ERRORS IN METER
```

Figure 3.16 The first run of the sample program.

STEP 8: EXECUTION, TESTING, AND DEBUGGING

The first run of the program, shown in Figure 3.16, has two syntax errors. The first one, at line 8, is the omission of the right parenthesis in the WRITE statement, and the second one, at line 11, is the use of the + sign instead of the = sign. Note that the first error message is printed after the comment lines. These errors are corrected and the program is resubmitted for execution.

The second run is shown in Figure 3.17. The program has several errors:

1. GAS USED must be positive, but the printout shows negative numbers. This is due to the formula

   GU = PMR − CMR

   which should be corrected to read:

   GU = CMR − PMR

   This is a logical error.

```
1             PROGRAM METER
2       C THIS PROGRAM GENERATES A REPORT FOR THE GAS USERS OF THE COMPANY.
3       C THE INPUTS ARE CUSTOMER ID, PREVIOUS METER READING (PMR),
4       C CURRENT METER READING (CMR), AND THE RATE FOR EACH CUSTOMER (RATE).
5       C AUTHOR:   T. H. SMALL, AUGUST 19, 19--
6       C
7       C HEADINGS
8             WRITE(6,10)
9       C
10      C TGU = TOTAL GAS USAGE          TAMT = TOTAL BILLINGS
11            TGU = 0.0
12            TAMT = 0.0
13      C
14      C READING THE DATA, THE START OF THE LOOP
15         15 READ(5,20, END=99) ID, PMR, CMR, RATE
16      C
17            GU = PMR - CMR
18            AMT = RATE * GU
19            WRITE(6,30) ID, GU, RATE, AMT
20      C
21      C CALCULATE TOTALS FOR ALL CUSTOMERS
22      C
23      C END OF THE LOOP
24            GO TO 15
25      C
26      C WRITING THE FOOTINGS
27         99 WRITE(6,40) TGU, TAMT
28      C
29      C THE FORMATS USED IN THE PROGRAM
30         10 FORMAT('1',35X,'XYZ COMPANY'///13X,'CUSTOMER ID',6X,'GAS USED',7X,
31            *'RATE',9X,'AMOUNT'///)
32         20 FORMAT(I5,1X,F6.2,1X,F6.2,1X,F5.3)
33         30 FORMAT(14X,I5,11X,F6.2,7X,F6.3,7X,F7.2)
34         40 FORMAT(///12X,'TOTAL GAS USED',3X,F7.2,7X,'TOTAL AMOUNT',2X,F7.2)
35            TGU = TGU + GU
36            TAMT = TAMT + AMT
37            END
```

```
                        XYZ COMPANY

      CUSTOMER ID    GAS USED       RATE        AMOUNT

         54985        ******        .350        -55.77
          4982        -91.25        .380        -34.68
         36872        -71.75        .330        -23.68
         99995        ******        .390        -92.10
         45632        ******        .330        -52.40
         33333        ******        .390        -48.17
         45678        ******        .380        -65.74



        TOTAL GAS USED    0.00    TOTAL AMOUNT    0.00
```

Figure 3.17 The second run of the sample problem.

116 *Programming Development and Execution*

2. The ****** in the GAS USED column is an indication of overflow—showing that the field is too small. However, the field being too small is a result of the negative numbers due to the previous error. Once these are gone, the field-size problem will be corrected automatically. Nevertheless, to be on the safe side, the field size has been increased to F7.2.
3. The printout shows that the total gas used and the total amount are equal to zero, whereas the totals should be positive numbers as several test data were made up. Careful review of the program and the flowchart showed that the formulas for calculating the totals were placed after the GO TO statement, whereas they ought to be before it in the loop. This is also a logical error.

The program is corrected and resubmitted for execution. The final program and the output for the test data are shown in Figure 3.18.

```
1              PROGRAM METER
2        C THIS PROGRAM GENERATES A REPORT FOR THE GAS USERS OF THE COMPANY.
3        C THE INPUTS ARE CUSTOMER ID, PREVIOUS METER READING (PMR),
4        C CURRENT METER READING (CMR), AND THE RATE FOR EACH CUSTOMER (RATE).
5        C AUTHOR:  T. H. SMALL, AUGUST 19, 19--
6        C
7        C HEADINGS
8              WRITE(6,10)
9        C
10       C TGU = TOTAL GAS USAGE          TAMT = TOTAL BILLINGS
11             TGU = 0.0
12             TAMT = 0.0
13       C
14       C READING THE DATA, THE START OF THE LOOP
15          15 READ(5,20, END=99) ID, PMR, CMR, RATE
16       C
17             GU = CMR - PMR
18             AMT = RATE * GU
19             WRITE(6,30) ID, GU, RATE, AMT
20       C
21       C CALCULATE TOTALS FOR ALL CUSTOMERS
22             TGU = TGU + GU
23             TAMT = TAMT + AMT
24       C
25       C END OF THE LOOP
26             GO TO 15
27       C
28       C WRITING THE FOOTINGS
29          99 WRITE(6,40) TGU, TAMT
30       C
31       C THE FORMATS USED IN THE PROGRAM
32          10 FORMAT('1',35X,'XYZ COMPANY'///13X,'CUSTOMER ID',6X,'GAS USED',7X,
33             *'RATE',9X,'AMOUNT'//)
34          20 FORMAT(I5,1X,F6.2,1X,F6.2,1X,F5.3)
35          30 FORMAT(14X,I5,11X,F7.2,7X,F6.3,7X,F7.2)
36          40 FORMAT(///12X,'TOTAL GAS USED',3X,F7.2,7X,'TOTAL AMOUNT',2X,F7.2)
37             END
```

```
                              XYZ COMPANY

         CUSTOMER ID      GAS USED       RATE        AMOUNT

            54985          159.34        .350         55.77
             4982           91.25        .380         34.68
            36872           71.75        .330         23.68
            99995          236.15        .390         92.10
            45632          158.80        .330         52.40
            33333          123.50        .390         48.17
            45678          173.00        .380         65.74



         TOTAL GAS USED    1013.79     TOTAL AMOUNT    372.53
```

Figure 3.18 The final run of the sample program.

STEP 9: DOCUMENTATION

The documentation of this program can be provided as follows:

1. The problem and the solution procedure (summary of steps 1–5).
2. The flowchart for the program, as shown in Figure 3.14.
3. The final run of the program (Figure 3.18).
4. Comments: This program is designed for a summary report where all the input data are keyed into the terminal just for this program. This program is not designed to print the bill for each customer, nor to store the data for later billing or processing.

EXERCISES

3.2 Describe the print position of each variable resulting from the following statements:

a.
```
         WRITE (6,20) A,B,C
20       FORMAT (11X,F6.2/11X,F6.2/11X,F6.2)
```
b.
```
         WRITE (6,10) A
         WRITE (6,10) B
         WRITE (6,10) C
10       FORMAT (11X,F6.2)
```
c.
```
         WRITE (6,30) X,Y,Z
30       FORMAT (21X,F6.2,5X,F6.2/25X,F6.2)
```
d.
```
         WRITE (6,30) U,P,Q
30       FORMAT ('1',57X,'XYZ COMPANY'/41X,
        *'U VARIABLE',5X,'P VARIABLE',5X,
        *'Q VARIABLE'///41X,F10.2,5X,F10.2,5X,F102.)
```
e.
```
         WRITE (6,40) APE,BOB,COP
40       FORMAT ('1',F6.2,10X,F6.2,5X,F5.1)
```
f.
```
5        READ(5,10,END=99)A
         WRITE(6,20)A
         GO TO 5
10       FORMAT(F6.2)
20       FORMAT('0',F7.2)
99       STOP
         END
```
g.
```
         WRITE (6,30) R
         WRITE (6,40) P
30       FORMAT (1X,F6.2)
40       FORMAT ('+',12X,F6.2)
```
h.
```
         WRITE (6,10) X
         WRITE (6,20) Y
10       FORMAT (1X,F6.2)
20       FORMAT ('0',F6.2)
```

i.
```
      WRITE (6,10) LENGTH, WIDTH, AREA
10    FORMAT ('1',I5/1X,F6.2/1X,F8.2)
```

3.3 Find all errors in the following statements or programs. Identify the errors as syntax, execution, or logical.

a.
```
      WRITE (6,20) A,B
20    (1X,F6.2,1X,F7.2)
```
b.
```
      READ (5,10) X,Y,Z,
10    FORMAT (F4.3,F5.1F6.1)
```
c.
```
      WRITE (6,20) I,J,K
20    FORMAT (F6.2,F4.1,F6.1)
```
d.
```
      READ (5,10) JOHN,CAR
      FORMAT (5X,F6.2,F5.1)
```
e.
```
      READ (5,100) X,Y,
100   FORMATE (F6.1,5X,F3.2)
```
f.
```
10    READ (5,100,) WGE,RTE
      PAY-WGE,RTE
      WRITE (6,200) WAGE,RATE,PAY
100   FORMAT (2F6.2)
200   FORMAT (1X,F4.2,1X,F5.2,1X,F3.2,)
      GO TO 10
      END
```
g.
```
      THIS PROGRAM CLACULATES THE SALARY OF THE
      EMPLOYEES.
5     REED (5,100) W,R
      SALARY=W+R
      WRGHT (6,10) W,R,SALARY
      GO TO 5 0
      END
```
h.
```
      READ (5,10) A
      FORMAT (1X,F6.2)
```
i.
```
      WRITE (6,20) B
20FORMAT (1X,F7.2)
```
j.
```
      READ (5,10) A,B
      OCCUPATION=A+B
      A*B=PAYRATE
      TOTAL=TOTAL+A
      WRITE (6,20) A,B,OCCUPATION,PAYRATE,TOTAL
      FORMAT (2F6.2)
      GO TO READ
      COMMENT: THIS IS THE END
      END
```

k.
```
50 0    READ (5,10END=99) X,Y
        N=0
        TOT=0
        Z=X+Y/2.0
        TOT=TOT+Z
        WRITE (6,20) AVG,X,Y
        GO TO 50
        WRITE (6,10) TOT
        10FORMAT (1X,F10.2)
        STOP
        END
```

l.
```
        TAX = 0
5       AMOUNT = 26,000.00
        RATE = TAX/AMOUNT
        GO TO 5
        END
```

m.
```
        READ(5,30)A, PAY,
```

n.
```
        TEMP = (F - 32.0)(X - Y)
```

3.4 If you have a computer system available:

a. Find out the following information about the computer to which you have access:

(1) What is the name of the computer model?
(2) Who is the manufacturer?
(3) Is any terminal (CRT or teletype) available to the users?
(4) Are both batch processing and on-line processing available to the users?
(5) What versions of FORTRAN compiler are available?

b. Ask someone to show you how a keypunch machine works, then practice punching a card with the keypunch.

3.5 If you have a computer system available, punch or key the program shown in Figure 3.15 exactly as it appears. Find out the necessary control statements for the system that you will be using. Make up three or four data records. (Don't forget the exact location of the numbers, as explained in Step 3 of the sample program.) Prepare a batch by putting the source program, the control statements, and the data together. The general order is shown in Figure 3.2; however, you should find out the specific control statements for the system that you will be using. Submit the batch for execution. Your next step, of course, will be to pick up your printout. If you find any error, it should be the result of some typing error. Correct any errors and resubmit the deck for execution. You may have to do this several times until the program is correct.

3.6 Suppose you would like to find the average heights and weights of the students in the FORTRAN course. Develop and design a program for this purpose. Follow the programming cycle explained in this chapter.

3.7 Suppose you would like to calculate the average percentages of the price increases of certain products in a grocery store. A simple way is to record the prices of certain products for one period, and the prices of the same products in another period, say after 6 months, and then calculate the average percentage price increase. Develop and design a program for this purpose. Follow the programming cycle explained in this chapter.

3.8 Suppose you buy a car on the installment plan for T dollars to be paid off in N years; if the interest rate is R (expressed in percentages), your monthly payment (P) is:

$$P = T \times \frac{R/1200}{1 - (1 + R/1200)^{-12N}}$$

Develop and write a program that reads T, R, and N, and calculates the monthly payment. Print the input data as well as the results.

Hint: Change the formula to a shorter form by using different variables such as:

$$X = R/1200$$
$$Y = 1 + X$$
$$Z = 1/(Y**(12*N)) \quad \text{This is the same as: } (1+R/1200)^{-12N}$$
$$P = T*X/(1-Z)$$

SUMMARY OF CHAPTER 3

You have learned the following terms and should be able to explain each one briefly:

Execution	Executable statement	Compilation
Machine language	Nonexecutable statement	FORTRAN IV
Higher-level language	Debugging	ANS FORTRAN
Source program	JCL cards	WATFOR
Object program	Compiler	FORTRAN 77
		File, record, field

You have also learned the following:

1. There are basically two ways a program can be run:
   a. Running the program with punched cards
   b. Keying the program into a terminal
2. Certain columns of a line are assigned to a particular item of a FORTRAN statement:

Columns	Purpose
1	for placing a C to signify a comment, for internal documentation
1–5	for the statement number
6	for placing a character indicating the continuation of the previous line
7–72	for the statement
73–80	blank, or may be used for any identification

3. 1X (or 'ƀ'), '1', '+', and '0' at the beginning of a write FORMAT are called carriage control characters. They control the position of the

paper. Further, using a slash (/) anywhere in the format causes the printer (or card reader) to start a new line (or card).
4. A program may have to be rerun several times before it is completely correct. A program can have three kinds of errors:
   a. syntax error
   b. execution error
   c. logical error
5. A flowchart is a pictorial presentation of the necessary steps for solving a problem. You should be familiar with the flowcharting symbols and techniques.
6. A pseudocode represents the necessary step for solving a problem in a narrative form.
7. A logic-chart shows the logical structure of a program.
8. An algorithm is the precise detail of a method of solving a problem.
9. A programmer must understand the problem to be solved and plan the solution procedures. The programming cycle explained in the text should help you in planning your programs.

SELF-TEST REVIEW

3.1 What is the difference between card-oriented batch processing and on-line processing?

3.2 How many columns are there on a typical punched card?

3.3 What is the difference between the control cards and the source program? Are control cards part of the FORTRAN language?

3.4 Certain columns of a line are reserved for certain purposes. Complete the following statements:

   a. Column 1 is reserved for _____.
   b. Columns 1–5 are reserved for _____.
   c. Column 6 is reserved for _____.
   d. Columns 7–72 are reserved for _____.
   e. Columns 73–80 are reserved for _____.

3.5 What is the difference between machine language and a higher-level language?

3.6 What is the difference between the source program and the object program?

3.7 What is a compiler? What is compilation?

3.8 FORTRAN 77 is the newest version of standard FORTRAN, developed and published in 1978. True or false?

3.9 The process of writing statements of a program is called _____.

3.10 Removing errors from a program is called _____.

3.11 A program can have three kinds of errors. What are they? Explain each briefly with an example.

3.12 A programmer must know the solution procedure of a problem before programming. True or false? Discuss.

3.13 Several planning steps were explained as part of the programming cycle. Explain how each step is important in solving a problem.

3.14 Explain the following:

    a. An algorithm
    b. A flowchart
    c. A pseudocode

3.15 Which of the following statements has a syntax error?

    a. AMT$ = A * P
    b. RAED (5,10) X,Y
    c. TOT = TOT + −B
    d. 20 FORMAT (1X,F6.2
    e. DISCOUNT = AMT * 3%
    f. All of the above

☐ Answers

The answers to the review questions can be found on the following pages in the chapter:

3.1 82, 83

3.2 82

3.3 83

3.4 84, 91

3.5 95, 96

3.6 96

3.7 96

3.8 97

3.9 85

3.10 98

3.11 98

3.12 108

3.13 108, 109

3.14 102, 104, 105

3.15 f.

Chapter 4

Decision Making, Comparing, and Branching

THE LOGICAL IF STATEMENT

    Important Notes about the Logical IF Statement

PROBLEMS REQUIRING THE IF STATEMENT

    Finding the Largest Number
    Terminating a Loop
        Header Record
        Trailer Record

PROGRAMMING STYLE AND STRUCTURED IF BLOCK

    GO TO Style—Style One
    GO TO-less Style—Style Two
    Structured IF—Style Three
        Rules of Structured IF
        Advantages of Using Structured IF

CONDITIONAL BRANCHING

    Arithmetic IF Statement
    Computed GO TO Statement

A SAMPLE PROBLEM

When writing programs, we often need to take some action based on the outcome of the comparing quantities or variables. For example, sometimes it is necessary to find out which quantity is larger, whether a value is less than zero, or whether a variable is equal to another. This chapter explains several alternative statements in FORTRAN for comparison and decision making. Understanding these statements and their applications is very important because the appropriate usage of conditional statements directly affects one's problem-solving capability and style of programming.

THE LOGICAL IF STATEMENT

The *logical IF statement* can be used for comparing two variables or two values. Example 4.1 shows the general form of the statement. Let's look at the example before explaining it further.

EXAMPLE 4.1

Problem: The normal discount rate is 5% unless the amount is greater than $10,000.00, in which case the discount rate is 10 percent. Write a program which reads a value (an amount), and calculates the discount. Print the amount, the discount rate, and the discount. Assume there is more than one record. Explain the programming procedure (the plan) and draw the flowchart before writing the program.

Plan:
Start
READ AMT, at the end of data STOP
   Assume: RATE = .05
   IF AMT > 10,000.0 then: RATE = .10
   Calculate the discount: DIS = RATE * AMT
   WRITE AMT, RATE, DIS
GO TO READ for reading additional data
END

Flowchart: Shown in Figure 4.1.

Program:

```
          PROGRAM DSCONT
10        FORMAT(F8.2)
100       READ (5,10,END = 95) AMT
          RATE = .05
          IF (AMT .GT. 10000.0) RATE = .10
          DIS = RATE*AMT
          WRITE (6,50) AMT,RATE,DIS
          GO TO 100
95        STOP
50        FORMAT(2X,'AMOUNT=',F8.2,1X,'RATE=',F4.2,1X,
         *'THE DISCOUNT=',F7.2)
          END
```

The Logical IF Statement 125

```
                    ┌─────────┐
                    │  START  │
                    └────┬────┘
                         ▼
              ┌──────────────────┐
         ┌───▶│    READ AMT     /
         │    └────────┬─────────┘
         │             ▼
         │      ╱ END OF ╲   YES    ┌─────────┐
         │     ⟨  DATA?   ⟩────────▶│  STOP   │
         │      ╲        ╱   95     └─────────┘
         │         │ NO
         │         ▼
         │   ┌───────────┐
         │   │ RATE = .05│
         │   └─────┬─────┘
         │         ▼
         │      ╱ AMT > ╲   YES   ┌───────────┐
         │     ⟨ 10000?  ⟩───────▶│ RATE = .10│
         │      ╲       ╱         └─────┬─────┘
         │         │ NO                 │
         │         ▼                    │
         │   ┌───────────┐              │
         │   │   DIS =   │◀─────────────┘
         │   │ RATE*AMT  │
         │   └─────┬─────┘
         │         ▼
         │   ┌──────────────┐
         │  / WRITE AMT,   /
         └──/ RATE, DIS   /
            └─────────────┘
```

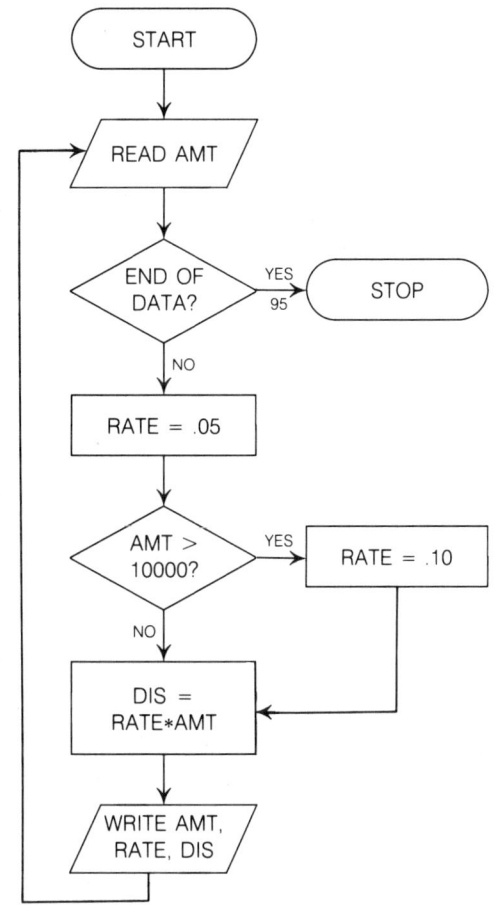

Figure 4.1 The flowchart for Example 4.1.

Sample data:

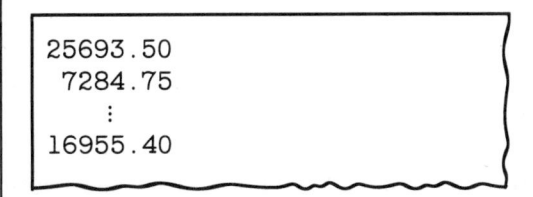

```
25693.50
 7284.75
   ⋮
16955.40
```

Notes:

1. The IF statement makes a comparison between the variable AMT and 10,000.0, to check whether AMT is *greater than* (GT) 10,000.0. If so, the value of RATE will be changed to 10% (note that it was 5%). If not, the statement connected to the IF will be ignored; that is, there will be no change in the value of RATE. In any case, the next statement below the IF statement will be executed.
2. The statements in the loop are indented for readability. As discussed in Chapter 3, blank spaces after column 7 do not affect the statements.
3. The flowchart for this example, shown in Figure 4.1, helps to understand the logic of the program. Note that the IF statement in the program corresponds to the decision symbol in the flowchart.

126 Decision Making, Comparing, and Branching

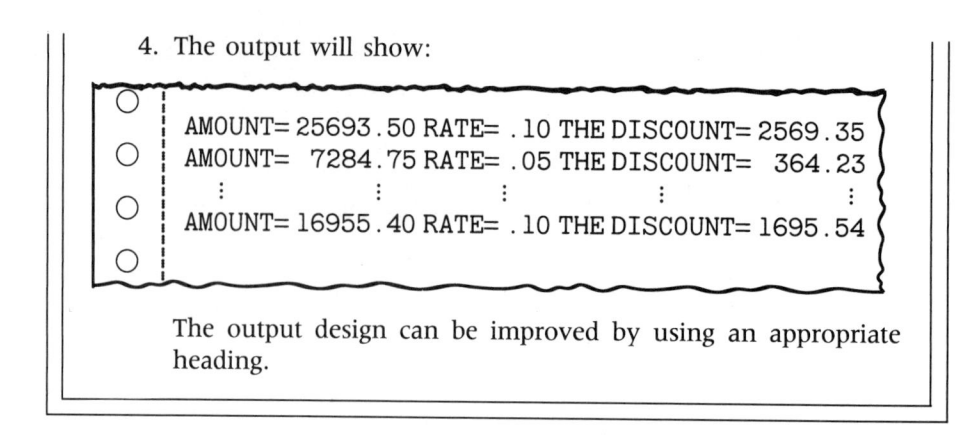

> 4. The output will show:
>
> ```
> AMOUNT= 25693.50 RATE= .10 THE DISCOUNT= 2569.35
> AMOUNT=  7284.75 RATE= .05 THE DISCOUNT=  364.23
>    ⋮           ⋮        ⋮            ⋮
> AMOUNT= 16955.40 RATE= .10 THE DISCOUNT= 1695.54
> ```
>
> The output design can be improved by using an appropriate heading.

The symbol .GT., the abbreviation for the phrase "greater than," in the above example is called a *relational operator*. A variety of such comparisons are permitted by FORTRAN. The most commonly used are shown in Table 4.1.

The following are some examples of logical IF statements:

```
IF (A .GT. B) GO TO 100
IF (JOHN .GT. 9) DIS = A*P
IF (I .NE. 4) SUM = SUM + B
IF (FUN .EQ. CC) GO TO 100
IF (A .EQ. B) C = D
IF (N .LT. 57) GO TO 5
IF (BAD .NE. GOAL) GOAL = ACE
IF (X .EQ. 0.0) STOP
IF (A + B .EQ. 50.0) GO TO 10
IF (X*Y .LE. X + Y) GO TO 50
IF (4*A*C + B**2 .LT. 0.0) GO TO 100
```

Thus, the general form of the logical IF statement is:

IF (logical expression) statement
 ↑   ↑   ↑
 1   2   3

and has three components:

1. The word IF.
2. A logical expression enclosed within a pair of parentheses. The outcome of this logical expression will be either true or false.
3. An executable FORTRAN statement. This statement will be performed if the expression in the parentheses is true; otherwise it will be ignored.

Table 4.1 Relational Operators in FORTRAN

RELATIONAL OPERATOR	MEANING	EQUIVALENT
.GT.	greater than	$>$
.GE.	greater than or equal to	$>=$ or $\geq$
.EQ.	equal to	$=$
.NE.	not equal to	$\neq$ or $\neg =$
.LT.	less than	$<$
.LE.	less than or equal to	$\leq$ or $<=$

In either case, execution normally proceeds to the next statement after IF (unless there is a GO TO statement).

RULE 4.1
The statement to the right of the parentheses of an IF statement is executed if and only if the expression in the parentheses is true.

☐ Important Notes about the Logical IF Statement

1. It is good programming practice to choose the variables in a simple logical expression with the same mode. For example,

   `IF (I .LE. 5.5) GO TO 56`

   should be avoided because of its mixed real and integer modes. This point is especially important when you use .EQ. to compare two values to see if they are *equal*. Because of the internal storage structure of a computer, the stored value of a number in real form may not be equal to the stored value of the same number in integer form.

2. A simple logical expression can be constructed by using a relational operator with variables, constants, or arithmetic expressions, but it must be capable of being evaluated as either true or false. For example,

   `IF (A + B) WRITE (6,10) X`

   and

   `IF (55 .LT. A .GT. 10.0) GO TO 50`

   are not valid, because the expression is not capable of being evaluated as either true or false.

3. The statement to the right of the IF must be an executable statement. For example, it cannot be a FORMAT statement.

Take a look at the following solved problems for more examples of the use of these notes.

SOLVED PROBLEMS

4.1 Write a logical FORTRAN expression for each of the following cases. Indicate whether each outcome is true or false.

  a. 51 greater than 5
  b. 10 less than or equal to 9
  c. 100 greater than or equal to 100
  d. 15953 greater than or equal to 15952
  e. 1.0 less than or equal to .99999

☐ Answers

  a. (51 .GT. 5), true
  b. (10 .LE. 9), false

c. (100 .GE. 100), true
d. (15953 .GE. 15952), true
e. (1.0 .LE. .99999), false

4.2 Write a logical expression for each of the following situations:

a. If ISUM is greater than or equal to 100, then GO TO statement 33.
b. If MAD is not equal to LOAD, LOAD becomes equal to 1000.
c. If STK is less than or equal to 593.5, then STK becomes equal to 593.5.
d. If X is less than zero, then X becomes equal to zero.
e. If PARK is greater than or equal to 65, then PARK becomes equal to zero.
f. If NOON is not equal to 12.0, then NOON becomes equal to 12.0.
g. If N is equal to 25, then write P and Q.
h. If WAGE is equal to 0.0, STOP.
i. If N + 1 is less than 65, GO TO statement number 5.
j. If 3X + Y is less than 5A + B * C, STOP.

☐ **Answers**

a. IF (ISUM .GE. 100) GO TO 33
b. IF (MAD .NE. LOAD) LOAD = 1000
c. IF (STK .LE. 593.5) STK = 593.5
d. IF (X .LT. 0.0) X = 0.0
e. IF (PARK .GE. 65.0) PARK = 0.0
f. IF (NOON .NE. 12) NOON = 12
g. IF (N .EQ. 25) WRITE (6,33) P, Q
h. IF (WAGE .EQ. 0.0) STOP
i. IF (N + 1 .LT. 65) GO TO 5
j. IF (3 * X + Y .LT. 5 * A + B * C) STOP

4.3 Do you think the following statements are correct? If not, identify the error.

a. IF (K = S) GO TO 60
b. IF (MAD EQ LOAD) MAD = J
c. 99 IF (K .EQ. 9) GO TO 99
d. IF (P .NE. 9.5) WRITE (6, 10) X, Y
e. IF (Q .NE. 9.5) 10 FORMAT (1X, F8.2)
f. IF (4 * Z + 3 * Y) GO TO 10
g. IF (I + 1 .LE. 6.5) GO TO 100
h. IF (A + B .EQ. I + J) STOP
i. IF (A) GO TO 10
j. IF (X + Y − Z) A = B

☐ **Answers**

a. The relational operator is not used correctly.
b. The periods before and after the relational operator (EQ) are missing.
c. It is inappropriate to transfer the control from a statement to itself, although syntactically it is correct.
d. This is correct.

e. The FORMAT statement cannot be placed to the right of an IF statement because it is not an executable statement.
f. The relational operator is missing.
g. Mixed modes should be avoided, although syntactically it is correct.
h. Mixed modes must be avoided in this case because of .EQ.
i. Same as **f.**: the expression cannot be evaluated as true or false.
j. Same as **i.** and **f.**

PROBLEMS REQUIRING THE IF STATEMENT

The IF statement is useful in any problem which involves any kind of conditional situation: conditional branching, termination, comparing, sorting, decision making, or rule making. The following are some examples:

☐ Finding the Largest Number

EXAMPLE 4.2

■ **Problem:** Develop an algorithm, draw a flowchart, and write a program which reads two real numbers and prints the larger number.

Flowchart: Shown in Figure 4.2.

Program:

```
        PROGRAM COMPAR
10      FORMAT (F5.2, 2X, F5.2)
150     READ (5,10,END = 99) A, B
        GREAT = A
        IF (B .GT. A) GREAT = B
        WRITE (6,20) GREAT
        GO TO 150
99      STOP
20      FORMAT (2X, 'THE LARGER NUMBER IS=', F5.2)
        END
```

Sample data:

92.3	61.5
-8.3	-5.5
-3.0	0.0
⋮	
+5.5	6.0

Notes:

1. At the beginning, it is assumed that A is the greater number. Then, a comparison is made between A and B. If B is greater

130 *Decision Making, Comparing, and Branching*

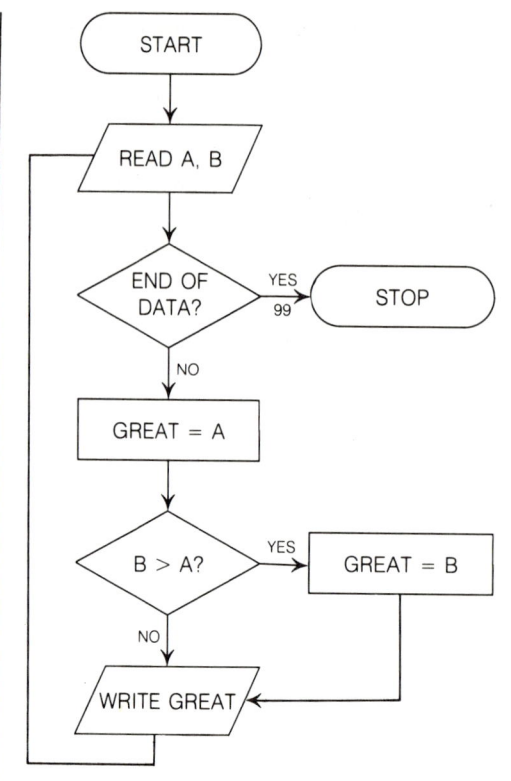

```
                    START
                      │
                      ▼
                  READ A, B
                      │
                      ▼
                 END OF     YES
                 DATA?  ────────▶  STOP
                      │   99
                    NO│
                      ▼
                  GREAT = A
                      │
                      ▼
                   B > A?    YES
                      │    ────────▶  GREAT = B
                    NO│                  │
                      ▼                  │
                 WRITE GREAT ◀───────────┘
                      │
                      └──── (loop back to READ A, B)
```

Figure 4.2 Flowchart for Example 4.2.

than A, the value of GREAT will be changed to B; otherwise GREAT remains A.

2. The variable name LARGE was not chosen, to avoid an integer mode.
3. The output will show

```
THE LARGER NUMBER IS=  92.30
THE LARGER NUMBER IS=  -5.50
THE LARGER NUMBER IS=   0.00
          ⋮
THE LARGER NUMBER IS=   6.00
```

4. The output design is deficient because it does not give a picture of the input. If the WRITE-FORMAT is changed to

```
       WRITE(6,20)A,B,GREAT
20     FORMAT(2X,'THE LARGER OF',1X,F5.2,1X,'AND',
      *1X,F5.2,1X,'IS',1X,F5.2)
```

then the output will show

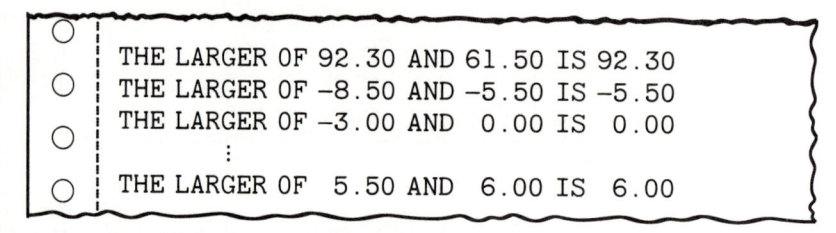

```
THE LARGER OF 92.30 AND 61.50 IS 92.30
THE LARGER OF -8.50 AND -5.50 IS -5.50
THE LARGER OF -3.00 AND  0.00 IS  0.00
          ⋮
THE LARGER OF  5.50 AND  6.00 IS  6.00
```

EXAMPLE 4.3

■ **Problem:** Write a program which reads a series of numbers, each of which is on a separate line. Find and print the largest number among the data. Draw the flowchart before writing the program.

Flowchart: Shown in Figure 4.3.

Program:

```
        PROGRAM LARGE
        GREAT = 0.0
40      FORMAT (F5.2)
50      READ (5,40,END = 99)A
            IF (A .GT. GREAT) GREAT = A
        GO TO 50
99      WRITE (6,60) GREAT
        FORMAT (11X, 'THE LARGEST NUMBER IS', F6.2)
        STOP
        END
```

Notes:

1. At the beginning, it is assumed that the greatest number is equal to zero. Each time that a new number (A) is read, it will be compared with GREAT (the largest number up to that point).

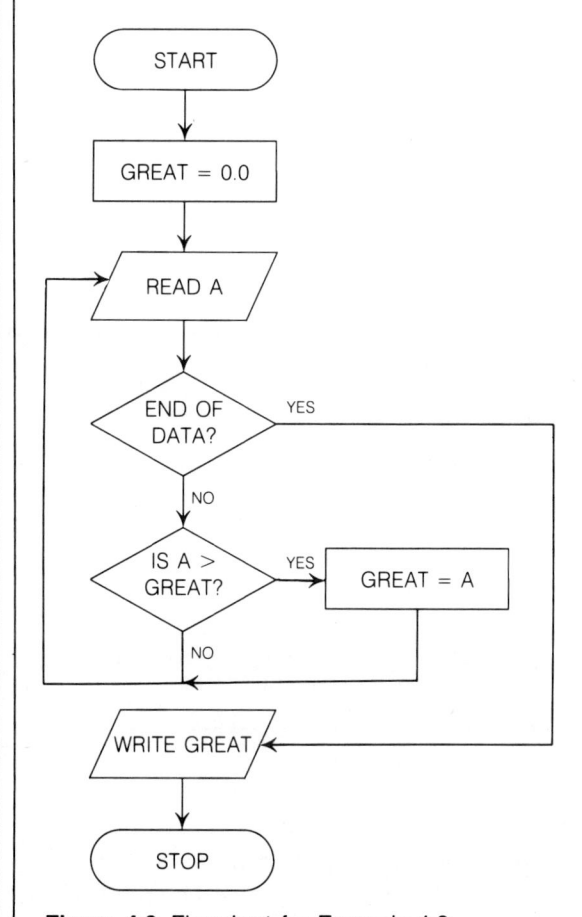

Figure 4.3 Flowchart for Example 4.3.

132 Decision Making, Comparing, and Branching

> If it is greater than GREAT, then GREAT will be changed to the new number; otherwise it remains the same, and the program continues until all the data are read, when the largest number will be printed.
> 2. It is assumed that all the numbers are greater than zero. (Otherwise, GREAT = 0.0 at the beginning of the program must be changed accordingly.)

☐ Terminating a Loop

An infinite loop created by a GO TO statement can be terminated by an IF statement. This section explains several techniques.

EXAMPLE 4.4

■ **Problem:** Calculate and print the weekly pay of the employees. Each data record contains wage rate and hours worked. There are exactly 65 records.

Program:

```
        PROGRAM LOOP2
        N = 0
        WRITE (6, 100)
100     FORMAT (11X, 'WAGE RATE', 5X, 'HOURS WORKED',
       *5X, 'PAY')
10      FORMAT (2F5.2)
5       READ (5, 10)WAGE,HOURS
            N = N + 1
            PAY = WAGE*HOURS
            WRITE (6,20)WAGE, HOURS, PAY
        IF (N .LT. 65) GO TO 5
95      WRITE (6,40)
20      FORMAT (13X, F6.2, 9X, F6.2, 7X, F8.2)
40      FORMAT (21X, 'END OF REPORT')
        END
```

Notes:

1. The statement N = N + 1 (the *counter block*) counts the number of times that the program goes through the loop, thus counting the number of data records.
2. The IF statement stops the loop after 65 times: If N (the number of repetitions or loops) is less than 65, the control will be transferred to the beginning of the program; otherwise (that is, if the number of loops is equal to or greater than 65), the loop will be terminated.

Header Record

The program in Example 4.4 works only if there are exactly 65 records in the data file. If there are more or less records, the program must be modified. We would like, however, to write a program which works with any number of records without being changed. Alternatively, a better method is to have *the number* of data records placed on a line and placed at the beginning of the data file. This record, which is called a *header record,* can then be read at the beginning of the program. The following example demonstrates this technique.

EXAMPLE 4.5

■ **Problem:** As in Example 4.4, except that the number of records (whatever it may be) is to be given on a line at the beginning of the data file.

Program:

```
      PROGRAM LOOP3
      N = 0
      WRITE (6,100)
100   FORMAT (11X, 'WAGE RATE', 5X, 'HOURS WORKED',
     *5X, 'PAY')
10    FORMAT (I2)
      READ (5, 10)K
C K IS THE NUMBER OF DATA RECORDS
15    READ (5, 20)WAGE,HOURS
      N = N + 1
      PAY = WAGE*HOURS
      WRITE (6, 30) WAGE,HOURS,PAY
      IF (N .LT. K) GO TO 15
      WRITE (6, 40)
40    FORMAT (21X, 'END OF REPORT')
20    FORMAT (2F5.2)
30    FORMAT (13X, F6.2, 9X, F6.2, 7X, F8.2)
      END
```

Notes:

1. On the first data record, the header record, a number such as 65 is placed to show how many data records will follow. Then the first READ statement reads that number (K).
2. If N (the number of loops) is less than K, the loop will be repeated; otherwise the loop will be terminated.

Trailer Record

The program in Example 4.5 is independent of the number of data. In order to change the number of data we need only change the first data record, the header record. However, the disadvantage of such a method is that the number of data must be precisely counted and entered.

Another method for terminating a loop is by placing a number that is out of the range of our regular data on the last data record in the data file. We take this number to mean "end of data," and an IF statement can terminate the loop after reading the number. This record is called a *trailer card, trailer record,* or *sentinel value.*

For instance, in the previous example we could place a negative number instead of the wage rate on a line at the end of the data file. This negative number could then be recognized as the end of the data by an IF statement. The following program is another example.

134 *Decision Making, Comparing, and Branching*

EXAMPLE 4.6

■ **Problem:** Calculate the average of the scores of students in a FORTRAN course. Each score is placed on a separate line, and on the last line -99.0 is placed as a trailer record to indicate the end of the data.

Program:

```
        PROGRAM LOOP4
        N = 0
        SUM = 0.0
10      READ (5,20) SCORE
        IF (SCORE .LT. 0.0) GOTO 50
            N = N + 1
            SUM = SUM + SCORE
        GO TO 10
50      AVG = SUM/N
        WRITE (6,30) AVG
30      FORMAT (11X, 'THE AVERAGE IS =', F6.2)
20      FORMAT (F5.2)
        END
```

Notes:

1. The READ statement reads each number. If the number is negative, it signals that all data have been read; the loop will then be terminated.
2. N = N + 1 counts how many data records have been read so far. Note that it does not count the last datum (the trailer record), because we need N to calculate the average.
3. The test for the sentinel value is done before any calculation in the loop.
4. The data are read in the loop, and the sum is cumulated also in the loop. The average is calculated and printed after the loop.

■

SOLVED PROBLEMS

■ **4.4** Write the following statements in FORTRAN:

   a. If N (the number of loops, say) still is not equal to K (the number of data lines), GO TO the beginning of the loop, statement number 5.
   b. If the number just read for X is negative, terminate the loop. The statement right after the loop is labeled 95.
   c. If the value of variable A is negative, make it positive.
   d. If X is not equal to Y, make it equal.
   e. There are two variables, A and B. Calculate the square of the smaller number and the cube of the larger number.
   f. If N is less than K, go to statement number 5.
   g. If N is greater than M, stop execution.

h. If X^3 is less than $3X$, go to statement number 50.
i. If $X - Y > X * Y$, GO TO statement number 100.
j. If A is equal to 1000, GO TO statement 60.

☐ **Answers**

a.
```
         IF (N .LT. K) GO TO 5
```
b.
```
         IF (X .LT. 0.0) GO TO 95
```
c.
```
         IF (A .LT. 0.0) A = -A
```
d.
```
         IF (X .NE. Y) X = Y
```
e.
```
         IF (A .LT. B) GO TO 10
         SS = B*B
         Q = A*A*A
         GO TO 20
10       SS = A*A
         Q = B*B*B
20       ...
  ⋮
```
f.
```
         IF (N .LT. K) GO TO 5
```
g.
```
         IF (N .GT. M) STOP
```
h.
```
         IF (X**3 .LT. 3*X) GO TO 50
```
i.
```
         IF (X - Y .GT. X*Y) GO TO 100
```
j.
```
         IF (A .EQ. 1000.0) GO TO 60
```

■ **4.5** Rewrite Example 4.1, using a GO TO statement after IF to transfer the control to a block of several statements for discount calculation.

☐ **Answer**
```
         PROGRAM, REPEAT
10       FORMAT (F8.2)
130      READ (5, 10, END = 95) AMT
         IF (AMT .LT. 10000.0) GO TO 50
             RATE = .10
             DIS = RATE*AMT
         GO TO 100
```

```
        50              RATE = .05
                        DIS = RATE*AMT
       100              WRITE (6, 52) AMT,RATE,DIS
                        GO TO 130
        32      FORMAT (11X, 'AMOUNT=', F9.2, 'RATE=',
               *F4.2, 'DISCOUNT=', F8.2)
        95      STOP
                END
```

■ **4.6** Write a program which finds and prints the smallest number among a series of data. Each number is placed on a separate line.

☐ **Answer**

```
                PROGRAM SMLEST
                SMALL = 9999.99
C IT IS ASSUMED THAT THE NUMBERS ARE SMALLER
C THAN 9999.99
        40      FORMAT (F6.2)
        50      READ (5, 40, END = 99) X
                IF (X .LT. SMALL) SMALL = X
                GO TO 50
        99      WRITE (6,60) SMALL
        60      FORMAT (11X, 'THE SMALLEST NUMBER IS',
               *F8.2)
                END
```

■ **4.7** Factorial N means $1 \times 2 \times 3 \times \cdots \times N$. For example, factorial 5 is $1 \times 2 \times 3 \times 4 \times 5 = 120$. Write a program which calculates factorial N, where N is to be read from a data line.

☐ **Answer**

```
                PROGRAM FACTOR
        10      FORMAT (I2)
                READ (5, 10) N
                I = 0
                IFACT = 1
        20      I = I + 1
                IFACT = IFACT*I
                IF (I .LT. N) GO TO 20
                WRITE (6, 30) N, IFACT
        30      FORMAT (11X, 'FACTORIAL OF', I3, 'IS' I10)
                END
```

■ PROGRAMMING STYLE AND STRUCTURED IF BLOCK

As you have noticed, a program can be written in different ways to solve the same problem. Programming is not merely technique; it is an art, science, and skill as well. When two programmers write a program to solve the same problem, chances are the designs and algorithms of the two will not be alike. This will be more apparent as your programs become longer and more complicated.

Structured programming is a technique that enables one to develop a well-organized program. It requires a style that makes the program easy to understand, has logic that is not difficult to follow, and has a greater probability of being error free.

One statement that tends to make a program unstructured is frequent use of the GO TO statement. Structured FORTRAN blocks have been developed to prevent such disorganization. In the following section, three styles are presented. In the first and second, logical IF and GO TO statements are used. In the third style, you will be introduced to one of the *structured* FORTRAN blocks—the IF-THEN-ELSE structure.

☐ GO TO Style—Style One

EXAMPLE 4.7

■ Problem: Calculate the discount for an amount where the discount rate is

- 5% if the amount is up to $10,000,
- 10% if the amount is up to $20,000,
- 20% if the amount is more than $20,000.

Program:

```
       PROGRAM GOTOST
C STYLE ONE
10     FORMAT (F7.2)
5      READ (5,10,END = 99) AMT
         IF (AMT .LT. 10000.0) GO TO 50
         IF (AMT .LT. 20000.0) GO TO 100
             RATE = .20
             DIS = RATE*AMT
         GO TO 200
50           RATE = .05
             DIS = RATE*AMT
         GO TO 200
100          RATE = .10
             DIS = RATE*AMT
200      WRITE(6,250) RATE,DIS,AMT
         GO TO 5
250    FORMAT (11X,'AMOUNT=',F9.2,'THE RATE=',
      *F4.2, 'THE DISCOUNT=',F8.2)
99     STOP
       END
```

Notes:

1. Logical IF and GO TO statements are used to branch to a particular block of statements.
2. Some of the statements are indented for readability of the logic of the program. This does not affect the statements because the blank spaces are ignored by the computer, as explained in Chapter 3.

138 Decision Making, Comparing, and Branching

> 3. The program is broken down into blocks. Whether a block is executed depends on whether a logical IF statement is true or false. For example, the block which starts with statement 50 is executed when AMT is less than 10000.0 and the discount rate is 5%.

You are allowed to write only one statement after IF (statement A in Figure 4.4a). In programs in which you have to write several statements after IF, you must branch to a block of statements by a GO TO statement (Figure 4.4b). This is an example of how a GO TO statement can be used too frequently.

(a)   (b)

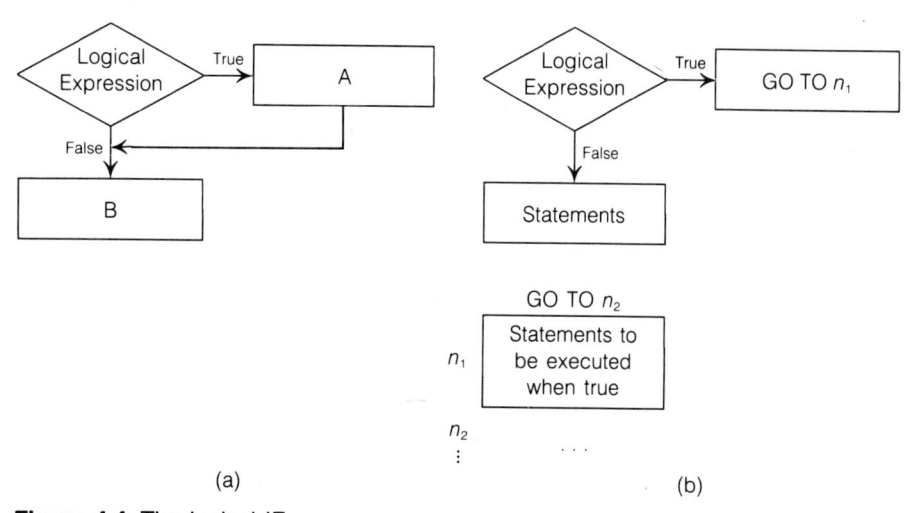

Figure 4.4 The logical IF.

GO TO-less Style—Style Two

The following example demonstrates another way to write a program for the same problem.

> **EXAMPLE 4.8**
>
> ■ **Problem:** Same as Example 4.7.
>
> **Program:**
>
> ```
>         PROGRAM LESSGO
> C STYLE TWO
> 10      FORMAT (F7.2)
> 5       READ (5, 10, END = 99) AMT
>         RATE = .05
>                 IF (AMT .GT. 10000.0) RATE = .10
>                 IF (AMT .GT. 20000.0) RATE = .20
>         DIS = RATE*AMT
>         WRITE (6,250) RATE,DIS,AMT
>         GO TO 5
> 99      STOP
> 250     FORMAT (11X,'AMOUNT=',F8.2, 'THE RATE=',
>        *F4.2, 'THE DISCOUNT=', F8.2)
>         END
> ```

> **Note:** The style is the same as that used for Example 4.1. The discount rate is assumed to be .05 at the beginning. If the amount is greater than 10000.0, the rate will be changed to .10 (and if not, the rate remains at .05). If the amount is greater than 20000.0, the rate will be changed to 20%.

Style one, which has many GO TO statements, seems easy to understand. Style two is simpler, yet the logic is not as easy to follow as style one's; which style is chosen depends on the programmer.

Structured IF—Style Three

In this style, an IF-THEN-ELSE structure is used. Let's take a look at the same problem before explaining this method.

EXAMPLE 4.9

■ **Problem:** Same as Example 4.7.

Program:

```
      PROGRAM STRCTR
C STYLE THREE, STRUCTURED FORM
10    FORMAT (F7.2)
5     READ (5,10,END = 99)AMT
      IF (AMT .LT. 10000.0) THEN
                   RATE = .05
                   DIS = RATE*AMT
        ELSE
          IF (AMT .LT. 20000.0) THEN
                   RATE = .10
                   DIS = RATE*AMT
          ELSE
                   RATE = .20
                   DIS = RATE*AMT
          END IF
      END IF
      WRITE (6,250) RATE,DIS,AMT
      GO TO 5
99    STOP
250   FORMAT (11X,'AMOUNT=',F9.2,'THE RATE=',F4.2,
     *'THE DISCOUNT=',F8.2)
      END
```

Notes:

1. When the logical expression is true (e.g., if the amount is less than $10,000.0), then statements between THEN and the corresponding ELSE statement will be executed, and the rest of the statements up to END IF are ignored. Otherwise (i.e., if false), the statements after the ELSE statement will be executed.
2. Some of the statements are indented for better understanding of the pattern of the logic.
3. Each IF-THEN-ELSE block has a corresponding END IF statement.
4. The logic-chart for this algorithm is shown in Figure 4.5.

140 Decision Making, Comparing, and Branching

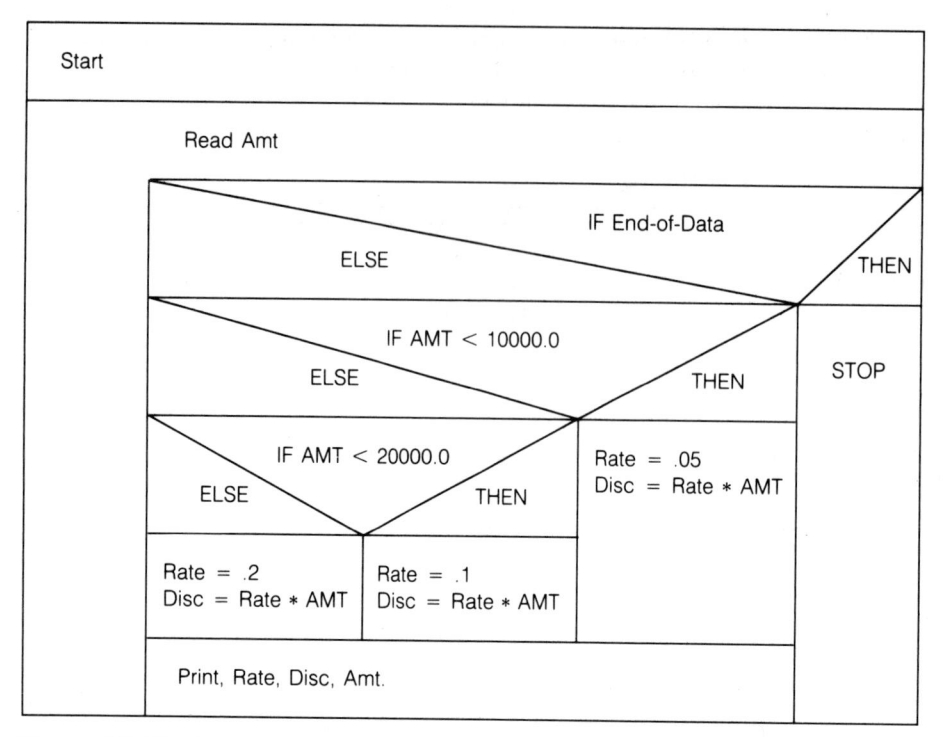

```
┌─────────────────────────────────────────────────────────────┐
│ Start                                                        │
├─────────────────────────────────────────────────────────────┤
│                    Read Amt                                  │
├──────────────────────────────────────────────────┬──────────┤
│                                  IF End-of-Data  │          │
│              ELSE                          THEN  │          │
│  ┌───────────────────────────────────────────┐   │          │
│  │                 IF AMT < 10000.0          │   │          │
│  │       ELSE                    THEN        │   │   STOP   │
│  │  ┌─────────────────────────┬───────────┐  │   │          │
│  │  │      IF AMT < 20000.0   │ Rate = .05│  │   │          │
│  │  │  ELSE            THEN   │ Disc = Rate * AMT │          │
│  │  │  ┌──────────┬────────┐  │           │  │   │          │
│  │  │  │Rate = .2 │Rate=.1 │  │           │  │   │          │
│  │  │  │Disc=Rate*│Disc=   │  │           │  │   │          │
│  │  │  │  AMT     │Rate*AMT│  │           │  │   │          │
│  ├──┴──┴──────────┴────────┴──┴───────────┘  │              │
│           Print, Rate, Disc, Amt.                            │
└─────────────────────────────────────────────────────────────┘
```

Figure 4.5 The logic-chart for Example 4.9.

The general form of an IF-THEN-ELSE block is

        IF (logical expression) THEN
            ⋮
            statements
            ⋮
        ELSE
            ⋮
            statements
            ⋮
        END IF

As explained previously, when the logical expression is true, the statements between THEN and ELSE will be executed and the statements between ELSE and END IF will be ignored. When the logical expression is false, then the statements after ELSE will be executed and those between the THEN and ELSE will be ignored. The END IF marks the range of the IF block. Indenting the statements in the range of the THEN and ELSE is necessary for clarity and readability.

Rules of Structured IF

1. Each block should have the words THEN, ELSE, and END IF (an exception is explained in rule 2).
2. If there is no statement after ELSE, the word ELSE *can* be omitted. For example, the following block is valid:

    IF (   ) THEN
       ⋮
    END IF

3. The forms ENDIF and END IF are both permissible.
4. No statement may be written directly after the word THEN or ELSE; instead the statement must begin on a new line.
5. An exception to rule 4 is that another IF statement may start directly after the word ELSE. The structure is then called an *ELSE-IF* block. In this case, the corresponding END IF *must* be omitted. That is, the ELSE-IF block does not need END IF. An example is:

```
┌─IF(A .LE. 500.0) THEN
│              COM = A*.01
│  ELSE┌─IF (A .LE. 1000.0) THEN
│      │          COM = A*.02
│      ├─ELSE
│      └          COM = A*.03
└─END IF
```

To remember this, you may note when an IF statement does not start a new line, it does not need the corresponding END IF. The following picture is another example:

```
IF(    )THEN
  ⋮
ELSE IF(    )THEN
     ⋮
     ELSE IF(    )THEN
          ⋮
          ELSE
     ⋮
END IF
```

6. An IF block can be inside another IF block, called a *nested block*. Each IF block in a nested block must be a complete block. That is, it must have the corresponding THEN, ELSE, and END IF (except for an ELSE-IF block). Furthermore, the inner IF block must be completely inside of the range of either the THEN or the ELSE of the outer IF. The following is an example:

```
IF (A .LT. 100.0) THEN
             IF (KODE .EQ. 1) THEN
                        DIS = A*.02
             ELSE
                        DIS = A*.03
             END IF
ELSE
             IF (KODE .EQ. 1) THEN
                        DIS = 5.0 + A*.02
             ELSE
                        DIS = 7.0 + A*.03
             END IF
END IF
```

Figure 4.6 on page 142 shows some valid nested IF blocks.

7. Indentation is not only helpful but often necessary to facilitate understanding the pattern of the logic in an IF block.
8. A GO TO statement can be inside an IF block to transfer the control out of the block, but control cannot be transferred into the IF block.

142 Decision Making, Comparing, and Branching

```
┌ IF (   ) THEN
│   ⋮
├ ELSE
│   ⋮
└ END IF


┌ IF (   ) THEN
│   ⋮
└ ENDIF


┌ IF (   ) THEN
│   ⋮
├ ELSE IF (   ) THEN
│   ⋮
│   ┌ ELSE
│   │   ⋮
└ END IF


┌ IF (   ) THEN
│   ⋮
├ ELSE
│   ┌ IF (   ) THEN
│   │   ⋮
│   ├ ELSE
│   │   ┌ IF (   ) THEN
│   │   │   ⋮
│   │   ├ ELSE
│   │   │   ⋮
│   │   └ END IF
│   └ END IF
└ END IF


┌ IF (   ) THEN
│   ┌ IF (   ) THEN
│   │   ┌ IF (   ) THEN
│   │   │   ⋮
│   │   ├ ELSE
│   │   │   ┌ IF (   ) THEN
│   │   │   │   ⋮
│   │   │   ├ ELSE
│   │   │   │   ┌ IF (   ) THEN
│   │   │   │   │   ⋮
│   │   │   │   ├ ELSE
│   │   │   │   │   ⋮
│   │   │   │   └ END IF
│   │   │   └ END IF
│   │   └ END IF
│   └ END IF
└ END IF


┌ IF (   ) THEN
│   ┌ IF (   ) THEN
│   │   ⋮
│   ├ ELSE IF (   ) THEN
│   │   ⋮
│   └ END IF
├ ELSE
│   ┌ IF (   ) THEN
│   │   ⋮
│   ├ ELSE
│   │   ┌ IF (   ) THEN
│   │   │   ⋮
│   │   ├ ELSE
│   │   │   ⋮
│   │   └ END IF
│   └ END IF
└ END IF
```

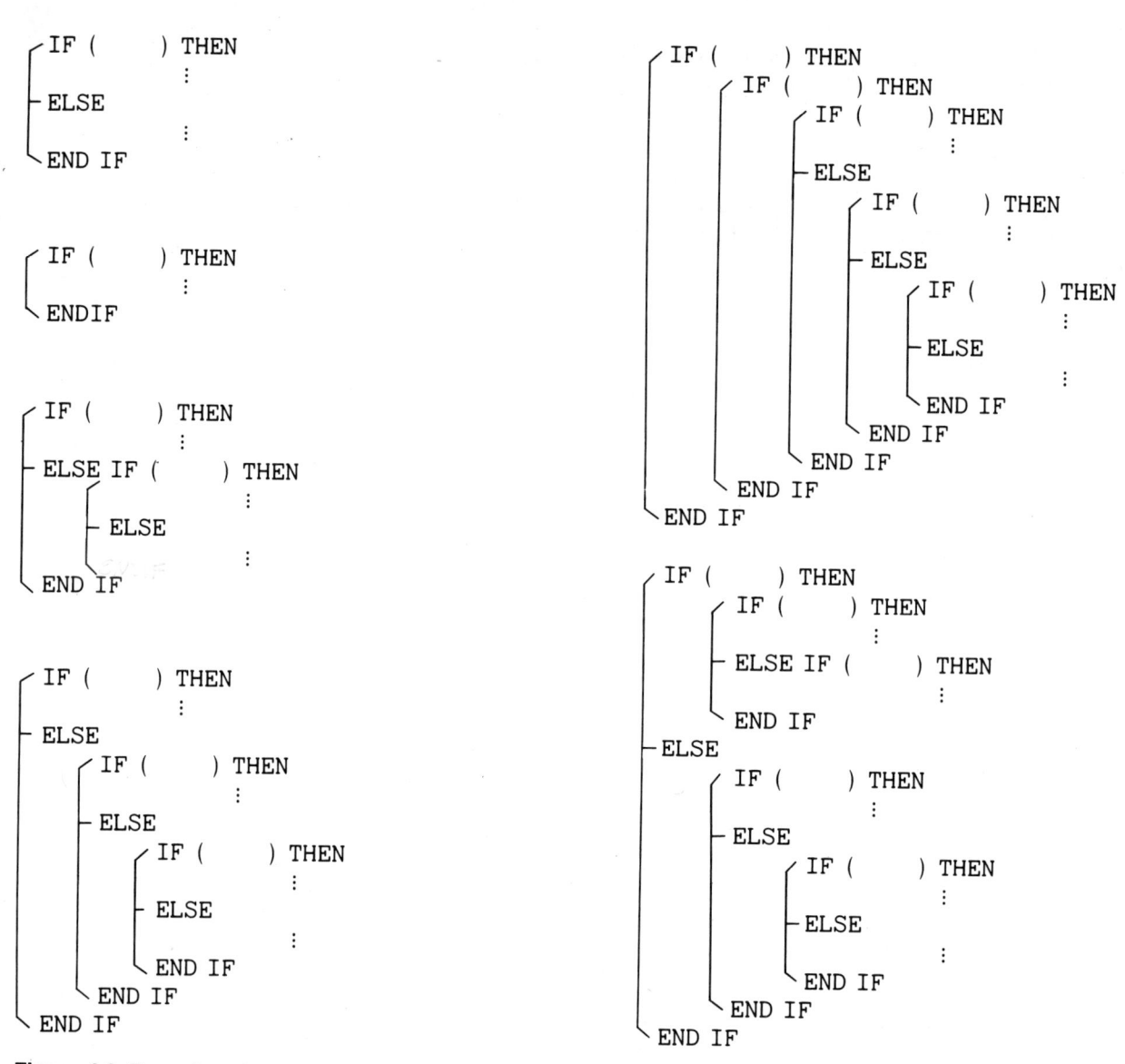

Figure 4.6 Examples of valid nested IF blocks.

The following is an example of a permissible and an impermissible case.

```
        Permissible                    Impermissible
        IF (   ) THEN                  IF (   ) THEN
          ⋮                              ⋮
        ELSE IF (   ) THEN             ELSE IF (   ) THEN
          ⋮            30                 A=B+C
          GO TO 80                       ⋮
          ⋮                            ELSE
        ELSE                             ⋮
          ⋮                            END IF
        END IF                           ⋮
   80   ...                           GO TO 30
          ⋮
```

Advantages of Using Structured IF

The structured IF block, or generally structured programming, makes a program

- Easier to understand by following the logic
- Easier to code
- Easier to review at a later time
- Less prone to error and less time consuming
- Easier to modify and document

Therefore, it is highly recommended to use the structured IF in a program rather than other kinds of IF statements. Furthermore, it is strongly advisable to avoid using the GO TO statement in a program if at all possible.

SOLVED PROBLEMS

4.8 Write a structured IF statement equivalent to the following:

a.
```
        IF (A .LT. B) GO TO 100
            X = A*B
            Y = T - X
            WRITE (6,10) A,B,X,Y
            GO TO 110
100     X = C*D
        WRITE (6,20) C,D,X
110     ...
```

b.
```
        IF (N .LT. I) N = I
```

c.
```
        IF (HOUR .GT. 40.0) GO TO 50
            GROSS = HOUR*RATE
            GO TO 60
50          GROSS = (HOUR - 40.0)*1.5*RATE
     *          + 40*RATE
60      ...
```

d.
```
        IF (A .LT. 1000.0) GO TO 10
        IF (A .LT. 2000.0) GO TO 20
            RATE = R
            GO TO 60
10      RATE = P
            GO TO 60
20      RATE = Q
60      ...
```

e.
```
        IF (X .GT. 50000.0) GO TO 40
        IF (X .GT. 40000.0) GO TO 50
        GO TO 60
```

144 Decision Making, Comparing, and Branching

```
40      RATE = 13.0
        IF (I .EQ. 1) RATE = 15.0
        IF (I .EQ. 2) RATE = 14.0
        GO TO 60
50      RATE = 10.0
        IF (I .EQ. 1) RATE = 12.0
        IF (I .EQ. 2) RATE = 11.0
60      ...
```

☐ **Answers**

a.
```
        IF (A .LT. B) THEN
            X = C*D
            WRITE (6,20) C,D,X
        ELSE
            X = A*B
            Y = T - X
            WRITE (6,10) A,B,X,Y
        END IF
110     ...
```

b.
```
        IF (N .LT. I) THEN
            N = I
        ENDIF
```
(Note: ELSE is omitted.)

c.
```
        IF (HOUR .GT. 40.0) THEN
            GROSS = (HOUR - 40.0)*1.5*RATE +
     *       40*RATE
        ELSE
            GROSS = HOUR*RATE
        ENDIF
60      ...
```

d.
```
        IF (A .LT. 1000.0) THEN
            RATE = P
        ELSE IF (A .LT. 2000.0) THEN
            RATE = Q
        ELSE
            RATE = R
        END IF
60      ...
```

e.
```
        IF (X .GT. 50000.0) THEN
            IF (I .EQ. 1) THEN
                RATE = 15.0
            ELSE IF (I .EQ. 2) THEN
                RATE = 14.0
            ELSE
                RATE = 13.0
```

```
                    END IF
          ELSE IF (X .GT. 40000.0) THEN
              IF (I .EQ. 1) THEN
                    RATE = 12.0
              ELSE IF (I .EQ. 2) THEN
                    RATE = 11.0
                 ELSE
                    RATE = 10.0
              END IF
          END IF
   60     ...
```

4.9 Correct the following IF blocks:

a.
```
          IF (A .LT. 5.0) THEN
                KODE = 1
                GO TO 50
          ELSE IF (A .LT. 10) THEN
                    KODE = 2
                    GO TO 50
             ELSE
                    KODE = 3
                    GO TO 50
          END IF
   50     ...
```

b.
```
          IF (X .EQ. Y) THEN
                P = 3.5
          ELSE
              IF (X .LT. Y) THEN
                    P = 2.5
              ELSE
                    P = 4.5
          END IF
```

c.
```
          IF (A .GE. 90.0) THEN
              IGRADE = 4
          ELSE IF (A .GE. 80.0) THEN
                 IGRADE = 3
             ELSE
                 IGRADE = 2
             END IF
          END IF
```

d.
```
          IF (A .LE. 10.0) THEN X = A
          ELSE Y = Z
          END IF
```

e.
```
          IF (X .GE. Y)
                RATE = .05
```

```
                ELSE
                    RATE = .01
f.

            GO TO 25
                :
            IF (SPEED .GT. 65.0) THEN
                FINE = SPEED*1.5
            ELSE
     25         FINE = 0.0
            END IF
```

Answers

a. The GO TO 50 statements are not necessary. The control will be transferred automatically to the statement after the END IF (statement 50) in any case.
b. The inner IF block needs an END IF statement. Generally any IF which starts a new line must have an END IF.
c. The inner IF block does not need an END IF statement. (It is an ELSE-IF statement.)
d. There must not be any statement right after the words THEN and ELSE. X = A and Y = Z must start a new line.
e. The IF block needs the words THEN and END IF.
f. Control cannot be transferred into an IF block.

CONDITIONAL BRANCHING

With the logical IF statement, or with the IF block, you can write almost any program which involves comparison, decision making, and conditional branching. But FORTRAN also allows you additional choices. Arithmetic IF and computed GO TO statements are additional features of FORTRAN used for this purpose.

Arithmetic IF Statement

The *arithmetic IF statement* is used mostly for conditional branching. It checks the sign of an arithmetic expression, and then transfers the control to a different part of the program depending on the sign of the arithmetic expression, whether it is negative, zero, or positive. Let's take a look at an example:

EXAMPLE 4.10

■ **Problem:** Read X and Y from a card and print the following:

THE ANSWER IS NEGATIVE   if   X − Y is negative,
THE ANSWER IS ZERO   if   X − Y is zero,
THE ANSWER IS POSITIVE   if   X − Y is positive.

Program:

```
        PROGRAM ARIF
10      FORMAT (2F5.1)
        READ (5,10) X,Y
C
        IF(X - Y) 20,30,40
C
20      WRITE (6,25)
25      FORMAT (11X,'THE ANSWER IS NEGATIVE')
        GO TO 99
C
30      WRITE (6,35)
35      FORMAT (11X,'THE ANSWER IS ZERO')
        GO TO 99
C
40      WRITE (6,45)
45      FORMAT (11X,'THE ANSWER IS POSITIVE')
99      STOP
        END
```

Note: IF (X − Y) 20,30,40 works like a GO TO statement; it means the following:

if X − Y is negative, GO TO 20
if X − Y is zero, GO TO 30
if X − Y is positive, GO TO 40

The general form of an arithmetic IF statement is:

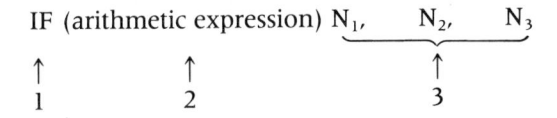

IF (arithmetic expression) N_1, N_2, N_3

↑   ↑   ↑
1   2   3

and has three parts:

1. The word IF.
2. An arithmetic expression in parentheses such as (3 * X * Y − B), (A + B), or (X − Y). The sign of the expression will be evaluated as negative, zero, or positive.
3. Exactly three statement labels, separated by commas. The statement works as (see Figure 4.7)

GO TO N_1 if the value of the arithmetic expression is negative,
GO TO N_2 if the value of the arithmetic expression is zero,
GO TO N_3 if the value of the arithmetic expression is positive.

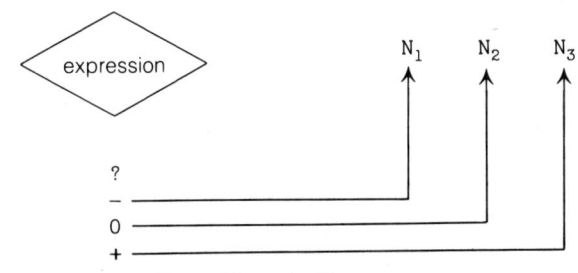

Figure 4.7 The arithmetic IF.

148 Decision Making, Comparing, and Branching

The statements N1, N2, and N3 can be the same or different. However, all of them must be executable statements. The following are examples of correct arithmetic IF statements:

```
IF (3*A - B) 30, 20, 20
IF (5*Y - 2*X/3*Y) 100, 100, 90
IF (ROOT) 10, 30, 10
IF (2*Y) 90, 3, 5
IF (VALUE) 95, 95, 100
IF ((A*B) - (C*D)) 5, 10, 20
```

The following are examples of incorrect arithmetic IF statements with the reasons:

IF (A - B), 10, 20, 30	Comma after parentheses is not needed
IF (C - D) 10 10 30	Labels must be separated by commas
IF (3 - X) GO TO 10, 30, 30	GO TO is not needed
IF A - Z, 50, 60, 70	Parentheses missing
IF 10, 20, 30 (P - Q)	Labels should go after (P - Q)
IF (V - W) 10, 20, 30, 40	A fourth label is not permissible.
IF (A .LT. Y) 10, 20, 30	Logical IF and arithmetic IF cannot be mixed.

A segment of a program can be written with either a logical or an arithmetic IF statement. The following are some examples:

```
           Logical IF                        Arithmetic IF

      IF(A .GE. B) GO TO 20             IF (A - B) 10,20,20
          RATE = .08                        RATE = .08
          GO TO 30                          GO TO 30
   20     RATE = .05                 20     RATE = .05
   30     ...                        30     ...


      IF(X .LT. Y) GO TO 5              IF(X - Y) 5, 10, 10
          W = X*Y                   10      W = X*Y
          GO TO 30                          GO TO 30
    5     W = X*Y - .5*(X - Y)       5      W = X*Y - .5*(X - Y)
          V = 3*X*Y                         V = 3*X*Y
   30     ...                        30     ...


      IF(V .EQ. W) GO TO 20             IF(V - W) 10, 20, 10
          V = W                     10      V = W
   20     WRITE(6,30)                20     WRITE (6,30)
   30     FORMAT (11X,'EQUAL')       30     FORMAT (11X,'EQUAL')

      IF(A .LE. 0.0)STOP                IF(A) 100,100,50
   50     ...                        50     ...
    ⋮                                100    STOP
```

Using an arithmetic IF makes a program unstructured and is not recommended.

Computed GO TO Statement

The *computed GO TO statement* is used for branching (like GO TO) when the control must be transferred to one of several statements, depending on different conditions of an integer variable. Compare this with a simple GO TO statement which transfers the control to only one statement. The following is an example:

EXAMPLE 4.11

Problem: The grade code for each student in a FORTRAN course is placed on a line. Each code is either 1, 2, 3, 4, or 5, where 1 is an E, 2 is a D, 3 is a C, 4 is a B, and 5 is an A. There are several data lines. The last one, with a negative number, is a trailer record. Write a program which reads each code, calculates the number of each grade code, and prints each grade code. Print the number of each grade code at the end.

Program:

```
      PROGRAM COMPGD
          NE = 0
          ND = 0
          NC = 0
          NB = 0
          NA = 0
5     READ (5,100) KODE
      IF (KODE .LT. 0) GO TO 99
C
          GO TO (10,20,30,40,50),KODE
C
10        NE = NE + 1
          GO TO 70
C
20        ND = ND + 1
          GO TO 70
C
30        NC = NC + 1
          GO TO 70
C
40        NB = NB + 1
          GO TO 70
C
50        NA = NA + 1
C
70        WRITE (6,110)KODE
      GO TO 5
99    WRITE (6,120)NE,ND,NC,NB,NA
C
100   FORMAT(I1)
110   FORMAT (11X,'GRADE CODE=',I2)
120   FORMAT (11X,'NO OF ES=',I2,'NO OF DS=',I2,
     *'NO OF CS=',I2,'NO OF BS=',I2,'NO OF AS=',I2)
      END
```

150 *Decision Making, Comparing, and Branching*

> **Notes:**
>
> 1. The GO TO (10, 20, 30, 40),KODE transfers the control to the block which starts with:
>
>   statement 10 if KODE = 1 (to count the number of E's)
>   statement 20 if KODE = 2 (to count the number of D's)
>   statement 30 if KODE = 3 (to count the number of C's)
>   statement 40 if KODE = 4 (to count the number of B's)
>   statement 50 if KODE = 5 (to count the number of A's)
>
> 2. At the end of the data, the number of A's, B's, C's, D's, and E's will be printed by statement 99.

The general form of a computed GO TO statement is

GO TO $(N_1, N_2, N_3, ..., N_n), I$

where $N_1, N_2, N_3, ..., N_n$ are integers representing a statement label, and I is an integer variable. The placement of parentheses is critical, but the comma after the parentheses is optional. FORTRAN 77 allows using a real variable or an arithmetic expression, and the omission of the comma after the parentheses.

The computed GO TO statement works like:

GO TO N_1 if I is equal to 1
GO TO N_2 if I is equal to 2
GO TO N_3 if I is equal to 3
GO TO N_4 if I is equal to 4
$\vdots$
GO TO N_n if I is equal to n

If the value of I is less than one or greater than n, execution continues with the statement following the computed GO TO. For example:

```
GO TO (400,6,280,99),M
```

means

GO TO 400 if $M = 1$
GO TO 6 if $M = 2$
GO TO 280 if $M = 3$
GO TO 99 if $M = 4$

If the value of M is less than 1 or greater than 4 in this case, the computed GO TO statement is ignored and processing continues to the next statement.

The computed GO TO statement is very useful when it is necessary to establish several independent blocks for different values of a variable. Each block can be a combination of statements built for the particular value of that variable. Such blocking is common in applications where different types of calculations must be performed for different values of a variable, especially when the variable takes on sequential integer values (like a code).

SOLVED PROBLEMS

4.10 Find and correct the errors in the following computed GO TO statements:

a.
```
        IF I = 1, GO TO 50
        IF I = 2, GO TO 60
        IF I = 3, GO TO 70
        ELSE GO TO 80
```

b.   GO TO 10,20,3,40,40,K

c.
```
        IF K = 2 GO TO 25 ELSE IF K = 1 GO TO
        10 ELSE PROCEED
```

d.   GO TO (K,J,K,N)L

e.   IF (5,10,15) KAY

f.   GO TO (10,20,30,40) L - 3

g.   GO TO (10,20),ROOL

h.
```
5       GO TO (5,10,15)L
```

i.   GO TO (90,50,300) X/K

(identify the statement to which the control will be transferred if X = 8.8, and K = 3).

☐ **Answers**

a.
```
        GO TO (50,60,70),I
80      ...
```

b.   GO TO (10,20,3,40,40),K

c.   GO TO (10,25),K

d. A variable cannot be used in the parentheses.

e.   GO TO (5,10,15), KAY

f. This is correct in FORTRAN 77. Note that the expression will be evaluated first. For example, if L = 6, then the control will be transferred to statement 30.

g. A real variable, such as ROOL, can be used only in FORTRAN 77.

h. It is inappropriate to transfer the control from a statement to itself, although syntactically it may be correct.

i. This is correct in FORTRAN 77, and the control transfers to statement 50 if X = 8.8 and K = 3 (the integer value of X/K will be equal to 2).

■ **4.11** Rewrite the following statements with a computed GO TO statement.

a.
```
        IF (K .EQ. 4) GO TO 50
        IF (K .EQ. 5) GO TO 40
```

152 Decision Making, Comparing, and Branching

```
                    IF (K .EQ. 3) GO TO 30
                    IF (K .EQ. 2) GO TO 20
                    IF (K .EQ. 1) GO TO 10
              60    ...
```

b.
```
                    IF (KODE .EQ. 1) THEN
                         GO TO 10
                    ELSE IF (KODE .EQ. 2) THEN
                         GO TO 20
                    ELSE IF KODE .EQ. 3) THEN
                         GO TO 30
                    ELSE
                         GO TO 40
                    END IF
```

c.
```
                    IF (I .EQ. 50) GO TO 5
                    IF (I .EQ. 51) GO TO 10
                    IF (I .EQ. 52) GO TO 20
```

d.
```
                    IF (ISCORE .LE. 6) GO TO 60
                    IF (ISCORE .EQ. 7) GO TO 70
                    IF (ISCORE .EQ. 8) GO TO 80
                    IF (ISCORE .EQ. 9) GO TO 90
                    IF (ISCORE .EQ. 10) GO TO 100
```

e.
```
                    IF (SCORE .LT. 60.0) GO TO 60
                    IF (SCORE .LT. 70.0) GO TO 70
                    IF (SCORE .LT. 80.0) GO TO 80
                    IF (SCORE .LT. 90.0) GO TO 90
                    IF (SCORE .LT. 100.0) GO TO 100
```

☐ **Answers**

a.
```
                    GO TO (10,20,30,50,40),K
              60    ...
```

b.
```
                    GO TO (10,20,30),KODE
              40    ...
```

c.
```
                    J = I - 49
                    GO TO (5,10,20), J
```
or
```
                    GO TO (5,10,20)I - 49
```
Note the technique for making J equal to 1, 2, 3.

d.
```
                    J = ISCORE - 5
                    GO TO (60,70,80,90,100),J
```
or
```
                    GO TO (60,70,80,90,100) ISCORE - 5
```

e.
```
    ISCORE = SCORE/10
    J = ISCORE - 4
    GO TO (60,70,80,90,100),J
```
or
```
    GO TO (60,70,80,90,100) SCORE/10 - 4
```
Note the technique for making J equal to 1, 2, 3, 4, and 5. (ISCORE will be an integer.)

A SAMPLE PROBLEM

The following problem has been developed to demonstrate the application of the material which has been covered so far. We will use the problem-solving concepts developed in Chapter 3.

STEP 1: THE PROBLEM

XYZ Company would like to have a summary report showing the weekly pay and insurance deductions for each employee. Further analysis shows that:

- The insurance deductions depend upon the number of dependents.
- The hours worked and wage rate data are available.
- The overtime (over 40 hours) is paid 1.5 times the regular rate.
- The total pay and total deductions are desirable at the end of the report.

Thus, the objective is to write a program to generate such a report.

STEP 2: INPUT-OUTPUT FORMULATION

The OUTPUT: After careful analysis, it is determined that the variables listed in Table 4.2 are necessary for the output.

The INPUT: In order to calculate the pay, deductions, pay after deductions, and totals, we need the input information shown in Table 4.3.

Table 4.2 Output Analysis Form for the Sample Program

ITEM	SYMBOLIC NAME OF VARIABLE	TYPE	FIELD LENGTH (COLUMNS)
1. Employee's identification number	ID	Integer	5
2. Hours worked	HOURS	Real	7
3. Wage rate	WAGE	Real	7
4. Pay for the period	PAY	Real	6
5. Code for number of dependents	KODE	Integer	1
6. Deduction	DEDUCT	Real	5
7. Pay after deduction (net pay)	PAD	Real	7
8. Total pay to all	TOTPAY	Real	9
9. Total deductions	TOTDED	Real	9
10. Total pay after deductions	TOTPAD	Real	9

154 Decision Making, Comparing, and Branching

Table 4.3 Input Analysis Form for the Sample Program

ITEM	SYMBOLIC NAME OF VARIABLE	TYPE	FIELD LENGTH (COLUMNS)
1. Employee's identification number	ID	Integer	5
2. Hours worked	HOURS	Real	5
3. Wage rate	RATE	Real	5
4. Code for number of dependents	KODE	Integer	1

STEP 3: INPUT-OUTPUT LAYOUT

The OUTPUT: The output is designed as shown in Figure 4.8, on page 155.
   The INPUT: It is determined that the input data will be entered as follows:

1. Employee's ID number   Columns 1–5
2. Hours worked for the period   Columns 7–11
3. Wage rate   Columns 13–17
4. Code for number of dependents Column 19

Observe that columns 6, 12, and 18 have no assignment and will be left blank. This will make the data easier to read. These blanks must be considered when coding the input format. A negative number (−9999) will be entered instead of the employee's ID number as the last item of input to indicate the end of data. Figure 4.9, on page 156, shows a sample of the input data which will be used by the program.

STEP 4: PROCESS DESIGN

The program is broken down into the following modules and submodules:

1. Housekeeping
    a. Printing the heading
    b. Initializing the variables
2. Reading the input information and terminating upon processing a negative number
3. Calculating
    a. Calculating the pay
    b. Calculating the deduction
    c. Calculating the pay after deduction (the net)
    d. Calculating the totals
4. Printing the information
    a. Writing the data about each employee, using the loop
    b. Writing the totals after the loop
    c. Printing END OF REPORT at the end of the program

The process requirements for the calculation module are as follows:

1. To calculate the pay (PAY) for the period, the following formulas are used:

   a. PAY = HOURS * RATE

      if the hours worked are 40 hours or less;

   b. PAY = RATE * 40 + 1.5 * (HOURS − 40) * RATE

      if the hours worked are over 40 hours.

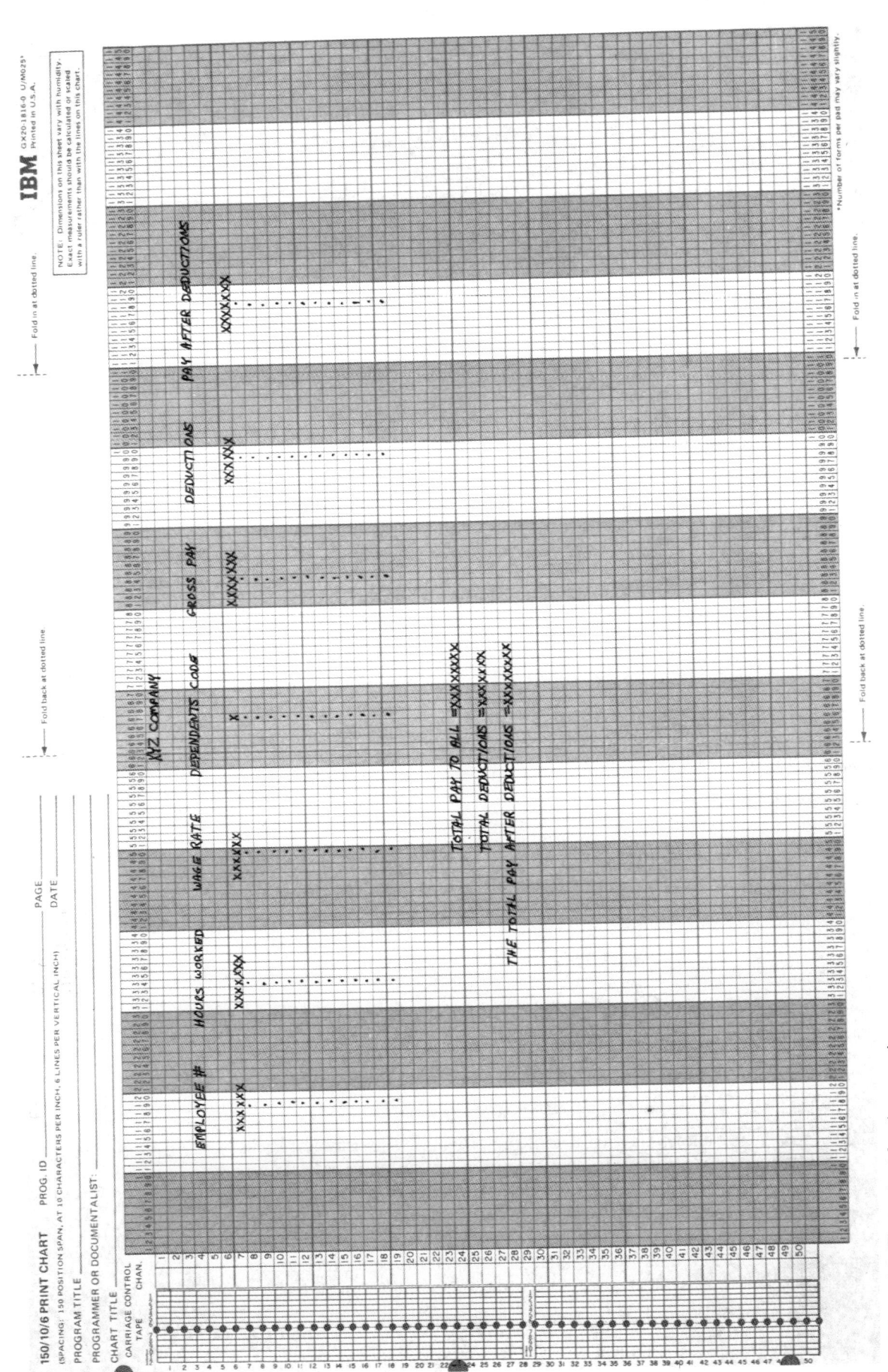

Figure 4.8 The output for the sample program.

156 Decision Making, Comparing, and Branching

```
56298   38.50   12.50   2
69820   52.50    9.50   1
56210   29.25   11.50   3
62890   40.00   15.78   2
26850   60.00    6.75   1
66268   72.00    8.36   3
78256   58.25    5.20   3
-9999
```

Figure 4.9 Sample input data.

2. It turns out that the deductions depend directly on the number of dependents; thus, KODE is designed to be equal to the number of dependents. The deductions are:

   12% of the pay if KODE is 3 or more: DEDUCT = .12 * PAY
   11% of the pay if KODE is 2:   DEDUCT = .11 * PAY
   10% of the pay if KODE is 1:   DEDUCT = .10 * PAY

3. The pay after deduction will be calculated by the following formula:

   PAD = PAY − DEDUCT

4. The totals can be calculated by

   TOTPAY = TOTPAY + PAY
   TOTDED = TOTDED + DEDUCT
   TOTPAD = TOTPAD + PAD

   in the loop created for reading the data. The loop will be terminated upon reading a negative number, namely −9999 for the ID number.

STEP 5: FLOWCHART

The flowchart for this problem is shown in Figure 4.10.

STEP 6: CODING

The coding for the program is shown in Figure 4.11, page 158.

STEP 7: RUN PREPARATION

The program statements, as well as the control statements, are entered at the terminal. After the program is reviewed carefully for typing errors, it is submitted for execution.

STEP 8: EXECUTION, TESTING, AND DEBUGGING

Figure 4.12 on page 159 shows the first run of the program. The listing of the program indicates a syntax error after line 18. Careful review of 10 FORMAT (line 14) shows that the left quotation mark for 'HOURS WORKED' is missing. This is corrected. However, since only 72 columns may be used for a FORTRAN statement, it is necessary to move the trailing comma down to the next line after inserting the quotation mark. Notice that the compiler does not recognize the missing apostrophe; instead it prints three messages indicating possible syntax errors.

The second run of the program, shown in Figure 4.13 on pages 160–161, does not have any syntax errors, but it has several logical errors.

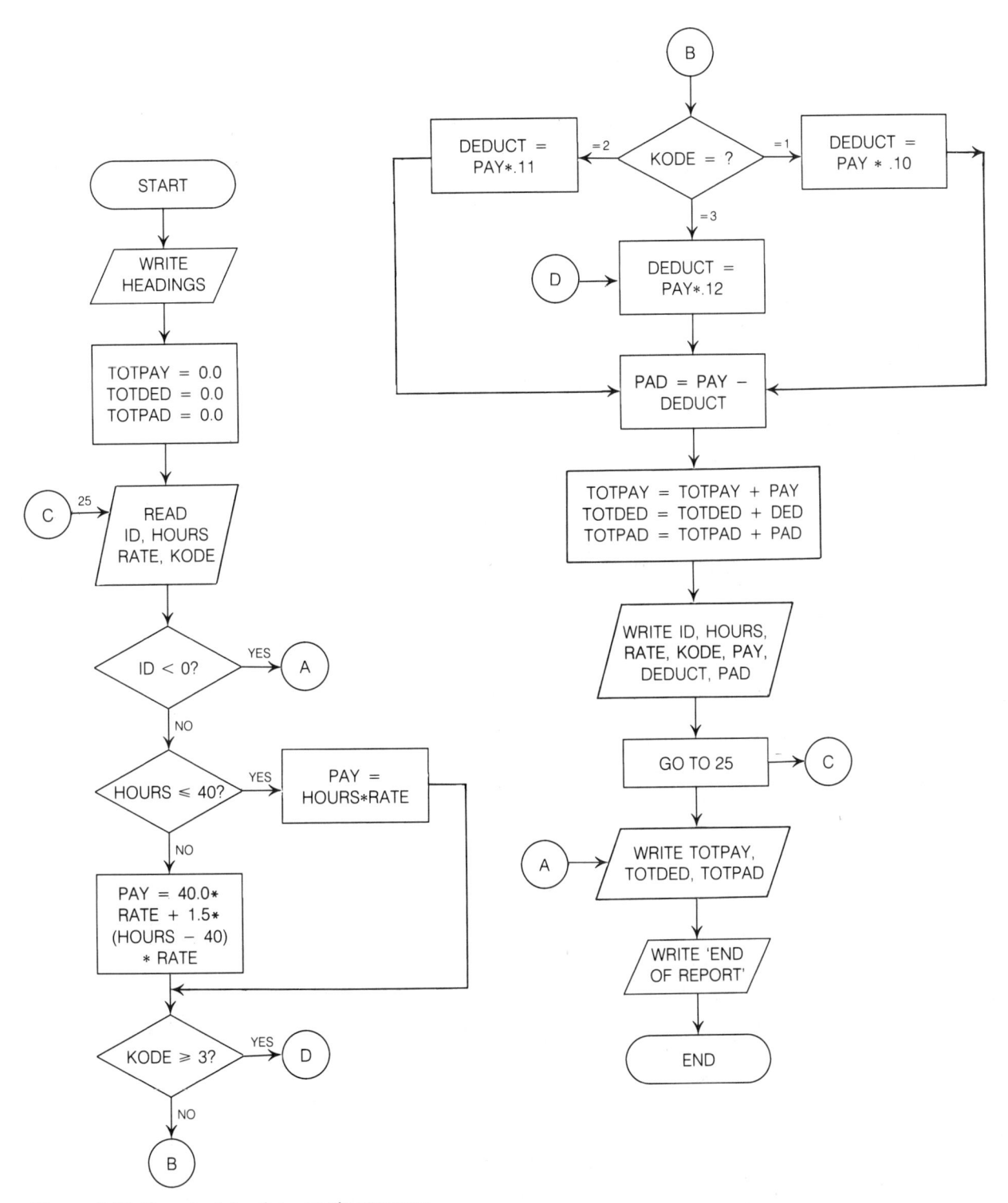

Figure 4.10 Flowchart for the sample program.

First of all, the figures do not align with the headings. The format for the WRITE statements must be revised to correspond with the output layout (shown in Figure 4.8). In 10 FORMAT the title centering is revised and 5X is inserted between HOURS WORKED and WAGE RATE. This was omitted when the comma was inserted to correct the syntax error. The 16 at the beginning of 90 FORMAT was mistakenly keyed as /6. This is corrected. 100 FORMAT is also corrected to show the footings better. The most important logical error

158 Decision Making, Comparing, and Branching

```
              PROGRAM WAGES
C     AUTHOR: RICH NEWMAN     DATE: AUGUST 8, 19--
C     THIS PROGRAM GENERATES A REPORT SHOWING THE PAY AND INSURANCE DEDUCTIONS
C     FOR A GROUP OF EMPLOYEES
C
C     VARIABLES:   ID = EMPLOYEE NO.              DEDUCT = DEDUCTIONS
C                  HOURS = NO. OF HOURS WORKED    PAD = PAY AFTER DEDUCTIONS
C                  RATE = WAGE RATE               TOTPAY = TOTAL PAY
C                  KODE = NO. OF DEPENDENTS       TOTDED = TOTAL DEDUCTIONS
C                  PAY = GROSS PAY                TOTPAY = TOTAL PAY AFTER DEDUCTIONS
C     PRINT THE HEADINGS, STARTING ON A NEW PAY
          WRITE(6,10)
10        FORMAT('1',59X,'XYZ COMPANY'///14X,'EMPLOYEE #',5X,'HOURS WORKED'
      1          'WAGE RATE',5X,'DEPENDENTS CODE',5X,'GROSS PAY',5X,
      2          'DEDUCTIONS',5X,'PAY AFTER DEDUCTION'//)
C
C     INITIALIZE THE TOTALS TO ZERO
          TOTPAY = 0.0
          TOTDED = 0.0
          TOTPAD = 0.0
C
C     THIS IS THE BEGINNING OF THE LOOP
25        READ(5,20)ID,HOURS,RATE,KODE
20        FORMAT(15,1X,F5.2,1X,F5.2,1X,11)
C
C     READING A NEGATIVE ID NUMBER TERMINATES THE LOOP
          IF (ID) 900,30,30
C
C     CALCULATE THE PAY (OVER 40 HOURS IS PAID ONE AND A HALF TIMES THE RATE)
30        IF (HOURS .LE. 40.0) THEN
              PAY = HOURS * RATE
          ELSE
              PAY = 40.0*RATE + 1.5*(HOURS - 40)*RATE
          ENDIF
C
C     CALCULATE THE DEDUCTIONS: 10%, 11%, OR 12% OF THE PAY DEPENDING ON THE CODE
          IF (KODE .GE. 3) GO TO 70
          GO TO (50,60), KODE
50        DEDUCT = PAY*.10
          GO TO 80
60        DEDUCT = PAY*.11
          GO TO 80
70        DEDUCT = PAY*.12
C
C     CALCULATE THE PAY AFTER DEDUCTIONS (NET PAY)
80        PAD = PAY - DEDUCT
C
C     CALCULATE THE TOTALS
          TOTPAY = TOTPAY + PAY
          TOTDED = TOTDED + DEDUCT
          TOTPAD = TOTPAD + PAD
C
C     WRITE THE INFORMATION (DATA AND CALCULATIONS)
          WRITE (6,90) ID,HOURS,RATE,KODE,PAY,DEDUCT,PAD
90        FORMAT(/6X,I5,9X,F7.2,9X,F6.2,14X,I1,13X,F7.2,9X,F5.2,13X,F7.2/)
C
C     END OF THE LOOP
          GO TO 25
900       WRITE (6,100) TOTPAY, TOTDED, TOTPAD
100       FORMAT (//61X,'TOTAL PAY TO ALL =',F9.2//75X,'TOTAL DEDUCTIONS =',F9.2//
     *           81X,'THE TOTAL PAY AFTER DEDUCTIONS',F9.2)
          WRITE (6,150)
150       FORMAT (///56X,'END OF REPORT')
          END
```

Figure 4.11 Coding for the sample program.

```
         PROGRAM WAGES      73/172  OPT=0              FTN 5.1+528    82/06/17. 19.29.22     PAGE    1


         1              PROGRAM WAGES
         2        C AUTHOR: RICH NEWMAN      DATE: AUGUST 8, 1990
         3        C THIS PROGRAM GENERATES A REPORT SHOWING THE PAY AND THE INSURANCE DEDUCTIONS
         4        C FOR A GROUP OF EMPLOYEES
         5        C
         6        C VARIABLES: ID = EMPLOYEE NO.                 DEDUCT = DEDUCTIONS
         7        C            HOURS = NO. OF HOURS WORKED       PAD    = PAY AFTER DEDUCTIONS
         8        C            RATE  = WAGE RATE                 TOTPAY = TOTAL PAY
         9        C            KODE  = INDICATES NO. OF DEPENDENTS  TOTDED = TOTAL DEDUCTIONS
        10        C            PAY   = GROSS PAY                 TOTPAD = TOTAL PAY AFTER DEDS.
        11        C
        12        C PRINT THE HEADINGS, STARTING ON A NEW PAGE
        13              WRITE (6,10)
        14           10 FORMAT ('1',59X,'XYZ COMPANY'///14X,'EMPLOYEE #',5X,'HOURS WORKED',
        15             1          5X,'WAGE RATE',5X,'DEPENDENTS CODE',5X,'GROSS PAY',5X,
        16             2          'DEDUCTIONS',5X,'PAY AFTER DEDUCTION'//)
        17        C
        18        C INITIALIZE THE TOTALS TO ZERO
FATAL   *         TERMINAL DELIMITER _' MISSING
FATAL   *         UNKNOWN EDIT DESCRIPTOR _H
FATAL   *         UNBALANCED PARENS
        19              TOTPAY = 0.0
        20              TOTDED = 0.0
        21              TOTPAD = 0.0
        22        C
        23        C THIS IS THE BEGINNING OF THE LOOP
        24           25 READ (5,20) ID, HOURS, RATE, KODE
        25           20 FORMAT (I5,1X,F5.2,1X,F5.2,1X,I1)
        26        C
        27        C READING A NEGATIVE ID NUMBER TERMINATES THE LOOP
        28              IF (ID) 900, 30, 30
        29        C
        30        C CALCULATE THE PAY, (OVER 40 HOURS IS PAID ONE AND A HALF TIMES THE RATE)
        31           30    IF (HOURS .LE. 40.0) THEN
        32                    PAY = HOURS * RATE
        33                 ELSE
        34                    PAY = HOURS * RATE * 1.5
        35                 ENDIF
        36        C
        37        C CALCULATE THE DEDUCTIONS: 10%, 11%, OR 12% OF THE PAY DEPENDING ON THE CODE
        38              IF (KODE .GE. 3) GO TO 70
        39              GO TO (50,60), KODE
        40           50 DEDUCT = PAY * .10
        41              GO TO 80
        42           60 DEDUCT = PAY * .11
        43              GO TO 80
        44           70 DEDUCT = PAY * .12
        45        C
        46        C CALCULATE THE PAY AFTER DEDUCTIONS, (NET PAY)
        47           80 PAD = PAY - DEDUCT
        48        C
        49        C CALCULATE THE TOTALS
        50              TOTPAY = TOTPAY + PAY
        51              TOTDED = TOTDED + DEDUCT
        52              TOTPAD = TOTPAD + PAD
        53        C
        54        C WRITE THE INFORMATION, (DATA AND CALCULATIONS)
        55              WRITE (6,90) ID, HOURS, RATE, KODE, PAY, DEDUCT, PAD
        56           90 FORMAT (/6X,I6,9X,F7.2,9X,F6.2,14X,I1,13X,F7.2,9X,F5.2,13X,
        57             1          F7.2/)
        58        C
        59        C END OF THE LOOP
        60              GO TO 25
        61        C
        62        C WRITE THE TOTALS
        63          900 WRITE (6,100) TOTPAY, TOTDED, TOTPAD
        64          100 FORMAT (//61X,'TOTAL PAY TO ALL =',F9.2//75X,'TOTAL DEDUCTIONS =',
        65             1         F9.2//81X,'THE TOTAL PAY AFTER DEDUCTIONS',F9.2)
        66              WRITE (6,150)
        67          150 FORMAT (///56X,'END OF REPORT')
        68              END

         PROGRAM WAGES      73/172  OPT=0              FTN 5.1+528    82/06/17. 19.29.22     PAGE    2

--VARIABLE MAP--(LO=A)
-NAME---ADDRESS--BLOCK-----PROPERTIES-------TYPE---------SIZE      -NAME---ADDRESS--BLOCK-----PROPERTIES-------TYPE---------SIZE

 DEDUCT    0B                               REAL                    PAY       0B                               REAL
 HOURS     0B                               REAL                    RATE      0B                               REAL
 ID        0B                               INTEGER                 TOTDED    0B                               REAL
 KODE      0B                               INTEGER                 TOTPAD    0B                               REAL
 PAD       0B                               REAL                    TOTPAY    0B                               REAL

--STATEMENT LABELS--(LO=A)
-LABEL-ADDRESS-----PROPERTIES----DEF   -LABEL-ADDRESS-----PROPERTIES----DEF   -LABEL-ADDRESS-----PROPERTIES----DEF

  10      0B       FORMAT          14     50      0B                     40     90      0B       FORMAT          56
  20      0B       FORMAT          25     60      0B                     42    100      0B       FORMAT          64
  25      0B                       24     70      0B                     44    150      0B       FORMAT          67
  30      INACTIVE                 31     80      0B                     47    900      0B                       63

--ENTRY POINTS--(LO=A)
-NAME---ADDRESS--ARGS---
 WAGES    0B      0

--I/O UNITS--(LO=A)
-NAME--- PROPERTIES-------------

 TAPE5   FMT/SEQ
 TAPE6   FMT/SEQ

--STATISTICS--
 PROGRAM-UNIT LENGTH         0B = 0
 CM STORAGE USED         60600B = 24960
 COMPILE TIME             0.268 SECONDS

     3  FATAL    ERRORS IN WAGES
```

Figure 4.12 The first run of the sample program.

160 Decision Making, Comparing, and Branching

```
T
1              PROGRAM WAGES      73/172  OPT=0              FTN 5.1+528        82/06/19. 09.51.41        PAGE    1

      1              PROGRAM WAGES
      2       C AUTHOR: RICH NEWMAN    DATE: AUGUST 8, 1990
      3       C THIS PROGRAM GENERATES A REPORT SHOWING THE PAY AND THE INSURANCE DEDUCTIONS
      4       C FOR A GROUP OF EMPLOYEES
      5       C
      6       C VARIABLES:  ID = EMPLOYEE NO.                DEDUCT = DEDUCTIONS
      7       C             HOURS = NO. OF HOURS WORKED       PAD = PAY AFTER DEDUCTIONS
      8       C             RATE = WAGE RATE                  TOTPAY = TOTAL PAY
      9       C             KODE = INDICATES NO. OF DEPENDENTS TOTDED = TOTAL DEDUCTIONS
     10       C             PAY = GROSS PAY                   TOTPAD = TOTAL PAY AFTER DEDS.
     11       C
     12       C PRINT THE HEADINGS, STARTING ON A NEW PAGE
     13              WRITE (6,10)
     14       10     FORMAT ('1',59X,'XYZ COMPANY'///14X,'EMPLOYEE #',5X,'HOURS WORKED'
     15             1       ,'WAGE RATE',5X,'DEPENDENTS CODE',5X,'GROSS PAY',5X,
     16             2       'DEDUCTIONS',5X,'PAY AFTER DEDUCTION'//)
     17       C
     18       C INITIALIZE THE TOTALS TO ZERO
     19              TOTPAY = 0.0
     20              TOTDED = 0.0
     21              TOTPAD = 0.0
     22       C
     23       C THIS IS THE BEGINNING OF THE LOOP
     24       25    READ (5,20) ID, HOURS, RATE, KODE
     25       20    FORMAT (I5,1X,F5.2,1X,F5.2,1X,I1)
     26       C
     27       C READING A NEGATIVE ID NUMBER TERMINATES THE LOOP
     28              IF (ID) 900, 30, 30
     29       C
     30       C CALCULATE THE PAY, (OVER 40 HOURS IS PAID ONE AND A HALF TIMES THE RATE)
     31       30    IF (HOURS .LE. 40.0) THEN
     32                  PAY = HOURS * RATE
     33              ELSE
     34                  PAY = HOURS * RATE * 1.5
     35              ENDIF
     36       C
     37       C CALCULATE THE DEDUCTIONS:  10%, 11%, OR 12% OF THE PAY DEPENDING ON THE CODE
     38              IF (KODE .GE. 3) GO TO 70
     39              GO TO (50,60), KODE
     40       50    DEDUCT = PAY * .10
     41              GO TO 80
     42       60    DEDUCT = PAY * .11
     43              GO TO 80
     44       70    DEDUCT = PAY * .12
     45       C
     46       C CALCULATE THE PAY AFTER DEDUCTIONS, (NET PAY)
     47       80    PAD = PAY - DEDUCT
     48       C
     49       C CALCULATE THE TOTALS
     50              TOTPAY = TOTPAY + PAY
     51              TOTDED = TOTDED + DEDUCT
     52              TOTPAD = TOTPAD + PAD
     53       C
     54       C WRITE THE INFORMATION, (DATA AND CALCULATIONS)
     55              WRITE (6,90) ID, HOURS, RATE, KODE, PAY, DEDUCT, PAD
     56       90    FORMAT (/6X,I6,9X,F7.2,9X,F6.2,14X,I1,13X,F7.2,9X,F5.2,13X,
     57             1       F7.2/)
     58       C
     59       C END OF THE LOOP
     60              GO TO 25
     61       C
     62       C WRITE THE TOTALS
     63       900   WRITE (6,100) TOTPAY, TOTDED, TOTPAD
     64       100   FORMAT (//61X,'TOTAL PAY TO ALL =',F9.2//75X,'TOTAL DEDUCTIONS =',
     65             1       F9.2//81X,'THE TOTAL PAY AFTER DEDUCTIONS',F9.2)
     66              WRITE (6,150)
     67       150   FORMAT (///56X,'END OF REPORT')
     68              END
```

Figure 4.13 The second run of the sample program (continued on next page).

was found in calculating the pay. For example, if an employee worked 60 hours, and the wage rate is 6.75, the gross pay must be

$$40 \times 6.75 + 1.5 \times 20 \times 6.75 = 472.50;$$

however, the output shows the total pay equal to $607.50. This kind of error is difficult to detect without test data. With the test data, one would notice immediately that there is something wrong with one of the formulas. The formulas are checked, and an error is found in the calculation of gross pay with overtime (Figure 4.13, line 34). This formula is corrected to read

$$PAY = 40.0 * RATE + 1.5 * (HOURS - 40.0) * RATE$$

```
1         PROGRAM WAGES      73/172  OPT=0                      FTN 5.1+528       82/06/19. 09.51.41      PAGE     2

  --VARIABLE MAP--(LO=A)
   -NAME---ADDRESS--BLOCK-----PROPERTIES-------TYPE---------SIZE    -NAME---ADDRESS--BLOCK-----PROPERTIES-------TYPE---------SIZE

    DEDUCT   237B                             REAL                   PAY      236B                            REAL
    HOURS    233B                             REAL                   RATE     234B                            REAL
    ID       232B                             INTEGER                TOTDED   230B                            REAL
    KODE     235B                             INTEGER                TOTPAD   231B                            REAL
    PAD      240B                             REAL                   TOTPAY   227B                            REAL

  --STATEMENT LABELS--(LO=A)
   -LABEL-ADDRESS-----PROPERTIES----DEF     -LABEL-ADDRESS-----PROPERTIES----DEF     -LABEL-ADDRESS-----PROPERTIES----DEF

      10    112B    FORMAT       14           50    45B                  40           90   137B    FORMAT      56
      20    133B    FORMAT       25           60    51B                  42          100   146B    FORMAT      64
      25     14B                 24           70    55B                  44          150   162B    FORMAT      67
      30    INACTIVE             31           80    60B                  47          900    74B                63
```

```
1                                              XYZ COMPANY

           EMPLOYEE #   HOURS WORKED   WAGE RATE   DEPENDENTS CODE    GROSS PAY    DEDUCTIONS    PAY AFTER DEDUCTION

             56298         38.50         12.50           2              481.25        52.94            428.31

             69820         52.50          9.50           1              748.13        74.81            673.31

             56210         29.25         11.50           3              336.38        40.37            296.01

             62890         40.00         15.78           2              631.20        69.43            561.77

             26850         60.00          6.75           1              607.50        60.75            546.75

             66268         72.00          8.36           3              902.88        *****            794.53

             78256         58.25          5.20           3              454.35        54.52            399.83

                                            TOTAL PAY TO ALL =   4161.68

                                                     TOTAL DEDUCTIONS =   461.16

                                                        THE TOTAL PAY AFTER DEDUCTIONS  3700.52

                                            END OF REPORT
```

Figure 4.13 (continued)

Finally, the ****** for the deduction figure is an indication of too small a field. This also could be the result of the previously mentioned error. The field descriptor is changed from F5.2 to F6.2 anyway. The final run and the output are shown in Figure 4.14.

STEP 9: DOCUMENTATION

This is a very simple, straightforward program. The documentation would be prepared as follows:

1. Summary of steps 1 through 4.
2. The flowchart.
3. The coded program which includes internal comments.
4. A report about how to place the data, how to use the program, and the limitations of the program.

162 Decision Making, Comparing, and Branching

```
1              PROGRAM WAGES
2      C AUTHOR:  RICH NEWMAN     DATE:  AUGUST 8, 1990
3      C THIS PROGRAM GENERATES A REPORT SHOWING THE PAY AND THE INSURANCE DEDUCTIONS
4      C FOR A GROUP OF EMPLOYEES
5      C
6      C VARIABLES:  ID = EMPLOYEE NO.               DEDUCT = DEDUCTIONS
7      C             HOURS = NO. OF HOURS WORKED     PAD = PAY AFTER DEDUCTIONS
8      C             RATE = WAGE RATE                TOTPAY = TOTAL PAY
9      C             KODE = INDICATES NO. OF DEPENDENTS  TOTDED = TOTAL DEDUCTIONS
10     C             PAY = GROSS PAY                 TOTPAD = TOTAL PAY AFTER DEDS.
11     C
12     C PRINT THE HEADINGS, STARTING ON A NEW PAGE
13             WRITE (6,10)
14       10    FORMAT ('1',62X,'XYZ COMPANY'///14X,'EMPLOYEE #',5X,'HOURS WORKED'
15          1        ,5X,'WAGE RATE',5X,'DEPENDENTS CODE',5X,'GROSS PAY',5X,
16          2        'DEDUCTIONS',5X,'PAY AFTER DEDUCTIONS'//)
17     C
18     C INITIALIZE THE TOTALS TO ZERO
19             TOTPAY = 0.0
20             TOTDED = 0.0
21             TOTPAD = 0.0
22     C
23     C THIS IS THE BEGINNING OF THE LOOP
24       25    READ (5,20) ID, HOURS, RATE, KODE
25       20    FORMAT (I5,1X,F5.2,1X,F5.2,1X,I1)
26     C
27     C READING A NEGATIVE ID NUMBER TERMINATES THE LOOP
28             IF (ID) 900, 30, 30
29     C
30     C CALCULATE THE PAY, (OVER 40 HOURS IS PAID ONE AND A HALF TIMES THE RATE)
31       30    IF (HOURS .LE. 40.0) THEN
32                 PAY = HOURS * RATE
33             ELSE
34                 PAY = 40.0 * RATE + 1.5 * (HOURS - 40.0) * RATE
35             ENDIF
36     C
37     C CALCULATE THE DEDUCTIONS:  10%, 11%, OR 12% OF THE PAY DEPENDING ON THE CODE
38             IF (KODE .GE. 3) GO TO 70
39             GO TO (50,60), KODE
40       50    DEDUCT = PAY * .10
41             GO TO 80
42       60    DEDUCT = PAY * .11
43             GO TO 80
44       70    DEDUCT = PAY * .12
45     C
46     C CALCULATE THE PAY AFTER DEDUCTIONS, (NET PAY)
47       80    PAD = PAY - DEDUCT
48     C
49     C CALCULATE THE TOTALS
50             TOTPAY = TOTPAY + PAY
51             TOTDED = TOTDED + DEDUCT
52             TOTPAD = TOTPAD + PAD
53     C
54     C WRITE THE INFORMATION, (DATA AND CALCULATIONS)
55             WRITE (6,90) ID, HOURS, RATE, KODE, PAY, DEDUCT, PAD
56       90    FORMAT (16X,I6,9X,F7.2,9X,F6.2,14X,I1,13X,F7.2,8X,F6.2,13X,
57          1         F7.2/)
58     C
59     C END OF THE LOOP
60             GO TO 25
61     C
62     C WRITE THE TOTALS
63      900    WRITE (6,100) TOTPAY, TOTDED, TOTPAD
64      100    FORMAT (//50X,'TOTAL PAY TO ALL =',F8.2//50X,'TOTAL DEDUCTIONS =',
65          1         F7.2//36X,'THE TOTAL PAY AFTER DEDUCTIONS =',F8.2)
66             WRITE (6,150)
67      150    FORMAT (///61X,'END OF REPORT')
68             END
```

XYZ COMPANY

EMPLOYEE #	HOURS WORKED	WAGE RATE	DEPENDENTS CODE	GROSS PAY	DEDUCTIONS	PAY AFTER DEDUCTIONS
56298	38.50	12.50	2	481.25	52.94	428.31
69820	52.50	9.50	1	558.13	55.81	502.31
56210	29.25	11.50	3	336.38	40.37	296.01
62890	40.00	15.78	2	631.20	69.43	561.77
26850	60.00	6.75	1	472.50	47.25	425.25
66268	72.00	8.36	3	735.68	88.28	647.40
78256	58.25	5.20	3	350.35	42.04	308.31

TOTAL PAY TO ALL = 3565.48

TOTAL DEDUCTIONS = 396.12

THE TOTAL PAY AFTER DEDUCTIONS = 3169.36

END OF REPORT

Figure 4.14 The listing and the final output of the sample program.

EXERCISES

4.1 Find any errors in the following statements:

- **a.** IF (A .EQ. N) GO TO 30
- **b.** IF (U = G) STOP
- **c.** IF (X .LT. 30.5) 10 FORMAT (1X,F10.1)
- **d.** IF,(AREA .EQ. 0.0) GO TO 9
- **e.** IF (M .LT. 5), RATE = .05
- **f.** IF (X - Y) GO TO 30
- **g.** IF (A .LT. B) 20, 30, 40
- **h.** IF (X GT Y) A = B*C
- **i.** IF (I .EQ. 95.5) GO TO 55
- **j.** IF (X OR Y .LT. 5.6) A = B
- **k.**
  10 IF (U .EQ. Q) FORMAT (1X,'EQUAL')
- **l.** IF (X .LT. POP) RATE = 0.0, STOP
- **m.** IF (M .EQ. X) GO TO 9
- **n.** IF,(V .LT. W), WRITE (6,50) V,W
- **o.** IF (A .LT. 10.0) G = .05, N = N + 1
- **p.** IF (AMT .LT. 1000.0) THEN RATE = .05
- **q.** IF (X .LT. 99.99) RATE = 5.5, GO TO 99
- **r.** IF (X .GT. Y) THEN X = Y
      ELSE X = A
  END IF
- **s.**
  IF (A .EQ. B) THEN
      V = W + X
      GO TO 5
  ELSE
      X = Y
  5 END IF
- **t.** IF (X .LE. Y) THEN U = W + X, DIS = R*A
      P = P + Q ELSE
      U = X + V END IF
- **u.** GO TO (M,N,L),I
- **v.** GO TO, 10,20,30,I
- **w.** IF (A) GO TO (20,30,40) K
- **x.** IF (100,200,300), I*K

4.2 Find the logical errors in the following programs, or program segments.

a.

```
         PROGRAM DDD
10       FORMAT (F5.1)
         READ (5,10) A
         IF (A .LT. 1000.0) DISC = A*.05
         DISC = A*.1
         WRITE (6,20) A,DISC
20       FORMAT (11X,'A=',F6.1,'DISCOUNT =',F4.2)
         END
```

b.

```
         PROGRAM SMALL
20       FORMAT (2F6.2)
         READ (5,20) A,B
         IF (A .LT. B) WRITE (6,30) A
         WRITE (6,30) B
30       FORMAT (11X,'THE SMALLER NO. IS', F7.2)
         END
```

c.

```
         IF (XXX)100,200,300
100      WRITE (6, 10) A
200      WRITE (6, 20) B
300      WRITE (6, 30) C
```

d.

```
         GO TO (10,20,30),I
10       A = B*C
20       A = Y + X
30       A = X*Y
```

e.

```
         IF (A .LE. 10.0) GO TO 32
         B = A*A
32       B = X*X
         WRITE (6,33) B
```

f.

```
         IF (AMT .LT. 1000.0) RATE = .05
         RATE = .08
         DIS = AMT*RATE
         WRITE (6,10) AMT, RATE, DIS
```

4.3 Assume K = 2, A1 = 2.0, A2 = 5.0, and A3 = −4.0; to which statement number will the control be transferred in each of the following cases:

a.

```
         IF (A1 .LT. A2) GO TO 10
         GO TO 30
```

b.

```
         IF (A3 .LT. A1) GO TO 20
         GO TO 40
```

c.
```
    IF (A1 + A2 .GT. A3) GO TO 10
    GO TO 50
```
d.
```
    IF (A2 .LT. A1 + A3) GO TO 30
    GO TO 60
```
e.   `IF (A1 + A2) 10,20,30`

f.   `IF (A1 + A3) 40,50,60`

g.   `IF (A1 + A2 + A3) 70,80,90`

h.   `IF (A1 - A2 - A3) 5,10,25`

i.   `IF (A1*A2 + A3) 9,10,11`

j.
```
       GO TO (20,10,30),A3
    40 ...
```
k.
```
    L = K*K
    GO TO (50,50,70,80),L
```
l.
```
    M = ((A1 + A2)/K) - 2
    GO TO (90,80,70,60,50) M
```
m.   `GO TO (110,90,60,130) (A1 + A2)/K`

n.
```
    IF (A1 .GT. A2) GO TO 25
    GO TO 80
```
o.
```
    IF (A3 .LT. A1) GO TO 10
    GO TO 20
```

PROGRAMMING EXERCISES

Write a complete FORTRAN program for each of the following problems. Assume more than one data record in each of the problems; use appropriate loops, headings, spaces, and FORMATs.

4.4 Write a program which calculates the grade code for students in a course. The input is the student's ID number and the average of the exam scores. The grade is determined as:

Average	Code
0 to 60	0 (for F)
more than 60 to 70	1 (for D)
more than 70 to 80	2 (for C)
more than 80 to 90	3 (for B)
more than 90 to 100	4 (for A)

Print the ID, average, the grade code for each student, and the number of students with each grade code at the end. Write the segment of the program which calculates the grade code with:

a. Logical IF statement,
b. IF-THEN-ELSE statements,

166 Decision Making, Comparing, and Branching

      c. Arithmetic IF statement,
      d. Computed GO TO statement,

    and then compare them. Which method do you find easiest?

4.5 Write a program which reads the IDs and test scores for several students as input. Find and print the ID and the score for the student who had the largest score.

4.6 Write a program which reads values for four variables (all on one line) *A, B, C,* and *D,* and prints the following:

    EQUAL   if $A/B = C/D$,
    NOT EQUAL   if $A/B = C/D$,
    UNDEFINED   if either *B* or *D* is equal to zero.

4.7 Write a program that inputs the value of a variable, and then outputs the absolute value of that variable. The *absolute value* of a number is the number without regard to its sign. For example, the absolute value of -3 is 3, and the absolute value of 4 (i.e., $+4$) is 4.

*4.8 Write a program using arithmetic IF, to calculate the roots of the equation:

$$AX^2 + BX + C = 0$$

Hint: *A, B,* and *C* are the input (real numbers). If $A = 0$ (and $B \neq 0$) the root of the equation is:

$$X = -C/B$$

Otherwise the equation has two roots. The roots are calculated by

$$X_1 = \frac{-B + \sqrt{B^2 - 4AC}}{2A}$$

$$X_2 = \frac{-B - \sqrt{B^2 - 4AC}}{2A}$$

Notice that if $B^2 - 4AC$ is negative, the equation does not have a real root, and if $B^2 - AC$ is zero, the equation has two identical roots

$$X = -\frac{B}{2A}$$

4.9 There are several data records; each contains the following information about students in a course:

    Column 1–6   ID number
    Column 8–12   Test score #1
    Column 14–18   Test score #2

Write a program which reads the information and prints the following for each student:

1. a sequence number
2. the students ID
3. the test scores
4. his/her average
5. the grade code (like Exercise 4.4)

Print also the ID number and the average of the student who had the highest average and the student who had the lowest average at the end.

* Answers to starred exercises are provided at the end of the chapter.

Make sure to design the input and output, and draw a flowchart, before writing the program.

4.10 Write a program which reads a real number as an amount A, then calculates and prints the discount based on the following rates:

Amount	Discount Rate
A ≤ 200	6%
200 < A ≤ 500	8%
500 < A ≤ 1000	10%
A > 1000	12%

Design the output.

*4.11 Write a program which reads several data records. On each record there is a value less than 100.0. The first data record is a header record indicating the number of data. Calculate and print the average, the largest, the smallest, and the variance of the data. The variance (VAR) can be calculated by the following formula:

$$VAR = \frac{\text{sum of the squares of the numbers}}{\text{number of data}} - (\text{average})^2$$

4.12 The tuition in a college is calculated as follows:

Up to 11 hours	$35 per hour
12 to 16 hours	$420
Over 16 hours	$420 + $30 per hour over 16 hours

Write a program which reads the ID of the students and the number of hours taken, and calculates the tuition. Print the ID and the tuition. Use the following input data to test the program.

ID	Number of Hours Taken
56291	10
96213	12
32152	15
93215	9
32131	18

Write the segment of the program which calculates the tuition with three styles explained in the text:

a. GO TO style
b. GO TO-less style and
c. structured style

4.13 The commission paid to a salesperson is calculated as:

sales ≤ $1000	2% of sales
$1000 < sales ≤ $5000	3% of sales
$5000 < sales ≤ $10,000	4% of sales
sales > $10,000	5% of sales

Write a program which reads the ID of a salesperson and the amount sold. Calculate and print the ID, sales, commission rate, and commission for each sale. Assume there are several data records. Terminate the program if the ID is 99999.

4.14 The tax withholding is calculated based on the following table.

Gross (Yearly)	Tax Rate
Less than 5000	0
$5000 to 10,000	5% of excess
$10,000 to 15,000	8% of excess
$15,000 to 20,000	12% of excess
$20,000 to 30,000	17% of excess
more than $30,000	25% of excess

Notice that the tax deduction is progressive; that is, if for example the employee's income is $18,000, the tax is

TAX = 5000.0*0.0 + 5000.0*.05 + 5000.0*.08 + 3000.0*.12

Write a program which calculates the tax deduction for the employees of a company. Input: the employee's ID number and the yearly gross income. Output: ID number, gross income, tax deduction, and net pay to each employee. Print the total of the gross incomes, deductions, and net pay at the end. Use a trailer record to terminate the loop.

4.15 Write a program that reads in ID number, hours, and wage rate. Calculate the pay. Overtime (over 40 hours) is paid 1.5 times the regular rate. Print the ID number, hours, wage rate, and the pay to each employee. Print an asterisk next to the employee's pay if the number of hours worked is more than 60 (to get management's attention). Also, print the total pay to all employees at the end. Use a trailer record for terminating the loop.

4.16 A construction company pays employees weekly according to the following formula:

PAY = HOURS*RATE + EXTRA   if HOURS ≤ 40,
PAY = 40.0*RATE + 1.5*(time over 40 hours)*RATE + EXTRA
                                                                    otherwise

The EXTRA is based upon the number of years the employee has worked for the company:

0–5 years	$50
6–10 years	$75
11–20 years	$100
Over 20 years	$125

Write a program which generates the pay report. The input is the employee's number, hours worked, number of years employed, and base wage rate. The output is the input information and the pay to each employee. Print also the total pay, the total overtime, and the total pay for overtime at the end. There are less than 200 employees. The last record contains −9999 for the employee's number.

4.17 Write a program for the library that will count how many books were checked out altogether and how many from each section in a day. The input is the book number, the section code (1, 2, 3, 4, or 5), and the due date (punched as 02-02-89). The output is:

1. The book number, the section code, and the due date for each book.
2. The total number of books checked out.
3. The number of books checked out from each section.

Hint: Read and print the date as three separate integer numbers with hyphens between them.

4.18 Write a program for students' grade information.

Input:
- **a.** Student's ID number — Columns 1–7
- **b.** A code for sex — Column 9
  - 1 for male
  - 2 for female
- **c.** A code for class — Column 11
  - 1 for undergraduate
  - 2 for graduate
- **d.** Scores of four exams:
  - Exam #1 — Columns 13–17
  - Exam #2 — Columns 19–23
  - Exam #3 — Columns 25–29
  - Exam #4 — Columns 31–35

Output: For each student, in report form, a sequence number followed by the ID, the sex code, the class code, the exam scores, the average of the scores, and the grade code based on:

- 4 (for A)   between 90–100
- 3 (for B)   between 80–90
- 2 (for C)   between 70–80
- 1 (for D)   between 60–70
- 0 (for F)   less than 60

and at the end of the report:

The number of A's, B's, C's, D's, and E's.
The grand average for all students.
The average for the graduate students.
The average for the undergraduate students.

4.19 The end-of-year bonus to the employees of the GHM Company is based upon the following plan:

5% of the income for single employees (KODE = 1)
7% of the income for married, no children (KODE = 2)
9% of the income for married with children (KODE = 3)

In addition, the following lump sum is added to the bonus:

$1,000 if the income is less than $10,000
$750 if the income is 10,000 to 20,000
$500 if the income is more than $20,000

Write a program which generates the bonus report and the totals. Design the inputs and outputs. Follow the general programming cycle explained in chapter three. You may use IDs to identify the employees.

4.20 The Consumer Gas Company charges its customers by the number of units used (hundreds of cubic feet) during the billing period according to the following schedule:

1. For residential use (KODE = 1):

0 to 50.00 units at $.55 per unit
more than 50 to 100.00 units at $.49 per unit
more than 100 to 500.00 units at $.39 per unit

more than 500 to 800.00 units at $.30 per unit
more than 800 units at $.26 per unit.

2. For industrial use (KODE = 2):

0 to 400 units at $.49 per unit
more than 400 to 1000 units at $.35 per unit
more than 1000 units at $.29 per unit

The charges are progressive. That is, if a residential customer used, for example, 120 units, the charge is

CHARGE = 50 × .55 + 50 × .49 + 20 × .39

Write a program to generate a report for customer's charges. Design the input and outputs. You may use customer's ID, last readings, and current readings as input. Make sure to print the totals at the end.

SUMMARY OF CHAPTER 4

You have learned:

1. An IF statement can be used for comparison purposes. Several forms of the IF statement are as follows:
   a. The logical IF statement has the general form

   IF (X .GT. Y)   an executable FORTRAN statement

   where X or Y can be a variable, a constant, or an arithmetic expression. We can use not only .GT. (for greater than), but also other *relational operators* such as .LT., .GE., .LE., .EQ., and .NE. The statement to the right of the parentheses will be executed only while the expression in the parentheses remains true. If it is not true, the statement will be ignored. Only one statement can be put on the right of the parentheses. If it is necessary to write several statements, one must use a GO TO statement to branch to a block of statements.
   b. The structured IF block has the general form of

   IF (a logical expression) THEN
       ⋮
   ELSE
       ⋮
   ENDIF

   If the expression in the parentheses is true, the statements between THEN and ELSE will be executed; otherwise the statements after ELSE will be executed.
   c. The arithmetic IF statement has the general form of

   IF (an arithmetic expression) N_1, N_2, N_3

   where N_1, N_2, and N_3 are statement labels. Control will be transferred to statement N_1, N_2, or N_3 according to whether the expression is negative, zero, or positive, respectively.
2. A computed GO TO statement has the general form of

   GO TO (N_1, N_2, N_3, ..., N_n), I

The control will be transferred to $N_1, N_2, N_3, ..., N_n$ according to whether the value of I (a variable or an arithmetic expression) is 1, 2, 3, ..., or n. The computed GO TO is suitable when the control must be transferred to a different part of a program, depending on the value of a variable. It is especially useful when the variable is a code which can take sequential values, and when different calculations must be done for each value of the code.

3. A header record is a record at the beginning of the file which shows how many data records will follow. A trailer or sentinel value is a record at the end of the data file, indicating the end of the data by a number which is out of the range of the regular data. Either a header or a trailer record can be used to terminate a loop.

4. Some typical examples of the use of IF or computed GO TO statements include the following:

   - Calculating discounts for different rates, commissions for different criteria, grades for different standards, and the outcomes of an equation for different values
   - Terminating loops
   - Finding the largest or the smallest number among a set of data
   - Solving other problems which require comparison and decision making

5. You also learned that:

   - A good programming style is one which is easy to follow and understand.
   - Using GO TO statements too frequently makes a program unstructured and hard to follow.
   - Indentation is necessary for the readability of a program.
   - Structured programming, utilizing structured IF blocks, makes a program more readable, easier to code, and less time consuming.

SELF-TEST REVIEW

4.1 Find the errors in the following program segments or statements:

a. IF (A = B) GO TO 10

b. IF (A .LE. X) 10 FORMAT (11X,F6.2)

c. IF (Z EQ B) WRITE (6,20)Z

d. IF (C .NE. G), STOP

e. IF (I .EQ. 5.5) STOP

f. IF (A .LE. 100.0) RATE = .05, DIS = A*RATE

g.
```
        IF (A .EQ. B) X = A*B
            WRITE (6,10) X
            GO TO 30
        ELSE
            Y = A + B
            WRITE (6,20) Y
            GO TO 30
```

```
            END IF
30      ...
    h.
            IF (X .LT. Y) THEN
                    P = RATE
            ELSE
                IF(Z .EQ. 9.5) THEN
                        P = Q
                ELSE
                        P = S
            END IF
    i.
            IF (M .EQ. I) GO TO 45
                ⋮
            IF (A .LE. X) THEN
                    G = A*B
45                  N = N + 1
            ELSE
                    P = Q
            END IF
30      ...
```

j. IF (AMT) GO TO 10,20,30

k. IF (A*5*B) 10,20,30,40

l. IF (X .EQ. Y) 10,20,30

m. GO TO (10,20,30) IF I = 1,2,3

n. IF (10,20,30), K

4.2 Write the segment of a program which calculates D = A * RATE, where

RATE is .10 if A is less than 200
RATE is .20 if A is up to 300
RATE is .30 if A is more than 300

Use:

a. Logical IF statement
b. Structured IF block
c. Computed GO TO
d. Arithmetic IF

4.3 A company pays salary commissions to its salespersons based upon the following plan:

$500 plus 6% of the amount sold if the amount is less than $5000,
$750 plus 7% of the amount sold if the amount is less than $10,000 (but more than $5000),
$1000 plus 8% of the amount sold if the amount is more than $10,000.

Write a complete program, using headings, appropriate loops, and a trailer record, to calculate and print the salary commission for the salespersons of the company. Print totals at the end. Follow the general programming cycle explained in Chapter 3. Use the salesperson's ID and monthly amount as input.

☐ Answers

4.1

    a. The correct form is: IF (A .EQ. B) GO TO 10
    b. FORMAT is not an executable statement.
    c. The periods before and after the relational operator EQ is missing.
    d. There must not be a comma after the parentheses.
    e. Both variables in the parentheses should have the same mode.
    f. You cannot write two statements next to the parentheses.
    g. The word THEN is missing. The statement X = A * B must start on a new line. GO TO 30 is not necessary.
    h. The word ENDIF for the inner IF is missing.
    i. Control cannot be transferred into an IF block.
    j. The GO TO must not be there.
    k. There must be only three statement numbers after the parentheses.
    l. A mixture of logical IF and arithmetic IF cannot be used.
    m. The correct form is: GO TO (10,20,30), I
    n. The correct form is: GO TO (10,20,30), K

4.2

    a.

```
           IF (A .LE. 200.0) GO TO 10
           IF (A .LE. 300.0) GO TO 20
                 RATE = .30
           GO TO 50
  10             RATE = .10
           GO TO 50
  20             RATE = .2
  50       D = A*RATE
            :
```

    b.

```
           IF (A .LE. 200.0) THEN
                       RATE = .10
           ELSE IF (A .LE. 300) THEN
                       RATE = .20
                ELSE
                       RATE = .30
           END IF
           D = A*RATE
            :
```

    c.

```
           IA = A/100
           GO TO (10,20,30),IA
  10             RATE = .10
           GO TO 50
  20             RATE = .20
           GO TO 50
  30             RATE = .3
  50       D = A*RATE
            :
```

d.

```
        IF (A - 200.0) 10,10,300
300     IF (A - 300.0) 20,20,30
30         RATE = .3
        GO TO 50
10         RATE = .10
        GO TO 50
20         RATE = .2
50      D = A*RATE
           ⋮
```

4.3 The coded program follows:

```
      PROGRAM CMSION
C XYZ SALARY-COMMISSION REPORT
C
C AUTHOR S.H. WHITE, AUGUST 29, 19--
C VARIABLES ARE: ID: SALESPERSON'S ID
C                AMT: AMOUNT SOLD BY SALESPERSON
C                SAL: SALARY-COMMISSION
C                TAMT: TOTAL AMOUNT
C                TSAL: TOTAL SALARY
C
C HEADINGS
      WRITE (6,10)
10    FORMAT ('1',49X,'XYZ COMPANY'//25X,'SEQ #',3X,
     *'SALESMAN ID #',4X,'THE AMOUNT',5X,
     *'SALARY-COMMISSION'/)
      TAMT = 0.0
      TSAL = 0.0
      N = 0
20    FORMAT (I6,1X,F7.1)
5     READ (5,20) ID, AMT
C IF AMT IS EQUAL TO -99.9, THE LOOP IS TERMINATED
      IF (AMT .EQ. -99.9) GO TO 99
      N = N + 1
      IF (AMT .LE. 5000.0) THEN
            SAL = 500.0 + AMT*.06
      ELSE IF (AMT .LE. 10000.0) THEN
            SAL = 750.0 + AMT*.07
      ELSE
            SAL = 1000.0 + AMT*.08
      ENDIF
      WRITE (6,30) N,ID,AMT,SAL
      TSAL = TSAL + SAL
      TAMT = TAMT + AMT
      GO TO 5
99    WRITE (6,40) TAMT,TSAL
40    FORMAT (35X'TOTAL AMOUNT',3X,F8.1/54X
     *'TOTAL SALARY',3X,F8.2)
30    FORMAT (27X,I3,6X,I7,9X,F8.1,9X,F9.2)
      END
```

ANSWERS TO SELECTED EXERCISES

4.8

```
            PROGRAM ROOTS
C THIS PROGRAM CALCULATES THE ROOTS OF A QUADRATIC
C EQUATION; A, B, C, ARE GIVEN AS INPUT
C
10          FORMAT (3F5.2)
            READ (5,10) A,B,C
            IF (A) 20,100,20
20          DD = B**2 - 4.0*A*C
            IF (DD) 30, 40, 50
30          WRITE (6,35)
35          FORMAT (21X,'NO REAL ROOTS')
            GO TO 150
40          X = -B/(2*A)
            WRITE (6,45) X
45          FORMAT (21X,'THERE ARE TWO IDENTICAL ROOTS OF',
            * F6.2)
            GO TO 150
50          X1 = (-B + DD**.5)/(2.0*A)
            X2 = (-B - DD**.5)/(2.0*A)
            WRITE (6,60) X1,X2
60          FORMAT (11X,'THE FIRST ROOT IS',F6.2,
            * 'THE SECOND ROOT IS',F6.2)
            GO TO 150
100         X = -C/B
            WRITE (6,110) X
110         FORMAT (11X,'THE ROOT OF THE EQUATION IS',
            *F6.2)
150         STOP
            END
```

Note: This program needs appropriate comments, loops, and spacing.

4.11

```
            PROGRAM VARIAN
            SUM = 0.0
            SS  = 0.0
            READ ( 5,10) N
5           READ ( 5,20)X
            SUM = SUM + X
            SS  = SS + X**2
            K = K + 1
40          IF (K .LT. N) GO TO 5
            AVG = SUM/N
            VAR = SS/N - AVG**2
            WRITE ( 6,50) VAR
            STOP
10          FORMAT (I2)
20          FORMAT (F6.2)
50          FORMAT (6X, 'THE VARIANCE IS', F6.2)
            END
```

Note: This program prints only the variance.

Chapter 5
Looping

THE DO LOOP

    DO-Loop Flowcharting
    Rules of a DO Loop
      The Control Variable
      The CONTINUE Statement
      Normal and Abnormal Termination of a DO Loop
      Transfer within and from a DO Loop
      DO Loop and IF Block
      Summary of the Rules about DO Loops
    Nested DO Loops

SELF-TEST REVIEW

As discussed before, automatic repetition of a segment of a program several times is called looping. In all aspects of programming, loops are essential. Every higher-level language offers some methods of looping to facilitate this important feature. The DO loop is the most common method of looping in FORTRAN. We discuss the DO loop in this chapter.

THE DO LOOP

The following section demonstrates two methods of looping. In Example 5.1, a GO TO statement is used. In Example 5.2, the DO loop is presented.

EXAMPLE 5.1

Problem: Write a program which prints a table of two columns. The first column shows the numbers from 1 to 100; the second column shows the square of the numbers in the first column.

Program:

```
          PROGRAM GOTOLP
          N = 0
   55     N = N + 1
          IF(N .GT. 100) GOTO 99
              M = N**2
              WRITE (6,33) N,M
          GO TO 55
   99     STOP
          FORMAT (11X, 'N=', I3, 'M=', I6)
          END
```

Notes:

1. The statement N = N + 1 generates the sequence numbers from 1 to 100. It is used to control the number of executions of the loop.
2. The loop is executed 100 times; an IF statement is used to stop the loop after 100 times. If N, the control variable, is less than 100, the loop is repeated. When N becomes greater than 100, the loop will be terminated.
3. Four statements are essential parts of the loop: (1) initializing the control variable (N = 0), (2) incrementing the control variable (N = N + 1), (3) testing and transferring (IF ... GO TO), and (4) closing the loop (GOTO 55).
4. The range of the loop is from statement 55 up to the GOTO 55 statement.
5. The statements in the loop are indented for readability. As discussed in chapter three, indentation does not affect the logic of the program.

Now Example 5.1 is repeated, but this time a DO loop is used. Compare the following program with the previous one.

> **EXAMPLE 5.2**
>
> ■ **Problem:** The same as Example 5.1: Print the numbers from 1 to 100 and their squares.
>
> **Program:**
>
> ```
>         PROGRAM DLOOP1
>         DO 89 N = 1, 100
>             M = N**2
>             WRITE (6, 33) N, M
> 89      CONTINUE
> 33      FORMAT (11X, 'N=', I3, 'M=', I6)
>         END
> ```
>
> **Notes:**
>
> 1. The two statements DO and CONTINUE substitute for four statements in Example 5.1: (1) N = 0, (2) N = N + 1, (3) IF ... GO TO, and (4) GO TO 55.
> 2. The DO statement performs two important tasks:
>   a. executes a loop for a certain number of times, in this example 100 times.
>   b. increments values of N by 1, starting from 1 up to 100 (there is no need for initializing N to zero). This integer variable is used for the control of the loop.
> 3. The range of the loop is from the DO statement up to the CONTINUE statement, statement 89.
> 4. The statements in the DO loop are indented for readability.

The DO loop in the example looks like the following and has eight important components:

```
        1   2   3   4   5   6   7
        ↓   ↓   ↓   ↓   ↓   ↓   ↓
        DO  89  N   =   1   ,  100
            ⋮
89    · CONTINUE
        ↑
        8
```

1. The word DO.
2. A statement number right after the word DO. This number indicates the end of the loop, the CONTINUE statement.
3. A variable, called the *control variable* or the *index variable*. The control variable is used to control the loop, and its value varies each time the loop is executed.
4. An "=" sign.
5. The initial value for the control variable.
6. A comma.
7. The final value of the control variable.
8. The word CONTINUE, which shows the range of the loop.

Needless to say, components 1, 4, 6, and 8 have to be written exactly as they are in the example, but components 2, 3, 5, and 7 can be chosen by the programmer, based on the need of the programming.

The control variable in the previous example was incremented by one. That is, the value of the control variable was increased by one each time that the loop was executed. The following example shows how the increment can be other than one.

EXAMPLE 5.3

Problem: Write a program which prints a table of two columns. The first column shows *odd* numbers from 11 to 99; the second column shows the square of the numbers in the first column.

Program:

```
         PROGRAM DLOOP2
         DO 10 J = 11, 99, 2
              K = J**2
              WRITE (6, 53) J, K
10       CONTINUE
53       FORMAT (5X, I3, 5X, I6)
         STOP
         END
```

Notes:

1. The starting value of the control variable J is equal to 11, and the loop terminates when the value of J is more than 99.
2. The increment of the control variable is 2, i.e., the value of J will increase by 2 each time that the loop is executed.

Thus, the general form of a DO statement is:

DO *n*, index = m_1, m_2, m_3

where *n* is a statement label, the first comma is optional, the index is a control variable, m_1 is the starting value of the control variable, m_2 is the final value of the control variable, and m_3 is the increment value of the control variable.

Furthermore, a DO statement does the following tasks:

1. Starts a loop by means of the word DO.
2. Establishes the range of the loop, which is from the word DO up to statement *n*.
3. Reexecutes the segment of the program the desired number of times.
4. Increments the control variable by m_3 after each time the loop is executed. If m_3 is 1, its specification can be omitted.
5. Tests the control variable. If the new value is less than or equal to the upper limit, execution will be repeated; if it is greater than the upper limit, the next statement after statement *n* will be executed.

The following are two more examples using a DO loop:

EXAMPLE A

The following program prints numbers 10 through 99 with an interval of 3.

```
         PROGRAM DLOOP3
         DO 125 N = 10, 99, 3
              WRITE (6, 10) N
```

```
125     CONTINUE
10      FORMAT (21X, I3)
        END
```

EXAMPLE B

The following program reads two numbers and prints (1) a sequence number, (2) the data, (3) the product of the data, and (4) the sum of the products at the end of the program. The loop is executed 49 times.

```
        PROGRAM DLOOP4
        DO 120 M = 1, 49
        READ (5, 30) A, B
            C = A*B
            SUM = SUM + C
            WRITE (6, 35) M, A, B, C
120     CONTINUE
        WRITE (6, 50) SUM
30      FORMAT (2F5.1)
35      FORMAT (21X, I2, 5X, F6.2, 5X, F6.2, 5X, F8.2)
50      FORMAT (21X,'THE SUM OF THE MULTIPLICATIONS IS =',
       *F10.2)
        END
```

☐ DO-Loop Flowcharting

We use two symbols to indicate the DO loop, as shown in Figure 5.1a. The first symbol shows the beginning of the loop, and the second one shows the end of the loop. For example, the flowchart for the previous example is shown in Figure 5.2a. An ⌈-shaped box can be used to show a DO loop in a logic-chart (Figure 5.1b).

☐ Rules of a DO Loop

There are certain rules concerning the use of DO loops with which you should become familiar. These rules will be explained through examples in the following section.

The Control Variable

The control variable is tested each time that the DO statement is executed. If m_3 is positive, a DO loop will be executed as long as the value of the control variable is equal to or less than its final value. For example, the following loop will be executed once:

```
        DO 10 I = 1, 1, 1
            WRITE (6, 30) I
10      CONTINUE
```

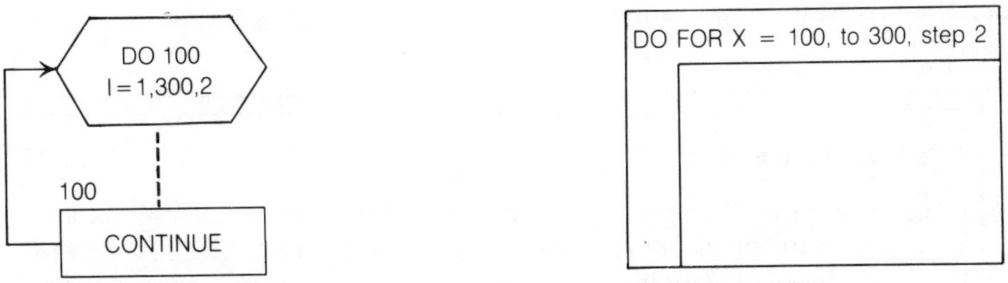

(a) DO flowchart

(b) DO logic-chart

Figure 5.1 A DO loop with a logic-chart

182 Looping

(a) Flowchart   **(b)** Logic-chart

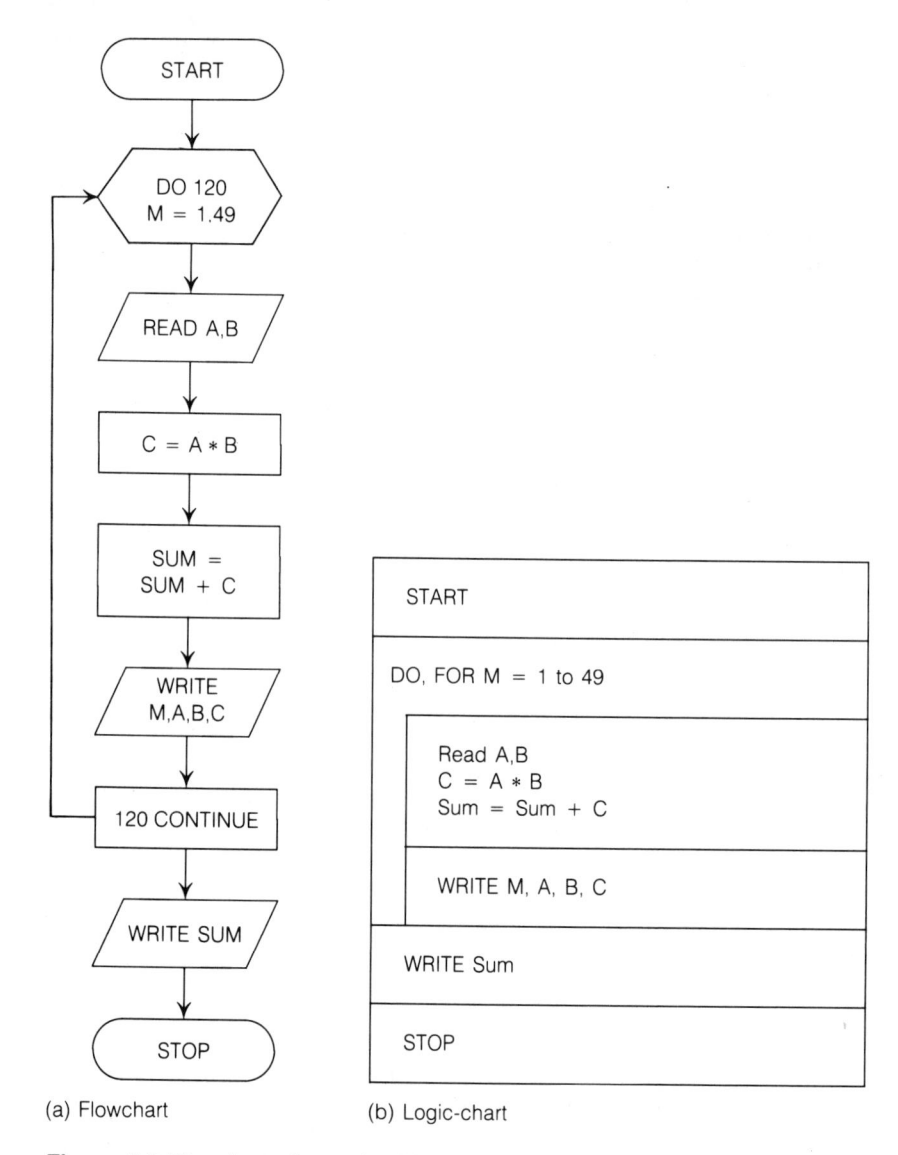

Figure 5.2 The charts for a simple example of a DO loop.

but the following loop will never be executed, because the initial value of the control variable is more than its final value:

```
        DO 10 I = 2, 1, 1
            WRITE (6, 30) I
10      CONTINUE
```

Furthermore, the initial, final, and increment values are not limited to being constants (numbers). All or any of the values can be variables if they are defined before the loop. The following is an example.

EXAMPLE 5.4

■ **Problem:** There are several employee data records, each containing the number of hours worked and the wage rate. The first record is a header record indicating the number of employee records. Draw a flowchart and write a complete program which reads the data and prints: (1) the sequence number, (2) the employee number,

(3) the hours worked, (4) the wage rate, and (5) the pay to each employee. Print the total pay of all employees at the end of the report.

The Flowchart: The flowchart is shown in Figure 5.3, below.

Program:

```
      PROGRAM DLOOP4
C PRINTING THE HEADING
      WRITE (6,5)
```

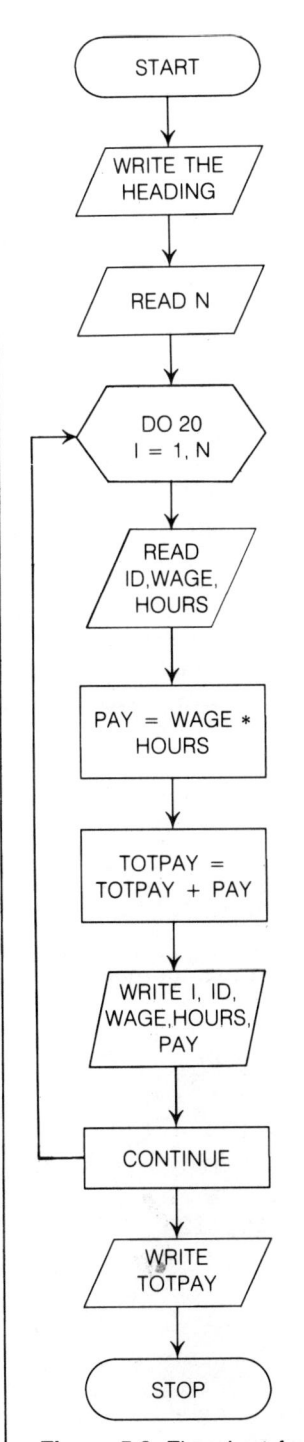

```
                START
                  ↓
            WRITE THE
             HEADING
                  ↓
              READ N
                  ↓
              DO 20
             I = 1, N
                  ↓
               READ
             ID, WAGE,
              HOURS
                  ↓
           PAY = WAGE *
              HOURS
                  ↓
            TOTPAY =
          TOTPAY + PAY
                  ↓
            WRITE I, ID,
           WAGE, HOURS,
               PAY
                  ↓
            CONTINUE
                  ↓
              WRITE
             TOTPAY
                  ↓
               STOP
```

Figure 5.3 Flowchart for Example 5.6.

```
        5       FORMAT (16X,'SEQ. NO.',2X,'EMPLOYEE #',
                *5X,'HOURS',5X,'RATE',25X,'PAY')
C READING THE HEADER RECORD
                READ (5, 10) N
C
                TOTPAY = 0.0
                DO 20 I = 1, N
                    READ (5, 30) ID, WAGE, HOURS
                    PAY = WAGE*HOURS
                    TOTPAY = TOTPAY + PAY
                    WRITE (6, 40) I, ID, WAGE, HOURS, PAY
        20      CONTINUE
                WRITE (6, 50) TOTPAY
        10      FORMAT (I2)
        30      FORMAT (I5,1X,F5.2,1X,F5.2)
        40      FORMAT (18X,I2,6X,I6,7X,F6.2,3X,F6.2,3X,
                *F6.2)
        50      FORMAT (9X, 'TOTAL PAY TO ALL EMPLOYEES =',
                *F7.2)
                STOP
                END
```

Notes:

1. The first READ statement reads the number of records (N).
2. The final value of the control variable in the DO statement is set to N, where the value of N has been previously defined.

In FORTRAN 77, the increment value of a DO statement can be negative as well as positive. If the increment value is positive, the control variable *increases* in value after each loop. This also implies that the initial value of the control variable must be *less* than the final value. But if the increment value is negative, the control variable decreases in value, and the loop will be executed while the value of the control variable is equal to or *greater* than the final value. The following example demonstrates this.

EXAMPLE 5.5

■ **Problem:** Write a program that prints the numbers from 99 to 1.

Program:

```
            PROGRAM DCRES1
            DO 200 I = 1, 99, 1
                J = 100 - I
                WRITE (6, 10) J
    200     CONTINUE
    10      FORMAT (1X, I2)
            STOP
            END
```

or

```
            PROGRAM DCRES2
            DO 200 I = 99, 1, -1
                WRITE (6,10) I
```

```
        200    CONTINUE
        10     FORMAT (1X,I2)
               STOP
               END
```

Note: If the increment value is negative, the control variable decreases in value after each loop, and the final value must be less than or equal to the initial value.

Two more important rules about the control variable are:

1. The value of the control variable cannot be changed within the range of the loop. For example,

```
        DO 10 K = 1, 10, 2
            K = K + 5
10      CONTINUE
```

is not valid.

2. The value of the control variable after a DO loop is completely executed is not always equal to the final value. The value can, however, be predicted and used in FORTRAN 77 compilers. For example, the following program:

```
        DO 100 J = 1, 10, 2
            ⋮
100     CONTINUE
        WRITE (6, 20) J
20      FORMAT (1X, I2)
        END
```

will print 11 as the final value of J. Note that the values of J are 1, 3, 5, 7, 9, and 11. When the value is 11, the loop will not be executed but will be terminated instead.

The CONTINUE Statement

The CONTINUE statement does not have any effect on the loop except to set the end of its range. It is only a dummy statement, and can even be omitted in certain cases, as demonstrated in the following example.

EXAMPLE 5.6

■ **Problem:** Read two numbers from a line, and print the sequence number, the data, and the average of the two numbers. There are 100 lines. Print the grand average at the end.

Program:

```
        PROGRAM NOCONT
        DO 10 J = 1, 100
            READ (5,20) A,B
            AVG = (A + B)/2.0
            SUM = SUM + AVG
10      WRITE (6,30) J, A, B, AVG
        GAVG = SUM/100.0
        WRITE (6,40) GAVG
```

```
    20    FORMAT (2F5.2)
    30    FORMAT (11X,'#',I2,'A =',F6.2,'B =',F6.2,
         *'AVERAGE =', F6.2)
    40    FORMAT (11X,'THE GRAND AVERAGE IS =',F6.2)
          END
```

Notes:

1. The CONTINUE statement is omitted.
2. The range of the loop is from the word DO up to the statement number 10, the WRITE statement.

The CONTINUE statement does nothing. It can be used anywhere in a program with no effect, but it is a convenient ending to a DO loop. Nevertheless, a CONTINUE statement *must be used* if the last statement in a DO loop is (1) a GO TO statement, (2) an arithmetic IF statement, (3) another DO statement, (4) a nonexecutable statement (such as FORMAT), or (5) any statement which contains any of the aforementioned statements. The following rule makes it easy to remember when a CONTINUE statement can be omitted.

> **RULE 5.1**
> *The CONTINUE statement can be omitted if the last statement in the loop is either (1) an arithmetic expression, or (2) an input or output (READ or WRITE) statement.*

It is advisable nevertheless to always end a DO loop with a CONTINUE statement. Some of the reasons are:

- It prevents the possible error of ending a DO loop with one of the nonallowable statements.
- It serves as a delimiter to a DO loop; this helps to standardize the form of the DO loop in a program.
- It provides flexibility to add or delete a statement in a DO loop at a later time.

> **RULE 5.2**
> *Always use a CONTINUE statement as the last statement in a loop.*

Normal and Abnormal Termination of a DO Loop

A DO loop can be terminated without being completely executed. The following is an example.

> **EXAMPLE 5.7**
>
> ■ **Problem:** Test scores for a FORTRAN course are punched on cards (one score per card). The last card is a trailer card with the number −99.9. There are less than 110 cards. Write a program which calculates the average of the scores.

Program:

```
            PROGRAM DLOOP7
            S = 0.0
            L = 0
10          FORMAT (F5.2)
            DO 90 J = 1, 110
                READ (5, 10) A
                IF (A .EQ. -99.9) GO TO 95
                L = L + 1
                SUM = SUM + A
90          CONTINUE
95          AVE = SUM/L
            WRITE (6, 20) AVG
20          FORMAT (6X, 'AVERAGE =', F6.2)
            STOP
            END
```

Notes:

1. The DO loop will be terminated when the value −99.9 is read.
2. L = L + 1 keeps track of the number of scores. We need this to calculate the average.
3. SUM = SUM + A adds up the scores.
4. The average is calculated and printed after the loop is terminated.

There are two ways to terminate a DO loop:

1. *Normal termination:* when the DO loop is terminated after complete execution, where the control will be transferred to the first executable statement after the loop. In this case, after the loop is completely executed, if the increment value is positive, the value of the control variable is greater than its final value; if the increment value is negative, the value of the control variable is less than its final value.
2. *Abnormal termination:* when a GO TO or IF ... GO TO statement transfers the control outside of the loop, regardless of the value of the control variable at the time.

The current value of the control variable after abnormal termination stays in the memory, and can be used in other points further on in the program. For example, the program

```
        DO 100 J = 1, 50
            IF (J .EQ. 3) GO TO 150
        CONTINUE
150     WRITE (6,20) J
20      FORMAT (1X,I2)
        END
```

will print 3, the current value of J. As another example, the average in Example 5.9 could be calculated by

```
    AVG = SUM/(J - 1)
```

Here (J − 1) is used because the last card (the trailer card) must not be counted in calculating the average.

188 *Looping*

Transfer within and from a DO Loop

The control can always be transferred from one point in the DO loop to another point in the same loop. But

> **RULE 5.3**
> *Control cannot be transferred from a point outside a DO loop into the loop.*

Of course the control can be transferred to the beginning of a DO loop from any point in the program at any time. However, remember that by doing this, the control variable will be set to its initial value. For this reason, when you need to transfer control from a point inside the DO loop to the beginning of the loop, you must branch to the CONTINUE statement first; otherwise you may loop indefinitely. As an example, assume we want to write a program which prints numbers 1 to 20 but avoids printing 10. Consider the following loop for this purpose:

```
50      DO 100 I = 1, 20
            IF (I .EQ. 10) GO TO 50
            WRITE (6,70) I
100     CONTINUE
```

This loop is infinite because whenever the control is transferred to the beginning, the control variable will be set equal to 1 (the initial value), and it never reaches 20 (the final value). The loop can be corrected to

```
50      DO 100 I = 1, 20
            IF (I .EQ. 10) GO TO 100
            WRITE (6,70) I
100     CONTINUE
```

The following picture shows some of the valid and invalid uses of GO TO in a DO loop.

Valid

```
10      ┌─DO 50 I = 1, 60
        │     ⋮
        │   ┌─IF (I.EQ.19) GO TO 50
        │   │  ⋮
50      └─CONTINUE ←┘

        ┌─DO 100 I = 1, 200
        │     ⋮
        │   GO TO 30──┐
        │     ⋮       │
30      │   A = B + C←┘
        │     ⋮
100     └─CONTINUE

        ┌─DO 99 K = 1, 300
        │     ⋮
        │   GO TO 50──┐
        │     ⋮       │
99      └─CONTINUE    │
              ⋮       │
50          X = A + B←┘
```

Invalid

```
10      ┌─DO 50 I = 1, 60
        │  ↑  ⋮
        │  └─IF (I.EQ.19) GO TO 10
        │     ⋮
50      └─CONTINUE

            GO TO 30──┐
              ⋮       │
        ┌─DO 100 J = 1, 200
        │     ⋮       │
30      │   A = B + C←┘
        │     ⋮
100     └─CONTINUE

        ┌─DO 99 K = 1, 300
        │     ⋮
50      │   A = B + C←┐
        │     ⋮       │
99      └─CONTINUE    │
              ⋮       │
            GO TO 50──┘
```

```
        ,GO TO 70                              ,GO TO 70
       (    :                                 (    :
    70  ┌DO 50 N = 1, 500              ┌DO 70 N = 1, 500
        │   :                          │   :
    50  └CONTINUE              70  ▼  └CONTINUE
    60   X = Y + Z ◄─────┐              GO TO 60 ─────┐
             :                             :          │
        ┌DO 90 M = 1, 300 │            ┌DO 90 M = 1, 300 │
        │   :             │            │   :             │
        │IF(...) GO TO 60─┘       60   │X = P + Q ◄──────┘
        │   :                          │   :
    90  └CONTINUE                 90   └CONTINUE
```

DO Loop and IF Block

The DO loop and the IF-THEN-ELSE statements can be used together as long as they do not interfere with each other's range. This requires one of them to be completely within the range of the other. The following pictures of valid and invalid uses of DO and IF-THEN-ELSE best explain this:

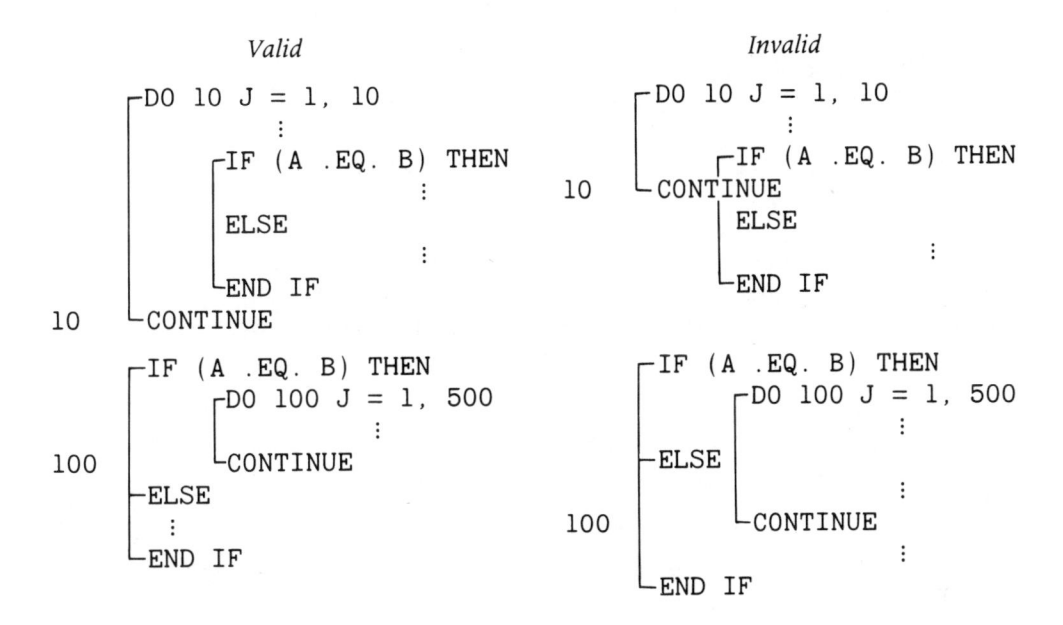

```
              Valid                              Invalid

        ┌DO 10 J = 1, 10                   ┌DO 10 J = 1, 10
        │   :                              │   :
        │   ┌IF (A .EQ. B) THEN             │   ┌IF (A .EQ. B) THEN
        │   │   :                      10  └CONTINUE
        │   │ELSE                          │ELSE
        │   │   :                          │   :
        │   └END IF                        └END IF
    10  └CONTINUE
        ┌IF (A .EQ. B) THEN                ┌IF (A .EQ. B) THEN
        │   ┌DO 100 J = 1, 500              │   ┌DO 100 J = 1, 500
        │   │   :                           │   │   :
   100  │   └CONTINUE                      ├ELSE
        ├ELSE                               │   :
        │   :                          100  │   └CONTINUE
        └END IF                             │   :
                                            └END IF
```

Summary of the Rules about DO Loops

1. The initial value and the final value of the control variable must be present in the DO statement. However, if the increment value is not present, it will be assumed equal to one.
2. In FORTRAN 77, any of the initial, final, and increment values of the control variable (1) can be positive, negative, or zero (except the increment value), and (2) can be a variable or an expression as long as its value is defined.
3. If the increment value in a DO statement is positive, the control variable must be increasing in value. That is, the final value of the control variable must be greater than or equal to the initial value. If the increment is negative, the control variable will be decreasing in value,

and its final value must be less than or equal to the initial value. The increment may not be zero.

4. The value of the control variable cannot be changed inside the DO loop. For example,

```
        DO 9 K = 1, 10
            K = K + 2
            WRITE (6, 20) K
9       CONTINUE
```

is not valid, since K, the control variable, is changed in the range of the loop.

5. If the increment value is positive, a DO loop will be executed as long as the current value of the control variable is less than or equal to its final value.

6. If the increment value is negative, a DO loop will be executed as long as the current value of the control variable is greater than or equal to its final value.

7. If any of the initial, final, or increment values in a DO statement is not defined, the loop will never be executed.

8. The last statement in a DO loop can be either (a) a CONTINUE statement, (b) an arithmetic statement, or (c) an input or output statement. Other statements (such as an IF statement, another DO loop, or a GO TO) are not valid as the last statement in a DO loop. It is recommended to use a CONTINUE statement always as the last statement in a DO loop.

9. It is possible to transfer control *out* of the loop by using a GO TO or IF ... GO TO statement. Control, however, cannot be transferred into a DO loop.

10. To skip the remaining statements within a DO loop, control must always be transferred to the end of the loop (the CONTINUE statement) rather than directly to the beginning (the DO statement). The following is an example:

```
        DO 55 JAY = 1, 10
            IF (JAY .EQ. 5) GO TO 55
            WRITE (6, 20) JAY
55      CONTINUE
```

11. The value of the control variable after the DO loop is completely executed (normal termination) is not always equal to its final value. For instance,

```
        DO 20 N = 1, 6, 2
            ⋮
20      CONTINUE
        WRITE (6, 30) N
30      FORMAT (1X, I2)
        END
```

prints 7 for N. Nevertheless, if control is transferred to the outside of a loop in the case of abnormal termination, the control variable retains its current value.

12. A DO loop and an IF-THEN-ELSE block can be used together, but they must not interfere with each other's range.

For more examples of these rules, look at the following solved problems.

SOLVED PROBLEMS

5.1 Detect the syntax errors, if any, in the following statements:
   a. FOR I = 1, 150 DO 60
   b. DO N = 99, 1, 100, 1
   c. DO 90 N = 20
   d. DO, 56, I = 1, 150
   e. FOR I = 1 TO 100 STEP 2, DO
   f. DO 99, NON = 5,25
   g. DO 85 I = 95, 5
   h. DO 50 INDEX = −1, 100, N**2
   i. DO 80 N = 0, 100, 0
   j. DO 5 K = −5, 100, −2
   k. DO 60 ICONT = 1, N, M
   l. DO 55 J = 2, 100, −1
   m. DO 65, K = 2*N**3/3 , 5*M*N/L , 2*I
   n. DO 99 X = 1, 200

☐ **Answers**

   a. The form of the DO statement must be DO 60 I = 1, 150
   b. The form of the DO statement is not correct. It could be corrected to read DO 99 N = 1, 100, 1.
   c. There must be at least two values after the equal sign: the initial and final values of the control variable.
   d. The comma after the word DO must be omitted. The comma after 56 is optional.
   e. The form must be:

   DO 10 I = 1, 100, 2

   f. This is correct in FORTRAN 77. Note that the comma after 99 is optional.
   g. The initial value must be less than the final value because the increment value is one. (This DO statement is, however, syntactically correct, but the loop will not be executed.)
   h. This is correct.
   i. The increment value of the control variable cannot be zero.
   j. The initial value of the control variable must be greater than the final value (the increment is negative).
   k. Correct, if N and M are previously defined.
   l. Either the increment must be positive, or the initial value must be more than the final value.
   m. This is correct in FORTRAN 77.
   n. The control variable is real, but its values are integers. (This may create execution errors.)

5.2 Are the following DO loops correct? If not, describe the error.

   a.
```
      DO 35 JAY = 1, 55
         ⋮
35    CONTINUE
```

b.

```
        DO 300 K = 1, 100, 2
            ⋮
300     IF (N .LT. 50) GO TO 30
```

c.

```
        DO 100 IX = 1, 200, 2
            ⋮
100     WRITE (6, 20) X, Y
```

d.

```
        DO 200 K = 1, 20
            ⋮
            K = K + 3
            ⋮
200     CONTINUE
```

e.

```
        DO 50 L = 1, 30, 5
        WRITE (6,50) L
50      FORMAT (1X,I2)
```

f.

```
        DO 60 I = 1, 100, 2
            ⋮
            IF (I .EQ. 10) GO TO 9
            ⋮
60      CONTINUE
            ⋮
9       X = X*Y
```

g.

```
            ⋮
        IF (I .EQ. 10) GO TO 25
            ⋮
        DO 90 J = 1, 100, 2
            ⋮
25          K = I + 5
            ⋮
90      CONTINUE
```

h.

```
250     I = J ** 2
            ⋮
        DO 250 K = 1, 50
```

i.

```
10      DO 900 I = 1, 200
            ⋮
            IF (A .EQ. B) GO TO 10
            ⋮
900     CONTINUE
```

j.
```
 50    DO 99 M = 1, 500
          ⋮
       IF (M .EQ. 5) GO TO 50
          ⋮
 99    CONTINUE
```

☐ **Answers**

    a. Correct.
    b. Incorrect: the last statement of a DO loop cannot be an IF ... GO TO statement.
    c. Correct.
    d. Incorrect; the value of the control variable cannot be changed in the range of the loop.
    e. Incorrect; the last statement in a DO loop cannot be a FORMAT statement.
    f. Correct.
    g. Incorrect; control cannot be transferred into a loop.
    h. Incorrect; the DO statement must be the first statement in the loop.
    i. Control must be transferred to the CONTINUE statement (99) instead of to the beginning (10).
    j. Same as **i**: the IF statement must be corrected to

        IF(M .EQ. 5) GO TO 99

■ **5.3**

    a. Is the following program correct?

```
       PROGRAM DLOOP
       N = 0
       I = 0
       DO 35 I = 1, 99
       N = N + 1
       WRITE (6,10) N
 35    CONTINUE
 10    FORMAT(11X, I3)
       END
```

    b. What does the output from the following program represent?

```
       PROGRAM SSS
       IS = 0
       DO 10 L = 1, 20
       IS = IS + L*L
 10    CONTINUE
       WRITE (6, 25) IS
       FORMAT (11X, I5)
       END
```

☐ **Answers**

    a. Yes, but I = 0 is redundant. Also, the statements N = 0 and N = N + 1 are redundant because I does the same job.

b. The sum of the squares of numbers 1, 2, 3, ..., 20. That is, IS = $1^2 + 2^2 + 3^2 + 4^2 + \cdots + 20^2$.

5.4 Write a program which prints a table of two columns, the first column showing numbers from N to 1, while the second column shows their squares. N is to be read at the beginning of the program. Assume the increment value of a DO statement cannot be negative.

☐ **Answer**

```
        PROGRAM DECRES
        WRITE (6, 5)
5       FORMAT (9X, 'THE NO', 4X, 'THE SQUARE')
10      FORMAT (I2)
        READ (5, 10) N
        DO 50 I = 1, N
          J = N + 1 - I
          M = J*J
50      WRITE (6, 20) J, M
20      FORMAT (11X, I3, 7X, I6)
        STOP
        END
```

5.5 As mentioned before in Chapter 4, factorial N (denoted by $N!$) means $1 \times 2 \times 3 \times 4 \times \cdots \times N$. Write a program which reads N and prints the factorial of N.

☐ **Answer**

```
        PROGRAM FACT
10      FORMAT (I2)
        READ (5, 10) N
        NF = 1
        DO 20 I = 1, N
          NF = NF*I
20      CONTINUE
        WRITE (6, 30) N, NF
30      FORMAT (1X, 'FACTORIAL OF', I3,
       *'IS =', I8)
        STOP
        END
```

5.6 Write a program to calculate $1 + 1/2 + 1/3 \ldots 1/N$. N is given as an input.

☐ **Answer**

```
        PROGRAM SERIES
1       READ (5, 10) N
10      FORMAT (I2)
        S = 0.0
        DO 20 J = 1, N
          S = S + 1.0/J
20      CONTINUE
        WRITE (6,30) N, S
```

```
            30      FORMAT (6X, 'FOR N =', I4,
                   *'THE SUM IS = ', F6.4)
                    STOP
                    END
```

■ **5.7** Write a program which prints numbers 1 through 99, three numbers in a row, that is,

```
1   2   3
4   5   6
:   :   :
```

☐ **Answer**

```
            DO 200 I = 1, 99, 3
                    J = I + 1
                    K = I + 2
           200      WRITE (6, 20) I, J, K
            20      FORMAT (21X, I3)
                    END
```

☐ **Nested DO Loops**

Sometimes it is necessary to have one or more loops inside another loop. Loops arranged in such a way are called *nested loops*. The following problem is an example.

EXAMPLE 5.8

■ **Problem:** Given a deck of 100 cards, each bearing a number N, calculate and print the factorial of N for each value (i.e., $1 \times 2 \times 3 \times \cdots \times N$; see Solved Problem 5.5).

Program:

```
            PROGRAM FACT
            DO 10 I = 1, 100
                    READ (5, 20) N
                    NF = 1
                        DO 30 J = 1, N
                            NF = NF*J
            30          CONTINUE
                    WRITE (6, 40) N, NF
            10      CONTINUE
            20      FORMAT (I2)
            40      FORMAT (11X, 'FACTORIAL OF', I2,
                   *'IS =', I10)
                    STOP
                    END
```

Note: There are two loops in the nested loop. The outer loop causes 100 cards to be read; the inner loop calculates and prints the factorial of each number read. The logic-chart for this problem is shown in Figure 5.4.

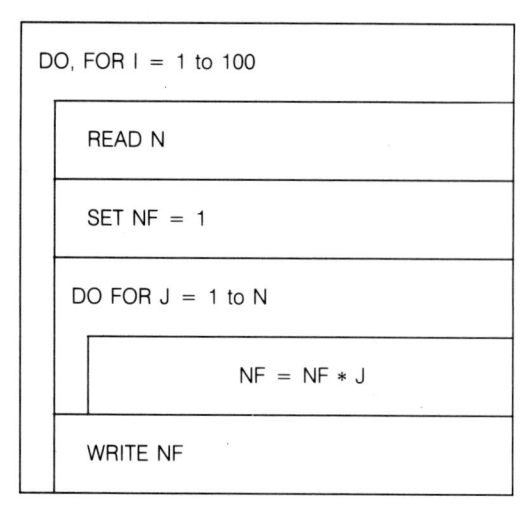

```
┌─────────────────────────────────────┐
│  DO, FOR I = 1 to 100               │
│   ┌─────────────────────────────┐   │
│   │         READ N              │   │
│   ├─────────────────────────────┤   │
│   │        SET NF = 1           │   │
│   ├─────────────────────────────┤   │
│   │   DO FOR J = 1 to N         │   │
│   │   ┌─────────────────────┐   │   │
│   │   │     NF = NF * J     │   │   │
│   │   └─────────────────────┘   │   │
│   ├─────────────────────────────┤   │
│   │        WRITE NF             │   │
│   └─────────────────────────────┘   │
└─────────────────────────────────────┘
```

Figure 5.4 The logic-chart for Example 5.8.

The rules of a DO loop discussed previously apply to any loop in a nested loop. Furthermore, the ranges of the loops must not interfere with each other. That is, the inner loop must lie completely inside the outer loop. For example,

```
          DO 10 I = 1, 30
              ⋮
              DO 20 J = 1, 40
                  ⋮
10        CONTINUE
              ⋮
20        CONTINUE
```

is not valid, because the loops overlap. Figure 5.5 shows some valid nested loops.

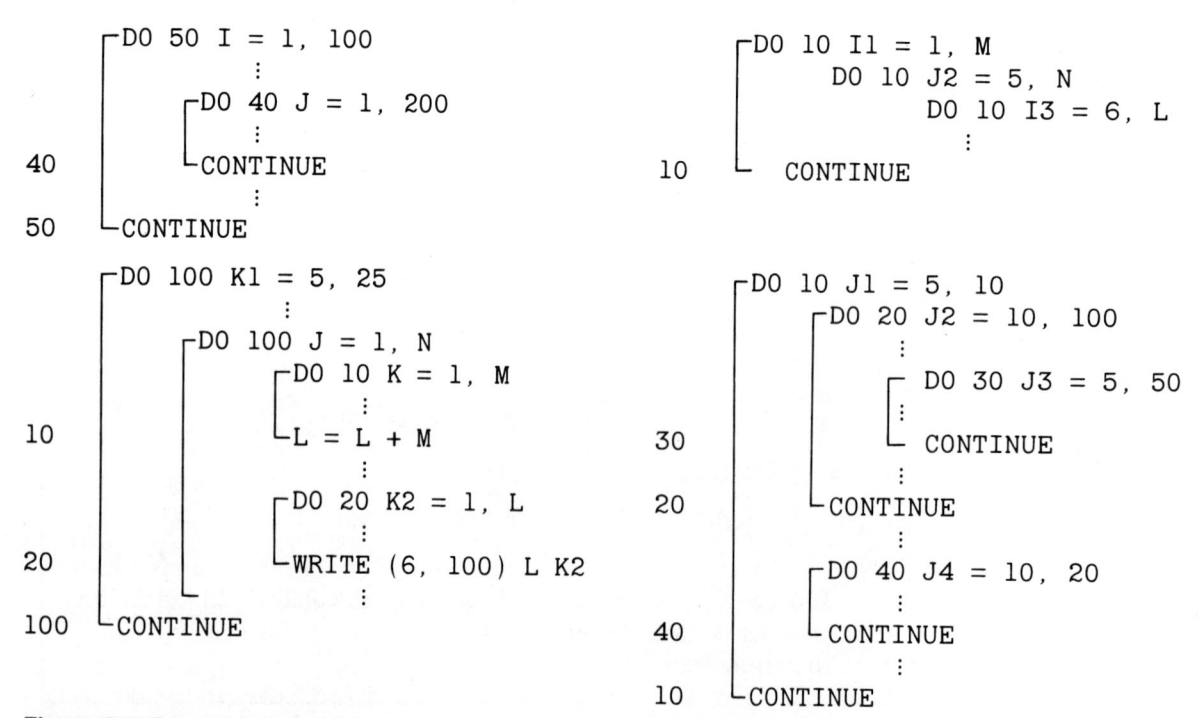

```
        ┌─DO 50 I = 1, 100                    ┌─DO 10 I1 = 1, M
        │   ⋮                                 │    DO 10 J2 = 5, N
        │ ┌─DO 40 J = 1, 200                  │      DO 10 I3 = 6, L
        │ │   ⋮                               │         ⋮
   40   │ └─CONTINUE                     10   └─ CONTINUE
        │   ⋮
   50   └─CONTINUE

        ┌─DO 100 K1 = 5, 25                   ┌─DO 10 J1 = 5, 10
        │   ⋮                                 │ ┌─DO 20 J2 = 10, 100
        │ ┌─DO 100 J = 1, N                   │ │   ⋮
        │ │ ┌─DO 10 K = 1, M                  │ │  ┌ DO 30 J3 = 5, 50
        │ │ │   ⋮                             │ │  │   ⋮
   10   │ │ └─L = L + M                  30   │ │  └ CONTINUE
        │ │   ⋮                               │ │   ⋮
        │ │ ┌─DO 20 K2 = 1, L             20  │ └─CONTINUE
        │ │ │   ⋮                             │   ⋮
   20   │ │ └─WRITE (6, 100) L K2             │ ┌─DO 40 J4 = 10, 20
        │ │   ⋮                               │ │   ⋮
  100   └─CONTINUE                       40   │ └─CONTINUE
                                              │   ⋮
                                         10   └─CONTINUE
```

Figure 5.5 Some valid nested loops.

Correct   *Incorrect*

```
            DO 90 I = 1, 30                            DO 90 I = 1, 30
                :                                          :
                DO 80 J = 1, 100                           DO 80 J = 1, 100
                    :                                          :
                    DO 70 K = 1, 200                           DO 70 J = 1, 200
                        :                                          :
70                  CONTINUE                 70             CONTINUE
                    :                                          :
80              CONTINUE                     80         CONTINUE
                :                                          :
90          CONTINUE                         90     CONTINUE

            DO 100 I = 1, 200                          DO 100 I = 1, 200
                :                                          :
                DO 300 K = 1, 400                          DO 300 K = 1, 400
                    :                                          :
300             CONTINUE                     300        CONTINUE
                :                                          :
                DO 400 K = 1, 150                          DO 400 I = 1, 130
                    :                                          :
400             CONTINUE                     400        CONTINUE
                :                                          :
100         CONTINUE                         100    CONTINUE

            DO 50 L = 1, 100                           DO 50 L = 1, 100
                :                                          :
                DO 200 J = 1, 300                          DO 200 J = 1, 300
                    :                                          :
                    DO 300 K = 1, 400                          DO 300 K = 1, 400
                        :                                          :
300                 CONTINUE                 300            CONTINUE
                    :                                          :
200             CONTINUE                                    DO 400 J = 1, 200
                :                                              :
                DO 400 J = 1, 200            400            CONTINUE
                    :                                          :
400             CONTINUE                     200        CONTINUE
                :                                          :
50          CONTINUE                         50     CONTINUE
```

Figure 5.6 Using the same control variable in nested loops.

You may use the same CONTINUE statement with several loops. However, you may not use the same variable as the control variable for two nested loops unless one of the loops is completely executed before the other starts. Figure 5.6 shows where you can use the same control variable for two loops.

The following is another example of a nested loop.

EXAMPLE 5.9

■ **Problem:** There are about 40 data records, each containing an amount. The interest on the amount (AMT) is 5% and is compounded yearly. Calculate the total (amount and interest) for each amount deposited in 1985 and withdrawn in 1995. The last record is a trailer record with the number 9999.99 placed on it.

Program:

```
        PROGRAM NEST
        DO 95 J = 1,40
            READ (5, 10) AMT
            IF (AMT .EQ. 999.99) GO TO 900
            TOTAMT = AMT
                DO 90 J = 1985, 1995
                    XINT = TOTAMT*.05
                    TOTAMT = TOTAMT + XINT
90              CONTINUE
            WRITE (6, 20) AMT, TOTAMT
95      CONTINUE
900     STOP
10      FORMAT (F5.2)
20      FORMAT (11X, 'TOTAL AMOUNT AND INTEREST ON',
       *F6.2, 'IS =', F7.2)
        END
```

Notes:

1. The outer loop (the first loop) attempts to read 40 data records. But the trailer record (the record with 9999.99) terminates the loop.
2. The inner loop calculates the interest on an amount, and the total amount and interest.

SOLVED PROBLEMS

5.8 Detect the errors, if any, in the following loops:

a.

```
        DO 50 I = 1, 30
            ⋮
            DO 60 J = 1, 20
                ⋮
50      CONTINUE
            ⋮
60      CONTINUE
```

b.

```
        DO 60 J = 1, 300
            ⋮
            DO 70 J = 35, 500
                ⋮
70      CONTINUE
            ⋮
60      CONTINUE
```

c.

```
          DO 100 M = 1,90
               ⋮
              IF (A .EQ. B) GO TO 50
               ⋮
              DO 200 N = 100, 200
               ⋮
50            I = N + M
               ⋮
200       CONTINUE
               ⋮
100   CONTINUE
```

d.

```
          DO 100 I = 1, 600
               ⋮
              DO 200 J = 2, 80
               ⋮
                IF (J .GT. K) GO TO 50
               ⋮
200       CONTINUE
               ⋮
              DO 300 J = 1, 200
               ⋮
300       CONTINUE
               ⋮
100   CONTINUE
               ⋮
50    ...
```

e.

```
          DO 200 N = 1, 50
               ⋮
              IF (N .EQ. J) GO TO 30
               ⋮
                  DO 300 M = 50, 200
               ⋮
300           CONTINUE
               ⋮
30        S = S + X
               ⋮
200   CONTINUE
```

f.

```
          DO 500 L = 25, 300, 2
               ⋮
              DO 600 N = 1, 400
               ⋮
                  IF (N .EQ. J) GO TO 400
               ⋮
600       CONTINUE
               ⋮
400       S = S + A
               ⋮
500   CONTINUE
```

Answers

a. Incorrect; the loops overlap.
b. Incorrect; the same control variable is used for both loops.
c. Incorrect; control cannot be transferred into a loop (the inner loop).
d. This is correct.
e. Correct; note that control is transferred from a point inside a loop to another point inside the same loop.
f. This is correct.

5.9 Write a program which prints the following sequence of numbers:

```
1    1
1    2
1    3
2    1
2    2
2    3
3    1
3    2
3    3
```

Answer

```
           PROGRAM XX
           DO 100 I = 1,3
              DO 100 J = 1,3
                 WRITE (6,10)I,J
100        CONTINUE
10         FORMAT (11X,I2,19X,I2)
           END
```

5.10 Write a program which reads 60 data records. On each record there is a number N. Calculate $1 + 2 + 3 + \cdots + N$ for each record.

Answer

```
           PROGRAM SUM
20         FORMAT (I2)
           DO 10 I = 1, 60
              READ (5, 20) N
              ISS = 0
                 DO 30 J = 1, N
30                  ISS = ISS + N
              WRITE (6, 40) N,ISS
10         CONTINUE
40         FORMAT (6X, 'FOR N=', 'THE SUM IS =',
           *I6)
           STOP
           END
```

Note: ISS = 0 before the inner loop is necessary to set ISS equal to zero for each new N.

EXERCISES

5.1 Identify errors in the following DO statements:

a.

30   DO 10 JAN = 10, 15, -1

b.

      DO 60 JAY = -5, 10, 0

c.

      DO 10 KAY = 50, 15, 2

d.

      DO 60 NO = 1, N*M , 0

e.

      DO 70 R = 5, N

f.

      DO 80 J = 1, P, Q

g.

      FOR J = 1, P, Q

h.

      DO, I, 99 = 1, 10 STEP 2

5.2 Identify errors in the following loops:

a.

5   DO 60 J = 1, N
        ⋮
60   IF (I .GE. J) GO TO 5

b.

20   A = A + K
        ⋮
      DO 20 J = 1, K

c.

      DO 99 I = 1, 10, 2
        ⋮
      I = I + 1
        ⋮
99   CONTINUE

d.

      DO 50 SS = 1, 30, 2
        ⋮
50   CONTINUE

e.

      DO 90 INPUT = 1, 30, 3
        ⋮

```
55    J = J + 3
       ⋮
90    CONTINUE
       ⋮
      GO TO 55
```

f.
```
      I = 0
      DO 100 K = 1, 95, 1
          I = I + K
          WRITE (6,100) K, I
100   FORMAT (6X, I3, 5X, I2)
```

g.
```
      DO 60 K = 1, 30, 2
          DO 60 K = 31, 60, 2
           ⋮
60    CONTINUE
```

h.
```
      DO 8 L = 90, 260, 2
       ⋮
      IF (L .EQ. 96) GO TO 10
       ⋮
      DO 8 M = 90, 260, 2
       ⋮
10    N = I + S
       ⋮
8     CONTINUE
```

i.
```
5     DO 80 INDEX = 1, 50
       ⋮
      IF (INDEX .LT. N) GO TO 5
       ⋮
80    CONTINUE
```

j.
```
50    DO 65 KAY = 1,100
          DO 65 JAY = 1,200
              IF (JAY .EQ. 9) GO TO 50
           ⋮
65    CONTINUE
```

k.
```
      DO 89 LOAD = 9, 90
       ⋮
          DO 60 MAY = 1,100
           ⋮
35            DO 6 MAY = 100, 200
               ⋮
6             CONTINUE
60        CONTINUE
89    CONTINUE
```

PROGRAMMING EXERCISES

Write a complete FORTRAN program for each of the following problems. Use a DO loop for looping, appropriate headings, formats, and spaces.

5.3 Print

    THIS IS A FORTRAN COURSE

thirty times.

5.4 Add the integers from 1 to N, where N is to be read from a data record.

5.5 Print the odd numbers from 1 to N with five numbers to a line. The input is N, and the output will look like this:

    1   3   5   7   9
   11   13   15   17   19
    :   :   :   :   :

5.6 Print a table of three columns. The first column shows the numbers from 1 to N, the second column shows the squares of the numbers, and the third column shows the cubes of the numbers. N is to be read as an input. Place asterisks (*) around the table. The printout then will look like:

```
******************************************
*   NUMBER   *   SQUARE   *    CUBE    *
******************************************
*     1      *     1      *     1      *
*            *            *            *
*     2      *     4      *     8      *
*            *            *            *
*     :      *     :      *     :      *
*            *            *            *
******************************************
```

5.7 The sine of X can be calculated approximately by summing the first N terms of the infinite series:

$$\sin X = X - \frac{X^3}{3!} + \frac{X^5}{5!} - \frac{X^7}{7!} + \cdots$$

Write a program which calculates and prints $\sin X$. X and N will be read in as input data. (! is a factorial sign.)

5.8 Combination of N and M can be calculated by:

$$C = \frac{N!}{M!(N - M)!}$$

where C is the number of combinations of N items taken M at a time (the number of ways that a subset of size M can be selected from a set of size N), and

$$N! = 1 \times 2 \times 3 \times \cdots \times N,$$
$$M! = 1 \times 2 \times 3 \times \cdots \times M,$$
$$(N - M)! = 1 \times 2 \times 3 \times \cdots \times (N - M)$$

Write a program which calculates and prints C, where N and M are to be read as input (with N larger than M). Keep the values of M and N low (below 10) to keep their factorials from becoming too large.

5.9 Write a program which calculates the average of a set of numbers. The number of data (N) is to be read at the beginning, followed by the data. Each number is on a separate line.

*5.10 Write a program which calculates the variance of a set of data. Each number is punched on a separate card. The first card is a header card indicating the number of data (N). The variance (VAR) can be calculated by

$$\text{VAR} = \frac{\text{sum of squares of numbers}}{N} - (\text{average})^2$$

5.11 Write a program which calculates the mean, variance, and largest and smallest member of a set of numbers. Each number is on a separate card. The first one shows the number of data.

5.12 Write a program which prints the numbers and the squares of the numbers from M to N with an increment of I. The starting value M, the final value N, and the increment I are to be read from a data card. For example, if M = 6, N = 20, and I = 2, the printout will show:

NUMBERS	SQUARES
6	36
8	64
10	100
⋮	⋮
20	400

5.13 Write the previous program (Exercise 5.12) so that it accepts several sets of data (instead of one), and prints the numbers and their squares for each. Use a trailer record to indicate the end of data. Assume there are less than 10 sets of data.

5.14 Write a program which calculates the volume of a cylinder for heights 10,20,30 and radii 1,2,3,4,5,6,7,8,9,10.

5.15 If X dollars are deposited in an account with the interest rate R, compounded annually, the total money after N years is

$$T = X(1 + R)^N.$$

Write a program for calculating the total of principal and interest for a deposit of X dollars after N years, at different interest rates. The input is X and N. The output is a table of two columns. The first column shows the different interest rates starting from .03, with an increment of .01, up to the point where the money is doubled (i.e., T = 2X). The second column shows T.

5.16 Write a program which finds values of Y for values of X in the equation $Y = 5X^2 + 3X - 2$ from A to B in increments of C. A, B, and C are to be read from a data record.

5.17 If a component has an average life of T hours, the probability P of running without failure for at least A hours can be calculated as

$$P = \frac{1}{2.718^{A/T}}.$$

Write a program which calculates the probability P for a component. The input is the product number and the average life T for the component. The output is a table of two columns. The first column shows different values of A with an interval of 2.5 from 1 up to $5*T$, and the second column shows the corresponding probabilities.

5.18 Write a program that will balance the checking accounts for the customers of a bank. The input is the account number (columns 1–6) and the current balance (columns 8–14) on one line, followed by deposits (which are positive) or withdrawals (which are negative) in columns 1–6 on separate lines. A "deposit" of 0.00 indicates the end of deposits and withdrawals for the customer. The number of withdrawals and deposits for a customer is less than 10. There are less than 500 customers. Number 00000 for the ID number indicates the end of the data. Design your own output. Make sure to print the totals at the end. (Hint: You can use two DO loops. Read the account number and the current balance in the first loop, and read the deposits and withdrawals in the second one).

5.19 Write a complete program which calculates the average of N exam marks for students. The number of exams varies from student to student. The ID number and the number of exams for a student are recorded on a line and are followed by the exam marks for the student (each exam mark on a separate line). There are about 100 students. The ID number -9999 indicates the end of the data. At the end, print the grand average and the number of students.

5.20 Assume that you have bought a car, and your monthly payment is PAY. Write a program which prints a table of your payments. The output should contain:

1. The sequence numbers as the period (or the transaction date).
2. The payments (transaction amount).
3. The interest.
4. The payments applied to the principal (amortization payment).
5. The remaining balance.

The input should contain:

1. The beginning balance (BAL).
2. The percentage interest rate (R).
3. The monthly payments (PAY).
4. The number of periods to be included in the table (N).

You can use the following formulas:

monthly interest rate, XR = R/12/100
interest for each period, XINT = BAL * XR
payment applied to principal, AP = PAY − XINT
(new BAL) = (previous BAL) − AP

For example, if

BAL = 6589.5
PAY = 139.5
R = 13.5
N = 3

Then the table is:

```
PERIOD    PAYMENT    INTEREST    APPLIED    BALANCE
   1       139.5      74.1319    65.3681    6524.13
   2       139.5      73.3965    66.1035    6458.03
   3       139.5      72.6528    66.8472    6391.18
```

SUMMARY OF CHAPTER 5

You have learned:

1. The DO loop is one of the features of FORTRAN for looping. The general form of the DO loop is:

   DO n INDEX = m_1, m_2, m_3
   $\vdots$
   n CONTINUE

   Here n is the statement number for the CONTINUE statement, or the last statement in the loop. INDEX is a variable which controls the loop. Its value starts from m_1 and increases by m_3 each time that the loop is executed until the value is greater than m_2. The most important rules for the DO loop are:
   a. The initial, final, and increment values of the control variable must be defined.
   b. The last statement in a DO loop can be one of the following:
      i. A CONTINUE statement
      ii. A READ, or WRITE statement
      iii. An arithmetic statement
   c. Control cannot be transferred into a loop, but control can be transferred out of a loop.
   d. When using a DO loop and IF block together, their ranges must not interfere with each other.
2. Loops inside loops are called nested loops. The ranges of nested loops must not interfere with each other. The rules of a DO loop apply to each nested loop.

SELF-TEST REVIEW

5.1 Find all the errors in the following statements:
   a. FOR I = 1 TO 100
   b. MAIN = 1,99,DO 100
   c. DO N = 1, 200, 2
   d. DO 99 JAY = 1
   e. DO M5 = 1,M,3
   f. DO 100, K = 1,M,3
   g. DO 999 R = 1.0, 30.0, 2.5
   h. DO 55 I = M, N, 2*K + 1
   i. DO 69 L = 99, 1, 2
   j. DO 90 L = 1, 80, 0

5.2 Find the errors in the following loops:
a.
```
          DO 80 I = 1, 99
              ⋮
             I = I + 2
              ⋮
80        CONTINUE
```
b.
```
10        DO 99 IND = 1939, 1989,1
              ⋮
          IF (IND .EQ. 1959) GOTO 10
              ⋮
99        CONTINUE
```
c.
```
          DO 100 J = 1,100
              ⋮
100       IF (J .EQ. M) GO TO 10
```
d.
```
          DO 89 K = 1,100
              ⋮
            IF (K .LE. 30)THEN
              ⋮
89        CONTINUE
              ⋮
          ELSE
              ⋮
             J = 2*K
              ⋮
          ENDIF
```
e.
```
          DO 30 I = 1, 59
              ⋮
          DO 45 K = 1, 99
              ⋮
30        CONTINUE

45        CONTINUE
```
f.
```
          DO 95 J = 1, 90
              ⋮
            IF (J .EQ. 35)GOTO65
              ⋮
              DO 30 K = 1, 80
65                M = 3*J
                  ⋮
30                CONTINUE
              ⋮
95        CONTINUE
```

5.3 If the yearly interest on an amount is *R*, and is compounded yearly, we would like to calculate the total money accumulated on an investment installed from an initial year to a final year. Write a program for this purpose. Input:

The initial investment, columns 1–7.
The interest rate, columns 9–11.
The initial year, columns 13–16.
The final year, columns 18–21.

Output: The input information as well as the total money. Assume there are several sets of data (less than 100). The last record is a trailer record with a negative number as the initial amount.

☐ **Answers**

5.1 a. The form is not correct.
  b. The form should be DO 100 MAIN = 1,99
  c. The statement label for the CONTINUE statement is missing.
  d. The final value of the control variable is missing.
  e. Same as **c**.
  f. This is correct in FORTRAN 77.
  g. This is also correct in FORTRAN 77.
  h. This is also correct in FORTRAN 77.
  i. The initial value of the control variable is more than its final value (the increment is positive).
  j. The increment value cannot be zero.

5.2 a. The value of the control variable cannot be changed in the DO loop.
  b. The control must be transferred to statement 99 instead of to 10.
  c. The last statement in a DO loop cannot be IF ... GO TO statement.
  d. The DO loop and IF block are overlapping.
  e. The loops are overlapping.
  f. The control cannot be transferred into a DO loop (to the inner loop).

5.3

```
         PROGRAM INTRST
         WRITE ( 6,10)
10       FORMAT ('1',5X,'THE AMOUNT',2X,'THE INTEREST RATE',
        *2X,'THE BEGINNING YEAR',2X,'THE END YEAR',
        *4X,'TOTAL'   )
         DO 200 I = 1, 100
             READ ( 5,30) AMT, R, IN1, IN2
             IF(AMT .LT. 0.0) GOTO 99
             TOT = AMT
                 DO 40 K = IN1, IN2
                     XINT = TOT*R
                     TOT = TOT + XINT
40               CONTINUE
             WRITE (6, 50) AMT, R, IN1, IN2, TOT
200      CONTINUE
99       STOP
30       FORMAT (F7.2,1X, F3.2, 1X, I4, 1X, I4)
50       FORMAT ('0',6X,F8.2,9X,F3.2,16X,I5,12X,I5,
        *6X,F10.2)
         END
```

ANSWERS TO SELECTED EXERCISES

5.10

```
      PROGRAM VARIAN
      SUM = 0.0
      SS  = 0.0
      READ ( 5,10) N
      DO 40 K = 1, N
          READ ( 5, 20)X
          SUM = SUM + X
          SS  = SS + X**2
40    CONTINUE
      AVG = SUM/N
      VAR = SS/N - AVG**2
      WRITE ( 6,50) VAR
      STOP
10    FORMAT (I2)
20    FORMAT (F6.2)
50    FORMAT (6X, 'THE VARIANCE IS=', F9.2)
      END
```

Chapter 6

Data Types and More about Input and Output

DATA TYPES

    REAL and INTEGER Statement
    Character Data
        CHARACTER Statement
        Character Variables
        Character Input and Output; A-field
        Character Constants
        Substring Reference
        Character Expressions
        Important Notes about Alphanumerics

DATA STATEMENT

OTHER FEATURES OF FORMAT

    Repeating Field Descriptors
        FORMAT Scanning
        T Descriptor
        H Descriptor
    Printing an Apostrophe
    Reading and Writing Literals
    The Slash
        Summary of Rules about the Slash

LIST-DIRECTED INPUT AND OUTPUT

    List-Directed Input
    List-Directed Output
    Important Notes About List-Directed READ and WRITE

FORMATTED PRINT STATEMENT

The first part of this text (Chapters 1–5) was designed to introduce you to the basic programming techniques and FORTRAN in general. In the following chapters we will discuss additional features of FORTRAN.

In this chapter you will learn how to use:

- REAL and INTEGER statements to explicitly define variables
- Character data
- DATA statements to initialize variables
- Other FORMAT features, such as automatic repetition and tabbing
- READ, WRITE, and PRINT statements without a FORMAT
- PRINT statements with FORMAT.

FORTRAN 77 standards are followed; however, it is sometimes necessary to check some of the variations and special features of the system which you are using. The *FORTRAN Reference Manual*, available in your computer center, is a good reference source for this purpose. The manual can answer most of your questions about the variations in your compiler. Of course, you can always ask your instructor or a member of the computer staff for assistance.

DATA TYPES

There are several types of data in FORTRAN:

- Real
- Integer
- Character
- Double precision
- Logical
- Complex

Real and integer data were discussed in Chapter 2. To begin this chapter, we discuss only the REAL and INTEGER statements, which are useful for defining a variable as real or integer. Character data then will be discussed in detail. Double precision, logical, and complex data will be discussed in Chapter 11.

☐ REAL and INTEGER Statements

As explained in Chapter 2, if the first letter of a variable's symbolic name is I, J, K, L, M, or N; the variable identifies integer data. If the first letter is anything else, the variable identifies real type data. This is an *implicit* rule: the symbolic name of a variable automatically identifies its type. Sometimes, however, it is not desirable to follow this rule. In this case, the REAL and INTEGER statements are used to define the type of the variable. The form of the statement is:

REAL var1, var2, . . .

or

INTEGER var1, var2, ...

The following are some examples:

```
REAL INTRST
REAL I,J,MONEY,SUM
```

```
      REAL LARGE,KAY,NUMBER,MULT
      INTEGER COUNT,A,TIME,X
```

Notice that declaring a variable which is already implicitly defined is also valid, although not necessary (such as REAL SUM). Such type statements are nonexecutable. They provide information about the type of data to be processed. Such information is necessary for compiling a program. Thus, it must be placed before any executable statements—at the beginning of the program, immediately after the PROGRAM statement.

The REAL and INTEGER statements are very useful for explicitly defining the mode of variables. For example, the length of a room is unlikely to be an integer. Therefore, if LENGTH is to be used as the name of this variable, the statement

```
      REAL LENGTH
```

is necessary to define the variable as real. Similarly, *I* is used often in physics to indicate a current, and normally it is not an integer quantity. Therefore the statement

```
      REAL I
```

is useful to define I as real.

☐ Character Data

So far our programs have had all numeric data. However, for some applications we need to input or output characters or symbols, such as names, addresses, and so forth. For instance, in some programs we need to input (or output) the name of the customer for billing purposes, or the name of the student for grading purposes. In FORTRAN this type of data is called *alphanumeric, alphameric,* or *character* (also *string* or *literal*). It may contain letters, numerals, and special characters such as commas, dollar signs, periods, or blanks.

CHARACTER Statement

A CHARACTER statement is used to define a variable as a character type. The form of the CHARACTER statement is:

```
        CHARACTER   Variable-name   *   N
            ↑            ↑          ↑   ↑
            1            2          3   4
```

It has four parts:

1. The word CHARACTER, which must start in or after column 7, and must be at the very beginning of the program.
2. A variable name.
3. The symbol *.
4. The length of the character data; *if a length is not specified, the default length is one.*

More than one variable can be declared by a CHARACTER statement. Thus, the general form of the CHARACTER statement is:

```
      CHARACTER Var1 * N1 , Var2 * N2 , Var3 * N3 , . . . .
```

For example, the statement

    CHARACTER INDEX*16 , NAME*12 , ADRES*24 , SOCSEC*11 , GRADE

declares that INDEX, NAME, ADRES, SOCSEC, and GRADE are character-type variables, and that data for INDEX has a maximum of 16 characters; NAME, 12; ADRES, 24; SOCSEC, 11; and GRADE, only one. If the length specifications of several variables are the same, the character statement can be specified as follows:

    CHARACTER*N Var1 , Var2 , Var3 , ...

For example, the statement

    CHARACTER*12 A , B , C

indicates that the variables A, B, and C are character types, and that the length of data for all of them is 12.

Combination of the above character statements is also permissible. For example, the statement

    CHARACTER*12 A , C , B , D*16

declares that the data for variables A, B, and C have a maximum of 12 characters each, and that the data for variable D have a maximum of 16 characters.

In summary, any of the following forms are permissible for the CHARACTER statement:

    CHARACTER M*12 , N*8 , SSN*11
    CHARACTER*12 M , J , K
    CHARACTER*12 A , B , SSN*11 , M , N*8

Don't forget to separate the variables with commas (but put no comma after the word CHARACTER).

The CHARACTER statement is a type statement and is not executable. It only specifies the type of data and their length. It is needed for execution of any statement containing alphanumeric data. Therefore, *the CHARACTER statement must be placed before any executable statements*, at the very beginning of the program, immediately after the program statement.

Character Variables

A character variable is a variable which has been defined as such by the CHARACTER statement. For example, the statement

    CHARACTER A*3, NAME*15

declares that A and NAME are character type variables, and that A is 3 and NAME is 15 characters long.

Character Input and Output; A-Field

For reading or writing alphanumeric data, a variable name is chosen in the same way as for real or integer data. But we use an *A-field* to describe the length of the field.

> **RULE 6.1**
> *A-field for alphanumeric data:* The descriptor An is used for showing the length of an alphanumeric field, where n is the number of characters occupied by the field.

The A format is comparable to the I format. For instance, A4 is used to read a field which contains four characters. The following field descriptors provide examples:

Data	Field descriptor
PAUL	A4
APOLLOS	A7
4811 E. MICHIGAN	A16
E	A1
MT.VER,MI	A9

Notice that any characters, including blanks, commas, and periods, can be included in the field. The following example demonstrates how alphanumeric data can be read or written.

EXAMPLE 6.1

Problem: There are several records, each containing individual-student information: name (maximum of 12 characters), ID number, and two test scores. Write a program which reads the information, calculates the average of the scores, and prints the information.

Program:

```
            PROGRAM CHAR1
            CHARACTER ANAME*12
5           READ (5,10,END = 99) ANAME,ID,SCORE1,SCORE2
10          FORMAT (A12,1X,I5,1X,F5.2,1X,F5.2)
            AVG = (SCORE1 + SCORE2)/2.0
            WRITE (6,20) ANAME,ID,SCORE1,SCORE2,AVG
20          FORMAT(2X,A12,2X,I5,2X,F6.2,2X,F6.2,2X,F6.2)
            GO TO 5
99          STOP
            END
```

Each data record must look like this:

```
JOHN HIGGINS 62593 95.50 87.50
```

The output will look like this:

```
  JOHN HIGGINS     62593     95.50     87.50     91.5
```

Notes:

1. A heading would improve the output.
2. The CHARACTER statement at the beginning of the program declares that:

   a. the variable ANAME is of character type, and
   b. the length of the field is 12 characters.

Character Constants

A character string enclosed in a pair of apostrophes forms a character constant. An assignment statement may be used to assign a character constant to a character-type variable. Example:

```
CHARACTER*3 ANAME
ANAME = 'PAT'
```

Substring Reference

A specific part of a character string can be specified in the following form:

$$\text{VAR}\begin{pmatrix}\text{first} & \text{last}\\ \text{position} & : \text{position}\end{pmatrix}$$

where VAR is a character type variable.

For example, if A is defined as character variable, and its value is

```
A = 'MY FAIR LADY'
```

then we may refer to the following substrings of A anywhere in the program:

Code	Value
A(4:7)	'FAIR'
A(1:2)	'MY'
A(2:2)	'Y'
A(5:)	'AIR LADY'
A(:5)	'MY FA'

Character Expressions

The *character operator* // represents concatenation.

For example, the result of 'XY'//'Z' is the string 'XYZ'. A *character expression* can be formed by using one or more character operators and character operands (i.e., character constants, symbolic names of a character variables, etc.). The following is an example:

```
'A'//'B'//'CDE'//'FGHI'
```

The value of this example is

```
'ABCDEFGHI'
```

The following are further examples:

EXAMPLE A

If using

```
CHARACTER NAME*10,FNAME*5,LNAME*5
FNAME = 'JACK'
LNAME = 'BROWN'
NAME = FNAME//LNAME
```

then NAME is equivalent to

```
NAME = 'JACK BROWN'
```

EXAMPLE B

If using

```
CHARACTER TOT*13 , FIRST*2
FIRST = 'US'
TOT = FIRST//' OF AMERICA'
```

then TOT is equivalent to

```
TOT = 'US OF AMERICA'
```

Important Notes about Alphanumerics

When working with character data, the following notes deserve special attention.

1. Alphanumeric data cannot be added together or multiplied by each other in a mathematical sense. We can, however, sort them, compare them, concatenate them, and transfer the contents of one variable to another.
2. A character constant may be assigned to a character variable. For example,

   ```
   GRADE = 'B'
   NAME = 'JOHN'
   CODE = '34RN'
   STATE = 'S.C.'
   BLANK = ' '
   IF(SCORE .GE. 90.0)GRADE = 'A'
   ```

   are all valid, provided, of course, that *each variable is defined as character type*.
3. If *the length of the data* is less than the defined length, the data will be stored *left justified* with blanks on the right. That is, characters will be stored starting from the left. For example, in

   ```
   CHARACTER NAME*4
   NAME = 'JOE'
   ```

   Joe will be stored as follows:

   Variable: NAME
   Content: | J | O | E | |

   In the program segment:

   ```
   CHARACTER*12 N
   READ(5,10)N
   FORMAT(A3)
   ```

   with data

   ```
   JOE
   ```

   JOE will be stored as

   | J | O | E | | | | | | | | | |

   Note that the remaining spaces are automatically filled with blanks.
4. If the length of the data is greater than the specified length, then:

   a. The *right characters* will be truncated if an assignment statement is used. For example, in

```
      CHARACTER NAME*4
      NAME = 'JOSEPH'
```

Only four characters, JOSE, will be stored in the memory location called NAME.

b. The *leftmost* characters will be truncated if a READ statement is used. For example, in

```
      CHARACTER ADDR*4
      READ(5,10) ADDR
10    FORMAT (A11)
```

with data

> 48 MICHIGAN

only the last four characters will be stored.

This point is very important, particularly when the CHARACTER statement is missing. In the absence of the CHARACTER statement, only a limited number of characters are allowed to be read or written, depending on the system being used. Therefore, if part of the character data in your program is missing, check the corresponding CHARACTER statement.

5. The content of one variable can be copied into another by using an assignment statement. For example, if A = 'DEB', then B = A is valid, provided that the variable B is also defined as character type.

6. If numeric data are stored as characters, they will be treated like characters, not numbers. For example,

```
A = '25'
B = '8'
```

will store 25 in memory A, and 8 in memory B. However,

```
C = A*B
```

is not valid, because 8 and 25 are *characters* in this example, not numbers.

7. The character variables or character constants can be compared with each other in an IF statement. For example:

```
IF(NAME .LT. 'PAT')GO TO 50
```

is valid. Furthermore, the characters are stored in terms of numeric codes in the usual alphabetical order. (Thus the code for a letter in the latter part of the alphabet is more than the code for a letter in the early part.)

8. If a variable is defined as a character with a specified length, then its field width *n* can be omitted in the FORMAT statement when reading or writing the data. For example:

```
      CHARACTER X*12
      READ (5,10) X
10    FORMAT (A)
```

is valid, and the FORMAT is equivalent to

```
10    FORMAT (A12)
```

The following is another example of using character data in a program.

EXAMPLE 6.2

Problem: Write a complete program which reads the student's name, social security number, and three test scores. Calculate the average of the scores, and assign a letter grade (90–100: A, 80–90: B, etc.). Print all the information with appropriate headings.

Program:

```
      PROGRAM CHACTR
C THIS PROGRAM READS THE NAMES OF STUDENTS, THEIR
C SOCIAL SECURITY NUMBERS, AND THEIR THREE EXAM
C SCORES. IT PRINTS THE DATA AS WELL AS THEIR
C FINAL AVERAGE AND A LETTER GRADE. THE DATA IS
C PLACED AS FOLLOWS:
C COLUMNS   1-12   NAME
C COLUMNS  14-24   SOCIAL SECURITY NUMBER, INCLUDING
C '-' AS 999-99-9999
C COLUMNS  26-30   EXAM #1, COLUMNS 32-36   EXAM #2,
C COLUMNS  38-42   EXAM #3
C ****************************************************
C
      CHARACTER GRADE*1,NAME*12,SSN*11
C
C WRITING THE HEADING, STARTING A NEW PAGE
C
      WRITE (6,2)
2     FORMAT ('1',30X,'THE STUDENTS GRADE REPORT')
      WRITE (6,4)
4     FORMAT (6X,'NAME',10X,'SOC SEC NO',5X,'EXAM
     *#1',5X,'EXAM #2',5X,'EXAM #3',5X,'AVERAGE',
     *5X,'GRADE')
5     READ(5,10,END=99) NAME,SSN,EXAM1,EXAM2,EXAM3
C
C CALCULATING THE AVERAGE
C
         AVG = (EXAM1 + EXAM2 + EXAM3 )/3.0
C
C GRADE ASSIGNMENT
         IF(AVG .GE. 90.0)THEN
                    GRADE = 'A'
         ELSE IF(AVG .GE. 80.0)THEN
                    GRADE = 'B'
            ELSE IF(AVG .GE. 70.0)THEN
                       GRADE = 'C'
               ELSE IF(AVG .GE. 60.0)THEN
                          GRADE = 'D'
                  ELSE
                          GRADE = 'E'
         END IF
C PRINTING DATA
C
         WRITE(6,20)NAME,SSN,EXAM1,EXAM2,EXAM3,
     *AVG, GRADE
```

```
C
C MAKING A LOOP
      GO TO 5
99    WRITE(6,30)
30    FORMAT( 31X,'END OF REPORT')
10    FORMAT(A12,1X,A11,1X,F5.2,1X,F5.2,1X,F5.2)
20    FORMAT(3X,A12,5X,A11,5X,F6.2,5X,F6.2,5X,
     *F6.2,5X,F6.2,17X,A1)
      END
```

SOLVED PROBLEMS

6.1 What is the value of each variable read by each of the following READ statements?

a.
```
      CHARACTER*4 X,Y,Z,P
      READ(5,10)X,Y,Z,P,Q
10    FORMAT(4A4,F5.2)
```
Data:

```
PAT JOHNSON 96.256.298
```

b.
```
      CHARACTER*4
     *BOAT,COAT,ROAD,NAM,KAD,PAD,MAD
      READ(5,20)BOAT,COAT,ROAD,NAM,KAD,PAD,MAD
20    FORMAT(4A4,A2,A4,1X,A4)
```
Data:

```
IMPERIAL PALACE IN NEW CITY
```

c.
```
      CHARACTER ADRESS*14,ST*4
      READ(5,40)ADRESS,ST
40    FORMAT(A4, A6)
```
Data:

```
21PENNSYLVANIA AVENUE
```

Answers

a.

X = |P|A|T| |

Y = |J|O|H|N|

Z = |S|O|N| |

P = |9|6|.|2|

Q = 56.29

Note that P is considered as a character field, not as numeric data.

b.

BOAT = |I|M|P|E|

COAT = |R|I|A|L|

ROAD = | |P|A|L|

NAM = |A|C|E| |

KAD = |I|N| | |

PAD = | |N|E|W|

MAD = |C|I|T|Y|

c.

ADRESS = |2|1|P|E|

ST = |S|Y|L|V|

6.2 Are the following statements correct? If not, indicate the errors.

a.

```
            PROGRAM ABC
            READ(5,10)A,B
            S = A*B
            WRITE(6,15)S
10          FORMAT(2A4)
15          FORMAT(A4)
            END
```

b.

```
            CHARACTER X*4
            READ(5,15)X,Y
15          FORMAT(A4,F4.1)
            D = Y + X
            WRITE(6,25)D
25          FORMAT(1X,F4.1)
            END
```

Data:

```
95.5b82.5
```

c.
```
      CHARACTER*12 NAME,CODE,REAL NORM, JAY
```
d.
```
      INTEGER WAGE, NO
      CHARACTER*12 N*10,M*9,ADD*4
```
e.
```
      CHARACTER*12 AAA, BBB, CCC*4
      AAA = 'STORAGE'
          ⋮
      CCC = AAA(1:3)
      BBB = AAA(3: )
```
f.
```
      GRADE = B
```
g.
```
      NAME1 = 'CHUCK NEWMAN'
      NAME2 = 'PAT BROWN'
      IF(NAME1 .LT. NAME2)FIRST = NAME1
```
h.
```
      PROGRAM XYY
      CHARACTER   X*13
      X = 'UNITEDSTATES'
      Y = X( :6)
      Z = X(7: )
      X = X//' '//Z
```
i.
```
      IF (SCORE .GE. 90.0) GRADE = 'A'
```
j.
```
      CHARACTER*13 JJ*5 , KK*8, LONG
      READ (5, 10) JJ, KK
  10  FORMAT (A4 , A7)
      LONG = JJ//KK
```
k.
```
      CHARACTER K*8
      READ (5 , 20)K
  20  FORMAT (A13)
```
l.
```
      IF (NAME .EQ. PAT) GO TO 10
```
m.
```
      IF (KODE .EQ. '    ') GO TO 20
```
n.
```
      K = 'FINISH    '
      IF (K .EQ. 'FINISH') GO TO 30
```

o.
```
      CHARACTER K1,K2,K3,K4,K5,K6,K7,K8,
     *K9,K10,K11,K12,K13
      READS (5, 70)K1,K2,K3,K4,K5,K6,K7,
     *K8,K9,K10,K11,K12,K13
70    FORMAT (13A1)
      END
```

Data:

```
THIS IS JOHNY
```

☐ **Answers**

- **a.** Incorrect; the alphanumeric data cannot be multiplied in a mathematical sense. The CHARACTER statement is also missing from the program.
- **b.** Also incorrect, because numerical data and alphanumerical data cannot be added together; X is defined as alphanumeric data, while Y is defined as numeric data.
- **c.** The CHARACTER statement and REAL statement cannot be mixed.
- **d.** This is correct.
- **e.** This is also correct.
- **f.** If GRADE is a character-type variable and B is a character constant, then apostrophes are needed around B.
- **g.** This is correct if the type and length of the variables NAME1, NAME2, and FIRST have been specified.
- **h.** Incorrect; if the variables Y and Z had been defined as character type, then the statements would be correct.
- **i.** This is correct if the variable GRADE has been defined.
- **j.** This is correct.
- **k.** K is declared as eight characters long but is read with A13. Therefore, the leftmost five characters will be ignored. That is, if the datum is, for example,

```
WILLIAM WHITE
```

  only AM WHITE will be stored.
- **l.** If PAT is a variable's name, this is correct; but if it is a string constant, it should be in apostrophes:

    IF (NAME .EQ. 'PAT') GO TO 10

- **m.** This is correct if KODE has been defined.
- **n.** The statements are syntactically correct if the length of the variable K has been declared at the beginning of the program.
- **o.** This is correct. Note that the length is not declared, and thus 1 will be assumed.

■ **6.3** Write a program which reads a word consisting of seven letters such as FORTRAN, and prints it scrambled.

□ **Answer**

```
        PROGRAM SCRMBL
        CHARACTER L1,L2,L3,L4,L5,L6,L7
        READ(5,10)L1,L2,L3,L4,L5,L6,L7
10      FORMAT(7A1)
        WRITE(6,20)L3,L6,L2,L1,L4,L5,L7
20      FORMAT(1X,7A1)
        END
```

Data:

```
FORTRAN
```

6.4 What is the purpose of the following program segments?

a.
```
        CHARACTER NEWPAGE*1
        NEWPAG = '1'
           ⋮
        WRITE (6, 10) NEWPAG, RATE
10      FORMAT (A, F5.2)
```

b.
```
        CHARACTER DBLSPC*1
        DBLSPC = '0'
        DO 90 I = 1, 10
90         WRITE (6, 10) DBLSPC, I
20      FORMAT (A, I2)
```

c.
```
        CHARACTER*1 NEWLN, NOADVC
        NEWLN = ' '
        NOADVC = '+'
        WRITE (6,10) NEWLN
        WRITE (6,20) NOADVC
10      FORMAT (A, 10X, 'THIS IS')
20      FORMAT (A, 18X, 'THE ANSWER')
```

□ **Answers**

All three program segments incorporate the carriage control characters with variables.

a. Prints the value of the RATE at the beginning of a new page (carriage control '1').

b. Prints the numbers 1 through 10 on every other line (double space, carriage control '0').

c. Prints the heading

THIS IS THE ANSWER

starting in column 10 of a new line.

DATA STATEMENT

So far we have used an assignment statement to provide initial values for variables. Alternatively, we can use a *DATA statement* for this purpose. The form of the DATA statement is:

DATA var1/value/ , var2/value/ , ...

or

DATA var1,var2,var3, .../value 1 , value 2 , value 3 , .../

or generally

DATA var list/value list/ , var list/value list/, ...

For example, either of the statements

DATA A/5.0/, B/3.0/, SUM/0.0/, I/0/

or

DATA A,B,SUM/5.0,3.0,0.0/ , I/0/

assigns 5.0 to A, 3.0 to B, 0.0 to SUM, and 0 to I. The placement of commas is critical. Pay attention to the following when using DATA statements:

1. Once a variable is assigned a value in an assignment statement, a DATA statement cannot be used for reinitializing that variable.
2. For each variable in the list, a value must be specified in the corresponding value list. That is, the number of variables must be equal to the number of values in each list. A one-to-one correspondence exists between variables in the variable list and constants in the value list: the first item of the variable list corresponds to the first constant in the value list, the second item to the second constant, and so forth.
3. An asterisk (*) can be used as a repetition code in the value list. For example,

DATA A,B,C,D,E/5*0.0/

assigns 0.0 to A, B, C, D, and E.
4. If a value is a character constant, it must be placed within apostrophes.
5. The DATA statement is nonexecutable; normally it is placed after the type statements (REAL, INTEGER, etc.) and before the executable statements.
6. Using the DATA statement is more convenient, compact, and efficient (in terms of computer time) than using an assignment statement to initialize the variables.

SOLVED PROBLEMS

6.5 Find the errors in the following statements:

a.

DATA, A,B,C/3*4.0/ I,/0/K/1/

b.

DATA X,Y,Z/2*0.0/,P,Q/3*0.0/

c.

DATA NAME,SCORE/JIM,95.5/

d.

DATA ADRES/'49 E. BROADWAY'/

e.

DATA SUM/0.0/
SUM = SUM + X
DATA SUM/5.5/

f.

CHARACTER*8 A,B,C
DATA A,B,C/'KAREN','MARG','WASHINGTON'/

g.

DATA A,B,C/3*' '/

☐ **Answers**

  a. The commas are not used properly. The correct form is

  DATA A,B,C/3*4.0/, I/0/, K/1/

  b. The number of variables is not equal to the number of constants in each list.
  c. The string of characters JIM must be placed in apostrophes, and the variable NAME must have been defined.
  d. This is correct if the type and length of the variable ADRES is defined.
  e. Once SUM is assigned another value (SUM = SUM + X), the DATA statement cannot be used to reinitialize the SUM.
  f. This is correct, but notice that the constants for A and B will be stored left justified with blanks on the right, and WASHINGTON will be truncated to WASHINGT.
  g. This is correct if the variables A, B, and C are defined.

■

OTHER FEATURES OF FORMAT

☐ **Repeating Field Descriptors**

As briefly discussed in Chapter 2, the field descriptors can be repeated by prefixing the descriptor with a nonzero integer constant. This is called the repeat factor and specifies the number of descriptors required. For example,

    10   FORMAT (2F5.2,3I2)

is equivalent to

    10   FORMAT (F5.2,F5.2,I2,I2,I2).

The repeat factor can also be used for a *group of descriptors* by enclosing them in a pair of parentheses and prefixing them with the factor. Some examples follow.

EXAMPLE A

    20   FORMAT (F3.2,2(F5.1,2I3,F6.2))

is equivalent to

    20   FORMAT (F3.2,(F5.1,I3,I3,F6.2,F5.1,I3,I3,F6.2))

or

    20   FORMAT (F3.2,F5.1,I3,I3,F6.2,F5.1,I3,I3,F6.2)

EXAMPLE B

    30   FORMAT (3(F6.1,5X),10X,2(I3,3X))

is equivalent to

    30   FORMAT((F6.1,5X,F6.1,5X,F6.1,5X),10X,(I3,3X,I3,3X))

EXAMPLE C

    40   FORMAT (11X,4(/),1X,15('*'))

is equivalent to

    40   FORMAT (11X,////1X,'***************')

EXAMPLE D

    50   FORMAT (11X,2I3,2(/1X,F6.2))

is equivalent to:

    50   FORMAT (11X,I3,I3,(/1X,F6.2,/1X,F6.2))

EXAMPLE E

    60   FORMAT (11X,2(2(F8.2,5X,I2)))

is equivalent to

    60   FORMAT (11X,2(F8.2,5X,I2,F8.2,5X,I2))

or

    60   FORMAT (11X,F8.2,5X,I2,F8.2,5X,I2,F8.2,5X,I2,F8.2,5X,I2)

The last example shows the use of nested groups. The maximum number of inner parentheses allowed in nesting groups depends on the system being used.

The repeat factor is a very useful feature of the FORMAT statement allowing it to be made more compact and flexible.

FORMAT Scanning

> **RULE 6.2**
> *If the number of field descriptors in a FORMAT exceeds the number of variables in an input or output, the excess field descriptors are ignored.*

For example, in

```
         READ (5,10) A,B
10       FORMAT (F5.1,F6.2,F8.2)
```

F5.1 and F6.2 will be used for A and B, and F8.2 will be ignored.

> **RULE 6.3**
> *If the number of variables listed in an input or output list is more than the number of field descriptors in a FORMAT, then* the same FORMAT will be repeatedly scanned as many times as needed *until all variables are accommodated. However, for any rescanning, a new data record (a new card, or a new line) will be started.*

For example,

```
         READ (5,100) X,Y,Z,U,V
100      FORMAT (F5.1,F3.0)
```

causes the variables X,Y to be read with field descriptors F5.1 and F3.0 from the first line; Z,U with F5.1, F3.1 from the second line; and V with F5.1 from the third line.

If it is desirable to repeat only a part of the format, the repeated descriptors can be placed in parentheses as a group. If there is more than one pair of parentheses, the scanning begins with the rightmost pair.

> **RULE 6.4**
> *The format control starts at the* rightmost left parenthesis *and continues to the right until either the output list is exhausted or the* final *right parenthesis of the FORMAT statement is encountered.*

For example, in

```
         WRITE (6,200) A,B,C,D,E,F,G
200      FORMAT (11X,2(F6.2,2X),(3X,F8.2),F10.2)
```

the FORMAT is equivalent to

```
200      FORMAT (11X,(F6.2,2X,F6.2,2X),(3X,F8.2),F10.2).
```

The values of A, B, C, and D will be printed on the first line according to the field descriptors F6.2, F6.2, F8.2, and F10.2, respectively. The values of E and F will be printed on the second line after 2 spaces (3X includes the carriage control) according to F8.2 and F10.2. Finally the value of G will be printed on the third line, after 2 spaces, according to F8.2. Remember also that a carriage control character is needed for a new record when printing.

Format scanning can become complicated with nested groups. Usually it is neither necessary nor advisable to use such arrangements.

SOLVED PROBLEMS

6.6 Interpret the field descriptor and show the position of each variable in each of the following cases:

a.
```
        READ (5,100) A,B,C,D,E,F,G,H
100     FORMAT (F6.2,2(F8.1,1X),F6.1)
```

b.
```
        READ(5,200) A,B,C,D,E,F,G,H,X,Y,Z
200     FORMAT (F6.2,2(F3.1,1X),2(F5.1,2X,F3.1),
       *F8.1)
```

c.
```
        WRITE (6,300) X,Y,Z,U,P,Q,R,S,T
300     FORMAT ('1','RESULTS',/,(1X,3F9.2))
```

d.
```
        READ (5,400) G,H,I,J,K,L
400     FORMAT (1X,2(F8.2,1X),(I3,2X,I4,2X))
```

☐ **Answers**

a. The FORMAT is equivalent to

```
100     FORMAT (F6.2,(F8.1,1X,F8.1,1X),F6.1)
```

Variables are read as follows:

Variable	Field Descriptor	Line
A	F6.2	1
B	F8.1	1
C	F8.1	1
D	F6.1	1
E	F8.1	2
F	F8.1	2
G	F6.1	2
H	F8.1	3

b. The FORMAT is equivalent to

```
200     FORMAT (F6.2,(F3.1,1X,F3.1,1X),
       *(F5.1,2X,F3.1,F5.1,2X,F3.1),F8.1)
```

Thus variables A,B,C,D,E,F,G,H will be read from the first data line with the field descriptors shown in the expanded FORMAT, and X,Y,Z will be read from the second line with F5.1, F3.1, and F5.1.

c. RESULTS will be printed, followed by values of the variables on the next lines, three variables on a line. Note the 1X for carriage control.

230 *Data Types and More about Input and Output*

> **d.** The FORMAT is equivalent to
>
>   FORMAT (1X,(F8.2,1X,F8.2,1X),(I3,2X, *I4,2X))
>
> The variables G,H,I,J will be read from the first data line, and K,L will be read from the second line with I3 and I4 respectively.

T Descriptor

The T descriptor may be used in the FORMAT statement for tabbing. It informs the computer of the position where the data are to be read or printed. It works the same way as the tab key of a typewriter. The general form of this specification is:

    T*n*

where *n* is an integer constant. For example,

```
      WRITE(6,10)A,B,C
10    FORMAT(T25,F5.2,T45,F5.2,T60,F5.2)
```

prints the value of A starting at column 24 (T25 includes the carriage control), the value of B starting at column 44, and the value of C starting at column 59. The statements are the same as

```
      WRITE(6,10)A,B,C
10    FORMAT(25X,F5.2,15X,F5.2,10X,F5.2)
```

Since the first character on the output is used for carriage control, we have

> **RULE 6.2**
> *The starting position of the information to be printed is one less than the number appearing in the T specification.*

The T field can also be used in the READ statement. For instance,

```
      READ(5,30)X,Y,K
30    FORMAT(T5,F5.1,T20,F5.2,T30,I3)
```

reads a value for X starting at column 5 (no carriage control), for Y at column 20, and for K at column 30.

> **RULE 6.3**
> *The starting position for the information to be read is the number which appears in the T specification.*

By using the T descriptor, the data can be read or written in any order. For example, the statements

```
      READ(5,10)X,M,N
10    FORMAT(T59,F5.2,T12,I5,T42,I3)
```

read the value of X starting from column 59, the value of M, starting from column 12, and the value of N starting from column 42 of the same line.

Or, the statements

```
        WRITE(6,20)X,Y,K,L
20      FORMAT(T69,F5.1,T25,F5.2,T39,I5,T28,I3)
```

print X starting at column 68, Y starting at column 24, K starting at column 38, and L starting at column 27 of the same line.

The T format and X format can accomplish the same purpose. However, using X is a convenient way to space between data, and using T is an easier way to indicate the exact column that the data are being read from or written to.

■ SOLVED PROBLEMS

■ 6.7 Describe the location of the information which will be read or written:

a.
```
        READ(5,50)A,B,C,D,I
50      FORMAT(T25,3F5.2,5X,F5.2,T1,I2)
```

b.
```
        WRITE(6,10)
        WRITE(6,20)
10      FORMAT('1',T53,'XYZ COMPANY')
20      FORMAT(T36,'WAGE',T56,'HOURS',T76,
       *'PAY')
```

c.
```
        WRITE(6,45)SUM,A,B,N
45      FORMAT(T50,F6.2,T20,F5.1,T35,F5.2,
       *T6,I2)
```

d.
```
        WRITE(6,65)X,Y,ION
65      FORMAT(T35,A4,5X,A4,T2,I2)
```

☐ **Answers**

a. A, B, and C will be read starting at column 25, five columns each (no space between them). D will be read starting at column 46, and I will be read starting at column 1.

b. The heading XYZ COMPANY will be printed at the top of a new page starting at column 52; then the headings WAGE, HOURS, and PAY will be printed on the next line, with WAGE starting at column 35, HOURS starting at column 55, and PAY starting at column 75.

c. This will print first the value of N starting at column 5, second the value of A starting at column 19, then the value of B starting at column 34, and finally the value of SUM starting at column 49.

d. This prints the value of X starting at column 34, the value of Y starting at column 43, and the value of ION starting at column 1.

6.8 What is the value of each variable read by the following READ statements?

a.
```
        READ(5,35)X,Y,JAY,KAY
35      FORMAT(T10,F5.2,T2,F3.0,T1,I1,T16,I4)
```
Data:
```
5629321832565321234
```

b.
```
        READ(5,25)A,B,C,I,J
25      FORMAT(A4,T7,A4,2X,F6.2,I3,T25,I2)
```
Data:
```
21FIRST ST.,58293132585329
```

c.
```
        CHARACTER CODE*10,ST*6
        READ(5,55)CODE,ST,A,B
55      FORMAT(A9,T1,A6,T10,F6.2,T11,F3.0)
```
Data:
```
FINISHING3298523
```

☐ **Answers**

a.
```
X = 256.53
Y = 629.
JAY = 5
KAY = 1234
```

b.
```
A = 21FI
B = T ST
C = 5829.31
I = 325
J = 29
```

c.
```
CODE = FINISHING
ST = FINISH
A = 3298.52
B = 298.
```

H Descriptor

In Chapter 2 we learned how a heading or an explanation can be printed by placing it in a FORMAT statement. We used apostrophes for this purpose. An alternative to apostrophes is specifying the number of characters to be printed preceded by the letter H. The following example demonstrates this technique.

EXAMPLE 6.3

■ **Problem:** Print the heading XYZ COMPANY.

Program:

```
        PROGRAM HEAD
        WRITE(6,10)
10      FORMAT(T51,11HXYZ COMPANY)
        END
```

Note: 11H in the FORMAT indicates that 11 characters will follow. This is another alternative for printing a string of characters. This FORMAT has the same effect as

```
10      FORMAT(T51,'XYZ COMPANY')
```

The general form of the H descriptor is

$nHC_1C_2C_3 \ldots C_n$

where n is the number of characters immediately following the letter H.

Some examples of H fields and their equivalent in apostrophe form are:

```
11HXYZ COMPANY              'XYZ COMPANY'
13HTHE ANSWER IS            'THE ANSWER IS'
19HTHE VALUE OF A IS =      'THE VALUE OF A IS ='
```

H descriptors can also be used for carriage control characters. The following table shows the carriage control characters in both forms:

Apostrophe form	H form	Instruction
' ' (or 1X)	1H	Advance to new line
'1'	1H1	Advance to new page
'0'	1H0	Advance two lines
'+'	1H+	No advance (stay on same line)

The H descriptor is not allowed on input.

When using the H format, the exact number of characters must be counted and specified. Because of this, one may find apostrophes more convenient to use.

☐ Printing an Apostrophe

In the English language, an apostrophe can be used to:

1. indicate a possessive case, e.g., Tim's hat;
2. mark omissions of letters (contractions), e.g., he can't;
3. form certain plurals, e.g., three B's.

234 Data Types and More about Input and Output

For printing such apostrophes, either pairs of apostrophes or the H format can be used. The techniques are shown by the following example.

EXAMPLE 6.4

■ **Problem:** Write statements to print TOM'S ANSWER before printing the value of X.

Using Apostrophes:

```
        WRITE(6,20)X
20      FORMAT(11X,'TOM''S ANSWER',F8.2)
```

Note that this technique requires using two apostrophes after TOM; only one of them will be printed.

Using an H Field:

```
        WRITE(6,20)X
20      FORMAT(1X,12HTOM'S ANSWER,F8.2)
```

The following table shows further examples of both forms:

H form	Apostrophe form
10HJOHN'S HAT	'JOHN''S HAT'
20HTHE NUMBER OF A'S IS	'THE NUMBER OF A''S IS'
15HTHE A WON'T FIT	'THE A WON''T FIT'

☐ Reading and Writing Literals

We have learned to use apostrophes (or the H format) in the FORMAT to print literals. There are several more options in FORTRAN. The following example demonstrates some of them.

EXAMPLE 6.5

■ **Problem:** Write the heading THE CLASS REPORT at the beginning of a program.

Procedure: Any of the following techniques can be used:

Technique I:

```
        WRITE(6,10)
10      FORMAT(1X,'THE CLASS REPORT')
```

Technique II:

```
        WRITE(6,10)
10      FORMAT(1X,15HTHE CLASS REPORT)
```

Technique III:

```
        CHARACTER*4 A1,A2,A3,A4
        READ(5,10)A1,A2,A3,A4
```

```
      10    FORMAT(1X,4A4)
            WRITE(6,10)A1,A2,A3,A4
```

On the data line:

```
 THE CLASS REPORT
```

(Note the space at the beginning.)

Technique IV:

```
            CHARACTER A*16
            READ(5,10) A
      10    FORMAT(1X,A16)
            WRITE(6,10) A
```

On the data line:

```
 THE CLASS REPORT
```

Technique V:

```
            WRITE (6,10) 'THE CLASS ', 'REPORT'
      10    FORMAT (1X,A10,A6)
```

Note: You are already familiar with the techniques I, II, III, and IV. In technique V, the literals enclosed in apostrophes will be printed with the WRITE statement. A10 and A6 in the FORMAT indicate the length of each part. FORMAT (1X, A,A) can be used also. Here is another example:

```
            WRITE (6,10) 'A EQUALS=', A,B
      10    FORMAT (1X,A,F8.2,'B EQUALS=',F8.2)
```

☐ The Slash

As discussed in Chapter 3, a slash (/) in the FORMAT statement indicates that a new data record is to be read or written. When it appears in the FORMAT of a READ statement, it causes the reading device to read a new line (record); when it appears in the FORMAT of a WRITE statement, it causes a new line to be started. For instance,

```
            READ(5,33)X,Y
      33    FORMAT(F5.2/F5.2)
```

will read X from the first line and Y from the second line, and

```
            WRITE(6,40)X,Y
      33    FORMAT(1X,F10.3/1X,F9.2)
```

will print the value of Y on the next line under the value of X.

Slashes can be used to produce blank lines on the page. *n* slashes place the printer on the *n*th line. This means *n* slashes cause *n*-1 lines (or cards) to be skipped or left blank. Note, though, that if the slashes are at the beginning or at the end of a format, *n* lines (or cards) are skipped (because the parenthesis

236 *Data Types and More about Input and Output*

also causes the start of a new line). In short, the number of blank lines (cards) is one less than the number of slashes used in the middle of the FORMAT. But the number of blank lines is equal to the number of slashes used at the beginning or at the end of the FORMAT.

Summary of Rules about the Slash

1. One or several slashes can be used anywhere in a FORMAT statement, either at the beginning, in the middle, or at the end.
2. Both a slash and a comma are field separators in a format. However,
   - A comma means: next data item in the same line/card.
   - A slash means: next data item in the next line/card.

   Although it is valid to use a comma before or after a slash, it is not necessary.
3. n slashes at the beginning or at the end of a format causes n lines (or cards) to be skipped; but using n slashes elsewhere causes n-1 lines (or cards) to be skipped.
4. A carriage control character is necessary after the slash in a FORMAT for a WRITE. If several consecutive slashes are used, a carriage control character can be used only after the last one. There is no need for carriage control in the READ format.

For further examples, look at the following solved problems.

SOLVED PROBLEMS

6.9 What is the value of each variable read by the following statements?

a.
```
      READ(5,10)A,B,M,N
10    FORMAT(F5.2,F3.0//I3//I4)
```
Data:

Line 1	96.2595.6253.21
Line 2	2319321
Line 3	56239213
Line 4	85321321
Line 5	6812

b.
```
      READ(5,20)X,Y,K,L
20    FORMAT(/F4.1/2X,F3.1,I1////I2)
```
Data:

Line 1	6.258.932
Line 2	36.225893.4
Line 3	5.64329.321
Line 4	685963
Line 5	JOHN99.99
Line 6	APPLE325
Line 7	5832

c.
```
        READ(5,30)P,Q,R,KODE
30      FORMAT(//3F2.1//1X,I1//)
        READ(5,35)N
35      FORMAT(I2)
```

Data:

Line 1	66289
Line 2	58923
Line 3	328956
Line 4	562890
Line 5	6829
Line 6	32895
Line 7	96.28
Line 8	4252

d.
```
        READ(5,60)A,B,M
        READ(5,60)C,D,N
60      FORMAT(F6.3/F6.3,I2)
```

Data:

Line 1	632.558225
Line 2	8932.13289
Line 3	9385.648
Line 4	3128.369

☐ **Answers**

a.
A = 96.25	from line 1
B = 95.	from line 1
M = 562	from line 3
N = 6812	from line 5

b.
X = 36.2	from line 2
Y = 64.3	from line 3
K = 2	from line 3
L = 58	from line 7

c.
P = 3.2	from line 3
Q = 8.9	from line 3
R = 5.6	from line 3
KODE = 8	from line 5
N = 42	from line 8

d.
A = 632.55	from line 1
B = 8932.1	from line 2
M = 32	from line 2
C = 9385.6	from line 3
D = 3128.3	from line 4
N = 69	from line 4

6.10 Suppose A = 5.62, B = 1.08, I = 628. Show what the output would be (and the positions) when using the following WRITE and FORMAT instructions:

a.
```
       WRITE(6,20)A,B,I
20     FORMAT(1X,F4.2/1X,F4.2/1X,I3)
```

b.
```
       WRITE(6,30)I,A,B
30     FORMAT(I3/F4.2/F4.2)
```

c.
```
       WRITE(6,40)A,B,I
40     FORMAT(1X,F4.2//1X,F4.2//1X,I3)
```

d.
```
       WRITE(6,50)A,B,I
50     FORMAT(//1X,F4.2//1X,F4.2///1X,I3//)
       WRITE(6,55)I
55     FORMAT(1X,I3)
```

e.
```
       WRITE(6,60)A,B
60     FORMAT(1X,F4.2/F4.2)
       WRITE(6,65)I
65     FORMAT(I3)
```

☐ **Answers**

a.

Line 1   5.62
Line 2   1.08
Line 3   628

b.

Line 1   28
Line 2   .62

New page
Line 1   .08

Notice that the first character of each record (line) is taken as a carriage control.

c.

Line 1   5.62
Line 2
Line 3   1.08
Line 4
Line 5   628

d.

Line 1
Line 2
Line 3   ◯ | 5.62
Line 4   ◯ |
Line 5 | 1.08
Line 6   ◯ |
Line 7
Line 8   ◯ | 628
Line 9   ◯ |
Line 10
Line 11   ◯ | 628

e.

Line 1   ◯ | 5.62

New page

Line 1   ◯ | .08
Line 2   ◯ | 28

Again, notice that the first character is taken as a carriage control.

6.11 Write a program which prints the letter A like the following picture:

```
AAAAAAAAA
A       A
A       A
A       A
AAAAAAAAA
A       A
A       A
A       A
```

☐ **Answer**

This can be done by several techniques. Here are two:

```
      PROGRAM AAA
      WRITE(6,10)
      WRITE(6,20)
      WRITE(6,10)
      WRITE(6,20)
10    FORMAT(21X,9('A'))
20    FORMAT(21X,'A',7X,'A',,2(/21X,'A',
     *7X,'A'))
      END

      PROGRAM AAA
      CHARACTER*9 X,Y
      DATA X,Y/'AAAAAAAAA','A       A'/
      DO 100 I = 1,2
```

```
          100   WRITE(6,10)X,Y,Y,Y
          10    FORMAT(21X,A9)
                END
```

LIST-DIRECTED INPUT AND OUTPUT

FORTRAN allows the reading or writing of information to be controlled by the programmer's design as expressed in a FORMAT statement. FORTRAN 77 compilers, however, also accept READ, WRITE, or PRINT statements without the FORMAT statement. Input or output without FORMAT is referred to as *list-directed* or *format-free* input or output.

☐ List-Directed Input

The general form of the list-directed READ statement is:

```
READ*, Var1, Var2, ....
```

Notice the asterisk (*) and comma after the word READ. Example:

```
READ*, A, B, C, N
```

Another alternative for the list-directed READ is use of the general form of the formatted READ, with asterisks substituted for format numbers:

```
READ ( 5,*) A, B, C, N
```

If it is desired to use a file which is assigned to a device number other than 5, then the form:

```
READ (u,*) A,B,C,N
```

can be used, where u is the device number, which causes the data to be read from a specific file. If an asterisk is used instead of a device number, then the 'INPUT' file will be assumed. Example:

```
READ(* , *) A,B,C,N
```

In any case,

```
READ*,A,B,C,N
```

instructs the computer to read in three real numbers for A, B, C, and an integer number for N. The numbers are normally placed on one line, separated either by one or more spaces or by a comma. However, they can also be recorded on separate lines (cards), because the computer keeps on reading from successive lines (or cards) until all data are read. For example, the statement

```
READ*, LENGTH, WIDTH, HEIGHT, NUMBER
```

used with the two data lines

```
5625, 629.50      5625.2
262
```

will determine the value of LENGTH to be 5625, WIDTH to be 629.5, HEIGHT to be 5625.2, and NUMBER to be 262.

☐ List-Directed Output

The general form for the list-directed WRITE statement is

```
WRITE (6,*) A, B, C, M
```

Again, 6 is the output device number, which leads to the use of a specific output file. In practice this number depends on the file you are using.

An alternative to the list-directed WRITE statement is the PRINT statement, as it was used in Chapter 1. For example,

```
PRINT*, A, B, C, M
```

Notice again the asterisk and comma after PRINT. Hereafter, we use this form for our illustrations.

The above PRINT statements instruct the computer to print the values of three real variables A,B,C and the value of the integer M on one line. Notice also that every PRINT statement automatically produces a blank for carriage control as the first character of the record, and thus causes the printer to start on a new line. If the values of several variables are desired to be printed on different lines, several PRINT statements must be used. For example,

```
PRINT*,X, Y, Z
PRINT*,M, N, K
```

will print the values of X, Y, Z on one line and the values of M, N, K on the next line.

For the programmer who is not concerned about the exact location of the data, the format-free input and output are very useful. Also, they are often used by beginners who would like to write and run a simple program without being worried about how and where the data appear.

The format-free PRINT is used frequently by skilled programmers to debug a program, especially when it contains a logical error. The programmer uses several print statements in different parts of the program to examine the values of certain variables, to see if they are as expected. After finding the "bugs," those print statements can be dropped from the program.

If you are planning to use unformatted input or output, pay attention to the following notes.

☐ Important Notes about List-directed READ and WRITE

1. Either a comma or a space can separate numbers in the data line. For example, with

   ```
   READ *, X, Y, M
   ```

   either

   ```
   58.9,103.5,89
   ```

   or

   ```
   58.9    103.5    89
   ```

   is valid.

2. Two successive commas indicate that no value is to be read for the corresponding variable. For example, with

   ```
   READ *,X, Y, M
   ```

and the data

```
58.9,,89
```

no value for Y will be read.

3. A slash on the data line indicates the end of data. For example, in

READ *, A, B, C, D, I, J, K, L, M

with the data

```
9.1, 56.2,    12.5/
```

the first three values will be assigned to A, B, and C. No value for D, I, J, K, L, M will be read, because the slash indicates the end of the data.

4. Explanatory comments and simple mathematical expressions are allowed in the print statement. For example:

PRINT*, A, B, 'THE SUM IS', A + B

is a valid statement.

5. An asterisk (*) can be used to indicate a repetition factor on the data line. For example, with

READ*, A, B, C, D

the data line

```
4*3.5
```

indicates the value of A, B, C, and D as 3.5.

6. For reading a string of characters, the variable must have been defined as character type and the data must be placed between apostrophes. For example,

READ*, M, N

with the data

```
5, 'PAUL'
```

is valid if N has been defined as character type; this indicates that M = 5 and N = 'PAUL'.

■

FORMATTED PRINT STATEMENT

The FORMAT statement can be used with the PRINT statement as discussed in Chapter 2. The general form of the formatted PRINT statement is

PRINT fn, list

where fn is the format statement number. The following are some examples:

EXAMPLE A

```
        PRINT 10, A, B
 10     FORMAT (1X, F6.3, 5X, F5.2)
```

Note that 1X is for the carriage control.

EXAMPLE B

```
        PRINT 20, ' ', A, B
 20     FORMAT (A1, F6.3, 'IS A', 5X, F5.2, 'IS B')
```

The ' ' in the print statement and A1 in the FORMAT are for the carriage control.

EXAMPLE C

```
        PRINT 30,' ', A, 'IS A', B, 'IS B'
 30     FORMAT (A, F6.3, A, F5.2, A)
```

Note that the field length is omitted in the FORMAT. Thus the length is equal to the number of characters specified in the PRINT statement.

EXAMPLE D

```
        CHARACTER X*6, NEWPGE*1
        DATA NEWPGE, X, B, N/'1', 'FINISH', 62.5, 5/
        PRINT 40, NEWPGE, 'RECORD #', N, 'IS', B, X
 40     FORMAT (A, T10, A, I1, A, F6.2, A)
        END
```

EXAMPLE E

```
        A = 3.2
        PRINT 10, ' ', 'THE', 'VALU', 'E OF', ' A=', A
 10     FORMAT (A1, 4A4, F5.1)
```

SOLVED PROBLEMS

6.12 What value is read for each variable by the following statements? Assume the character variables have been defined.

a.

READ *, LENGTH, WIDTH, HEIGHT

Data:

```
32      , 2*12.5
```

b.

READ *, A, B, C, D, M, N

Data:

```
2*25.5,,,2*5
```

c.

READ *, X, Y, Z, G, H

Data:

```
59.5, , 2.5/
```

d.

READ *, M, N, P

Data:

```
5, 'JOHN', 'JOE'
```

e.

READ *, K, L, M, N

Data:

Line 1   360
Line 2   1359
Line 3   3
Line 4   069   .52

f.

READ *, A, B, N
READ *, X, M

Data:

Line 1   36.5,   2.6
Line 2   5   ,   80.5,   32
Line 3   69.5,32,6

☐ **Answers**

a.
   LENGTH = 32
   WIDTH = 12.5
   HEIGHT = 12.5

b.
   A = 25.5
   B = 25.5
   C,D: no value read
   M = 5
   N = 5

c.
> X = 59.5
> Y: no value read
> Z = 2.5
> G, H: no value read

d.
> M = 5
> N = ⎡J⎤⎡O⎤⎡H⎤⎡N⎤
> P = ⎡J⎤⎡O⎤⎡E⎤⎡ ⎤

e.
> K = 360
> L = 1359
> M = 3
> N = 069

f.
> A = 36.5
> B = 2.6
> N = 5
> X = 69.5
> M = 32

Note that only the first number from the second card is read. Because a new READ causes a new card to be read, the values of X and M will be read from the third card.

6.13 Find the errors, if any, in the following statements:

- **a.** READ A, B, D, C
- **b.** READ*, END = 99, S, R, X
- **c.** READ *, A B C
- **d.** READ(5,*)A,B,C
- **e.** WRITE (6,*)X, Y, Z
- **f.** WRITE *, A, B, C
- **g.** WRITE, X, Y, Z
- **h.** PRINT *, 'A IS =' A 'B IS =' B 'C IS =' C
- **i.** PRINT *, 'A =', A, 'B = ', B, 'C =', C
- **j.** PRINT *, 'A' = A, 'B' = B, 'C' = C
- **k.** READ *, N, A
  with data:

  ```
  KAREN RITA MORRISON, 36.5
  ```

- **l.** PRINT*, '4811 EAST MICHIGAN, MT. PLEASANT'

☐ **Answers**

- **a.** An asterisk and a comma are required after READ.
- **b.** An END option cannot be used with a format-free READ. The form READ (*,*, END = 99) S, R, X can be used.
- **c.** Commas are needed between the variables.
- **d.** This is correct.
- **e.** This is also correct.
- **f.** The word PRINT must be used instead of WRITE.
- **g.** The word PRINT must be used instead of WRITE. Also an asterisk is needed.
- **h.** Commas are needed between the strings and variables.

246 *Data Types and More about Input and Output*

    i. This is correct.

    j. The equal signs must be between the apostrophes:

```
PRINT*, 'A=', A, 'B =', B, 'C=', C
```

    k. The characters in the data line must be between apostrophes. Also, the length of the variable N should have been defined.

    l. This is correct.

6.15 Find the errors, if any, in the following formatted PRINT statements:

    a.
```
       PRINT 10, A, B, C
10     FORMAT (3F6.1)
```

    b.
```
       PRINT 30, 'THE VALUE OF THE VARIABLES
      *ARE'
30     FORMAT (1X, A31)
```

    c.
```
       PRINT 40, A, B, I, J
40     FORMAT (1X, 'A=',F5.2,'B=',F5.2,
      *'THE VALUE OF I,J, ARE =',2I6)
```

☐ **Answers**

    a. The carriage control is missing.
    b. This is correct.
    c. This is correct.

EXERCISES

☐ **Character Data**

6.1 Find the errors in each of the following statements:

    a. CHARACTER*9, N, M, J, K

    b. CHARACTER A*5 B*10 C*6

    c. CHARACTER*5 X*3 Y*9 Z*5

    d.
```
       CHARACTER*18 N, M*9
       READ(5,10) N,M,K
10     FORMAT(2A18,A9)
```

    e.
```
       CHARACTER*12 SSN, ADRES
       READ(5,100)SSN,ADRES
100    FORMAT(3A4,3A4)
```

f.
```
        READ(5,90) NAME,GRADE
90      FORMAT(3A4,F5.2)
```

g. NAME = JOHN

h. GRADE = B

i. IF(SCR .GT. 90.0) GRADE = A

j. DATA LNAME,FNAME/BROWN,JOHN/

k. DATA GRADE/'A','B','C','D'/

l. RE'

6.2 What is the value of each variable read by the following READ statements?

a.
```
        CHARACTER*4 A1,A2,A3,A4
        READ(5,35) A1,A2,A3,A4
35      FORMAT(4A4)
```

Data:

```
4811 E. MICHIGAN
```

b.
```
        CHARACTER*2 NAME1,NAME2,NAME3,NAME4
        READ(5,55) NAME1,NAME2,NAME3,NAME4
55      FORMAT(4A4)
```

Data:

```
MARGARET JONES
```

c.
```
        CHARACTER N1,N2,N3,N4,N5,N6,N7,N8,N9,N10
        READ(5,85) N1,N2,N3,N4,N5,N6,N7,N8,N9,N10
85      FORMAT(10A1)
```

Data:

```
FORTRAN 77
```

d.
```
        CHARACTER*9 M1, M2, M3
        READ(5,105) M1,M2,M3
105     FORMAT(3A9)
```

Data:

```
THIS IS A FORTRAN EXAMPLE
```

248 *Data Types and More about Input and Output*

6.3 Find the errors in the following programs:

   a.
```
              PROGRAM CHAR
              READ(5,10) NAME,DRESS,SOCSEC,SCORE
       10     FORMAT(3A12,F6.2)
              WRITE(6,20)NAME,DRESS,SOCSEC,SCORE
       20     FORMAT(1X,3A12,F10.2)
              END
```

   b.
```
              PROGRAM SUBCHR
              READ(5,25)NAME1,NAME2,NAME3,SCR1,SCR2,SCR3
              AVG = (SCR1 + SCR2 + SCR3)/3.0
              WRITE(6,35) NAME1,NAME2,NAME3,AVG
       25     FORMAT (A12,3F5.2)
       35     FORMAT(11X,A12,5X,F6.2,5X,F6.2,5X,F6.2)
              END
```

   c.
```
              PROGRAM SEVRL
              READ(5,100) NAME,SCR1,SCR2
              SUM = SCR1 + SCR2
              AVG = SUM/2.0
              WRITE(6,200)NAME,SUM,AVG
       100    FORMAT(3A4,2F5.2)
       200    FORMAT(11X,3A4,5X,F6.2,5X,F6.2)
              END
```

Data:

```
ROB JOHNSON 85.5099.5
```

☐ **T Descriptor**

6.4 What is the value of each variable read by the following READ statements?

   a.
```
              READ(5,80)A,B,C,I
       80     FORMAT(T10,F5.2,T2,F3.1,T5,F5.2,T1,I2)
```
Data:

```
56239832528123
```

   b.
```
              CHARACTER*4 A,B,C
              READ(5,90)A,B,C,I
       90     FORMAT(T12,3A4,T1,I2)
```
Data:

```
25123456789THIS IS A SURPRISE
```

c.
```
       CHARACTER*12 M,N,Y
       READ(5,110) M,N,X,Y
110    FORMAT(T5,2A4,F3.1,T2,A3)
```

Data:

```
TTTTABCDEFGH9.9
```

☐ **The Slash**

6.5 Suppose we have the following data file:

Line 1	623.59832652
Line 2	58932.8321536328
Line 3	93285.2328632852
Line 4	59823.83556329
Line 5	632885533.289645

What will be the value of each variable when each of the following program segments are executed with the above data?

a.
```
       READ(5,30)A,B,M
30     FORMAT(2F6.2,////I2)
```

b.
```
       READ(5,40)X
       READ(5,40)Y
       READ(5,40)Z
       READ(5,40)V
       READ(5,40)W
40     FORMAT(F6.2)
```

c.
```
       READ(5,50)X,Y,Z,V,W
50     FORMAT(F6.2/F6.2/F6.2/F6.2/F6.2)
```

d.
```
       READ(5,60)X,Y,M
       READ(5,70)A,B,N
60     FORMAT(1X,2F4.2/I3)
70     FORMAT(7X,F5.1/6X,F6.1/I3)
```

e.
```
       READ M
       READ(5,80)P,Q,R,S,T,U,M
80     FORMAT(4X,F7.2/F7.2/2F6.2/F4.1/2F4.2)
```

6.6 Suppose

A = 6.893
B = 325.3
I = 522

Show the position of each item, and what the output will be, when the following WRITE and FORMAT instructions are executed:

a.

```
        WRITE(6,50)A
        WRITE(6,50)B
        WRITE(6,60)I
50      FORMAT(1X,F7.3)
60      FORMAT(1X,I3)
```

b.

```
        WRITE(6,70)A,B,I
70      FORMAT(1X,F7.3/1X,F7.3/1X,I3)
```

c.

```
        WRITE(6,80)A,B,I
80      FORMAT(21X,'ABC CO.'///13X,'A',10X,'B',10X,
       *'I'///11X,F7.3,3X,F7.3,5X,I4)
```

d.

```
        WRITE(6,90)I,A,B
90      FORMAT(1X,I4///1X,F6.2///1X,F7.1)
```

☐ **List-directed Input and Output**

6.7 Find the errors in the following statements:

a. READ A,X,Y

b. READ P Q R

c. READ*,(END=55)A,B,C

d. READ, (5,*),X,A,B

e. REED*,U,P

f. WRITE X,Y,Z

g. PRINT,A = A, B = B, C = C

h. PRINT A B C

i.
```
CHARACTER NAME*4
READ*,NAME
```
Data:

```
JOHN
```

6.8 What value will be read for each variable when each of the following statements is executed?

a.

READ*,AA,BB,CC

Data:

```
5.6,,6.2,8.1
```

b.

```
READ*,I,J,K
```

Data:

```
2*5,3
```

c.

```
READ*, A,B,C,D,E
```

Data:

```
3*5.1/
```

d.

```
CHARACTER*4 NAME, ADRES
READ *, NAME, ADRES
```

Data:

```
'PAT','1ST'
```

e.

```
READ *, X, Y, Z
```

Data:

Line 1 62.5
Line 2 31.9
Line 3 6235.2

f.

```
READ*,P,Q,R
READ*,M,N,K
READ*,X,J,Y,I
```

Line 1 .95, 52.1, 393.5, 162
Line 2 72 86 52 89
Line 3 69.5 50
Line 4 95.2 70 80.5 60

g.

```
READ*,A,B,C,M,N
READ*,M,N,K
READ*,E,F,I,J
```

252 Data Types and More about Input and Output

```
Line 1   32.5, 80.9/
Line 2   8 , 3 /
Line 3   53.5 , 90.2
Line 4   5 , 8 , 3 , 30
```

☐ **Format Specification**

6.9 Find the errors in each of the following statements:

   a.
```
        READ(5,105) X,Y,Z,A
105     FORMAT(F10.2,20X,F8.2,20X,F6.2,10X,F8.2)
```
An 80-column punched card is used as an input record.

   b.
```
        WRITE(6,120) B,C,D,M
120     FORMAT(11X,'THE FIRST NO. =',F10.2,20X,'THE
       *SECOND NO. =',F10.2,20X,'THE THIRD NO. =',
       *F10.2,20X,'M =',I8)
```
Hint: For parts **a.** and **b.** pay attention to the length of the input or output record.

   c.
```
        READ(5,70) NAME, M, N
70      FORMAT(3A4,1X,A4,1X,A4)
```

   d.
```
        READ(5,80) A,B,C,D
        S = A + B + C + D
        WRITE(6,90) A,B,C,D,S
        END
90      FORMAT(5F6.3)
80      FORMAT(4F6.3)
        STOP
```

6.10 Describe the position and the field descriptor of the variables in the following statements:

   a.
```
        WRITE (6,100) X,Y,Z,A,B,C,D
100     FORMAT (1X,F6.1,6(1X,F5.2))
```
   b.
```
        WRITE (6,100) X,Y,Z,A,B,C,D
100     FORMAT (11X,F6.2)
```
   c.
```
        READ (5,200) X,Y,Z,D,F,G
200     FORMAT (F9.2,(F6.1))
```
   d.
```
        READ (5,150) P,Q,R,S,T,U,V,W
150     FORMAT (F6.1,2(F5.2,1X))
```

e.

```
          READ (5,300) A,B,C,D,E,F,G,H,I,J,K,L
300       FORMAT (8(F5.1,1X),(I2,I3))
```

f.

```
          WRITE (6,400) A1,A2,A3,A4,A5,A6,A7,A8,A9
400       FORMAT (5X,2(F6.2,2X),(1X,F3.1,F7.2),2X,F6.1)
```

g.

```
          READ (5,30) X1,X2,X3,X4,X5,X6,X7,X8,X9,X10,X11
30        FORMAT (F6.1,1X,2(3(F3.1,1X),F2.0,1X))
```

PROGRAMMING EXERCISES

Write a complete FORTRAN program for each of the following problems. Use appropriate headings, spaces, FORMATS, loops, and structured programming techniques.

6.11 Assume you are given a deck of data cards with information concerning the students in your FORTRAN course. The information about each student is punched as:

Name of the student, maximum of 16 characters, columns 1–16.
Course grade (A, B, C, D, or E), column 18.

Write a program which reads the data and prints them as a report. Search the names for your name; print '*' under your name in the report. NO MORE STUDENTS is punched on the last card in the name field. Terminate the DO loop by reading this card.

6.12 The data about grades for a course are recorded as follows:

Columns 1–10   Last name of student
Columns 11–20   First name
Column 22   Final grade: A, B, C, D, or E.

Write a program which reads the information and prints the following in report form:

The first name.
The last name.
The grade.

One of the students has changed her name from PAT JOHNSON to PAT WILLIAMS. Print her new name instead of the old name in the report. Print the following at the end of the report:

The average of the grades (on a scale of 1–4).
The numbers of A's, B's, C's, D's, and E's.

Use a DO loop for reading the data. There are less than 100 students. Terminate the loop with a trailer record with the word FINISH on it.

6.13 A part of a textbook is placed in a data file, line by line in columns 2–80. Altogether 93 lines are used. Write a program which reads the lines and prints the information as it was entered.

6.14 Design and develop a program which prints a series of names and addresses.

6.15 Write a program which reads a word such as your name, and prints it scrambled.

6.16 Write a program which reads a name with a given number of characters and prints each letter on a different line, in a diagonal form. For example, if it reads the name JOHNY, it prints

```
J
 O
  H
   N
    Y
```

Test the program with your name.

6.17 Write a program which prints the following picture:

```
      XXXXXXX
   X           X
X      MMM MMM    X
X       O   O     X
X                 X
X                 X
X       XXXXX     X
X        XXX      X
   X           X
      X     X
         XXX
```

6.18 Write a program which prints the letter E on a larger scale:

```
EEEEEEEEEEEE
EEEEEEEEEEEE
EEEE
EEEE
EEEEEEEE
EEEEEEEE
EEEE
EEEE
EEEEEEEEEEEE
EEEEEEEEEEEE
```

6.19 Write a program which calculates the pay for the workers who are in an incentive plan. Each worker's information is on three lines as follows:

Line 1:
    Name (maximum of 20 characters)   Columns 1–20
    Wage rate   Columns 22–26

Line 2:
    Hours worked for the week (regular)   Columns 1–5
    Overtime (over 40 hours, if any)   Columns 7–11

Line 3:
    Number of extra units produced   Columns 1–5
    Wage rate for each extra unit produced   Columns 7–11

There are less than 100 employees. The phrase NO MORE EMPLOYEE is entered in the name field on the last line. Overtime is paid at 1.5 times the regular rate. Print the input information—name, wage rate, hours worked, overtime, number of extra units produced, and rate of pay—and the total pay to each employee in a report form. Print the total pay to all employees at the end of the report.

6.20 Given is a data file of students' test scores as follows:

1. Student's name, maximum of 12 characters	Columns 1–12
2. Student's social security number, 11 characters	Columns 13–23
3. Student's test scores (in the form 99.99):	
Test 1	Columns 24–28
Test 2	Columns 29–33
Test 3	Columns 34–38

Write a program which calculates the average of the tests, the letter grades (100–90, A; 90–80, B; etc.), the grand average, and the numbers of A's, B's, C's, D's, and E's. Print the information. Design the output. Use a DO loop. There are less than 100 students. Terminate the loop by a trailer record.

6.21 Write a program which prints address labels.

Input:

1. First name and middle initial, maximum of 12 characters	columns 1–12
2. Last name, maximum of 12 characters	columns 13–24
3. Address, maximum of 24 characters	columns 25–48
4. City and state, maximum of 24 characters	columns 49–72
5. Zip code, 5 characters	columns 73–77

Output:

1. First and last name, starting at position 16.
2. Address, on the next line, starting at position 16.
3. City, state, and zip code, on the next line, starting at position 16.

Use a DO loop to read the data. Terminate the loop with a trailer record with FINISH in the name field. Assume there are less than 100 records.

6.22 Repeat the previous exercise, but print two address labels side by side. Be sure to design your output on a print chart before coding. (Hint: You should use two READ statements to read the two labels.)

6.23 The manager of a machine shop would like a list of the equipment in the shop. Write a program which reads the following information:

1. The name of each piece of equipment, maximum of 12 characters	columns 1–12
2. Date purchased:	
a. month	columns 14–15
b. year	columns 16–17
3. Cost when purchased	columns 19–25
4. A code for usage (A, B, or C)	column 27
5. Equipment location, 3 characters	columns 29–31

Print the input information for each piece of equipment. Print also how many pieces are available for each usage code at the end of the report. Use a DO loop for reading the data. The trailer record with NO MORE in the equipment name field terminates the loop. Assume there are less than 100 pieces of equipment.

6.24 Write a program which reports the total sales by the four divisions of a company: Detroit, MI; New York, NY; Miami, FL; and Columbia, SC.

There are three different products with respective codes AA, AB, BC.

Input:

1. Division name	columns 1–14
2. Product code	columns 15–16
3. Amount of sales	columns 18–27

Output:

1. Division name, product code, and amount of sales for all divisions as a report.
2. Total sales of product AA for all divisions.
3. Total sales of product AB for each division.
4. Total sales of product BC for each division.
5. Total sales for all divisions.

6.25 Write a program to report students' current GPA.

Input:

1. Student's name	columns 1–15
2. GPA, on a scale of 0.00 to 4.00	columns 17–20

Output:

1. Student's name, GPA, and the word HONOR for the students whose GPA is more than 3.3, or SUSPENDED for those whose GPA is less than 1.7.
2. Print the total number of students, the number of honor students, and the number of suspended students at the end of the report.

Use a DO loop for reading the data. Terminate the loop with a trailer record which says NO MORE STUDENTS. Assume there are less than 500 students.

6.26 A union would like to see some statistics about the salaries of the employees in a company. They have access to the following input data:

1. The names of the employees, maximum 20 characters	columns 1–20
2. Permanent-nonpermanent code: SP = permanent, SN = nonpermanent	columns 21–22
3. Marital status: S = single, M = married	column 23
4. Sex: M = male, F = female	column 25
5. Age	columns 27–30
6. Salary	columns 32–39

Write a program which reads this information and prints it in report form. Add a column to the report showing the category of the employees:

Category A if the salary is more than or equal to 20,000.00.
Category B if the salary is more than or equal to 10,000.00 but less than 20,000.00.
Category C if the salary is less than 10,000.00.

Print the following at the end of the report:

1. Average salary of all employees.
2. Average salary of males.
3. Average salary of females.
4. Average salary of the employees who are over sixty years old.
5. The number in each employee category (A, B, and C).

6.27 Write a program which prints weekly check stubs for employees of the XYZ company. The input is:

Employee's ID	Columns 1–5
Hours worked for the week	Columns 7–11
Wage rate	Columns 13–17
Employee's name	Columns 19–30
Employee's address (number and street)	Columns 32–51
Employee's address (city and state)	Columns 53–72

Let the deductions be 15% of the gross pay; time over 40 hours is paid at 1.5 times the regular rate. The last record is a trailer record with the number −9999 as the ID number. An example of the output:

```
***********************************************************
*                                                         *
*                    XYZ COMPANY                          *
*                                                         *
*    EMPLOYEE # 99999         JOHN DOE                    *
*                             4811 E. MICHIGAN ST.        *
*    HOURS WORKED XX.XX       MT. PLEASANT, MI 48858      *
*                                                         *
*    TOTAL PAY XXX.XX         DEDUCTIONS XXX.XX           *
*                                                         *
*                    NET PAY XXX.XX                       *
*                                                         *
***********************************************************
```

6.28 Rewrite the previous program so that it prints two check stubs side by side. (Hint: You should use two READ statements to read the information on two employees.)

SUMMARY OF CHAPTER 6

You have learned:

1. Data types:
   a. REAL and INTEGER statements can be used to define the type of a variable as real or integer, for example,

   REAL NET, MAY
   INTEGER COUNT, SEQ

   b. The CHARACTER statement must be used to declare the variable as character type and the length of the data for each variable. The following are examples of three forms of the CHARACTER statement:

   CHARACTER M*12 , N*12 , K*6
   CHARACTER*12 M,N
   CHARACTER*12 M,N,K*6

   c. The field descriptor for alphanumeric data is A. For example, A6 in the FORMAT indicates that the length of the data field is 6.
2. The DATA statement can be used to initialize the variables at the beginning of a program. The following is an example of the DATA statement:

DATA A,B,C,M,NAME/5.6,2*2.5,8,'JOE'/

3. FORMAT features:
   a. A group of field descriptors can be repeated by placing them in a pair of parentheses, preceded by an appropriate multiplier.
   b. If the number of variables in the input or output list is more than the number of field descriptors, the format will be repeated automatically until all of the variables in the list are exhausted. If there are any inner parentheses in the format, the scanning starts from the rightmost left parenthesis.
   c. If the number of field descriptors in the format is more than the number of variables in the list, the extra field descriptors are ignored by the compiler.
   d. The T descriptor can be used for tabbing. It shows the starting position of the data to be read or written.
   e. The H descriptor is an alternative method to using apostrophes for printing literals—headings, endings, or other descriptions—in the FORMAT of an output statement. The literal is preceded by nH (n equals the number of characters) rather than being enclosed in apostrophes.
   f. A slash can be used in a FORMAT to introduce a new record—a new line when printing or reading. Several consecutive slashes can be used to skip several data records. n slashes at the beginning, or at the end, of a FORMAT cause n lines to be skipped. n slashes elsewhere in the FORMAT cause n-1 lines to be skipped.
4. List-directed input and output: If the programmer is not concerned about the location of the data, he can use format-free READ, WRITE, or PRINT statements. The following are examples of each:

```
READ*, A, B, C . . .
READ(5,*)A,B,C . . .
WRITE (6,*) A,B,C . . .
PRINT *, A, B, C . . .
PRINT *, 'THE', 'VALUE', 'IS', X
```

The * after READ or PRINT is required.

5. Formatted PRINT: The FORMAT statement can be used with the PRINT statement. The following is an example:

```
      PRINT 10, ' ', 'X =', X, 'Y =', Y
10    FORMAT (A1,A2,F6.1,A2,F6.1)
```

SELF-TEST REVIEW

6.1 Find the errors, if any, in each of the following statements:

a. READ A, B, C,

b. PRINT*, A, B, C,

c. WRITE, A, B, C

d. CHARACTER*12, M,N,K*9

e. M = JOHNY PORT

f.
```
      CHARACTER*8 L,M
      READ(5,20) L,M
20    FORMAT(2A12)
```

g.
```
      READ(5,50)A,B,C
50    FORMAT(F3.2/T9,F3.0/5X,T9,F8.9)
```

h. `CHARACTER A*2, REAL MAY`

i. `DATA AG,BG,CG/0.0/, N/3,2,1/`

6.2 What is the value of each variable read by the following statements? Assume the character variables have been defined.

a.
```
READ *,X, Y, M
READ *,C, D
```

Data:

Line 1	52.123 69325.0
Line 2	'JOHNY' 82.4
Line 3	956.5 532.5

b.
```
      READ(5,40) NAME1,NAME2,NAME3,ISSN
40    FORMAT(3A4,A4)
```

Data:

```
RITA K. PAGE 562-63-2348
```

c.
```
      READ(5,50) M,N,P,Q
      READ(5,60) X,Y,NAME
50    FORMAT(T4,I2,T9,A4,1X,F3.2/T5,F5.1)
60    FORMAT(2F3.2//A4)
```

Data:

Line 1	9.3259328JIMY563
Line 2	693528.5329891.32
Line 3	.56.98JONES
Line 4	PAM
Line 5	ARNY

d.
```
READ (5,100) A1,A2,A3,A4,A5,A6
FORMAT (F3.1,2(F3.0,1X),(F4.0))
```

Data:

Line 1	5623198235689328512
Line 2	4267528
Line 3	59321832

6.3 Write a complete FORTRAN program for printing a report about customers' gas usage. Assume there are less than 100 users. The phrase NO MORE is entered in the trailer record.

Input:

1. Customer's name, maximum of 12 characters	columns 1–12
2. Customer's code, 7 characters	columns 14–20
3. Customer's type: R for residential, I for industrial	column 22
4. Last meter reading, 7 digits	columns 24–30
5. Current meter reading, 7 digits	columns 32–38

Calculation:

1. Calculate the charges at:
   a. 32.5¢ per unit for residential use.
   b. 29.9¢ per unit for industrial use.
2. Find the number of residential users and the number of industrial users.
3. Calculate the total amount of gas used.
4. Calculate total charges to all customers.

Output:

1. Print the sequence number and the input information in the same order as input for each customer. Print RESIDENTIAL if the code is R, and INDUSTRIAL if the code is I.
2. Print the totals at the end.

☐ **Answers**

6.1

a. Needs an asterisk and a comma after the word READ.
b. The comma after C must be omitted.
c. The word PRINT should be used instead of WRITE.
d. No comma is needed after 12.
e. Apostrophes are needed when assigning a constant to a character variable.
f. The length of the data is more than what is declared. (This is syntactically correct.)
g. This is correct.
h. The CHARACTER statement and the REAL statement cannot be mixed.
i. The number of variables does not correspond with the number of constants in each list.

6.2

a.
X = 52.123
Y = 69325.0
M = | J | O | H | N | Y |
C = 956.5
D = 532.5

b.
NAME1 = | R | I | T | A |
NAME2 = | | K | . | |

NAME3 = |P|A|G|E|
ISSN = |5|6|2|

c.

M = 25
N = |8|J|I|M|
P = 5.63
Q = 28.53
X = .56
Y = .98
NAME = |A|R|N|Y|

d.

A1 = 56.2
A2 = 319
A3 = 235
A4 = 8932
A5 = 4267
A6 = 5932

6.3

```
      PROGRAM CHRCTR
C AUTHOR KAREN BROWNSON, JULY 19--
C
C THIS PROGRAM PREPARES A REPORT ABOUT CUSTOMER'S
C GAS USAGE
C CUSTOMER'S NAME (COL 1-12), CUSTOMER'S CODE (COL
C 14-20), CUSTOMER'S TYPE CODE (COL 22), LAST
C METER READING (COL 24-30), CURRENT METER READING
C (COL 32-38) ARE PUNCHED ON CARDS
C CHARGES ARE:  32.5¢/UNIT FOR RESIDENTIAL, 29.9¢/
C UNIT FOR INDUSTRIAL USAGE
C
C ********** VARIABLE DICTIONARY **************
C NAME: NAME OF THE CUSTOMERS, KODE: CUSTOMERS CODE,
C LMR: LAST METER READING
C TYPE AND ATYPE: FOR RESIDENTIAL, INDUSTRIAL TYPE,
C CMR: CURRENT METER READING
C GASUSE:   GAS USAGE,
C CHARGE:   CHARGES TO EACH CUSTOMER
C NIND:  NO. OF INDUSTRIAL USERS,
C NRES: NO. OF RESIDENTIAL USERS
C TOTUSE = TOTAL GAS USE FOR ALL CUSTOMERS
C TOTCHG = TOTAL CHARGE TO ALL CUSTOMERS
C VARIABLE TYPE DECLARATION
      CHARACTER NAME*12,KODE*7,TYPE*1,ATYPE*11
      REAL LMR
C INITIALIZING THE VARIABLES
      DATA TOTUSE,TOTCHG,NIND,NRES,ATYPE/2*0.0,
     *2*0,' '/
C PRINTING THE HEADING
      WRITE (6,5)
```

```
    5       FORMAT (61X,'XYZ COMPANY'///11X,'SEQ #',10X,
           *'NAME',11X,'CODE',10X,'TYPE',12X,'LAST
           *READING',7X,'CURRENT READING',7X,'CHARGE')
C START OF THE LOOP, THE LOOP WILL BE TERMINATED
C BY READING 'NO MORE     ' IN THE NAME FIELD
            DO 90 N = 1, 500
                READ (5,10) NAME,KODE,TYPE,LMR,CMR
                IF (NAME .EQ. 'NO MORE') GO TO 99
                GASUSE = CMR - LMR
                TOTUSE = TOTUSE + GASUSE
                IF (TYPE .EQ. 'R') THEN
                        CHARGE = GASUSE * .325
                        NRES = NRES + 1
                        ATYPE = 'RESIDENTIAL'
                ELSE
                        CHARGE = GASUSE * .299
                        NIND = NIND + 1
                        ATYPE = 'INDUSTRIAL'
                END IF
                TOTCHG = TOTCHG + CHARGE
                WRITE (6,20) N,NAME,KODE,ATYPE,LMR,CMR,
           *CHARGE
   90   CONTINUE
C WRITING TOTALS
   99       WRITE (6,30) NIND,NRES,TOTUSE,TOTCHG
   10       FORMAT (A12,1X,A7,1X,A1,1X,F7.2,1X,F7.2)
   20       FORMAT (13X,I3,6X,A12,5X,A7,5X,A11,9X,F8.2,
           *13X,F8.2,10X,F8.2)
   30       FORMAT (11X,'NO. OF INDUSTRIAL USERS =',I3/
           *11X,'NO. OF RESIDENTIAL USERS =',I3//11X,
           *'TOTAL USAGE =',F9.2/11X,'TOTAL CHARGES =',
           *F9.2)
            END
```

Chapter 7
One-Dimensional Arrays

INTRODUCTION

SUBSCRIPTED VARIABLES

DIMENSION STATEMENT

REASON FOR ARRAYS: AN EXAMPLE

ARITHMETIC EXPRESSIONS WITH ARRAYS

ARRAY INPUT AND OUTPUT

    Array I/O with a DO Loop
    Implied DO Loop
    The Entire-List Method
    Mirror Printing

INITIALIZING THE ARRAY AND THE DATA STATEMENT

IMPORTANT NOTES ABOUT ARRAYS

SOME EXAMPLES OF ARRAY USE AND PROGRAMMING TECHNIQUES

INTRODUCTION

In data processing, it's necessary to read and store the data first. To do this, a variable name for each datum must be used, as in the following example:

EXAMPLE 7.1

Problem: Read and store ten test scores, then calculate the average of the scores. Print the scores and the average.

Program:

```
      PROGRAM ARRY1
      READ (5,10 ) A1
      READ (5,10 ) A2
      READ (5,10 ) A3
      READ (5,10 ) A4
      READ (5,10 ) A5
      READ (5,10 ) A6
      READ (5,10 ) A7
      READ (5,10 ) A8
      READ (5,10 ) A9
      READ (5,10 ) A10
      SUM = A1+A2+A3+A4+A5+A6+A7+A8+A9+A10
      AVG = SUM/10.0
      WRITE (6,20) A1
      WRITE (6,20) A2
      WRITE (6,20) A3
      WRITE (6,20) A4
      WRITE (6,20) A5
      WRITE (6,20) A6
      WRITE (6,20) A7
      WRITE (6,20) A8
      WRITE (6,20) A9
      WRITE (6,20) A10
      WRITE (6,20) AVG
10    FORMAT (F5.2)
20    FORMAT (11X,F6.2)
      END
```

Notes:

1. The variable names A1, A2, ... A10 are chosen to show test score 1, test score 2, and so on.
2. Because we would like to calculate the average after all data are read, we must choose different variable names. Even though the program

```
      SUM = 0.0
      DO 80 I = 1,10
          READ (5,10 ) A
          SUM = SUM + A
          WRITE (6,20) A
```

```
80      CONTINUE
        AVG = SUM/10.0
        WRITE (6,20) AVG
10      FORMAT (F5.2)
20      FORMAT (11X,F6.2)
        END
```

works fine, but it cannot be used when it is necessary to store all the information. (Reading a new item erases the previous one.)

Not only is the program in Example 7.1 unduly long, but also writing this kind of program is impractical if there are a lot of data. Imagine writing this type of program for 1000 records! Fortunately, FORTRAN allows us to write the same program as follows:

EXAMPLE 7.2

Problem: Same as Example 7.1.

Program:

```
        PROGRAM ARRAY2
        DIMENSION A(10)
        SUM = 0.0
        DO 50 I = 1,10
            READ(5,10 ) A(I)
50      CONTINUE
        DO 60 I = 1,10
            SUM = SUM + A(I)
60      CONTINUE
        AVG = SUM/10.0
        DO 70 I = 1,10
            WRITE(6,20) A(I)
70      CONTINUE
        WRITE (6,20) AVG
10      FORMAT (F5.2)
20      FORMAT (11X,F6.2)
        END
```

Notes:

1. A(I) is used for the name of the variables; when I changes from 1 to 2, to ..., to 10 in the DO loop, it creates A(1), A(2), ..., A(10). These variables correspond to A1, A2, ..., A10 in Example 7.1. (However, A1 is not the same as A(1); they are two different kinds of variables.)
2. The DIMENSION statement declares that there will be ten elements for A, namely A(1), A(2), A(3), ..., A(10).
3. The program is written in three small modules:
   a. reading the data,
   b. calculating the average,
   c. writing the data and the average.
   Although we could have used a single DO loop in the program,

writing a program in smaller modules is preferable. The value of this becomes apparent when your program becomes long and complicated.

4. The module for calculating SUM accomplishes the same function as follows:

```
SUM = 0.0
SUM = SUM + A(1)
SUM = SUM + A(2)
      ⋮
SUM = SUM + A(10)
```

SUBSCRIPTED VARIABLES

Instead of labeling a series of variables $A, B, C, ...,$ we can call them $A_1, A_2, A_3, ..., A_n$. These are called *subscripted variables*. In mathematics, subscripted symbols are used to denote the elements of a set, i.e., individual elements belonging to a group of related items.

In most programming languages, subscripts must be written in parentheses: A(1), A(2), A(3), ..., and the complete set of these variables is called an *array*; each subscripted variable is then an element of the array. Thus, an array is a group of elements, identified by a single name, joined to be treated as a whole.

> *An* array *refers to the collection of the elements of a group of variables. Each individual variable is called a* subscripted variable, *or simply an* element.

An array with elements specified by one subscript is called a *linear* or *one-dimensional* array. We may also have two-dimensional arrays with two subscripts such as B(2,3), or three-dimensional arrays with three subscripts such as C(2,3,4), etc. One-dimensional arrays will be discussed in this chapter, and multidimensional arrays in the next chapter.

By creating an array, the computer assigns a certain number of consecutive memory cells to the specified array name (Figure 7.1). Each individual element can be referred to by using an index—the subscript.

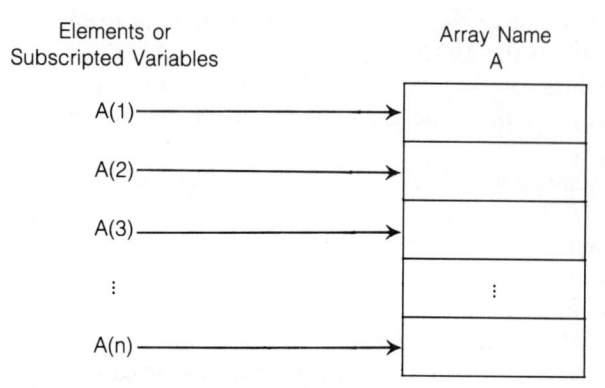

Figure 7.1 An array and its elements.

> *An* **array** *is a group of memory cells which collectively have the same name; each individual element can be located by an index—the subscript.*

The array and its number of elements must be declared by the DIMENSION statement.

■
DIMENSION STATEMENT

The DIMENSION statement defines an array and its size at the beginning of a program. It instructs the computer to reserve a certain number of consecutive memory locations for the elements of the array. More than one array can be declared in a single DIMENSION statement. The simple form of the DIMENSION statement for a one-dimensional array is

```
DIMENSION var1(n1), var2(n2), ....
```

where var1, var2,... are the array names, and $n1$, $n2$, ... are unsigned integer numbers representing the number of elements in each array. For example,

```
DIMENSION X(500), AMOUNT(100), JOY(50)
```

declares that X, AMOUNT, and JOY are arrays and they have 500, 100, and 50 elements, respectively. The DIMENSION statement must be placed before any executable statement and after a type statement.

Note that there is no comma after the word DIMENSION, but arrays are separated by commas. Also, the number of elements for each array (the array size) must be a positive integer. It cannot be a variable, a real number, or a negative number, except:

1. In subroutines, which will be discussed in Chapter 9, the array size can be an integer variable.
2. In FORTRAN 77, the boundaries of an array dimension can be defined, and either of them (the lower or the upper boundary) can be positive, negative, or zero as long as the upper boundary is greater than the lower boundary. The form of the DIMENSION statement in this case is

   DIMENSION var $\begin{pmatrix} \text{lower} \\ \text{bound} \end{pmatrix} : \begin{pmatrix} \text{upper} \\ \text{bound} \end{pmatrix}$, ...

For example:

```
DIMENSION XX(-1 : 4)
```

declares a one-dimensional array of six elements: XX(-1), XX(0), XX(1), XX(2), XX(3), and XX(4). Notice that the lower boundary is separated from the upper boundary by a colon. If the lower boundary is omitted, the value 1 will be assumed, and the form is the previous one.

All elements in an array must be of the same type (real, integer, character, etc.). Furthermore, the name of the array, which follows the rule for symbolic names in FORTRAN, determines the type of the array in the same way that the name of a variable determines its type. For example,

```
DIMENSION NET(300)
```

defines NET as an integer array with 300 elements.

Array names can also be declared by the type statement. For example,

```
INTEGER B,X
DIMENSION B(100), X(20)
```

identifies arrays B and X as integer, B having 100 elements and X having 20 elements. Also, the type statement and the DIMENSION statement can be combined. For instance, the previous example can be written as

```
INTEGER B(100), X(20)
```

Just as before, this declares B and X as integer arrays, with B having 100 elements and X having 20 elements. The following are further examples: The declaration

```
REAL MAY(300), NO(100), MEAN(500)
```

is equivalent to

```
REAL MAY, NO, MEAN
DIMENSION MAY(300), NO(100), MEAN(500)
```

The declaration

```
REAL M(1000), LARGE(250), L, J
```

is equivalent to

```
REAL M, LARGE, L, J
DIMENSION M(1000), LARGE(250)
```

The declaration

```
INTEGER GROSS(350), COUNT(500), SEQ
```

is equivalent to

```
INTEGER GROSS, COUNT, SEQ
DIMENSION GROSS(350), COUNT(500)
```

The declaration

```
CHARACTER*12 XX(300), NAME(250), B
```

is equivalent to

```
CHARACTER*12 XX, NAME, B
DIMENSION XX(300), NAME(250)
```

The declaration

```
CHARACTER*15 A(200), ADRES(100)*18, NAME
```

is equivalent to

```
CHARACTER*15, A,ADRES*18,NAME
DIMENSION A(200), ADRES(100)
```

The DIMENSION statement provides information about the maximum number of elements needed. The entire array, however, does not have to be used in a program. The following is an example:

```
      DIMENSION M(1000)
      DO 80 I = 1, 15
         READ (5,50) M(I)
         WRITE (6,60) M(I)
80    CONTINUE
```

```
50      FORMAT (I4)
60      FORMAT (1X,I5)
        END
```

When you need to declare the size of an array, estimate the maximum number of the necessary elements, but avoid excessive sizes because they make inefficient use of memory.

SOLVED PROBLEMS

7.1 Write a DIMENSION statement to declare the following arrays:

  a. Arrays COUNT with 30 elements and SEQ with 1000 elements.
  b. Arrays TABLE which can contain 65 elements and LIST which can contain 110 real numbers.
  c. Array NAME which can contain the names of the students in a FORTRAN course (about 62 students), where each name can be up to 20 characters.
  d. Arrays NAME with 100 elements, each up to 12 characters; ADRES with 100 elements, each up to 20 characters; ID with 100 integer elements; and SCORE with 100 real elements.

☐ **Answers**

  a.
  ```
  DIMENSION COUNT(30), SEQ(1000)
  ```
  b.
  ```
  REAL TABLE(65), LIST(110)
  ```
  c.
  ```
  CHARACTER*20 NAME(65)
  ```
  d.
  ```
  CHARACTER NAME(100)*12, ADRES(100)*20
  DIMENSION ID(100), SCORE(100)
  ```

7.2 Find any errors in the following statements:

  a. DIMENSION XXX(−5), YY(M), Z(56.0)
  b. DIM YR(60), Q(70), ION(300)
  c. INTEGER DIMENSION COIN(100), RX(70), I(300), XROSS
  d. REAL, MONEY(300),
  e. REAL CHARACTER*12 N(30), M
  f. DIMENSION POOR(100) TEST(200)
  g. DIMENSION (500)
  h. DIMENSION X(50 + 2), Y(N + 1)
  i. DIMENSION XXX(1000), YYY(−5 : 0), ZZZ(0 : 10),
     *   AAA(−7 : −2)

☐ **Answers**

a. The array size cannot be a negative number, a variable, or a real constant. Note: If you are using FORTRAN 77, the bounds can be negative, but the upper bound must be greater than the lower bound. In this example, the lower bound for the array XXX is omitted; thus it will be assumed to be 1, which is more than −5. In any case XXX(−5) is not correct.
b. The word DIMENSION must be spelled correctly.
c. INTEGER and DIMENSION words cannot be used simultaneously.
d. The commas after the word REAL and after the array name must be omitted.
e. REAL and CHARACTER words cannot be used simultaneously.
f. A comma must separate the arrays.
g. The name of the array is missing.
h. The array dimension cannot be 50 + 2 or N + 1.
i. This is correct.

REASON FOR ARRAYS: AN EXAMPLE

Sometimes the nature of the problem requires using an array, as in Example 7.3:

EXAMPLE 7.3

■ **Problem:** Draw a flowchart and write a program which reads student test scores, and prints the scores which are below average. There are 45 test scores.

Flowchart: Shown in Figure 7.2.

Program:

```
        PROGRAM ARRAY3
        DIMENSION SCORE(45)
        DATA SUM/0.0/
        DO 30 J = 1,45
            READ(5,10) SCORE(J)
30      CONTINUE
        DO 40 K = 1, 45
            SUM = SUM + SCORE(K)
40      CONTINUE
        AVG = SUM/45.0
        DO 50 L = 1,45
            IF (SCORE(L) .LT. AVG) WRITE (6,20)
       *        SCORE(L)
50      CONTINUE
10      FORMAT (F5.2)
20      FORMAT (11X,F6.2)
        END
```

Reason for Arrays: An Example

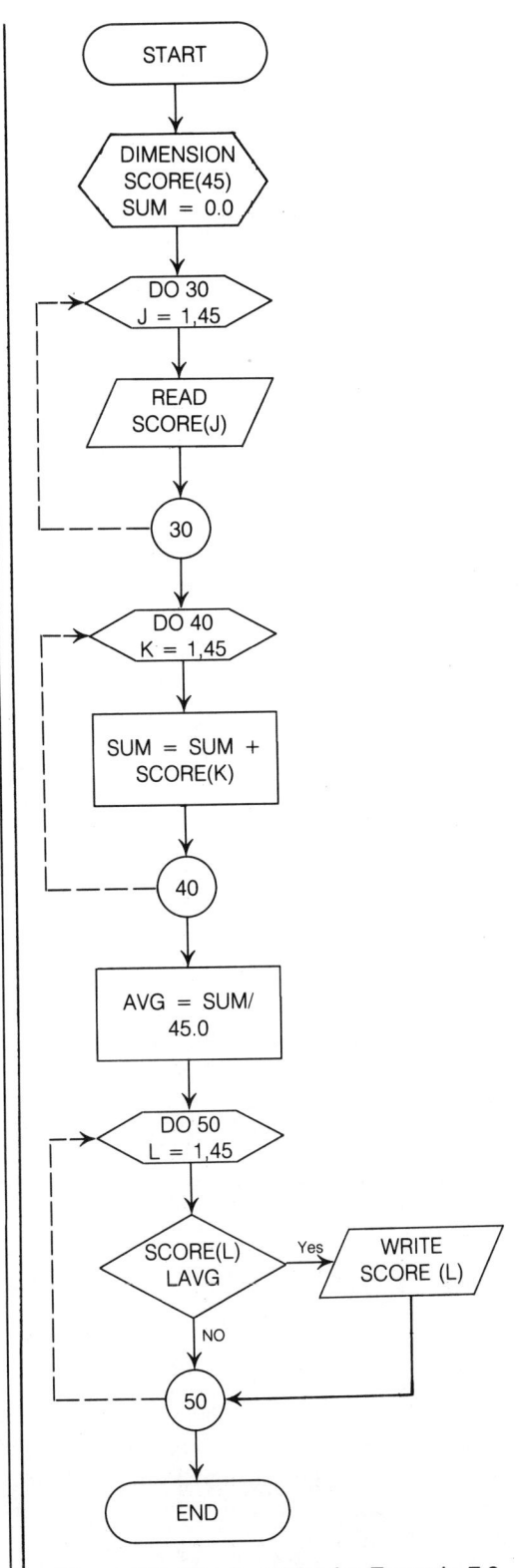

```
                START
                  │
                  ▼
          ╱DIMENSION╲
         ╱ SCORE(45) ╲
         ╲ SUM = 0.0 ╱
          ╲         ╱
                  │
                  ▼
           ╱ DO 30 ╲
      ┌──▶╲ J = 1,45╱
      │    ╲       ╱
      │        │
      │        ▼
      │    ╱ READ  ╱
      │   ╱ SCORE(J)╱
      │        │
      │        ▼
      │      ( 30 )
      └────────┘
               │
               ▼
          ╱ DO 40 ╲
     ┌──▶╲ K = 1,45╱
     │        │
     │        ▼
     │   ┌─────────┐
     │   │SUM = SUM +│
     │   │ SCORE(K) │
     │   └─────────┘
     │        │
     │        ▼
     │      ( 40 )
     └────────┘
              │
              ▼
         ┌─────────┐
         │AVG = SUM/│
         │  45.0   │
         └─────────┘
              │
              ▼
          ╱ DO 50 ╲
     ┌──▶╲ L = 1,45╱
     │        │
     │        ▼
     │      ╱╲
     │     ╱  ╲   Yes   ┌─────────┐
     │    ╱SCORE(L)╲───▶│  WRITE  │
     │    ╲ LAVG  ╱      │SCORE (L)│
     │     ╲    ╱       └─────────┘
     │      ╲ ╱              │
     │      NO│              │
     │        ▼              │
     └─────▶( 50 )◀──────────┘
              │
              ▼
             END
```

Figure 7.2 The flowchart for Example 7.3.

> **Notes:**
> 1. The program's modules are:
>   a. reading the data,
>   b. calculating SUM and AVERAGE,
>   c. writing the scores which are less than average.
> 2. The integer variable representing the subscript does not have to be the same for each module.

ARITHMETIC EXPRESSIONS WITH ARRAYS

The array name refers to all elements of the complete array. However, the array name followed by a subscript refers to a specific element of the array, and can be considered as a single variable.

Any calculation must be performed with the individual elements of the array, that is, the subscripted array name. The following are some examples (assume A, B, and C are defined as arrays):

```
C(6) = B(2)*A(9)
A(5) = A(5) + A(4)
C(10) = .04*A(7) + 2*B(5)
```

A variable can represent the subscript in a program as long as its value has been defined. For example, if the values of I and J are defined, we can then write

```
A(I) = B(J) + C(I)
```

or

```
C(J) = A(J)*B(J)
```

A DO loop is often used for manipulating the elements of an array. The following is an example:

> ### EXAMPLE 7.4
>
> ■ **Problem:** Write a program, using arrays, which reads and prints the information about the merchandise ordered by customers of a department store. Input:
>
> Customer's name (12 characters),
> ID number (integer, 5 digits),
> Number of units ordered,
> Price of each unit.
>
> The sales tax is 4% of the amount ordered. Print the input information, tax, and total amount of the order for each customer in report form. Print also the detailed information about the customers whose total amount of the order is above the average. There are less than 500 customers. The first data record indicates the number of customers with orders.

Program:

```
      PROGRAM ORDER
      INTEGER ID(500)
      REAL UNORD(500), PRICE(500), AMTORD(500),
     *TAX(500), TORD(500)
      CHARACTER*12 NAME(500)
      DATA TOT/0.0/
C
      READ(5,10)N
C
      DO 50 I = 1,N
           READ(5,20) NAME(I), ID(I), UNORD(I),
     *     PRICE(I)
50    CONTINUE
C
      DO 60 J = 1,N
           AMTORD(J) = UNORD(J)*PRICE(J)
           TAX(J) = .04*AMTORD(J)
           TORD(J) = AMTORD(J) + TAX(J)
           TOT = TOT + TORD(J)
60    CONTINUE
C
      DO 70 K = 1,N
           WRITE (6,30) NAME(K), ID(K), UNORD(K),
     *     PRICE(K), AMTORD(K), TAX(K), TORD(K)
70    CONTINUE
      WRITE (6,40) TOT
      AVG = TOT/N
      DO 80 L = 1,N
           IF(TORD(L) .GT. AVG)WRITE(6,30)NAME(L),
     *     ID(L),UNORD(L),PRICE(L),AMTORD(L),
     *     TAX(L),TORD(L)
80    CONTINUE
C
10    FORMAT(I3)
20    FORMAT(A12,I5,F2.0,F5.2)
30    FORMAT(11X,A12,2X,I6,2X,F3.0,2X,F6.2,2X,
     *F8.2,2X,F6.2,2X,F8.2)
40    FORMAT(///61X,'TOTAL',7X,F10.2///)
C
      END
```

Notes:

1. The DO loop is used for the manipulation of the individual elements of arrays. The first READ statement reads N (the number of data); then each loop will be repeated N times.
2. The amount ordered (AMTORD), tax (TAX), and total amount of an order (TORD) for each customer are each calculated in arrays. The appropriate arrays are declared at the beginning of the program.
3. The DIMENSION statement (the type statements work the same as the DIMENSION statements) defines the maximum number of elements, 500 in this case. Only N elements (N ≤ 500) are read and processed.

> 4. This program needs appropriate headings for both the customer's report and the report of the customer whose order is above the average.

An arithmetic expression can also represent the subscript. The following are examples (assume arrays are defined):

```
SUM(I + 1) = TOT(I - 1)
A(5*N + 2) = X(K + 3)
NET(3*M - 6) = MAY(2*M)
```

Earlier versions of FORTRAN would allow only certain simple forms of expressions as a subscript. But in FORTRAN 77, any arithmetic expression (simple or complicated, real or integer) can be a subscript. The following are examples:

```
XXX(3.0*R**2) = YYY(I + 1)
AEE((5.0*A - 3.0*B**2)/C) = 5.0*PPP(X + 3.0)
QUE((3*I**4 + 3)*(5.0)) = PIE(3.0*X)
```

If the subscript is an expression, it will be evaluated and converted if necessary to an *integer number* first. However, after conversion each value must not be less than the lower bound, or greater than the upper bound of the dimension.

It is important to emphasize that the individual elements of an array must be used in an expression. For example, the program segment

```
DIMENSION A(100)
     ⋮
A = 0.0
```

is not valid. If the intention is to set all the elements of the array equal to zero in this case, then the individual elements must be used in a loop, such as

```
      DO 10 I = 1,100
         A(I) = 0.0
10    CONTINUE
```

SOLVED PROBLEMS

■ 7.3 Find any errors in the following program segments.

a.
```
      DIMENSION X(10),Y(10),Z(10)
      DO 30 I = 1,10
         READ(5,10) X(I),Y(I)
            Z = X + Y
30    CONTINUE
```

b.
```
          DIMENSION X(100)
          DO 80 J = 5,105
80        X(J) = 5.0
```

c.
```
          DIMENSION X(30)
          DO 90 K = 1,30
90        X(K + 1) = K
```

d.
```
          DIMENSION IY(65)
          DO 90 L = 1,30
90        IY(L - 1) = L
```

e.
```
          READ (5,10) N
          DIMENSION X(N)
          DO 80 I = 1,N
              READ (5,20) X(I)
80        CONTINUE
```

f.
```
          REAL IS(500)
          DO 100 J = 1,500
              IS(J) = J + 1
100       CONTINUE
```

g.
```
          DIMENSION JOY(50)
          X = 0.0
          DO 60 M = 1,50
              X = X + 1.0
              JOY(M) = X
60        CONTINUE
```

h.
```
          DIMENSION SUM(100)
          SUM(1) = 0.0
          X = 0.0
          DO 60 I = 2,100
              X = X + 1.0
              SUM(I) = SUM(I - 1) + X
60        CONTINUE
```

i.
```
          I = 6
          X = P(2*I + 1)
```

j.
```
          DIMENSION A(100)
              ⋮
          IF (A .EQ. 0.0) GO TO 30
              ⋮
```

Answers

a. If one intends to add the elements of X and Y, then they should be subscripted. The arithmetic expression therefore must be corrected to

```
Z(I) = X(I) + Y(I)
```

b. The array X is defined to have up to 100 elements; therefore, the subscript J cannot go over 100.
c. First, the subscript cannot be greater than 30, whereas K + 1 will be 31 when K reaches 30. Secondly, the array X is of real type, but it has been assigned integer values (K).
d. When L starts with 1, L − 1 is zero, which is illegal.
e. The array size cannot be a variable (N), and it should be placed before executable statements.
f. The array IS is defined as real; hence it is inappropriate to assign integer values to it.
g. The array JOY is an integer array, but it is assigned real values.
h. This is correct.
i. This is correct if the array P has been defined.
j. Each individual element of the array A must be used in the IF statement, as in the following loop:

```
      DO 10 I = 1,100
         IF(A(I) .EQ. 0.0) GO TO 30
10    CONTINUE
```

7.4 Which of the following subscripts is invalid?

a. K(I)
b. A(I + 3)
c. K(3 * J + 1)
d. I(J + 3)
e. L((3 + J)/3 * X)
f. BB(ISUM/3.0)
g. X(SUB)
h. MAY(2 * FIX + 2)
i. TOT(I + J)
j. SUM(3 * IX ** 2 + 4 * 3)

Answers

All of them are valid in FORTRAN 77, provided the arrays are defined and the subscript, after conversion to an integer value, is not less than the lower bound or greater than the upper bound of the dimension.

ARRAY INPUT AND OUTPUT

The elements of an array can be read or written individually, such as:

```
      DIMENSION A(10)
      READ (5,100) A(1), A(2), A(3)
100   FORMAT (3F5.1)
```

However, this method is neither efficient nor practical, especially if the subscripts become large. Basically, there are three methods for array input and output:

1. Using a DO loop.
2. Using an implied DO loop.
3. Using the entire-list method.

These methods are explained in this section. Furthermore, *mirror printing* is a convenient way to check whether the input data have been read correctly, and the DATA statement is an efficient way to initialize the array elements. These techniques are also explained in this section.

☐ Array I/O with DO Loop

As we have seen in previous examples, the DO loop is a convenient way for the input or output of the elements of an array. The following is another example.

EXAMPLE 7.5

Problem: Read and write an array A composed of 100 elements. The data are on separate lines.

Program:

```
      PROGRAM ARRAY5
      DIMENSION A(100)
      DO 70 L = 1,100
          READ (5,10) A(L)
70    CONTINUE
      DO 80 K = 1,100
          WRITE (6,20) A(K)
80    CONTINUE
10    FORMAT (F4.1)
20    FORMAT (1X,F5.1)
      END
```

Note:

```
      DO 70 L = 1,100
          READ (5,10) A(L)
70    CONTINUE
```

performs the same function as

```
      READ(5,10) A(1)
      READ(5,10) A(2)
      READ(5,10) A(3)
            ⋮
      READ(5,10) A(100)
```

We could have used one DO loop for both reading and writing data in the previous program. It is, however, good programming practice to write a program in small modules (input can be considered as one module, and output as another; although these "modules" are too small in this example). Modular programming makes a program easier to code, understand, debug, and document.

Often a header or a trailer record is used to read an appropriate number of data. A trailer record is particularly useful when the number of data is unknown. However, when a trailer record is used, a counter block must keep track of the number of data for later use. The following is an example.

EXAMPLE 7.6

■ **Problem:** Write a program which reads the scores of the students in a course and calculates the average. Print the scores and the average. There are less than 100 students, but the exact number of students taking the test is unknown. The last line with a negative test score is a trailer record.

Program:

```
      PROGRAM ARRAY6
      DIMENSION SCORE(100)
      SUM = 0.0
      N = 0
      DO 50 I = 1,100
          READ(5,10) SCORE(I)
          IF(SCORE(I) .LT. 0.0) GO TO 80
          N = N + 1
50    CONTINUE
80    DO 60 J = 1,N
          SUM = SUM + SCORE(J)
60    CONTINUE
      AVG = SUM/N
      DO 70 K = 1,N
          WRITE(6,20) SCORE(K)
70    CONTINUE
      WRITE(6,20) AVG
10    FORMAT(F5.2)
20    FORMAT(11X,F6.2)
      END
```

Notes:

1. The trailer record with a negative number terminates the DO loop which reads the data.
2. The statement N = N + 1 is necessary to keep track of the number of elements read. Note that the N is used for the other DO loops.

☐ Implied DO Loop

The implied DO loop is a looping technique which can be used for array input and output. Let's take a look at an example before explaining this technique further.

EXAMPLE 7.7

Problem: Read 15 elements of the array B. The data are placed on one line. Print the data and the average on one line. Array B may have up to 100 elements.

Program:
```
      PROGRAM ARRAY7
      DIMENSION B(100)
      TOT = 0.0
C
      READ(5,10)(B(I),I = 1,15)
C
      DO 50 J = 1,15
50    TOT = TOT + B(J)
      AVG = TOT/15.0
C
      WRITE (6,20) (B(K),K = 1,15), AVG
10    FORMAT (15F4.1)
20    FORMAT (11X,16(2X,F5.1))
      END
```

Note:
```
      READ(5,10) (B(I),I = 1,15)
```
is equivalent to
```
      READ (5,10) B(1), B(2), B(3), B(4), ...,
     *B(13), B(14), B(15)
```

The implied DO loop can be used in the variable list of the READ, WRITE, and DATA statements. The general form of the implied DO loop is

   READ/WRITE (−, −) (array name (INDEX), INDEX = m1, m2, m3)

or

   READ/WRITE (−, −) (var1, var2, ..., INDEX = m1, m2, m3)

where INDEX is the control variable, and $m1$, $m2$, $m3$ are the initial, final, and increment value of the INDEX variable.

If the increment value is equal to 1, then $m3$ can be omitted. Placing parentheses around the entire loop and a comma after the "array index" are critical. The following are examples of using an implied DO loop.

EXAMPLE A
```
      K = 3
      READ (5,10) (A(I), I = 1,K)
10    FORMAT (3F5.1)
```
is equivalent to
```
      READ (5,10) A(1),A(2),A(3)
10    FORMAT (3F5.1)
```

EXAMPLE B

```
       READ (5,20) K,X,(B(J),J = 1,5),Y
20     FORMAT (I2,7F5.2)
```

is equivalent to

```
       READ (5,20) K,X,B(1),B(2),B(3),B(4),B(5),Y
20     FORMAT (I2,7F5.2)
```

EXAMPLE C

```
       READ (5,30) (X,Y(K),K = 1,5),Z
30     FORMAT (11F4.1)
```

is equivalent to

```
       READ (5,30) X,Y(1),X,Y(2),X,Y(3),X,Y(4),X,Y(5),Z
30     FORMAT (11F4.1)
```

Note that any variable in the parentheses is included in the DO loop.

EXAMPLE D

```
       READ (5,40) N,(A(I),I = 1,N)
40     FORMAT (I3/(15F4.1))
```

is equivalent to

```
       READ (5,30) N
30     FORMAT (I3)
       READ (5,40) A(1),A(2),...A(N)
40     FORMAT (15F4.1)
```

The following are additional examples of using an implied DO loop:

```
READ(5,100)(APE(J),J = 1,30),(POE(I),I = 1,100)
READ(5,300)(KAY(K),JAY(K),K = 1,30)
READ(5,400)(X(I),K(I),Y(I),I = 1,150,2)
WRITE(6,200)B,(X(I),I = 5,20),C
WRITE(6,300)(GROSS(K),K = 5,20)
WRITE(6,400)A,B,C,(D(K),K = I,J,K)
READ(5,90)(EXCH(NET),NET = 1,500,2)
WRITE(6,100)(PAT,MAT,JOE,I = 1,10)
```

The last statement prints the values of PAT, MAT, and JOE ten times each.

The rules of the DO loop discussed in Chapter 5 also apply to implied DO loops.

When using an implied DO loop, the FORMAT statement determines the location of the data and the number of records (or data cards). For example,

```
       READ(5,30)(A(I),I = 1,5)
30     FORMAT(5F5.1)
```

reads one record containing five values for A. But

```
       READ(5,40)(A(I),I = 1,5)
40     FORMAT(F5.1)
```

reads five records (or data cards), each containing a value for A. This was explained in chapter six.

An implied DO loop is compact and easy to use, also making it possible to print or read elements of an array on one line. When there are too many values to fit on one line, though, the computer automatically goes on to the next. An implied DO loop can also be used to write a simple variable more than one time in the loop.

☐ Entire-List Method

An entire array can be read or written by placing the array name, without subscripts, in an input or output list. The following is an example:

EXAMPLE 7.8

■ Problem: The array X has 15 elements. Write a program which reads and prints all elements of the array.

Program:

```
        PROGRAM ARRAY8
        DIMENSION X(15)
C
        READ(5,10)X
C
        WRITE(6,20)X
C
10      FORMAT(15F5.1)
20      FORMAT(1X,15F5.1)
        END
```

Note: Since X is defined by the DIMENSION statement as having 15 elements,

    READ(5,10)X

attempts to read the entire array and works the same way as

    READ(5,10) X(1),X(2),X(3),...,X(15)

Thus, by placing only the name of an array in the input or output list, the entire array will be read or written. This method is also called *array transmission*, or the *short-list* method. It is a short way to transmit *all the elements* of an array. However, if only part of an array needs to be input or output, then using the subscript and a DO loop is required.

Using the array name without a subscript is possible only in the input or output list. As was noted before, it is not permissible to use an array name without a subscript in arithmetic expressions.

The placement of the data also depends on the FORMAT statement when transmitting the entire array. For example,

```
        DIMENSION X(15)
        READ(5,10)X
10      FORMAT(15F5.1)
```

reads one record, containing 15 data, for the elements of the array X; but

```
        DIMENSION X(15)
        READ(5,20)X
20      FORMAT (F5.1)
```

282 One-Dimensional Arrays

Table 7.1 Comparing DO Loops

METHOD	DATA ON THE SAME LINE	DATA ON DIFFERENT LINES

Individual items	`DIMENSION A(4)`
	`READ(5,10)A(1),A(2),A(3)A(4)`
	`10   FORMAT(4F5.1)`
		`DIMENSION A(4)`
		`READ(5,10)A(1),A(2),A(3),A(4)`
		`10   FORMAT(F5.2)`

DO loop	—	`DIMENSION A(4)`
		`DO 90 I = 1,4`
		`    READ(5,10)A(I)`
		`90  CONTINUE`
		`10  FORMAT(F5.1)`

Implied DO	`DIMENSION A(4)`
	`READ(5,10)(A(I),I=1,4)`
	`10   FORMAT(4F5.1)`
		`DIMENSION A(4)`
		`READ(5,10)(A(I),I=1,4)`
		`10   FORMAT(F5.1)`

Entire list	`DIMENSION A(4)`
	`READ(5,10)A`
	`10   FORMAT(4F5.1)`
		`DIMENSION A(4)`
		`READ(5,10)A`
		`10   FORMAT(F5.1)`

reads fifteen records, each containing 1 datum, for the elements of X, and

```
      DIMENSION X(15)
      READ(5,30)X
30    FORMAT(5F5.1)
```

reads three records, each containing 5 data, for the elements of X.

Table 7.1 compares the input-output methods for an example.

☐ Mirror Printing

Incorrect results may be obtained in a program if the input data are not read correctly. To check the input data for errors, an extra WRITE (or PRINT) statement with identical variable list can be placed immediately following each READ statement to print the data after reading them. This technique is called *mirror printing* or *echo printing*. It is a good programming practice to use mirror printing in the first run of a program. Once it is confirmed that the input data are transmitting correctly, the extra WRITE statements can be easily removed.

Mirror printing is not only for arrays: it is advisable to use this technique with all the READ statements in any program. The list-directed PRINT statement is used in the following example to demonstrate the mirror-printing technique.

EXAMPLE 7.9

■ **Problem:** Write a program which reads a set of N data. Calculate the average. Print the data and the average.

Program:

```
      PROGRAM ARRAY9
      DIMENSION A(100)
      TOT = 0.0
      READ(5,10)N
      PRINT*,N
```

```
             DO 100 I = 1,N
                 READ(5,30)A(I)
                 PRINT*,A(I)
100          CONTINUE
             DO 200 J = 1,N
                 TOT = TOT + A(J)
200          CONTINUE
             AVG = TOT/N
             DO 300 K = 1,N
                 WRITE (6,60) A(K)
300          CONTINUE
             WRITE(6,80) AVG
10           FORMAT(I2)
30           FORMAT(F5.2)
60           FORMAT(11X,'DATA:',F6.2)
80           FORMAT(11X,'THE AVERAGE:',F6.2)
             END
```

Note: The print statements may be removed after confirming that the data have been read correctly.

INITIALIZING THE ARRAY AND THE DATA STATEMENT

Sometimes it is necessary to set elements of an array equal to some initial values. This can be done either by an assignment or by the DATA statement. The following is an example:

EXAMPLE 7.10

Problem: The array Y has been defined as

DIMENSION Y(5)

Show how the initial values can be set equal to zero at the beginning of a program.

Method I:
```
             Y(1) = 0.0
             Y(2) = 0.0
             Y(3) = 0.0
             Y(4) = 0.0
             Y(5) = 0.0
```

Method II:
```
             DO 40 I = 1,5
40           Y(I) = 0.0
```

Method III:
```
             DATA Y(1),Y(2),Y(3),Y(4),Y(5)/5*0.0/
```

Method IV:

        DATA (Y(I),I = 1,5)/5*0.0/

Method V:

        DATA Y/5*0.0/

Notes:

1. The DATA statement is a convenient way to set the initial values.
2. FORTRAN 77 permits the use of an implied DO loop in the DATA statement (method IV).
3. An unsubscripted array name may appear as an item in the variable list. However, a constant in the value list must be specified for each respective element of the array. The order in which the values are assigned is the storage order of the array.

The following are further examples of using a DATA statement:

```
      INTEGER A(10)
      DATA A(2)/5/,A(1),A(4),A(5)/3*1/

      DIMENSION X(100)
      DATA X/100*0.0/

      REAL MONEY(500)
      DATA (MONEY(J),J = 1,150)/150*1.0/,
     *(MONEY(K),K = 151,300)/150*0.0/

      DIMENSION KAY(510)
      DATA (KAY(K),K = 1,148)/148*0/,KAY(149),
     *KAY(150)/5,7/
```

SOLVED PROBLEMS

7.5 Detect errors in the following statements or program segments:

a.
```
          DIMENSION X(100)
          DO 90 I = 1,100
          READ(5,20) X(100)
90        CONTINUE
```

b.
```
          DIMENSION Y(50)
              ⋮
          DO 90 J = 1,50
          WRITE (6,30)Y
90        CONTINUE
```

c.
```
          DO 100 K = 1,5
          READ (5,50)B(K)
100       CONTINUE
50        FORMAT(5F6.1)
```

d.
```
      READ(5,30)X(L),L = 1,100,2
```
e.
```
      READ(5,40)(A(J)J = 1,30)
```
f.
```
      READ(5,50) (I,B(I),I = 1,50)
```
g.
```
      WRITE (6,30)(Y(M + 5),M = 1,10)
```
h.
```
      DIMENSION X(10)
      DATA X/0.0/
```
i.
```
      DIMENSION B(5),F(5)
      DATA (B(J),F(J),J = 1,5)/5*0.0/
```
j.
```
      WRITE (6,50)(I,X(I),I = 1,100)
```
k.
```
      READ(5,10)N,(WAGE(I),I = 1,N)
10    FORMAT(I2/(F5.1))
```
l.
```
      WRITE (6,10)(I,I = 1,30)
10    FORMAT(11X,30(I2,3X))
```

☐ **Answers**

a. The program segment reads the value of only one element, namely X(100), one hundred times. If the intention is to read the elements of the array X, then X(100) must be corrected to X(I).

b. The loop prints the entire array Y fifty times. However, if the intention is to write 50 elements of array Y in the DO loop, then Y(J) must be used with the WRITE statement in the loop.

c. The program segment is correct syntactically if the array B has been defined; however, if the intention is to read five values on one line as the FORMAT states (5F5.1), then an implied DO loop should be used.

d. A pair of parentheses is needed around the implied DO loop.

e. A comma is needed after A(J).

f. The value of I must not be changed within the DO loop by a READ statement.

g. This is correct.

h. The number of constants in the value list must be equal to the number of elements of array X.

i. There must be ten values in the value list of the DATA statement.

286 One-Dimensional Arrays

j. This is correct. It prints the integer numbers 1 through 100, as well as the content of 100 elements of the array X.
k. This is also correct. It reads N, and then N elements of the array WAGE, each on a different line.
l. This is also correct. It prints the sequence numbers from 1 to 30 on one line.

7.6 What would be the value of each element if the following READ statements are executed and the following data records are used?

Line 1	35892832105396568723524356
Line 2	932589614
Line 3	628315629
Line 4	3580153287
Line 5	298516231

a.
```
          DIMENSION A(50)
          DO 30 I = 1,5
30        READ (5,10) A(I)
10        FORMAT (F5.1)
```

b.
```
          DIMENSION B(5)
          READ(5,20) B
20        FORMAT (5F5.1)
```

c.
```
          DIMENSION C(5)
          READ (5,20) C
20        FORMAT (F5.1)
```

d.
```
          DIMENSION D(5)
          READ(5,30) D
30        FORMAT (5(F5.1/))
```

e.
```
          DIMENSION E(15)
          READ(5,10) N,(E(I),I = 1,N)
10        FORMAT (I1/15F3.1)
```

f.
```
          DIMENSION F(15)
          READ(5,10)M
10        FORMAT(I1)
          READ(5,30)(F(J),J = 1,M)
30        FORMAT(F3.1)
```

g.
```
          DIMENSION G(15), H(15)
          READ(5,40)(G(I),I = 1,3),(H(J),J =
         *12,15)
40        FORMAT(7F2.1)
```

h.

```
      DIMENSION G(15),H(15)
      READ(5,40)(G(I),H(I),I = 1,3)
40    FORMAT(6F4.1)
```

☐ **Answers**

a. Only five elements are read, as

A(1) = 3589.2
A(2) = 9325.8
A(3) = 6283.1
A(4) = 3580.1
A(5) = 2985.1

b.

B(1) = 3589.2
B(2) = 8321.0
B(3) = 5396.5
B(4) = 6872.3
B(5) = 5243.5

c.

C(1) = 3589.2
C(2) = 9325.8
A(3) = 6283.1
A(4) = 3580.1
A(5) = 2985.1

d.

D(1) = 3589.2
D(2) = 9325.8
D(3) = 6283.1
D(4) = 3580.1
D(5) = 2985.1

e. Three elements (N = 3) are read, as

E(1) = 93.2
E(2) = 58.9
E(3) = 61.4

f. Three elements (M = 3) are read, as

F(1) = 93.2
F(2) = 62.8
F(3) = 35.8

g.

G(1) = 3.5
G(2) = 8.9
G(3) = 2.8

and

H(12) = 3.2
H(13) = 1.0
H(14) = 5.3
H(15) = 9.6

288 One-Dimensional Arrays

h.

G(1) = 358.9
H(1) = 283.2
G(2) = 105.3
H(2) = 965.6
G(3) = 872.3
H(3) = 524.3

7.7 What do any of the following programs print? Also show the location of data.

a.
```
          PROGRAM SEQ
          DIMENSION IA(100)
          DO 90 I = 1,100
90        IA(I) = I
          WRITE(6,30) IA
30        FORMAT(10(6X,I4))
          END
```

b.
```
          PROGRAM FIVE
          DIMENSION A(100)
          DATA (A(J),J = 1,10)/10*5.5/
          WRITE (6,20)(A(I),I = 1,10)
20        FORMAT (1X,F4.1/)
          END
```

c.
```
          PROGRAM NUM
          DIMENSION X(100)
          DATA X(1),X(2),X(3),X(4)/
         *5.6,3.1,9.8,2.1/
          X(5) = 6.8
          DO 90 I = 1,5
90        WRITE (6,40) X(I)
40        FORMAT(1X,F4.1/)
          END
```

d.
```
          PROGRAM SQUARE
          DIMENSION I(100)
          DO 90 J = 1,100
90        I(J) = J*J
          WRITE(6,30)(K,I(K),K = 1,100)
30        FORMAT(11X,I3,5X,I6)
          END
```

e.
```
          PROGRAM CUTE
          DIMENSION TEST(100)
          READ(5,100) TEST
          WRITE(5,200)(TEST(I), I = 1,100)
100       FORMAT(10F5.2)
200       FORMAT(1X,100(21X,10(F6.2,3X)/1X))
          END
```

> ☐ **Answers**
>
> **a.** The program prints sequence numbers 1 through 100 with ten numbers to a line.
> **b.** The program prints 5.5 on every other line for ten times.
> **c.** The program prints 5.6, 3.1, 9.8, 2.1, and 6.8 each one on every other line.
> **d.** It prints two columns of numbers: integers from 1 to 100 in the first column, and the squares of those integers in the second column.
> **e.** The program reads 100 numbers, 10 per record. The data will be printed, 10 numbers per line, with three spaces between each number. The printed data will be centered on 132-column paper.

IMPORTANT NOTES ABOUT ARRAYS

1. The name and size of each array must be declared in a DIMENSION or type statement. The size must be a constant.
2. If the exact size of an array is not known, it should be declared larger than appears necessary. Obviously, one cannot use a subscript larger than the specified size. However, an excessive size is costly.
3. The DIMENSION statement is a nonexecutable statement; it specifies the number of memory locations needed for the array, and must be placed before the array is used. Normally it appears before any executable statement.
4. The type of the array elements is determined by the array name in the same way as the type of a variable is determined by its name (real, integer, character, etc.). All elements of an array must be of the same type.
5. The elements of an array are stored under the array name in order of the subscripts. The subscript is a reference number for a specific element. It is important to differentiate between the array subscript and the content of the element of an array. The subscript is only an index, a reference number, to select a particular element. The content of an element can be any data.
6. The name of a specific element is the name of the array followed by a subscript specification in parentheses. A typical subscript is an integer constant or variable. For example, if J is defined, then X(J) refers to the Jth element of the array X.
7. Zero, negative values, real variables, and real constants are allowed as subscripts only in FORTRAN 77, not in all versions of FORTRAN.
8. An arithmetic expression can be used as a subscript specification. The following are examples:

```
NAME (I + 3)
PAY (KAY - 2)
JOY (5*N)
AAA (3*I + 2)
```

9. The array name without the subscript refers to the complete array with all the elements. The array name cannot appear in an arithmetic expression, but it can appear in input or output list.

10. Elements of an array can be manipulated in the same way as a single variable is manipulated. In arithmetic expressions only one element (array name with the subscript) at a time must be used.
11. Array input and output can be accomplished by a DO loop, an implied DO, or the entire-list method. If all elements are to be input or output, the entire-list method is convenient; if only part of the array must be input or output, a DO loop is appropriate.
12. The DATA statement is a compact method to initialize the elements of an array.

SOME EXAMPLES OF ARRAY USE AND PROGRAMMING TECHNIQUES

Using arrays in a program offers more flexibility, challenge, and programming power. Some algorithms are impossible or extremely difficult to implement without using arrays. However, the programming techniques, data manipulations, and algorithms with arrays are somehow different than nonarray techniques.

In this chapter we will discuss some simple examples, and in Chapter 10 we will see applications of arrays in solving typical problems. To understand the technique used in the following examples, *it is imperative that you give enough thought to each problem before looking at the given solution procedure*. Write down your own solution method and then compare it with the text's method. The given procedure is one of several alternative algorithms—yours could even be a better one.

EXAMPLE 7.11 Storing a sum in an array

■ **Problem:** Write a program which calculates the sum of a series of 89 numbers in an array so that SUM(1) will contain the value of A(1), SUM(2) will contain A(1) + A(2), and so on. Print the subscripted data and the sum.

Program:

```
        PROGRAM SSS
        DIMENSION A(100), SUM(100)
        DO 90 I = 1,89
            READ(5,10) A(I)
90      CONTINUE
C
        SUM(1) = A(1)
        DO 100 J = 2,89
            SUM(J) = SUM(J - 1) + A(J)
100     CONTINUE
        DO 110 J = 1,89
            WRITE(6,20) J,A(J),J,SUM(J)
110     CONTINUE
10      FORMAT(F4.1)
20      FORMAT(2X,'A(',I2,')=',F5.1,3X,'SUM(',I2,
       *')=',F6.2)
        END
```

Data:

```
6.1
5.2
 ⋮
```

Output:

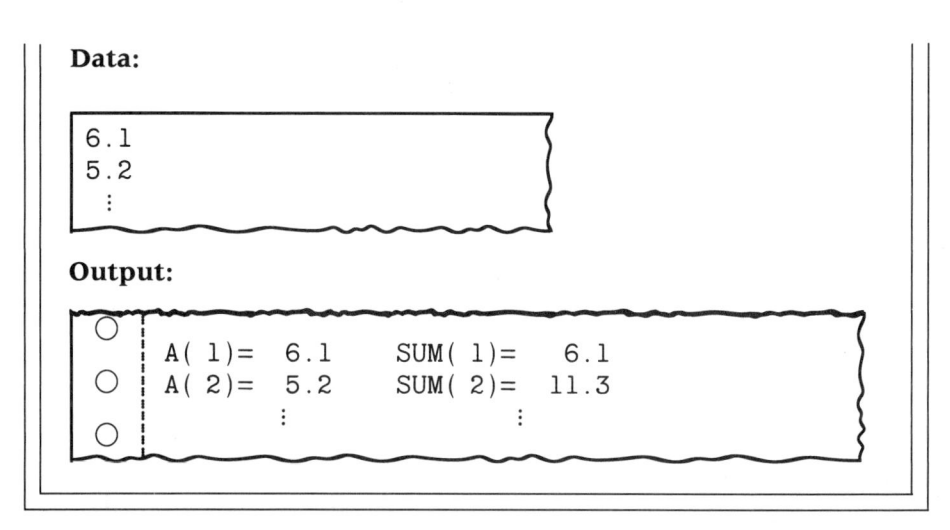

```
A( 1)=  6.1    SUM( 1)=   6.1
A( 2)=  5.2    SUM( 2)=  11.3
         ⋮                ⋮
```

EXAMPLE 7.12 Direct access to factorial

■ **Problem:** Write a program which stores factorial 1 through 10 in ten elements of an array FACT: factorial 1 in the first element, factorial 2 in the second, and so on. Then by reading N, an integer between 1 and 10, the program prints factorial N directly from the array FACT.

Program:

```
        PROGRAM FACTOR
        INTEGER FACT(10)
        FACT(1) = 1
        DO 100 I = 2,10
            FACT(I) = I*FACT(I - 1)
100     CONTINUE
        READ(5,10)N
        WRITE(6,20)N, FACT(N)
10      FORMAT(I2)
20      FORMAT(5X,'FACTORIAL',I3,'IS=',I10)
        END
```

EXAMPLE 7.13 Direct access to a table (table lookup)

■ **Problem:** Write a program which reads the ID, the number of hours worked, and the pay code (a number between 1 and 7) for employees during a week. The pay rate is based on the following table:

Pay Code	Pay Rate
1	5.50
2	6.50
3	7.75
4	9.10
5	10.90
6	12.90
7	14.00

Use arrays to calculate and print the weekly pay for about 100 employees. The first data card indicates the number of records.

Program:

```
      PROGRAM TABLE
C
      DIMENSION ID(100), HRS(100),KODE(100),
     *PAY(100), RATE(7)
      DATA RATE/5.5,6.5,7.75,9.10,10.90,12.90,
     *14.00/
C
      READ(5,10)N
C
      DO 30 I = 1,N
          READ(5,20) ID(I), HRS(I), KODE(I)
30    CONTINUE
C
      DO 40 J = 1,N
          KEY = KODE(J)
          PAY(J) = HRS(J)*RATE(KEY)
40    CONTINUE
C
      DO 50 I = 1,N
          WRITE (6,70) ID(I), HRS(I), PAY(I)
50    CONTINUE
10    FORMAT (I3)
20    FORMAT (I5, F5.2,I1)
70    FORMAT (11X, 'ID=', I6, 5X, 'HOURS WORKED=',
     *F6.2, 5X, 'PAY=',F6.2)
      END
```

Notes:

1. The term "table lookup" is used for the procedure of accessing data from a table. By knowing the key (pay code), the pay rate can be accessed directly.
2. The search key to find the appropriate pay rate from the pay table is KODE for each employee. Using RATE(KODE(J)) would also be valid.

EXAMPLE 7.14 Frequency Distribution

Problem: Write a program which reads a set of 100 integers from 1 to 10 into an array and then counts the number of times each number occurs in the set.

Program:

```
      PROGRAM FREQ
      INTEGER SET(100), COUNT(10)
      DATA COUNT/10*0/
C        READING THE NUMBERS
```

```
            READ(5,100) (SET(I), I=1,100)
100     FORMAT(I2)
C           COUNTING THE NUMBER OF OCCURRENCES
        DO 50 J = 1, 100
        K = SET(J)
50      COUNT(K) = COUNT(K) + 1
C           WRITING THE FREQUENCY
        DO 60 J = 1,10
60      WRITE (6,200) J, COUNT(J)
200     FORMAT(11X, 2(I3,10X))
        END
```

Notes:

1. Each element of the array COUNT counts the occurrences of the corresponding integer.
2. The numbers do not have to be in an array. The loop

```
        DO 10 I = 1,100
            READ*,SET
10          COUNT(SET) = COUNT(SET) + 1
```

will also work.

EXAMPLE 7.15 Indexing

Problem: The students' IDs and their test scores (between 50 and 99.99) are the input data. There are about 65 students; a negative test score indicates the end of data. Write a program which finds and prints the letter grade for each student, and prints the numbers of A's, B's, C's, D's, and E's.

Program:

```
        PROGRAM INDEX
        INTEGER COUNT(5)
        CHARACTER*1 GRADE(5), STGRD(65)
        DIMENSION ID(65), SCORE(65)
        DATA GRADE/'E', 'D', 'C', 'B', 'A'/,
     *       COUNT/5*0/,N/0/
        DO 50 I = 1,65
            READ(5,10) ID(I), SCORE(I)
            IF (SCORE(I) .LT. 0.0) GO TO 90
            N = N + 1
50      CONTINUE
90      DO 60 J = 1,N
            KEY = (SCORE(J)/10) - 4
            STGRD(J) = GRADE(KEY)
            COUNT(KEY) = COUNT(KEY) + 1
60      CONTINUE
        DO 70 K = 1,N
            WRITE(6,20) ID(K),STGRD(K)
70      CONTINUE
        WRITE (6,30) (COUNT(J),J = 1,5)
```

```
10      FORMAT (I5,F5.2)
20      FORMAT(11X,'ID#:',I6,5X,'GRADE:',A1)
30      FORMAT (6X,'NO. OF E''S=',I2,5X,
       *'NO. OF D''S=',I2,5X,'NO. OF C''S=',I2,5X,
       *'NO. OF B''S=',I2,5X,'NO. OF A''S=',I2)
        END
```

Notes:

1. Input data are read in the first module. The IF statement terminates the loop by reading a negative number. The statement N = N + 1 keeps track of the number of data records. N is needed for the next DO loops.

2. The array GRADE contains E,D,C,B,A. That is,

   GRADE(1) = 'E'
   GRADE(2) = 'D'
   GRADE(3) = 'C'
   GRADE(4) = 'B'
   GRADE(5) = 'A'

   The scores are converted to a key on a scale of 1 to 5 by the following formula:

   $$KEY = \frac{\text{score of each student}}{10} - 4$$

   Thus:

   if 90 ≤ SCORE ≤ 99.99,   KEY = 5,
   if 80 ≤ SCORE < 90.00,   KEY = 4,
   if 70 ≤ SCORE < 80.00,   KEY = 3,
   if 60 ≤ SCORE < 70.00,   KEY = 2,
   if 50 ≤ SCORE < 60.00,   KEY = 1.

   For example, if a student's score is 77, then

   $$KEY = \frac{77}{10} - 4 = 3 \text{ (integer)}$$

   and his or her grade is:

   GRAD(3) = 'C'

3. If scores below 50.0 were possible, then a statement such as

   IF(KEY .LT. 1)KEY = 1

   would be necessary after calculating KEY, to avoid an invalid subscript for the arrays GRADE and COUNT.

4. The array COUNT keeps track of the numbers of E's, D's, C's, B's, and A's as follows:

   COUNT(1) = COUNT(1) + 1   accumulates number of E's
   COUNT(2) = COUNT(2) + 1   accumulates number of D's
   COUNT(3) = COUNT(3) + 1   accumulates number of C's
   COUNT(4) = COUNT(4) + 1   accumulates number of B's
   COUNT(5) = COUNT(5) + 1   accumulates number of A's

   The initial values have been set to zero by the DATA statement.

EXAMPLE 7.16 Locating the smallest element in an array

Problem: Write a program which finds the position (subscript) of the smallest value among a set of data. There are less than 100 data, the first one being a header record. Print the smallest number and its subscript.

Program:

```
      PROGRAM LOCATE
      DIMENSION A(100)
      READ(5,10)N
C-----------------------------------
      READ(5,20) (A(I), I = 1,N)
C-----------------------------------
      K = 1
C ASSUMING THE SUBSCRIPT OF THE SMALLEST NO. (K) IS 1
      DO 200 J = 2,N
         IF(A(J) .LT. A(K)) K = J
200   CONTINUE
      WRITE(6,30) A(K), K
10    FORMAT (I3)
20    FORMAT (F5.1)
30    FORMAT (11X, 'THE SMALLEST NO. IS =', F6.1,
     *5X,'ITS SUBSCRIPT IS=', I2)
      END
```

Notes:

1. The data are read from different lines.
2. It is assumed at the beginning that the first value is the smallest number (K = 1). Each value is then compared with A(K) in the loop. If the value is smaller than A(K), then K will change to the subscript of the smaller value; otherwise, K will not change.

EXAMPLE 7.17 Inserting a value into an array

Problem: Suppose an array X is defined as having 100 elements, and data are already stored in N elements of the array (N < 100). We would like to insert a value D into the first element of array X while maintaining the existing data in the same order. Find a solution method and then write the program segment.

Solution Methods: In order to insert a value into the first element of array X, the other data must be shifted up the list by one element:

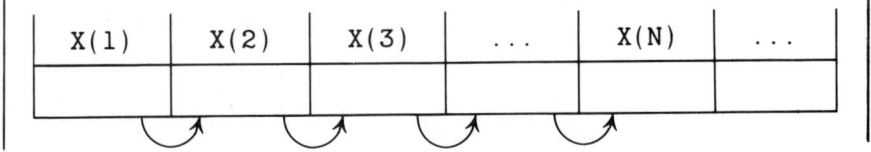

X(1)	X(2)	X(3)	...	X(N)	...

However, the program segment

```
      DO 50 K = 1,N
         X(K + 1) = X(K)
50    CONTINUE
```

will not work, because by moving the contents of the first element to the second, the original content of the second element will be erased. This will happen to the third, fourth, and finally all elements, and we end up assigning the value of A(1) to all elements of X. Three solution methods are presented below.

Method I:

1. Copy all elements of X into a new array Y, and
2. Transfer back the content of Y into the original array X, but one element up the list.

The program segment is:

```
      DO 50 I = 1,N
         Y(I) = X(I)
50    CONTINUE
      DO 60 J = 1,N
         X(J + 1) = Y(J)
60    CONTINUE
      X(1) = D
```

Note: This method requires 100 additional memory locations.

Method II: Store the content of X(1) into a temporary memory (TEMP) before assigning any value to it; store the original content of X(2) into the temporary memory (TEMP) before assigning X(1) to it; and follow this procedure for all N elements. However, when exchanging the content of each X with TEMP, another memory is needed; therefore, use two memories TEMP and D for the program segment. This method requires three steps:

1. Assign X_i into a temporary memory TEMP
2. Assign the value of D into X_i.
3. Assign TEMP into D (to be considered a new value)

Figure 7.3 shows the process. The program segment is as follows:

```
      DO 70 I = 1,N
         TEMP = X(I)
         X(I) = D
         D = TEMP
70    CONTINUE
```

Note: This method requires only two additional memory locations.

Method III: Shift the content of array X up the list by one element, but start with the last one. The program segment is

```
      DO 80 I = N,1, - 1
C
         X(I + 1) = X(I)
80    CONTINUE
      X(1) = D
```

Note: This method requires no additional memory.

Step 2   D   Step 3

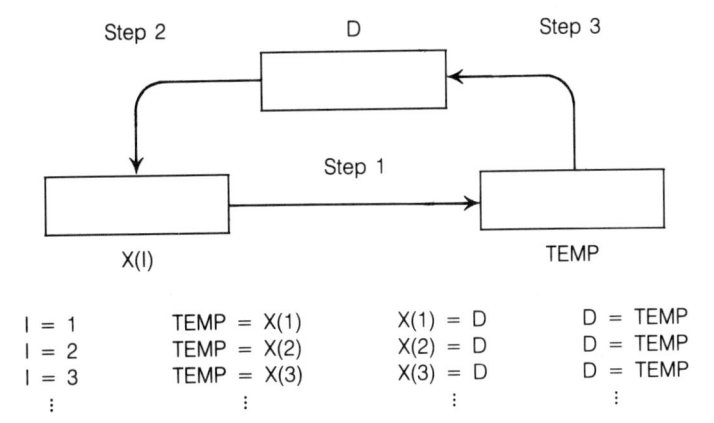

Step 1

X(I)   TEMP

I = 1	TEMP = X(1)	X(1) = D	D = TEMP
I = 2	TEMP = X(2)	X(2) = D	D = TEMP
I = 3	TEMP = X(3)	X(3) = D	D = TEMP

Figure 7.3 Interchanging the elements of an array.

EXAMPLE 7.18 Character Manipulation

■ **Problem:** A sentence is punched on a card (up to 80 columns). A period indicates the end of the sentence. Write a program which reads the sentence, prints the number of characters (including the spaces) in the sentence, and prints the sentence.

Program:

```
      PROGRAM CHAR
      CHARACTER*1 CARD(80)
C
      READ(5,10) CARD
C
      DO 20 I = 1, 80
          IF(CARD(I) .EQ. '.') GO TO 30
          N = N + 1
20    CONTINUE
30    WRITE(6,40)N,(CARD(J),J = 1,N)
10    FORMAT(80A1)
40    FORMAT('# OF CHARACTERS =',I2,5X,
     *'THE SENTENCE:'',80A1)
      END
```

Arrays are a convenient tool to manipulate characters, particularly when we would like (1) to break a field into several subfields, and (2) to read a field, one character at a time. The following are further examples:

EXAMPLE A

Write a program segment which reads the name of a student up to 20 characters, 4 characters at a time.

Here is the program segment:

```
      CHARACTER*4 NAME(5)
      READ(5,10) (NAME(I),I = 1,5)
10    FORMAT(5A4)
```

or

```
          CHARACTER*4 NAME(5)
          READ(5,10)NAME
10        FORMAT(5A4)
```

EXAMPLE B

Several sentences are punched on a card. Write a program segment which reads the sentences and counts how many commas they contain.

Here is the program segment:

```
          CHARACTER*1 CARD(80)
          READ(5,10) CARD
10        FORMAT(80A1)
          DO 50 I = 1,80
              IF(CARD(I) .EQ. ',') N = N + 1
50        CONTINUE
```

EXAMPLE C

Print a line of asterisks (132 columns).

Here is the program segment:

```
          CHARACTER*1 LINE(132)
          DATA LINE/132*'*'/
          WRITE(6,20) LINE
20        FORMAT(1X,132A1)
```

EXAMPLE D

Write a program segment which reads the name of a student, up to 20 characters, one character at a time.

Here is the program segment:

```
          CHARACTER*1 NAME(20)
          READ(5,10) NAME
10        FORMAT(20A1)
```

EXAMPLE E

Write a program segment which assigns the first to sixth characters of the 30th item of an array A into the variable B.

Here is the program segment:

```
          CHARACTER*12 A(100),B
              ⋮
          B = A(30)( :6)
```

EXERCISES

7.1 Identify any errors in the following DIMENSION statements.

    a. DIMENSION XXX(N),I(100),B(5)
    b. REAL DIMENSION A(100),B(2)
    c. INTEGER,I,J,K(100)
    d. DIMENSION A(−20),B(100)

e. DIMENSION (300),P(20)
  f. DIMENSION REAL I(30)
  g. CHARACTER*35,OIL(35)
  h. DIMENSION Q,R(100)
  i. DIM IX(10),Y(100)
  j. INTEGER PAY(0),RATE(-80)

7.2 Identify any errors in the following statements or loops:

  a.
```
      DIMENSION F(100), G(100)
         :
      G = F*F
```
  b.
```
      DIMENSION X(100), B(100)
         :
      DO 10 I = 1,100
         X(I + 1) = B(I**2)
 10   CONTINUE
```
  c.
```
      DIMENSION H(30),Q(30),R(30)
         :
      DO 90 I = 1,30
         H = Q(I) + R(I)
 90   CONTINUE
```
  d.
```
      DIMENSION Q(500),R(100)
         :
      DO 50 J = 1,500
         Q(J) = 2*R(J)
 50   CONTINUE
```
  e.
```
      DIMENSION X(100),Y(100)
      X = 0.0
      Y = 0.0
```
  f.
```
      DIMENSION ERROR(500)
         :
      IF(ERROR .EQ. 0.0) GO TO 10
```
  g.
```
      DIMENSION PAY(30),GROSS(30)
         :
      DO 100 I = 1,30
         PAY(I + 1) = GROSS(I)
100   CONTINUE
```

7.3 Identify any errors in the following statements:

  a. READ(5,10) A(I),I = 1,100
  b. READ(5,30) A,B,(C(J),DO 100J = 1,300)

c. WRITE(6,90) N,(Y(K),10 J = 1,100)
d. READ(5,50)(I,X(I),I = 1,100)
e. DATA (X(I),I = 1,10)/0.0/
f. WRITE(6,60)(P(K)K = 1,20),(Q(K)K = 1,20)
g. READ(5,100)(A(I),B(J),I,J = 1,100)

7.4 What are the purpose and the output of the following programs or program segments?

a.
```
         PROGRAM SQR
         DIMENSION K(100), L(100)
         DO 90 J = 1,50
            K(J) = J*J
            L(J) = J**3
90       CONTINUE
         WRITE(6,20)(N,K(N),L(N),N = 1,50)
20       FORMAT(11X,I2,5X,I6,5X,I10)
         END
```

b.
```
         DIMENSION B(100), A(101)
         READ(5,10) B
10       FORMAT(10F5.1)
         DO 100 I = 1,100
            A(I + 1) = B(I)
100      CONTINUE
```

c.
```
         DIMENSION B(100)
            ⋮
         DO 900 I = 1,100,2
            TEMP = B(I)
            B(I) = B(I + 1)
            B(I + 1) = TEMP
900      CONTINUE
```

d.
```
         PROGRAM SSS
         DIMENSION X(100)
         READ(5,10) X
10       FORMAT (10F6.1)
         SMALL = X(1)
         DO 70 K = 2,100
C
            IF(X(K) .LT. SMALL) SMALL = X(K)
C
70       CONTINUE
         WRITE(6,20) SMALL
20       FORMAT (11X,F7.1)
         END
```

e.
```
         DIMENSION B(100)
            ⋮
         DO 100 I = 1,N
```

```
              M = N - I + 1
              WRITE(6,20)B(M)
100     CONTINUE
```

f.

```
        PROGRAM FREQ
        DIMENSION N(10)
        DATA N/10*0/
        DO 100 I = 1,100
C READING THE DATA, EACH LESS THAN 99
        READ(5,10)A
        M = A/10
        N(M) = N(M) + 1
100     CONTINUE
10      FORMAT(I2)
        END
```

7.5 Find the logical errors in each of the following program segments:

a.

```
        DIMENSION XDATA(50),TOT(50)
        DATA TOT/50*0.0/
        READ(5,10)XDATA
10      FORMAT(F5.1)
C CALCULATING THE TOTAL OF XDATA
        DO 100 I = 1,50
        TOT(I) = TOT(I) + XDATA(I)
100     CONTINUE
           ⋮
```

b.

```
        INTEGER FACT(10)
        DATA FACT/10*1/
        READ(5,10)N
10      FORMAT(I1)
C CALCULATING FACTORIAL OF N, N < 10
        DO 100 I = 1,N
        FACT(I) = FACT(I)*I
100     CONTINUE
```

c.

```
C MOVING THE ELEMENTS OF ARRAY A UP BY ONE ELEMENT
        DO 500 I = 1,99
        A(I + 1) = A(I)
500     CONTINUE
```

PROGRAMMING EXERCISES

Write a complete FORTRAN program for each of the following problems, using appropriate arrays, headings, formats, and techniques. Use either a header or a trailer record to terminate the loops.

7.6 Write a program which reads 100 values into an array. Calculate the average of the numbers, then print the data and the average.

7.7 Write a program which reads N values of arrays A and B, each with 200 elements. Calculate the elements of the array C given by

```
C(I) = A(I) + B(I) + 2*A(I)*B(I)
```

Print all three arrays. N is to be read at the beginning of the program.

7.8 Write a program which reads N elements of the array EEE (N < 20). Then use an implied DO loop to print the values of N elements:

a. one element per line,
b. with all elements on the same line,
c. ten elements per line, three spaces between values.

N is to be read at the beginning of the program (use F4.1 format when reading the data).

7.9 The array DX has been defined to have 40 integer elements. Write the program segment, using the entire-list method, to print DX:

a. one element per line,
b. with all elements on the same line (use I3),
c. ten elements per line, five spaces between.

7.10 Write a program which reads 10 sales values into an array. The sales tax is 4%. Calculate the tax and totals for all sales. Print the sales, tax, and total:

a. In three columns such as

SALES	TAX	TOTAL
XXXXX	XXX	XXXXX
⋮	⋮	⋮

b. In three rows such as

SALES	XXX	XXX	XXX	...
TAX	XXX	XXX	XXX	...
TOTAL	XXX	XXX	XXX	...

Print the totals of all sales at the end of the report.

7.11 Write a program which reads 80 columns of a line and, after omitting all the commas in the line, prints the characters in a column.

7.12 Write a program which reads a word with four characters and prints it scrambled in several ways. Can you develop an algorithm which prints all possible combinations of the scrambled word?

7.13 Write a program which reads and prints a series of names (16 characters), Social Security numbers (11 characters), and addresses (24 characters). Assume that the computer being used cannot store a value with more than four characters.

7.14 Given a set of data cards which have an algebraic expression punched on each card. Write a program which reads each card, and prints a message indicating whether or not the left and right parentheses used in each expression are properly matched.

7.15 Several names are punched on a card (80 columns). There is a period after each name, but the number of characters in each name varies from 1 to 15. Write a program which reads the names, and prints each name backwards (the last letter first) on a separate line.

7.16 Write a program which calculates the sum of the squares of a series of numbers:

$$\text{SUM} = A_1^2 + A_2^2 + A_3^2 + A_4^2 + \cdots + A_n^2.$$

7.17 Write a program which calculates and stores the squares of integer numbers 1, 2, 3, ..., 40 in the array SQURE. Then have the program print the integer numbers and their squares as follows:

  a. In two columns, with the numbers in the first column, and their squares in the second column.

  b. In two lines (rows), with the numbers in the first line, and their squares in the second line. To make all the 40 elements fit in one line (132 columns), print:

  1 through 9 with two digits each (use I2),
  10 through 25 with three digits each (use I3),
  26 through 40 with four digits each (use I4).

  c. In four columns. In the first column are the numbers 1 through 20, and in the second column their squares. In the third column are the numbers 21 to 40, and in the fourth column their squares.

7.18 Write a program which reads N real numbers ($N < 100$) into an array and prints them in reverse order. N is to be read at the beginning of the program.

7.19 Suppose you have received a message in numerical codes. But you know every two digits represent a character: numbers 01 through 26 for A through Z respectively, and 27 through 40 for a space, period, comma, hyphen, and the digits 0 through 9 respectively. Write a program which reads the codes and prints the decoded message. Each message is less than 39 codes long, and 00 indicates the end of the message. Test the program by decoding the following message:

2515222701180527141528271514050 0

7.20 Suppose the data are stored in only N elements of an array A with 200 elements ($N < 200$). Write a program which inserts the value B in the Kth position of A ($K < N$), while maintaining the existing data in the same order.

7.21 The variance of a series of numbers can be found by the formula:

$$\text{VAR} = \frac{(A_1 - X)^2 + (A_2 - X)^2 + (A_3 - X)^2 + \cdots + (A_n - X)^2}{N}$$

where X is the mean of the numbers. Write a program which reads a set of data and calculates their variance. Print the data, the mean, and the variance.

7.22 In Example 7.16 in the text, the smallest number was located by an algorithm which found the subscript of the smallest number. Write a similar program which finds the smallest number among a set of data by comparing the values of each element with the smallest value (without considering the subscript).

7.23 Write a program which finds the largest number among a set of data. Print the data and the largest number.

7.24 Write a program which reads 200 scores of a test, then counts and prints the number of A's (90–100), B's (80–90), C's (70–80), D's (60–70), and E's (below 60).

7.25 The insurance premium for the employees of a company is as follows:

Number of dependents	Yearly premium
1	101.0
2	125.0
3	140.0
4	150.0
5	158.0
6	165.0
7	171.0
8	176.0
9 or more	180.0

Write a program which reads the following information about employees into arrays:

1. Name, 16 characters, columns 1–16.
2. Social Security number, 11 characters, columns 17–27.
3. Yearly salary, columns 29–36.
4. Number of dependents, columns 38–39.

Calculate the net pay after the premium deduction for the employees. (1) Print the information in report form. Print also the total pay, total deductions, and the total net at the end of the report. (2) Print the same report but only for employees whose salary is above $50,000.00.

7.26 The grade information for a FORTRAN course is available as:

The student's number	Columns 1–9
The student's name (12 characters)	Columns 11–22
The test score	Columns 24–28

Write a program which reads the information and then prints the student's number, name, test score, and letter grade, and a code AA if his or her score is above average. Print also a report of the names of the students in each grade category.

7.27 Write a program which reads a set of real numbers (less than 500 items). The first record is a header. Find and print the average, smallest, largest, range, and variance of the data (the range is the largest number minus the smallest number).

7.28 The personnel manager of an organization would like to have a report on the retirement situation of the employees in the organization. The following data are available:

Employee number	Columns 1–7
Name	Columns 9–20
Annual salary	Columns 22–27
Age	Columns 29–30
Number of years worked	Columns 32–33

The condition for retirement is that one should be over 65 years old and have worked more than 30 years. Write a program which reads the information, calculates the average salary, and prints the following about each employee in a report form:

1. Name, salary, age, number of years worked, a code for retirement (E for eligible, N for not eligible), and a code for salary situation (A for the salary being above average, B for the salary being less than or equal to the average).
2. The above information for the employees whose age is over 65.

7.29 Write a program which prints the number of students, categorized as freshman (1), sophomore (2), junior (3), and senior (4), who have successfully completed Engineering 101.

Input:

Student's class standing (1, 2, 3, or 4)	Column 1
Grade for the course (A, B, C, D, or E)	Column 3
Name	Columns 4–15

Output:

Four reports, each indicating the names and grades of students in each class.

A report indicating the number of students in each category who have passed the course (D or better), and the total number of students who have failed the course.

7.30 The following information is available about members in a book club:

1. Name, 16 characters, columns 1–16.
2. Account number, 7 digits, columns 18–24.
3. Address
   a. Number and street, 20 characters, columns 26–45.
   b. City, state, and zip code, 22 characters, columns 47–66.
4. Date of last bill (three fields, each two digits), columns 68–73.
5. Balance due (as 999.99), columns 75–80.

Write a program which reads the information and prints:

1. A report for the club managers containing the name, account number, address, date of last bill, and balance due for each member (on one line) in a report form.
2. Bills for the members, each including name, account number, address, and balance due.

Design the bill, and then print two bills side by side. An example of the bill follows:

```
***********************************************************
*                                                         *
*                     XYZ BOOK CLUB                       *
*                                                         *
*                   ACCOUNT # 5623518                     *
*                                                         *
*       RAY JOHNSON                                       *
*       562 EAST BROADWAY                                 *
*       MT. PLEASANT, MI 48858                            *
*                                                         *
*                            *********                    *
*                            *       *                    *
*               BALANCE DUE  * $56.25*                    *
*                            *       *                    *
*                            *********                    *
*                                                         *
*               HAVE A NICE DAY                           *
*                                                         *
***********************************************************
```

7.31 A FORTRAN instructor would like to have the final semester grade report for each student in a course. Design a program that will prepare the report as follows. The input consists of:

1. Student's name, 16 characters, columns 1–16.
2. Student's ID, 9 digits, column 17–25.
3. Class standing (1 = freshman, 2 = sophomore, 3 = junior, 4 = senior), column 26.
4. Male-female code (M or F), column 27.
5. Three test scores:

   a. Test 1, columns 28–32.
   b. Test 2, columns 33–37.
   c. Test 3, columns 38–42.

The output is composed of two parts:

1. The instructor's report, containing the student's name, ID, class standing, male-female code, three test scores, average of the tests, and the course grade (based on the average and a scale of 100–90: A; 80–90: B; and so on). Print the grand average and the number of A's, B's, C's, D's, and E's at the end of the report.
2. Students' reports, containing the above information for each student in a form that you design and that is ready to be cut and handed to students. Try to print two reports side by side.

SUMMARY OF CHAPTER 7

You have learned:

1. An array is a group of variables having the same name but different subscripts. In most programming languages the subscript must be placed in parentheses. Using an array gives the programmer flexibility and more programming power. Some algorithms are almost impossible to code without using an array.
2. The array size must be declared by a DIMENSION statement or by a type statement at the beginning of a program. The following are some examples:

   ```
   DIMENSION A(100), WAGE(450), I(300)
   REAL NET(500), K(300), I(-5:50)
   INTEGER COUNT(300), FACT(100)
   DIMENSION AAA(-1:100), BBB(0:100), CCC(200)
   ```

3. Any arithmetic expression must be performed on individual elements of an array. A DO loop is an efficient tool for this purpose.
4. The subscript can be an integer constant, variable, or arithmetic expression. The following is an example:

   ```
         DO 10 I = 1,100
            A(2*I + 3) = 5.5
   10    CONTINUE
   ```

   However, the value of the subscript must not go beyond the dimension size.
5. A DO loop can be used to read the elements of an array one by one, where the control variable of the DO loop is used to control the subscript of the array. The following is an example:

```
       ARRAY B(100)
       DO 20 I = 1,100
           READ(5,10) B(I)
20     CONTINUE
10     FORMAT (F5.1)
```

6. An implied DO loop is a compact method for reading the elements of an array from one line, or to print the elements on one line. The following is an example:

```
       DIMENSION B(10)
       READ(6,20) (B(J),J = 1,10)
20     FORMAT (10F5.1)
```

7. The name of the array without a subscript in an input or output list causes all the elements to be read or written. This is called the entire-list method or the short-list method. The following is an example:

```
       DIMENSION C(15)
       READ (5,20)C
20     FORMAT (15F5.1)
```

8. The FORMAT statement determines the location of the data in the implied DO loop or the entire-list method.

9. The DATA statement is an efficient method for initializing the elements of an array at the beginning of the program. The following are some examples:

```
       DATA X(5), X(6), X(7), X(8), X(9)/3*0.0,2*1.0/
       DATA (X(I), I = 1, 17)/17*0.0/
       DATA Y/100*0.0/
```

10. Mirror printing means using a WRITE (or PRINT) statement immediately after the READ statement to check whether the input data have been read correctly. The WRITE statement is normally removed after this check.

SELF-TEST REVIEW

7.1 Which of the following statements are incorrect in FORTRAN?
  a. DIMENSION ROAD(599),N(302)
  b. REAL INTER(532), MASH(282)
  c. INTEGER COL(390),PIE(30)
  d. DIMENSION CHARACTER*12 NAME(100)
  e. REAL NET(N)
  f. DIM A(100), B(200)

7.2 Which of the following subscripts are valid in FORTRAN 77?
  a. X(I + 2)
  b. Y (R + 1)
  c. YORK(3 * M ** 2 + 1)
  d. K(−3)
  e. GROSS(100/3)

7.3 Write a program segment which reads values of the array A with 100 elements. The data are on different lines. Use:
  a. A DO loop.

b. An implied DO.
c. The entire-list method.

7.4 Write a program segment which prints ten values from an array B on one line.

7.5 Are the following program segments correct? If they are not correct, identify the errors in each loop:

a.
```
      DIMENSION B(100)
      DO 10 I = 1,100
      DATA B(I)/0.0/
10    CONTINUE
```
b.
```
      DIMENSION X(500)
      DO 30 I = 1, 500
         X = 59.5
30    CONTINUE
```
c.
```
      INTEGER FATE (1500)
      DO 30 I = 1,1500
         FATE(I - 1) = I
30    CONTINUE
```

7.6 Suppose data are stored in 100 elements of an array A. Write a program segment which interchanges values of successive pairs of elements—that is, interchanges the values of A(1) and A(2), A(3) and A(4), and so on.

7.7 Write a complete FORTRAN program using arrays, headings, and formats for the inventory of a sport shop. Use the following information:

Input for each item:
1. Item number   columns 1–5
2. Number of items available columns 7–9
3. Minimum inventory level   columns 11–13
4. Ordering quantity   columns 15–17
5. Number of items sold   columns 19–21

Processing: The new inventory level for each item is the number of items available minus the number of items sold. If the inventory level is below the minimum level, the item must be ordered.

Output:
1. A report containing the item number, inventory level, minimum inventory level, and ordering quantity for each item.
2. The above information (as the first report) for the items which must be ordered in a second report.

There are less than 100 items. The first record is a header record indicating the number of items.

☐ Answers

7.1 **d., e.,** and **f.** are incorrect.

7.2 All of them are correct in FORTRAN 77.

7.3

a.
```
        DO 50 I = 1,100
            READ(5,10) A(I)
50      CONTINUE
10      FORMAT(F5.1)
```

b.
```
        READ(5,20)(A(I),I = 1,100)
20      FORMAT(F5.1)
```

c.
```
        READ(5,30) A
30      FORMAT(F5.1)
```

7.4
```
        WRITE(6,40)(B(I),I = 1,10)
40      FORMAT(10F5.1)
```

7.5

a. Incorrect; the DATA statement is not an executable statement.
b. Incorrect; the individual subscripted elements must be used in an assignment statement.
c. Incorrect; when I is equal to one, the subscript becomes zero.

7.6
```
        DO 90 I = 1,100,2
            T = A(I)
            B(I) = B(I + 1)
            B(I + 1) = T
90      CONTINUE
```

7.7
```
        PROGRAM INVENT
        INTEGER ITEMNO(100)
        REAL NOAVAL(100), MLEVEL(100),ORDQ(100),
       *NOSOLD(100),LEVEL(100)
C READING THE HEADER RECORD
        READ(5,10) N
C READING THE DATA
        DO 100 I = 1,N
            READ(5,20) ITEMNO(I),NOAVAL(I),MLEVEL
       *    (I),ORDQ(I),NOSOLD(I)
100     CONTINUE
C CALCULATE THE INVENTORY LEVEL
        DO 100 J = 1,N
            LEVEL(J) = NOAVAL(J) - NOSOLD(J)
110     CONTINUE
C WRITING THE HEADING
        WRITE(6,30)
C WRITING THE INFORMATION
        DO 120 K = 1,N
            WRITE(6,40) ITEMNO(K),LEVEL(K),
       *    MLEVEL(K),ORDQ(K)
120     CONTINUE
C WRITING THE ORDERING REPORT
C THE HEADING
```

```
              WRITE (6,25)
              WRITE(6,30)
C THE INFORMATION
              DO 130 L = 1,N
                  IF(LEVEL(L) .LE. MLEVEL(L)) WRITE(6,40)
     *              ITEMNO(L),LEVEL(L),MLEVEL(L),ORDQ(L)
130       CONTINUE
10        FORMAT(I2)
20        FORMAT(I5,4(F3.0,1X))
25        FORMAT(56X,'ORDER THE FOLLOWING ITEMS'//)
30        FORMAT(11X,'ITEM NUMBER',10X,
     *'INVENTORY LEVEL',10X,'MINIMUM LEVEL',10X,
     *'ORDERING QUANTITY'//)
40        FORMAT(13X,I6,19X,F4.0,20X,F4.0,20X,F4.0)
          END
```

Chapter 8
Multidimensional Arrays

INTRODUCTION

DIMENSION STATEMENT

ARITHMETIC EXPRESSIONS WITH MULTIDIMENSIONAL ARRAYS

ARRAY STORAGE

MULTIDIMENSIONAL-ARRAY INPUT AND OUTPUT
   The Entire-List Method

THE DATA STATEMENT

SUMMARY OF THE IMPORTANT RULES

EXAMPLES OF TWO-DIMENSIONAL ARRAYS AND PROGRAMMING TECHNIQUES

INTRODUCTION

A two-dimensional array is composed of related data organized into rows and columns. For example, the following table of scores can be considered as a two-dimensional array:

	QUARTER 1	QUARTER 2	QUARTER 3	QUARTER 4
TEAM 1	3	9	6	11
TEAM 2	2	7	13	0

Each element of a two-dimensional array is referred to by two subscripts. The first subscript shows the row number, the second one shows the column number. For example, SCORE(2,3) means the element in the second row and the third column of the array SCORE, which is 13.

Figure 8.1 depicts the elements of another two-dimensional array, with four rows and three columns.

To locate an element in the array, both the row number and the column number must be presented. For example, A(3,2) represents the element of the array A in the third row and second column. That element contains the real number 6.9 in Figure 8.1.

A three-dimensional array has three subscripts, a four-dimensional array has four subscripts, and so on. FORTRAN 77 compilers allow up to seven dimensions in an array. A three-dimensional array can be imagined as being several tables in different pages (or planes). For example, the names of the students and their test scores for different courses in Figure 8.2 can be considered as elements of a three-dimensional array. In a three-dimensional array, the first subscript shows the row number, the second subscript shows the column number, and the third subscript shows the page or plane number. Thus, X(5,3,2) shows the element in the 5th row, 3rd column, and 2nd page of the array X.

In mathematics a two-dimensional array is called a *matrix*, whereas a one-dimensional array is called a linear array or *vector*. Most of the examples in this chapter are designed for two-dimensional arrays.

A multidimensional array also must be defined in a DIMENSION statement before being used in a program.

	One column of the array A	
	A(1,1)	A(1,2)	A(1,3)
	5.6	8.6	3.2
	A(2,1)	A(2,2)	A(2,3)
One row of the array A	15.2	3.2	8.1
	A(3,1)	A(3,2)	A(3,3)
	15.1	6.9	5.8
	A(4,1)	A(4,2)	A(4,3)
	18.2	4.3	7.8

Figure 8.1 A two-dimensional array.

```
            Page 3              Course #3
                    Name    x  x  x
                      ⋮     ⋮  ⋮  ⋮

        Page 2              Course #2
                Name    x  x  x
                  ⋮     ⋮  ⋮  ⋮

    Page 1              Course #1
            Name    x  x  x
              ⋮     ⋮  ⋮  ⋮
```

Figure 8.2 Data for a three-dimensional array.

DIMENSION STATEMENT

The DIMENSION statement defines an array, its dimensions, and its size in each dimension. For example,

    DIMENSION TAB(4,3)

declares two-dimensional array TAB of four rows and three columns for a total of 12 elements (3 × 4);

    DIMENSION TABLE(5,3,6)

declares a three-dimensional array of five rows, three columns, and six planes, for a total of 90 elements.

When an array is defined, the compiler reserves a certain number of memory locations for the array. As with linear arrays, a type statement (REAL, INTEGER, CHARACTER, etc.) can also be used to define a multidimensional array. Several arrays can be defined in a single DIMENSION statement. Thus, the simple form of an array declaration statement is:

    DIMENSION var1(i,j,...,n), var2(i,j,...,n),...

or

    TYPE var1(i,j,...,n), var2(i,j,...,n),...

where var1, var2,... are FORTRAN symbolic names, and $i,j,...,n$ are unsigned integer constants representing the array sizes. The following are examples:

    DIMENSION B(50), X(5,20),N(100,2,5)

    DIMENSION TABLE(5,30), COUNT(30), IN(50,3)

    REAL MONTH(50,30) NET(100), INCOME(50,3)

    INTEGER A(20,3,5), SUM(5,5), GROSS(100,3,5,8)

    CHARACTER*20 ADDRES(100,10), SSN(100,3)*11

In FORTRAN 77, the boundaries of an array can be defined as follows:

$$\text{TYPE var}\left(\begin{array}{c}\text{lower}\\\text{bound 1}\end{array}:\begin{array}{c}\text{upper}\\\text{bound 1}\end{array},\begin{array}{c}\text{lower}\\\text{bound 2}\end{array}:\begin{array}{c}\text{upper}\\\text{bound 2}\end{array},...\right)$$

314 *Multidimensional Arrays*

Each boundary can be a negative constant or a positive one. However, the upper boundary must be greater than the lower boundary. If the lower boundary is omitted, a value of 1 will be assumed and the form is the same as given above. The following are examples:

```
DIMENSION A(-5:20,0:50), B(30,5),C(-5:54,6)
REAL MORE(-5:38,5:50), JOY(100,-10:5,0:10), NORM(1:100,1:200)
INTEGER COUNT(0:100),X(-5:-1,20:30), WHY(-10:0,100)
```

SOLVED PROBLEMS

8.1 Detect any errors in the following DIMENSION statements:

a. DIMENSION,A(10,5) INT(30,40)
b. REAL COR(30,5,80), MARK(30), LOAN(300,5), GROSS
c. DIMENSION AMOUNT(M,N), LARGE(N,K)
d. DIM AMT(5,500), SALARY(600,6)
e. DIMENSION POOR(50.0,100.0), MODEL(100.0,50.0)
f. INTEGER I(100,2),J(2,N), AVERAGE(110)
g. DIMENSION A(5,0,8), B(5,-8,2), B(5), C(10,20)
h. DIMENSION A(50,-10:0, -8:-5), B(5), C(10,20)

☐ Answers

a. The comma after the word DIMENSION must be omitted. But a comma is needed to separate the arrays.
b. This is correct.
c. The array size must be an integer constant. An integer variable can be used only in the subprograms, which will be explained in the next chapter.
d. The word DIMENSION must be spelled correctly.
e. The dimension size cannot be a real constant.
f. The dimension size cannot be a variable. Also, the name of the third array (AVERAGE) is too long.
g. The dimension size cannot be zero or a negative number.
h. This is correct.

ARITHMETIC EXPRESSIONS WITH MULTIDIMENSIONAL ARRAYS

As with one-dimensional arrays, the name of the array indicates all the elements together, and the name followed by subscripts indicates an individual element. Individual subscripted elements must be used with arithmetic expressions. A DO loop is a convenient tool for manipulating arrays. Normally, a nested DO loop is necessary for controlling two or more subscripts. The following are two simple examples.

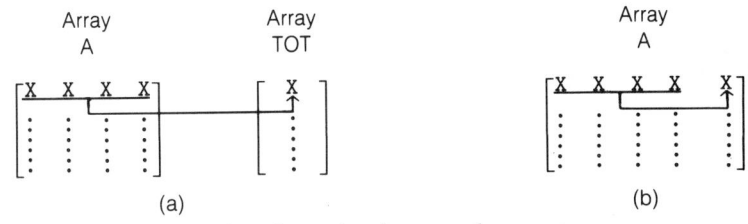

Figure 8.3 Adding the elements of a row of an array.

EXAMPLE A

Either of the following two program segments adds the elements of each row of the array A and stores the totals in a one-dimensional array TOT. A has four columns and fifty rows (Figure 8.3a):

```
      DO 100 I = 1,50
          TOT(I) = A(I,1) + A(I,2) + A(I,3) + A(I,4)
100   CONTINUE
```

or

```
      DO 100 I = 1,50
          TOT(I) = 0.0
          DO 100 J = 1,4
              TOT(I) = TOT(I) + A(I,J)
100   CONTINUE
```

EXAMPLE B

Either of the following program segments stores the sum of the first four elements of each row as the fifth element of the row (i.e., in the fifth column). The array A has five columns and fifty rows (Figure 8.3b):

```
      DO 200 I = 1,50
          A(I,5) = A(I,1) + A(I,2) + A(I,3) + A(I,4)
200   CONTINUE
```

or

```
      DO 200 I = 1,50
          A(I,5) = 0.0
          DO 200 J = 1,4
              A(I,5) = A(I,5) + A(I,J)
200   CONTINUE
```

Each subscript in a multidimensional array can be an integer constant, an integer variable, or an arithmetic expression, as explained for a one-dimensional array. The following program is an example of using a two-dimensional array.

EXAMPLE 8.1

■ **Problem:** Write a program which reads the scores at the end of each quarter of a football game and prints the final score. The scores for one occasion are:

	QUARTERS				
	1	2	3	4	FINAL
TEAM 1	3	9	6	11	?
TEAM 2	2	7	13	0	?

Program:

```
      PROGRAM MULTAR
      DIMENSION ISCORE(2,4), IFINAL(2)
C
      DO 90 I = 1,2
          READ(5,10) (ISCORE(I,J), J = 1,4)
90    CONTINUE
      DO 100 I = 1,2
          IFINAL(I) = ISCORE(I,1) + ISCORE(I,2) +
     *      ISCORE(I,3) + ISCORE(I,4)
100   CONTINUE
      DO 110 J = 1,2
          WRITE(6,20) (ISCORE (J,K), K = 1,4),
     *      IFINAL(J)
110   CONTINUE
10    FORMAT(4(I2,3X))
20    FORMAT(11X,5(I3,5X))
      END
```

Note: The calculation module can also be written as

```
      DO 100 I = 1,2
          IFINAL(I) = 0
          DO 100 J = 1,4
              IFINAL(I) = IFINAL(I) + ISCORE(I,J)
100   CONTINUE
```

SOLVED PROBLEMS

8.2 Find any errors in the following statements:

a.
```
      DIMENSION X(5,6)
      DO 10 I = 1,5
          DO 10 J = 1,6
              X = 0.0
10    CONTINUE
```

b.
```
      DO 50 J = 1,30
          DO 50 K = 1,300
              XXX(J + 5,K + 3) = 5.62
50    CONTINUE
```

c.
```
      (Y(1,J),J = 1,5) = 0.0
```

d.
```
      YORK(K,L**3,M) = 5.0
```

☐ **Answers**

a. The individual element must be used in an assignment statement. The assignment statement, then, must be corrected to X(I,J) = 0.0.
b. This is correct if the array XXX has been defined to have more than 35 rows and 303 columns.
c. The implied DO loop cannot be used in an assignment statement.
d. This is correct if the array YORK has been defined and the value of each subscript does not go beyond its DIMENSION size.

ARRAY STORAGE

The dimension statement tells the compiler to reserve

$N_1 \times N_2 \times N_3 \times \cdots$

elements for the array (where N_1, N_2, N_3, ... are the dimension sizes). For example, the statement

    DIMENSION A(5,30,2)

causes $5 \times 30 \times 2 = 300$ elements to be reserved for the array A.

The elements of an array are stored in a specific order. The internal storage order of a multidimensional array is somewhat different than it appears visually. A multidimensional array will always be converted into a linear (one-dimensional) array by the compiler, in the order of the subscripts, so that the first subscript value increases most rapidly, and the last subscript value increases least rapidly. In other words, the first column of the array will be stored first, the second column second, and so on. Figure 8.4 shows the internal storage pattern of a 3×4 array.

> **RULE 8.1**
> *A multidimensional array will always be stored in the form of one-dimensional arrays, one column under another.*

Figure 8.5 shows the storage pattern of a simple $3 \times 3 \times 3$ three-dimensional array.

MULTIDIMENSIONAL-ARRAY INPUT AND OUTPUT

As with a one-dimensional array, the elements of a multidimensional array can be read individually, but a more efficient method is by using a looping

318 *Multidimensional Arrays*

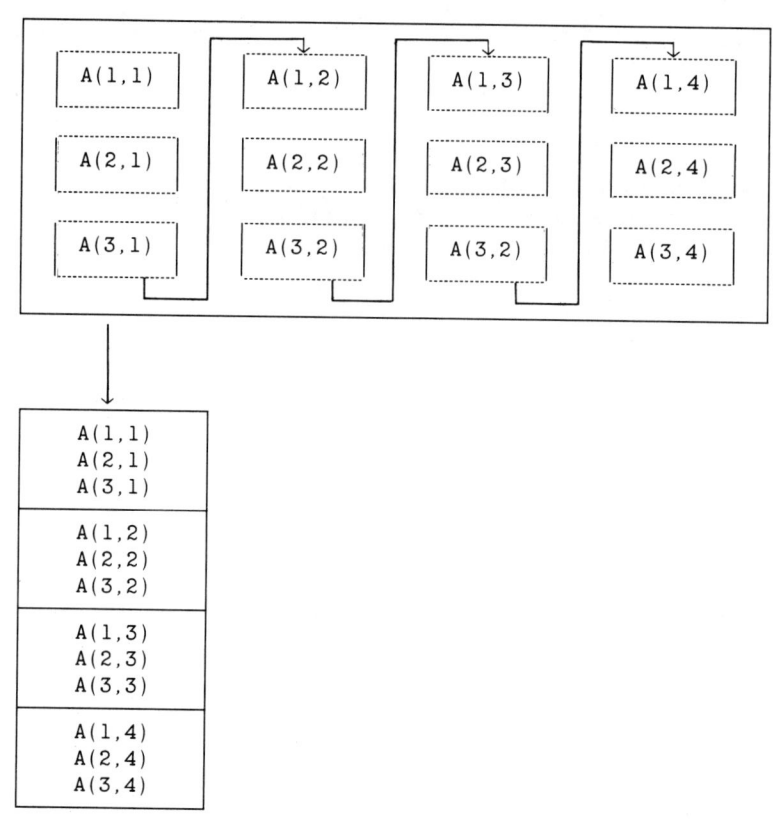

```
A(1,1)   A(1,2)   A(1,3)   A(1,4)
A(2,1)   A(2,2)   A(2,3)   A(2,4)
A(3,1)   A(3,2)   A(3,2)   A(3,4)
```

```
A(1,1)
A(2,1)
A(3,1)

A(1,2)
A(2,2)
A(3,2)

A(1,3)
A(2,3)
A(3,3)

A(1,4)
A(2,4)
A(3,4)
```

Figure 8.4 Storage order of a two-dimensional array.

technique. A nested DO loop is normally employed to input or output elements of a multidimensional array. The following is an example:

```
          DIMENSION X(50,3)
          DO 90 I = 1,50
              DO 90 J = 1,3
                  READ(5,10)X(I,J)
90        CONTINUE
10        FORMAT (F5.1)
```

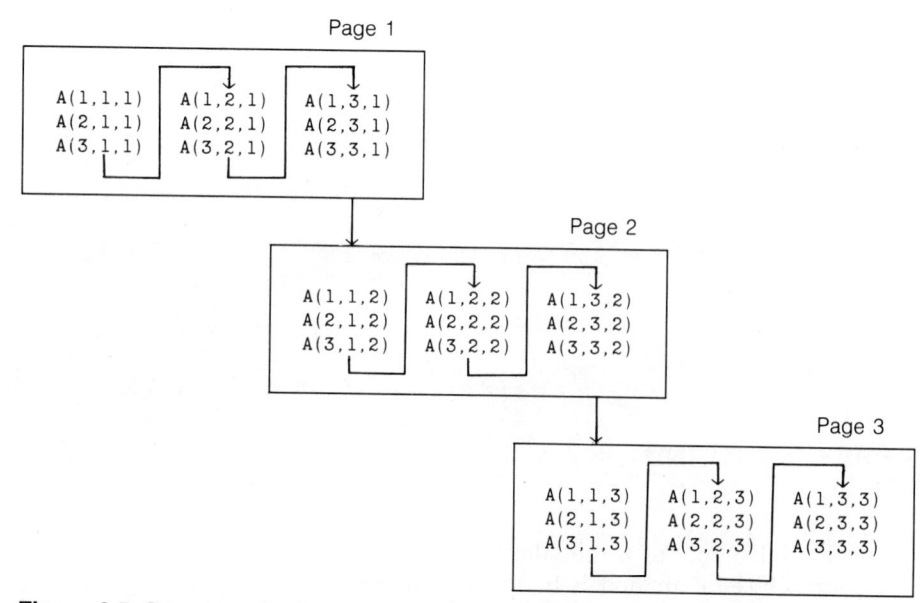

Page 1
```
A(1,1,1)  A(1,2,1)  A(1,3,1)
A(2,1,1)  A(2,2,1)  A(2,3,1)
A(3,1,1)  A(3,2,1)  A(3,3,1)
```

Page 2
```
A(1,1,2)  A(1,2,2)  A(1,3,2)
A(2,1,2)  A(2,2,2)  A(2,3,2)
A(3,1,2)  A(3,2,2)  A(3,3,2)
```

Page 3
```
A(1,1,3)  A(1,2,3)  A(1,3,3)
A(2,1,3)  A(2,2,3)  A(2,3,3)
A(3,1,3)  A(3,2,3)  A(3,3,3)
```

Figure 8.5 Storage order for a three-dimensional array.

Note that the control variable J in the inner loop changes more rapidly. This nested loop is equivalent to

```
READ (5,10)X(1,1)
READ (5,10)X(1,2)
READ (5,10)X(1,3)
READ (5,10)X(2,1)
READ (5,10)X(2,2)
READ (5,10)X(2,3)
            ⋮
READ (5,10)X(50,3)
```

The nested implied DO loop is also a convenient way to read the elements of an array. The following is an example:

```
DIMENSION X(50,3)
READ (5,10)((X(I,J),J = 1,3),I = 1,50)
```

Note again that the control variable in the inner loop (J) changes more rapidly. The above READ statement is equivalent to

```
READ (5,10)A(1,1),A(1,2),A(1,3),A(2,1),A(2,2),
*A(2,3), . . .,A(50,3)
```

The format determines the location of the data in this case. For example, when

```
10      FORMAT(F5.1)
```

is used, for the above example, each datum must be placed on a different line. But when

```
10      FORMAT(10F5.1)
```

is used, ten data must be placed on one line. The implied DO loop is particularly useful when the elements must be read from one line or printed on one line.

A regular DO loop and an implied DO loop are a good combination for reading or writing a multidimensional array *row by row* (row-wise), or *column by column*. The following is an example of reading the data row by row:

```
        DIMENSION Y (30,10)
        DO 100 L = 1,30
            READ(5,70)(Y(L,I),I = 1,10)
100     CONTINUE
70      FORMAT (10F3.1)
```

The following module prints the elements of the same array, one *column* per line:

```
        DO 200 J = 1,10
            WRITE(6,300)(Y(I,J),I = 1,30)
200     CONTINUE
300     FORMAT(1X,30F4.1)
```

The following program segment is an example of printing the elements of a three-dimensional array:

```
        DIMENSION TABLE(50,10,5)
        DO 200 K = 1,5
            DO 200 L = 1,50
                WRITE (6,20)(TABLE(L,M,K),M = 1,10)
200     CONTINUE
20      FORMAT (10F5.1)
```

The following are some more examples of the implied DO loop:

```
        READ(5,10) M,N, ((A(I,J),J = 1,N),I = 1,M)
10      FORMAT(2I2/(F5.2))
        READ(5,20) ((A(I,J),B(I,J),J = 1,30), I = 1,100)
20      FORMAT(2F6.2)
        READ(5,30)(A(I,J),J = 1,10),I = 1,30),((B(I,J),
       *J = 1,50),I = 1,100)
30      FORMAT(F5.1)
```

(Note that I and J in this last example can be used twice because the first nested loop is completed before the second nested loop starts.)

```
        READ(5,40)((A(I,J),J = 1,10),(B(I,J),J = 1,20),
       *I = 1,100)
10      FORMAT(30F2.0)
        READ(5,50)(((A(I,J,K),J = 1,5),I = 1,200),
       *K = 1,3)
10      FORMAT(5F8.1)
```

☐ The Entire-List Method

As explained before with one-dimensional arrays, the array name without a subscript in an input or output list causes the entire array to be read or printed. For example:

```
DIMENSION A(3,4)
READ(5,10)A
```

causes all 12 elements of the array to be read. The above READ statement is equivalent to

```
READ(5,10) A(1,1),A(2,1),A(3,1),A(1,2),A(2,2),
*          A(3,2),A(1,3),A(2,3),A(3,3),A(1,4),
*          A(2,4),A(3,4)
```

The FORMAT determines the location of the data. For example, by using

```
10      FORMAT(F5.1)
```

each element will be read from different lines, but by using

```
10      FORMAT(3F5.1)
```

the array will be read column by column. That is, each *column* (three elements) will be read from one line.

Note that the order of the elements read in the example is the same as the internal storage order. In fact, this is one of the reasons that a user should know about internal storage order.

RULE 8.2

The order of the elements of an array to be read or written in the entire-list method is the same order in which they are internally stored.

THE DATA STATEMENT

The DATA statement is an efficient method to initialize the elements of multidimensional arrays as well as one-dimensional arrays. Several examples follow:

EXAMPLE A
```
DATA A(1,1),A(1,2),A(1,3),B(2,5),B(2,6)/3*0.0,2*1.0/
```

EXAMPLE B
```
DIMENSION X(2,3)
DATA X/6*0.0/
```
This is equivalent to
```
DIMENSION X(2,3)
DATA X(1,1),X(2,1),X(1,2),X(2,2),X(1,3),X(2,3)/6*0.0/
```
Note that the array name without a subscript denotes all elements in the order in which they are stored (although the order does not matter in this example).

EXAMPLE C
```
DIMENSION Y(300,3)
DATA Y/900*0.0/
```

EXAMPLE D
```
DIMENSION DD(500,300)
DATA ((DD(I,J)J = 1,100),I = 1,450)/45000*0.0/
```

EXAMPLE E
```
DIMENSION P(1000), Q(10,3)
DATA P(3), Q(9,2)/2*5.0/
```

EXAMPLE F
```
DIMENSION X(6)
DATA X/1.,2.,3.,4.,5.,6./
```

SOLVED PROBLEMS

8.3 How many elements are in each of the following arrays?

  a. DIMENSION IX(53,2),XY(105,2)
  b. DIMENSION TA(1:10,1:10),KA(−1:20,−5:10)
  c. REAL COST(55,60,10),ION(−30:30,7,2)
  d. INTEGER III(30, 41:80),RO(50,−3:−1,−5:2,1:10)

Answers

  a. IX has 53 × 2 = 106, and XY has 210 elements.
  b. TA has 100, and KA has 352 elements (22 × 16 = 352).

c. COST has 33000, and ION has 854 elements (61 × 7 × 2 = 854).
d. III has 1200, and RO has 12000 elements (50 × 3 × 8 × 10 = 12000).

8.4 Identify any errors in the following DO loops:

a.
```
        READ(5,20) (XY(I,J),J = 1,10,
       *I = 1,100)
10      FORMAT (10F4.0)
```

b.
```
        DIMENSION M(50,5)
        READ (5,30)((M(L,K),K = 1,50),
       *L = 1,5)
30      FORMAT (5F5.1)
```

c.
```
        DIMENSION PART(5,2)
        DO 20 I = 1,2
            READ (5,10) (PART (I,J),J = 1,5)
20      CONTINUE
10      FORMAT (5F6.1)
```

d.
```
        READ (5,100) TABLE(I,J,K),I = 1,5,
       *J = 1,50,K = 1,3
```

e.
```
        DIMENSION A(50,5)
        DO 90 I = 1,50
            DO 90 J = 1,5
                READ (5,30) A(I,J)
90      CONTINUE
30      FORMAT (5F6.2)
```

☐ **Answers**

a. A pair of parentheses is necessary to separate the inner loop. The correct form is

```
        READ(5,10) ((XY(I,J),J = 1,10),
       *I = 1,100)
10      FORMAT(10F4.0)
```

b. When K changes from 1 to 2 to ... to 50, it creates M(L,1),M(L,2),...,M(L,50). However, the array M has been defined to have only 5 columns. Thus, either M or the loops should be corrected. If we leave the DIMENSION statement as is, the loops can then be corrected to read

```
        READ(5,30) ((M(L,K),K = 1,5),
       *L = 1,50)
```

c. The same problem as **b**.

d. A pair of parentheses should surround each loop. The correct form is

```
READ(5,100) (((TABLE(I,J,K),I = 1,5),
*J = 1,50),K = 1,3)
```

e. This is correct, and one item will be read per record. However, if the intention is to read five values to a record, as the FORMAT indicates, then an implied DO loop must be used, such as

```
      DIMENSION A(50,5)
      DO 90 I = 1,50
         READ(5,30) (A(I,J),J = 1,5)
90    CONTINUE
30    FORMAT(5F6.2)
```

■ **8.5** Explain whether each array is read row by row or column by column, in each of the following program segments:

a.
```
      DIMENSION AAA(40,5)
      DO 90 I = 1,40
         READ (5,100) (AA(I,J),J = 1,5)
90    CONTINUE
100   FORMAT (5F6.1)
```

b.
```
      DIMENSION XYZ(20,10)
      DO 80 I = 1,10
         READ (5,100)(XYZ(J,I),J = 1,20)
      CONTINUE
100   FORMAT (20F4.1)
```

c.
```
      DIMENSION IJK(8,4)
      READ (5,200) ((IJK(I,J),J = 1,4),
     *             I = 1,8)
200   FORMAT (4F6.2)
```

d.
```
      DIMENSION KAY(8,4)
      READ (5,300) ((KAY(I,J), I = 1,8),
     *             J = 1,4)
300   FORMAT (8F6.2)
```

e.
```
      DIMENSION AY(10,5)
      READ (5,100) AY
100   FORMAT (10F6.1)
```

f.
```
      DIMENSION AY(10,5)
      READ (5,100) AY
100   FORMAT (5F6.1)
```

g.

```
      DIMENSION X(10,3),Y(10,20)
      DO 80 I = 1,10
          READ (5,60) (X(I,J),J =
     *    1,3),(Y(I,J),J = 1,20)
80    CONTINUE
60    FORMAT (23F3.1)
```

☐ **Answers**

a. Five values for each row are read from a record. Thus, the array values are read row by row.

b. Twenty values for each column are read from each record. Thus, array values are read column by column.

c. The array values are read row by row, i.e., four values for each row are read from a record.

d. In contrast to the previous case, the array is read column by column, i.e., eight values for a column are read from each record.

e. The values are read column by column, i.e., ten values for a column are read from each record. Five records are needed to read the entire array.

f. The values are read column by column. However, five values for a column are read from a record, and then the next five values for the same column are read from the next record (two records per column). A total of ten records are needed to read the entire array.

g. Two arrays are read row by row; i.e., three values for each row of X and twenty values for each row of Y, a total of 23 values, will be read from one record. Ten records are needed.

■ **8.6** Can any of the following DO loops be written with the other DO loop form? If so, rewrite them.

a.

```
      DO 100 I = 1,300
          DO 100 J = 1,20
              READ(5,10) TABLE(I,J)
100   CONTINUE
10    FORMAT(F6.1)
```

b.

```
      DO 200 K = 1,30
          READ(5,20) (TABLE(K,L),L = 1,10)
200   CONTINUE
20    FORMAT(10F5.1)
```

c.

```
      READ(5,10) ((M(I,J),J = 1,10),
     *I = 1,3)
10    FORMAT(30I2)
```

d.

```
      READ(5,200) N,((A(K,L),L = 1,20),
     *                K = 1,N)
200   FORMAT(I2/(20F3.0))
```

☐ **Answers**

a.

```
      READ(5,10) ((TABLE(I,J),J = 1,20),
     *              I = 1,300)
10    FORMAT(F6.1)
```

accomplishes the same function.

b. A regular DO loop cannot be used for the inner loop (reading 10 values per record), but

```
      READ(5,20) ((TABLE(K,L),L = 1,10),
     *              K = 1,30)
20    FORMAT(10F5.1)
```

accomplishes the same function.

c. This cannot be accomplished by a regular DO loop.

d.

```
      READ(5,10) N
10    FORMAT(I2)
      DO 50 K = 1,N
          READ(5,200) (A(K,L),L = 1,20)
50    CONTINUE
200   FORMAT(20F3.0)
```

accomplishes the same function.

■ **8.7** Use a DO loop, instead of the entire-list method, to read each of the following arrays.

a.

```
      DIMENSION AAA(30,3)
      READ(5,10) AAA
10    FORMAT(30F2.0)
```

b.

```
      DIMENSION X(20,5),Y(10,3)
      READ(5,50) X,Y
50    FORMAT(10F5.2)
```

c.

```
      DIMENSION A(5,2),B(10)
      READ(5,20) A,B
20    FORMAT(10F5.2)
```

☐ **Answers**

a.

```
      DIMENSION A(30,3)
```

```
              DO 300 J = 1,3
                 READ(5,10) (AAA(I,J),I = 1,30)
300       CONTINUE
10        FORMAT(30F2.0)
```

b.
```
          DIMENSION X(20,5), Y(10,3)
          DO 10 J = 1,5
             READ(5,50)(X(I,J),I = 1,20)
10        CONTINUE
          DO 20 J = 1,3
             READ(5,50)(Y(I,J),I = 1,10)
20        CONTINUE
50        FORMAT(10F5.2)
```

or

```
          DIMENSION X(20,5),Y(10,3)
          READ(5,50) ((X(I,J),I = 1,20),
         *J = 1,5),((Y(I,J),I = 1,10),J = 1,3)
50        FORMAT (10F5.2)
```

c.
```
          DIMENSION A(5,2),B(10)
          READ (5,20) ((A(I,J),I = 1,5),
         *J = 1,2),(B(I),I = 1,10)
20        FORMAT (10F5.2)
```

8.8 Are any of the following statements correct?

a.
```
          DIMENSION A(10),B(5,2)
          DATA A,B/20*0.0/
```

b.
```
          DIMENSION X(20,5),Y(20,5)
          DATA X,Y/100*0.0/
```

c.
```
          DIMENSION D(5,10),E(10,5)
          DATA D(5,10),E(10,5)/100*0.0/
```

d.
```
          DIMENSION PAY(100,20)
          DATA ((PAY(I,J),J = 1,10),
         *     I = 1,50)/2000*0.0/
```

e.
```
          DIMENSION A(10,20),B(10,20)
          DATA ((A(I,J),B(I,J),J = 1,20),
         *     I = 1,10)/400*0.0/
```

☐ **Answers**

a. This is correct.
b. Incorrect; the number of values is not equal to the number of variables (200) in the variable list.
c. Incorrect; there are only two variables [two elements D(5,10),E(10,5)], but there are 100 values.
d. Incorrect; again the number of values is more than the number of elements (there are only 500 elements).
e. This is correct.

SUMMARY OF THE IMPORTANT RULES

The rules governing multidimensional arrays are basically the same as for one-dimensional arrays. The most important rules are:

1. The array's name, its dimension, and each dimension size must be defined at the beginning of a program by the DIMENSION statement.
2. In FORTRAN 77, the boundaries of an array dimension can be specified, and each boundary can be a negative constant as well as a positive one.
3. The array name refers to the whole array—all elements. Each element can be referred to by the array name followed by the subscript specification enclosed in parentheses. In an arithmetic expression, individual elements must be used.
4. Each subscript can be an integer constant, an integer variable, or a simple arithmetic expression. However, no subscript value can go beyond the DIMENSION boundaries.
5. Array elements are stored internally by columns: the first column first, the second column second, and so forth.
6. The array elements can be read or written by a DO loop, an implied DO loop, or the entire-list method. Normally, a combination of two or more DO loops is necessary to input or output a multidimensional array.
7. The array name in the input or output list denotes all elements of the array in the order in which they are stored internally.

EXAMPLES OF TWO-DIMENSIONAL ARRAYS AND PROGRAMMING TECHNIQUES

One- and two-dimensional arrays are the kinds most frequently used in programs. Employing arrays with three or more dimensions can become quite complicated, and their use is not as popular.

You can always convert a multidimensional array into one or more one-dimensional arrays. For example:

```
TABLE (50,3)
```

can be converted to

```
TABLE1(50), TABLE2(50), TABLE3(50)
```

or it can be converted to

```
TABLET(150)
```

Some simple examples of the use of two-dimensional arrays are presented in the following section. Be sure to study each problem thoroughly and have a written answer prepared before examining the solution supplied.

EXAMPLE 8.2 A Simple Example

■ **Problem:** Students of a FORTRAN course have taken three tests. There are less than 50 students. Write a program which reads and stores (1) the name and (2) the test results; calculates each student's average and the class average; and prints the information. The first record is a header record indicating the number of students taking the tests. The data will be similar to the following sample:

Name	Test 1	Test 2	Test 3
JOHN DOE	95.50	75.90	87.80
PAT JOY	83.00	92.50	85.0
⋮	⋮	⋮	⋮
JOYCE DOER	87.75	92.30	78.50

Program:

```
      PROGRAM MULT2
      CHARACTER*10 NAME(50)
      DIMENSION TEST(50,3), TOT(50), AVG(50)
      DATA SUM/0.0/
C READING THE HEADER RECORD
      READ (5,15) N
15    FORMAT (I2)
C READING THE DATA
      DO 100 I = 1,N
         READ (5,10) NAME(I),(TEST(I,J),J = 1,3)
100   CONTINUE
C CALCULATING THE AVERAGES
      DO 110 J = 1,N
         TOT(J) = 0.0
         DO 115 I = 1,3
115         TOT(J) = TOT(J) + TEST(J,I)
         AVG(J) = TOT(J)/3.0
         SUM = SUM + AVG(J)
110   CONTINUE
      GAVG = SUM/N
C WRITING THE DATA
      DO 120 K = 1,N
         WRITE (6,20) NAME(K), (TEST(K,J),J = 1,
     *      3),AVG(K)
120   CONTINUE
      WRITE (6,30) GAVG
10    FORMAT (A10, 3(F5.2,1X))
20    FORMAT (11X,A10,5X,4(F6.2,5X))
30    FORMAT(26X,'GRAND AVERAGE =',F6.2)
      END
```

Notes:

1. The test scores are stored in a two-dimensional array in which a row is reserved for each student's scores.
2. The total of the three test scores, and the average of the scores, are stored in the one-dimensional arrays TOT and AVG respectively.
3. The test data are read into the two-dimensional array with two DO loops. The first one reads each student in, and the second one, the implied DO loop, reads three test scores.

EXAMPLE 8.3 Table Manipulation

Problem: A school of engineering has seven departments, and each department has three secretarial positions. Write a program which reads the salaries of the secretaries by department and position, and prints a salary table including:

1. the total and average salaries in each department,
2. the total and average salaries in each position,
3. the total and average of all salaries,
4. the salaries which are below average.

The following shows the form of the table to be printed:

		Positions		Total	Average
Department	1	2	3	4	5
1	7890.00	9130.00	11500.00	?	?
2	.	.	.	?	?
3	.	.	.	?	?
4	.	.	.	?	?
5	.	.	.	?	?
6	.	.	.	?	?
7	.	.	.	?	?
8 (total)	?	?	?	?	?
9 (average)	?	?	?	?	?

Program:

```
      PROGRAM TABLE2
      DIMENSION TABLE (9,5)
      DATA TABLE/45*0.0/
C READING THE DATA
      DO 90 I = 1,7
          READ (5,10) (TABLE(I,J),J = 1,3)
90    CONTINUE
C CALCULATING ROW TOTALS AND AVERAGES
      DO 100 I = 1,7
          DO 110 J = 1,3
              TABLE(I,4) = TABLE(I,4) + TABLE
     *         (I,J)
110       CONTINUE
      TABLE(I,5) = TABLE(I,4)/3.0
100   CONTINUE
C CALCULATING COLUMN TOTALS AND AVERAGES
      DO 120 I = 1,3
          DO 130 J = 1,7
              TABLE(8,I) = TABLE(8,I) + TABLE
     *         (J,I)
130       CONTINUE
      TABLE(9,I) = TABLE(8,I)/7.0
120   CONTINUE
```

```
C CALCULATING THE GRAND TOTAL AND AVERAGE
      DO 140 I = 1,7
          TABLE(8,4) = TABLE(8,4) + TABLE(I,4)
          TABLE(8,5) = TABLE(8,5) + TABLE(I,5)
140   CONTINUE
      TABLE(9,4) = TABLE(8,4)/7.0
      TABLE(9,5) = TABLE(8,4)/21.0
C WRITING THE HEADING
      WRITE (6,200)
200   FORMAT (41X,'POSITION',35X,'TOTAL',13X,
     *'AVERAGE')
      WRITE (6,300)(I,I = 1,5)
300   FORMAT(25X,5(I1,17X))
C WRITING THE DATA
      DO 60 I = 1,9
          WRITE (6,20) I,(TABLE(I,J),J = 1,5)
60    CONTINUE
C WRITING BELOW AVERAGE SALARIES
      DO 70 K = 1,7
          DO 70 L = 1,3
          IF(TABLE(K,L) .LT. TABLE(9,5))WRITE(6,
     *30) TABLE(K,L)
70    CONTINUE
10    FORMAT(3(F8.2,1X))
20    FORMAT(11X,I2,8X,5(F10.2,10X))
30    FORMAT(21X,F10.2)
      END
```

Notes:

1. Row totals are assigned to the fourth column, and row averages to the fifth column of array TABLE.
2. Column totals are assigned to the eighth row, and column averages to the ninth row of array TABLE.
3. The grand total is assigned to element TABLE(8,4). This can be calculated by adding either the "column totals" or the "row totals."
4. The grand average is assigned to element TABLE(9,5). This can also be calculated by dividing the sum of the "column averages" by 7, or the sum of the row averages by 3.
5. TABLE(9,4) can also be calculated by:

```
          DO 150 J = 1,3
              TABLE(9,4) = TABLE(9,4) +
     *          TABLE(9,J)
150       CONTINUE
```

EXAMPLE 8.4 Table Lookup

■ **Problem:** A store sells eight products, each with three models: A, B, C. The unit price can be found from the following table:

Product Number	Model A	Model B	Model C
1	12.30	15.80	18.90
2	18.20	25.60	35.80
3	19.90	22.50	26.90
4	25.00	29.60	35.50
5	28.50	39.90	45.50
6	36.50	42.80	49.50
7	45.00	55.50	58.20
8	59.00	69.90	79.80

Write a program which reads (1) the product code and (2) the quantity ordered by the customers. Print the product code, quantity ordered, and amount of the order for each of the 50 customers. Each product code is composed of four digits, the first two (01 through 08) representing the product number, and the next two (01 through 03) representing models A, B, and C.

Program:

```
      PROGRAM LOOK
      DIMENSION PRICE(8,3),ORDER(50),AMT(50)
      INTEGER CODE(50,2)
      DATA PRICE/12.30,18.20,19.90,25.00,28.50,
     *36.50,45.00,59.00,15.80,25.60,22.50,29.60,
     *39.90,42.80,55.50,69.90,18.90,35.80,26.90,
     *35.50,45.50,49.50,58.20,79.80/
C READING THE DATA
      DO 70 I = 1,50
            READ(5,10) (CODE(I,J)J = 1,2),ORDER(I)
70    CONTINUE
C CALCULATING THE AMOUNT
      DO 80 J = 1,50
            M = CODE(J,1)
            N = CODE(J,2)
            AMT(J) = ORDER(J)*PRICE(M,N)
80    CONTINUE
C WRITING THE DATA
      DO 90 K = 1,50
            WRITE (6,20) (CODE(K,J),J = 1,2),ORDER(K),
     *      AMT(K)
90    CONTINUE
10    FORMAT(2I2,1X,F2.0)
20    FORMAT(11X,2I3,5X,F3.0,5X,F7.2)
      END
```

Note: The price table can be read in with a READ statement rather than a DATA statement.

EXAMPLE 8.5 Graphing a Histogram

■ **Problem:** Write a program which reads a set of 100 integers between 1 and 9. Count the number of times each of the numbers occurs in the set, and print a histogram (an asterisk for each occurrence of each digit). For example, if the number of occurrences of each number is given by

Number	Occurrences
1	3
2	4
3	10
4	20
5	25
6	18
7	10
8	6
9	4

then the program must print:

```
1 ***
2 ****
3 **********
4 ********************
5 *************************
6 ******************
7 **********
8 ******
9 ****
```

Program:

```
      PROGRAM HISTO
      INTEGER A(100),COUNT(9)
      CHARACTER*1 CHART(9,50)
      DATA COUNT, CHART/9*0,450*' '/
C READING THE DATA
      DO 50 I = 1,100
            READ (5,10) A(I)
50    CONTINUE
C COUNTING EACH OCCURRENCE
      DO 60 J = 1,100
            K = A(J)
            COUNT(K) = COUNT(K) + 1
60    CONTINUE
C FILLING THE ARRAY CHART WITH *
      DO 70 I = 1,9
            N = COUNT(I)
            DO 70 J = 1,N
                  CHART(I,J) ='*'
70    CONTINUE
C PRINTING THE CHART
      DO 80 K = 1,9
            WRITE(6,20) K,(CHART(K,J),J = 1,50)
80    CONTINUE
10    FORMAT (I3)
20    FORMAT (11X,I2,2X,50A1/)
      END
```

Notes:

1. A 9-by-50 two-dimensional character array is used to represent the graph. (A one-dimensional array could have been used to print a line of the graph also.)
2. The array is first filled with blanks (the DATA statement); then the appropriate number of asterisks (*) are stored in the array positions.

EXAMPLE 8.6 Searching

Problem: Assume integer data are stored in the array ISAM, which has 20 rows and 10 columns. Write a program segment which reads an integer and searches the array. As soon as it finds a number in the array equal to the number read, it prints the location (the row and the column) of the number and stops. If the program does not find the number at all, it prints a message.

Program Segment:

```
          :
      READ(5,10) NUMBER
      DO 100 I = 1,20
          DO 100 J = 1,20
              IF(ISAM(I,J) .EQ. NUMBER) THEN
                  WRITE (6,20) I,J
                  GOTO 99
              ENDIF
100   CONTINUE
      WRITE (6,30)
30    FORMAT(11X,'THE NUMBER NOT FOUND')
99    STOP
10    FORMAT(I3)
20    FORMAT(11X,'THE ROW IS=',I2,'THE COLUMN IS =',
     *I2)
          :
```

EXAMPLE 8.7 Matrix Multiplication

Problem: Assume the array A has M rows and N columns; B has N rows and one column (a linear array), and C has M rows

$$\begin{bmatrix} C_1 \\ C_2 \\ \vdots \\ C_M \end{bmatrix} = \begin{bmatrix} A_{1,1} & A_{1,2} & \cdots & A_{1,N} \\ A_{2,1} & A_{2,2} & \cdots & A_{2,N} \\ \vdots & \vdots & & \vdots \\ A_{M,1} & A_{M,2} & \cdots & A_{M,N} \end{bmatrix} \begin{bmatrix} B_1 \\ B_2 \\ \vdots \\ B_N \end{bmatrix}$$

Figure 8.6

and one column (also linear), as in Figure 8.6. Write the program segment which calculates C so that

```
C(1) = A(1,1)*B(1) + A(1,2)*B(2) + ··· + A(1,N)*B(N)
C(2) = A(2,1)*B(1) + A(2,2)*B(2) + ··· + A(2,N)*B(N)
  :
C(M) = A(M,1)*B(1) + A(2,2)*B(2) + ··· + A(M,N)*B(N)
```

Program Segment:

```
      DO 90 I = 1,M
         C(I) = 0.0
         DO 90 J = 1,N
            C(I) = C(I) + A(I,J)*B(J)
90    CONTINUE
```

EXAMPLE 8.8 Matrix Manipulation

■ **Problem:** Suppose array A has N rows and N columns. Write the program segment which interchanges the values of a_{ij} and a_{ji}. For example, if

$$A = \begin{bmatrix} 5 & 3 & 2 \\ 8 & 4 & 9 \\ 6 & 7 & 1 \end{bmatrix}$$

then the new array will be

$$A = \begin{bmatrix} 5 & 8 & 6 \\ 3 & 4 & 7 \\ 2 & 9 & 1 \end{bmatrix}$$

Program Segment:

```
      DO 10 I = 1,N
         DO 10 J = I,N
            TEMP = A(I,J)
            A(I,J) = A(J,I)
            A(J,I) = TEMP
10    CONTINUE
```

Note: The program segment

```
      DO 100 I = 1,N
         DO 100 J = 1,N
            A(I,J) = A(J,I)
100   CONTINUE
```

will not work because the value of an element must be stored in a temporary location (TEMP) before assigning any new value to it; otherwise, by storing a new value the previous one will be erased.

EXAMPLE 8.9 Matrix Manipulation

Problem: Assume the array A has N rows and M columns, and B has M rows and N columns. Write a program segment which inserts the elements of A into B so that the last column of A becomes the first row of B, the column before the last column of A becomes the second row of B, and so forth. In other words, if array A is turned 90 degrees, it becomes array B. Below is an example of A and B.

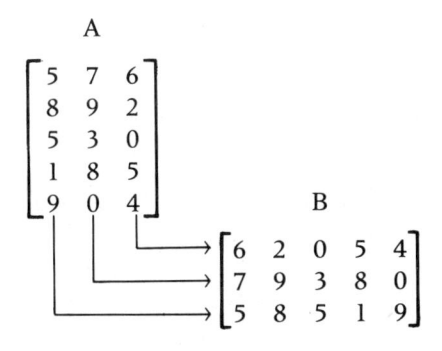

$$A = \begin{bmatrix} 5 & 7 & 6 \\ 8 & 9 & 2 \\ 5 & 3 & 0 \\ 1 & 8 & 5 \\ 9 & 0 & 4 \end{bmatrix}$$

$$B = \begin{bmatrix} 6 & 2 & 0 & 5 & 4 \\ 7 & 9 & 3 & 8 & 0 \\ 5 & 8 & 5 & 1 & 9 \end{bmatrix}$$

Program Segment:

```
        DO 10 I = 1,N
            DO 10 J = 1,M
            K = M - J + 1
            B(J,I) = A(I,K)        [Note: A(I,M - J + I)
 10     CONTINUE                    will also work]
```

Application: This technique can be used to print the graph produced by the program in Example 8.5 as the following figure:

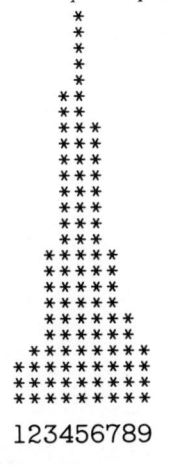

```
       *
       *
       *
       *
       *
       *
      **
      **
     ***
     ***
     ***
     ***
     ***
     ***
     ***
    *****
    *****
    *****
    *****
     ******
     ******
   ********
   ********
   ********
   ********
123456789
```

EXERCISES

8.1 Find all the errors in the following statements:

  a. DIMENSION CHARACTER*12 NAME(50,4)

  b. DIM GROSS(10,5,3),NET(100,3,2)

  c. DIMENSION X(M,N)

  d. DIMENSION Y(500,-3,0)

  e. DIMENSION Z(350.0,3.0)

  f. DIMENSION A(25,25)
  DATA I,J,K/3*0/
  A(I - J,K - 2) = 5.0

  g. DIMENSION X(10,10), A(100)
  DATA I,J/2*10/,A/100*0.0/
  X(I + 1,J + 2) = A(J**3)

  h. DIMENSION MARCH(100,100)
  DATA K,I/0,10/
  MARCH(K - 2,I*I) = 6

  i. READ(5,10)(A(I,J),J = 1,10,I = 1,20)

  j. WRITE(6,20)((AAA(K,L),K = 1,30)L = 1,5)

  k. WRITE(6,30)((N(I,J),I = 1,3 J = 1,300))

  l. DATA X(I,J)/50*3.0/

  m. DATA A(25,2)/50*0.0/

  n. DATA (B(I,J),I = 1,10,J = 1,30)/300*0.0/

8.2 Explain how the data will be read for arrays in each of the following program segments:

  a.
```
      DIMENSION A(20,10)
      DO 100 I = 1,20
          READ(5,30)(A(I,J),J = 1,10)
100   CONTINUE
30    FORMAT(10F4.0)
```

  b.
```
      DIMENSION B(20,20)
      DO 10 J = 1, 20
          READ(5,30)(B(I,J),I = 1,20)
10    CONTINUE
30    FORMAT(20F3.0)
```

  c.
```
      DIMENSION A(20,20)
      READ(5,40)((A(I,J),I = 1,20),J = 1,20)
40    FORMAT(20F3.0)
```

d.
```
          DIMENSION D(20,10)
          DO 10 J = 1,10
             DO 10 I = 1,20
                READ(5,50)D(I,J)
10        CONTINUE
50        FORMAT(F3.0)
```

e.
```
          DIMENSION E(20,20)
             READ(5,60)E
60        FORMAT(20F3.0)
```

f.
```
          DIMENSION F(20,10)
             READ(5,70)F
70        FORMAT(F3.0)
```

g.
```
          DIMENSION X(5,4),Y(10,2)
          READ(5,30) X,Y
80        FORMAT(10F5.1)
```

8.3 Find all the errors in the following loops:

a.
```
          DIMENSION PAT(50,2)
             ⋮
          DO 100 I = 1,100
             DO 100 J = 1,20
                READ(5,10)PAT(I,J)
100       CONTINUE
```

b.
```
          DIMENSION MEAN(100,20)
          DO 100 I = 1,100
             DATA(MEAN(I,J),J = 1,20)/20*0/
100       CONTINUE
```

c.
```
          DIMENSION GROSS(500,3)
          READ(5,10)((GROSS(I,J),J = 1,500),I = 1,3)
```

d.
```
          DIMENSION WORD(100,30)
          DO 100 I = 1,100
             (WORD(I,J),J = 1,30) = 0.0
10        CONTINUE
```

e.
```
          DIMENSION PPP(50,50)
          READ(5,10)PPP
             ⋮
          SUM = 0.0
```

```
          DO 100 I = 1,250
              SUM = SUM + PPP
100       CONTINUE
```

f.
```
          DIMENSION X(100,5),SUM(100)
          READ(5,10)X
          SUM = 0.0
             ⋮
          DO 100 I = 1,100
              DO 100 J = 1,5
                  SUM(I) = SUM + X(I,J)
100       CONTINUE
```

PROGRAMMING EXERCISES

Write a complete FORTRAN program for each of the following problems. Use appropriate arrays, headings, formats, and techniques.

8.4 The following table shows the number of patients who were treated in four emergency rooms of a hospital within three shift periods:

Shift	Room 1	Room 2	Room 3	Room 4
1	5	8	3	7
2	3	6	2	2
3	2	3	0	5

Write a program which reads the data into a two-dimensional array, and prints the data and the total number of the patients in each shift.

8.5 The following table shows the number of products shipped to three warehouses during four weeks:

Warehouse number	Week 1	Week 2	Week 3	Week 4
1	529	628	580	320
2	6290	5890	6120	4190
3	1928	2018	1980	1620

Write a program which reads the data into a two-dimensional array and prints:

1. the data in table form,
2. the total number of products which were shipped to each warehouse, and
3. the total number of products shipped each week.

8.6 Repeat the previous exercise, this time assume there are N warehouses. N is to be read as a header record ($N < 50$). Make up your own data.

8.7 Assume the array Q has three rows and four columns. Write the program segment which prints the values of the elements of Q:

a. Row by row; that is, the first row in the first line, the second row in the second line, and the third row in the third line.
b. Column by column; that is, the first column of the array in the first line, the second column in the second line, and the third column in the third line, and the fourth column in the fourth line.
c. With all values in one line.
d. With all values in one column.

8.8 The following table shows the number of votes, in thousands, for six states over three years.

State	Year 1	Year 2	Year 3
1	105.70	125.00	139.68
2	56.29	55.69	51.83
3	82.31	78.51	85.98
4	52.98	49.31	45.39
5	32.85	38.05	43.05
6	93.28	90.58	85.90

Write a program which reads the data into a two-dimensional array and:

a. Prints the data with a new row for the total votes in each year, and a new column for the total votes in each state.
b. Prints a new table which shows the percentages of votes in each state relative to the total number of votes cast in that year.

8.9 Consider a two-dimensional array

X(100,5)

Write a program which reads the values of the elements, row by row, and finds the largest value in the array.

8.10 The variance of a set of data in a two-dimensional array can be calculated by the formula:

$$\text{VAR} = \frac{(X_{11} - X)^2 + (X_{12} - X)^2 + (X_{13} - X)^2 + \cdots + (X_{mn} - X)^2}{N},$$

where

$X_{11}, X_{12}, X_{13}, \ldots, X_{mn}$ are the elements of the array
X is the mean (average)
N is the number of data

Write a program which reads a set of data consisting of 100 rows and 10 columns. Calculate the mean and the variance of the data. Print the data, mean, and variance.

8.11 Arrays A, B, C each have 10 rows and 10 columns. Write a program which reads the elements of A and B, and calculates the array C so that

$C_{ij} = A_{ij} + B_{ij}$,

that is, each element of C is the sum of the corresponding elements of A and B. Print all three arrays.

8.12 Assume that when a ball is dropped from a height of H feet, it bounces to 80% of its previous height, and it continues to bounce 80% each time until coming to rest. Assume $H = 120$ feet. Write a program which prints a graph with asterisks to show the height of the ball after each bounce up to 13 bounces. Print the graph:

a. using Example 8.5 as a model;
b. using the following figure. (Hint: Use the technique in Example 8.9 to turn the graph in **a.** 90 degrees.)

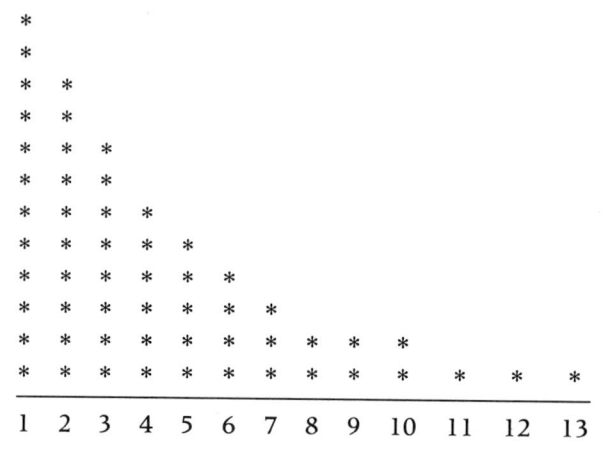

```
 *
 *
 *  *
 *  *
 *  *  *
 *  *  *
 *  *  *  *
 *  *  *  *  *
 *  *  *  *  *  *
 *  *  *  *  *  *  *
 *  *  *  *  *  *  *  *  *  *
 *  *  *  *  *  *  *  *  *  *  *  *  *
─────────────────────────────────────────
 1  2  3  4  5  6  7  8  9 10 11 12 13
```

8.13 Write a program which reads the students' scores (on a scale of 1 to 100) in a test. Count the numbers of A's, B's, C's, D's, and E's.

a. Print a histogram showing the number of A's, B's, C's, D's, and E's, such as the one shown in Example 8.5 in the text.
b. Print a histogram similar to the following graph:

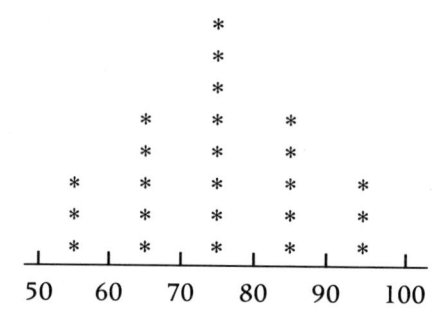

```
              *
              *
              *
        *     *     *
        *     *     *
  *     *     *     *     *
  *     *     *     *     *
  *     *     *     *     *
──┴─────┴─────┴─────┴─────┴─────┴──
 50    60    70    80    90   100
```

8.14 Write a program segment which reads the integer N and searches the two-dimensional array ITOT. If it finds a value in the array equal to N, it prints the location of the value and searches further until all the elements of array ITOT are searched. If it does not find any number equal to N, it prints a message. Count and print also the number of times that the number is found. Assume ITOT has 100 columns and 30 rows.

8.15 The following table shows the product weight produced by five machines in three shifts in a plant.

Machine	Shift 1	Shift 2	Shift 3
1	22.5	22.7	22.9
2	23.5	23.0	23.8
3	21.4	22.0	20.5
4	24.0	23.8	24.2
5	25.1	24.8	24.9

Write a program which reads the data into a two-dimensional array. Calculate:

a. the average weight of the products produced by each machine,
b. the average weight for each of the three shifts,
c. the grand average weight.

Print the data and averages.

8.16 A trucking company charges the customers a fixed amount per pound, based on the following table.

Zone code	Area code 1	Area code 2	Area code 3
1	.40	.55	.68
2	.60	.70	.85
3	.75	.78	.96
4	.90	1.05	1.10
5	.99	1.08	1.15

Write a program which reads the name of the customer, the charge code, and the weight of the item shipped. The charge code consists of two digits; the first one shows the zone, and the second one shows the area code. Assume there are less than 100 records for one truckload, the last one being a trailer record. Print the information as well as the shipping charge to each customer in report form. Print the total weights and the total charges at the end.

8.17 Suppose arrays A, B, and C have two rows and two columns.

a. Write the program segment which calculates array C so that

$$\begin{bmatrix} C_{11} & C_{12} \\ C_{21} & C_{22} \end{bmatrix} = \begin{bmatrix} A_{11} & A_{12} \\ A_{21} & A_{22} \end{bmatrix} \times \begin{bmatrix} B_{11} & B_{12} \\ B_{21} & B_{22} \end{bmatrix}$$

where

$C_{11} = A_{11}B_{11} + A_{12}B_{21}$
$C_{12} = A_{11}B_{12} + A_{12}B_{22}$
$C_{21} = A_{21}B_{11} + A_{22}B_{21}$
$C_{22} = A_{21}B_{12} + A_{22}B_{22}$

b. Can you generalize the idea to two $N \times N$ arrays?

8.18 Write a program which reads two arrays A and B of equal sizes. Then determine whether A is equal to B by comparing corresponding elements one by one. The two arrays are equal if all the corresponding elements are equal. Print the arrays and a message saying whether or not they are equal.

8.19 Referring to the previous problem, calculate the ratio of one of the elements of the array A to the corresponding element of B. Call the ratio C. Then compare each element of A with C multiplied by the corresponding element of B. If they are equal after multiplication, then A and C are *equivalent*. If the result of one comparison reveals inequality, the arrays are not equivalent (related by a ratio). Print the arrays, the ratio, and a statement of equality or inequality.

8.20 The hourly pay rate for the employees of the XYZ Company is based on the following table:

Job code	Dept. 1	Dept. 2	Dept. 3	Dept. 4	Dept. 5
1	4.80	5.00	5.30	5.80	6.30
2	5.30	5.50	5.90	6.60	7.30
3	5.80	6.05	7.55	7.55	8.40
4	6.35	6.65	8.40	8.50	9.60
5	6.95	7.30	9.20	9.60	10.90
6	7.70	8.00	10.10	10.70	12.30
7	8.50	8.80	11.20	12.00	14.90

The federal tax deduction rate is:

Weekly Income	Tax Rate
$100	0.0
More than $100 to $180	.03
More than $180 to $300	.07
More than $300 to $450	.12
More than $450 to $600	.18
More than $600	.25

Write a program which reads the input information and prints biweekly reports as follows:

Input:

	Columns
1. Name, 15 characters	1–14
2. Social Security number, 11 characters	15–25
3. Male-female code (M or F)	26
4. Single-married code (S or M)	27
5. Number of dependents	28–29
6. Pay code, 4 digits	30–33
a. Job code, 2 digits, such as 05	
b. Department code, 2 digits, such as 02	
7. Hours worked in week I	34–38
8. Hours worked in week II	39–43

Processing:

1. Gross pay: (hours worked) × (pay rate); time over 40 hours for each week is paid at 1.5 times the regular rate.
2. Federal tax deduction: [(weekly gross pay) − (number of dependents) × 40.0] × tax rate.
3. FICA deduction: (gross pay) × .0665.
4. State tax: (gross pay) × .045.
5. Total deductions: sum of all the deductions.
6. Net pay: (gross pay) − (total deductions).

Output:

1. A biweekly report including all the input information, the gross pay, all the deductions, and the net pay for each employee.
2. A summary report consisting of:
   a. Total gross pay.
   b. Total Federal tax deduction.
   c. Total FICA deductions.
   d. Total State tax deduction.
   e. Total of all deductions.
   f. Total net pay.
3. The checks and the check stubs for the employees. Design the checks and check stubs first.

SUMMARY OF CHAPTER 8

You have learned:

1. A multidimensional array is an array which has more than one subscript (dimension); for example, a table of data which has several rows and columns can be represented as a two-dimensional array. The first subscript of a multidimensional array shows the row number, the second shows the column number, the third shows the page or plane number, and so on.
2. A multidimensional array must be declared at the beginning of a program. Either a DIMENSION or a type statement can be used to define (1) the array, (2) the array's dimensions, and (3) the size in each dimension. The following statements are some examples:

   DIMENSION TOTAL(200,50,10), NICE(300,50,550)

   REAL GOOD(50,60,70), NONE(100,20)

   INTEGER PIE(60,5), GEE(100), GO

   DIMENSION PET(-100:5, 0:100), NOON(20,1:30), GET(100,10)

   (the last in FORTRAN 77 only).
3. The individual array elements (array name followed by the subscript in parentheses) must be used in an arithmetic expression or assignment statement.
4. The elements of a multidimensional array are internally stored in the order of the columns.
5. Array elements can be input or output by using:
   a. Regular DO loops, implied DO loops. Normally a combination of nested DO loops is necessary to input or output a multidimensional array. For example, the following statements read the elements of a two-dimensional array row by row:

   ```
         DO 10 I = 1,100
            READ(5,20)(A(I,J), J = 1,10)
   10     CONTINUE
   20     FORMAT( 10F5.1)
   ```

   b. The entire-list method; that is, by using only the array name in an input or output list. All the elements will then be read or written in the order that they are stored internally.
6. The DATA statement is a compact and efficient method to initialize the array elements. If using only the array name in the variable list of the DATA statement, all the elements, in the storage order, will be considered.
7. There are situations where a multidimensional array is a useful tool for solving problems. Table manipulation, table lookup, graphing a histogram, and matrix manipulation are some examples.

SELF-TEST REVIEW

8.1 Which of the following statements are correct?
  a. DIMENSION AAA(100,200), BBB(M,N)
  b. DIMENSION POOL(200,60,300), M(30)

c. REAL DIMENSION NET(100,20)
d. INTEGER AROW(30,60,0),P(100)
e. DIMENSION,A(100,20),B(2,300)
f. DIMENSION RARE(50,0,60,0),BAY(60)
g. DIMENSION X(1:50, 0:10, −1:+5), Y(60, −1:1, 0:10), Z(−5,5)

8.2 Correct the following program segments:
   a.
```
          DIMENSION B(100,10)
          DO 50 I = 1,100
              READ(5,20)B
50        CONTINUE
```
   b.
```
          DIMENSION X(100,200)
              ⋮
          IF(X.EQ.0.0)GO TO 50
              ⋮
50        ..................
```
   c.
```
          DIMENSION Y(5,20)
          DO 10 I = 1,5
              READ(5,50)(Y(J,I),J = 1,20)
10        CONTINUE
```

8.3 Identify any errors in the following statements:
   a. WRITE (6,10)((A(I,J),B(I,J),C(I,J)I = 1,10),J = 1,20)
   b. WRITE(6,20)(X(I,J,K),I = 1,10,J = 1,20,K = 1,30)
   c. DATA((X(K,L),K = 1,10),L = 1,30)/0.0/
   d. DATA (X(50,I),I = 1,100)/100*0.0/

8.4 Explain whether the elements of the arrays are read one by one, row by row, or column by column, in each of the following program segments:
   a.
```
          DIMENSION D(10,10)
          READ(5,10) ((D(I,J),I = 1,10),J = 1,10)
10        FORMAT(10F5.1)
```
   b.
```
          DIMENSION X(10,10)
          READ(5,20)((X(I,J),I = 1,10)J = 1,10)
20        FORMAT(F5.1)
```
   c.
```
          DIMENSION Y(20,20)
          READ(5,30)((Y(I,J),J = 1,20),I = 1,20)
30        FORMAT(20F3.0)
```
   d.
```
          DIMENSION Z(10,10)
          READ(5,40)Z
40        FORMAT(10F5.1)
```

e.
```
         DIMENSION ZZ(10,10)
         READ(5,50)ZZ
  50     FORMAT(F5.1)
```

8.5 What is the value of each element of the arrays A and B when using the following statements?

```
         DIMENSION A(2,3), B(3,2)
         DATA A,B/1.0,2.0,3.0,4.0,5.0,6.0,6*7.0/
```

8.6 Assume the array A has been defined as:

```
         DIMENSION A(10,15)
```

and the data are stored in the array. Write a program segment which prints the last row of the array first, the row before the last row second, and so on.

8.7 There are about 50 salesmen in a company. Write a program which prepares a "sales by the location" report as follows:

The input for each salesman is:

1. Salesman's name, 12 characters, columns 1–12.
2. Three sales amounts, columns 13–30.
3. Location code, A or B, column 31.

The output for each salesman is:

Each salesman's name, the amounts, total sales, and the commission (5% of the sales) in report form for each of the locations A and B.
Each sales amount which is above the average.
The total sales and commissions, at the end of the report.

The first record in the data is a header record. Use a two-dimensional array (five columns) for the sales and commissions.

☐ ANSWERS

8.1 Only **b.** and **g.** are correct.

8.2

a. One way to correct the loop is

```
         DIMENSION B(100,10)
         DO 50 I = 1, 100
             READ (5,20)(B(I,J),J = 1,10)
  50     CONTINUE
```

b. The IF statement should be placed in a loop such as

```
         DO 10 I = 1,100
             DO 10 J = 1,200
                 IF(X(I,J).EQ.0.0)GO TO 50
  10     CONTINUE
             ⋮
  50     ...
```

346 *Multidimensional Arrays*

c. One possible way to correct the loop is

```
      DO 10 J = 1,5
         READ(5,50)(Y(J,I),I = 1,20)
10    CONTINUE
```

8.3

a. This is correct.

b. Parentheses should enclose each loop:

```
WRITE(6,20)(((X(I,J,K),I = 1,10),J = 1,20),
*K = 1,30)
```

c. The number of values must be equal to the number of variables; the value list can be corrected to /300 * 0.0/

d. This is correct.

8.4

a. The values are read column by column (because the values for one column are read at a time).

b. The values are read one by one. (Pay attention to the FORMAT.)

c. The values are read row by row.

d. The values are read column by column.

e. The values are read one by one.

8.5

```
A(1,1) = 1.0      B(1,1) = 7.0
A(2,1) = 2.0      B(2,1) = 7.0
A(1,2) = 3.0      B(3,1) = 7.0
A(2,2) = 4.0      B(1,2) = 7.0
A(1,3) = 5.0      B(2,2) = 7.0
A(2,3) = 6.0      B(3,2) = 7.0
```

8.6

```
      DO 10 I = 1, 10
         M = 11 - I
         WRITE(6,30)(A(M,J),J = 1,15)
10    CONTINUE
```

or

```
      DO 10 I = 10,1,-1
         WRITE(6,30)(A(I,J),J = 1,15)
10    CONTINUE
```

8.7 The following program needs appropriate headings and comments:

```
      PROGRAM SALES
      CHARACTER*12 NAME(50), LOCATN(50)*1
      DIMENSION SALES (50,5)
      DATA SALES/250*0.0/,TOT,TCOM/2*0.0/
C
      READ(5,10)N
C
      DO 100 I = 1,N
         READ(5,20) NAME(I),(SALES(I,J),J = 1,3),
     *      LOCATN(I)
100   CONTINUE
```

```fortran
C
      DO 110 I = 1,N
         DO 120 J = 1,3
            SALES(I,4) = SALES(I,4) + SALES(I,J)
120      CONTINUE
         SALES(I,5) = .05*SALES(I,4)
         TCOM = TCOM + SALES(I,5)
         TOT = TOT + SALES(I,4)
110   CONTINUE
      AVG = TOT/(3.0*N)
C THE REPORT FOR LOCATION A
      DO 130 K = 1, N
         IF ( LOCATN .EQ. 'A' ) THEN
            WRITE(6,30) NAME(K), (SALES(K,J),
     *      J = 1,5 )
         ENDIF
130   CONTINUE
C THE REPORT FOR LOCATION B
      DO 135 K = 1, N
         IF ( LOCATN .EQ. 'B' ) THEN
            WRITE(6,30) NAME(K), (SALES(K,J),
     *      J = 1,5)
         ENDIF
135   CONTINUE
C PRINTING ABOVE AVERAGE SALES
      DO 140 L = 1,N
      DO 140 M = 1,3
         IF(SALES(L,M).GE.AVG)WRITE(6,40)SALES
     *   (L,M)
140   CONTINUE
C
      WRITE(6,50)TOT,TCOM
C
10    FORMAT(I2)
20    FORMAT(A12,3F6.2,A1)
30    FORMAT(11X,A12,5X,5(F8.2,5X))
40    FORMAT(11X,'ABOVE AVERAGE SALES =' F8.2)
50    FORMAT(11X, 'TOTAL SALES =',F10.2,10X,'TOTAL
     *COMMISSION = ', F8.2)
      END
```

Chapter 9
Subprograms

INTRODUCTION

SUBROUTINES

    Structure of a Subroutine
        The Name of the Subroutine
        Actual Arguments and Dummy Arguments
        Completeness and Independence of Subroutines
        The RETURN Statement
        Subroutines and Arrays
        Adjustable Array Size
    Summary of the Important Points about Subroutines

FUNCTIONS

    Introduction
    Library Functions
    Function Statement
    Function Subprograms
        Function Name
        Arrays and Functions
        The Difference Between an Array and a Function
        Summary of the Important Points about Function Subprograms
    Comparing Functions with Subroutines

COMMON BLOCK

    Unlabeled COMMON Block
    Labeled COMMON Block

ORDER OF STATEMENTS

INTRODUCTION

A *subprogram* is a program unit which allows us to assign a name to a set of instructions and "call" the name when it is time to execute those instructions. A subprogram communicates with the main program through its name and a series of parameters. There are at least two advantages in using subprograms:

1. When we write a long and complicated program, the process can become quite cumbersome. However, if the writing process can be broken down into smaller modules, it becomes more manageable. This is especially true when the modules can be written independently. By being able to manage a small module, we make the program simpler in its entirety and easier to work with. Using subprograms is a convenient modular approach where the subprogram unit is one of several modules of a larger unit.
2. Sometimes a series of instructions must be repeated several times at different places in a program. We can write those instructions once in a subprogram and "call" them whenever needed. Thus, by using a subprogram, we can avoid having to write the same sequence of instructions repeatedly throughout the program.

Subprograms in FORTRAN are of two types:

1. Subroutines
2. Functions

We will discuss these important features of FORTRAN in this chapter.

SUBROUTINES

☐ Structure of a Subroutine

A subroutine is a tool which allows us to assign a name to a series of instructions and separate them physically from the main program. The instructions can be executed by "calling" the name of the subroutine. Let's take a look at a simple example before explaining subroutines further.

EXAMPLE 9.1

■ Problem: Write a program which calculates and prints the sum and the average of three variables. Use a subroutine.

Program:

```
      PROGRAM DEMO1
C THE MAIN PROGRAM
      READ(5,10)A,B,C
C CALLING THE SUBROUTINE NAMED TOTAL
C
      CALL TOTAL(A,B,C,SUM,AVG)
C
      WRITE(6,20)SUM,AVG
```

```
      10      FORMAT(3F5.1)
      20      FORMAT(11X,2(F6.2,5X))
              END
C***************************************
C THE SUBPROGRAM
C***************************************
              SUBROUTINE TOTAL(A,B,C,SUM,AVG)
              SUM = A + B + C
              AVG = SUM/3.0
              RETURN
              END
```

Notes:

1. The instructions for calculating the sum and the average are performed by the SUBROUTINE named TOTAL.
2. The CALL statement is composed of the word CALL followed by the name of the subroutine, and variables to be transferred to and returned from the subroutine in parentheses.
3. When the CALL statement is encountered in the program, the values of the variables A, B, and C are passed to the subroutine; the instructions in the subroutine are executed, and the values of the variables SUM and AVG will be returned to the main program.
4. The subroutine returns the control to the main program when the RETURN statement is executed.
5. The subroutine is separated from the main program. It starts with the word SUBROUTINE and ends with an END statement.

The general form of a subroutine has seven important components as follows:

In the referencing program,

```
          ⋮
CALL  name   (a1,a2,a3,a4,...)
 ↑     ↑          ↑
 1     2          3
          ⋮
END
```

In the subroutine,

```
SUBROUTINE  name   (d1,d2,d3,d4,...)
    ↑        ↑          ↑
    4        2          5
              ⋮
           (instructions)
RETURN
  ↑
  6
END
  ↑
  7
```

The main components are:

1. The word CALL. The subroutine starts execution when the CALL statement is encountered in a program.

352 *Subprograms*

2. The name of the subroutine. This must be a FORTRAN name.
3. The list of variables, constants, and values in parentheses. Hereafter we call this list the *actual arguments* list.
4. The word SUBROUTINE followed by the name of the subroutine.
5. The list of the variables. Hereafter we call this list the *dummy arguments* list.
6. The word RETURN. This terminates the subroutine and returns the control back to the referencing program.
7. The END statement. This indicates the physical end of the subroutine.

A subroutine is normally placed after the END statement of the referencing program. A program may have more than one subroutine. The deck setup is shown in Figure 9.1.

The Name of the Subroutine

The name of a subroutine must follow the FORTRAN rule for a symbolic name: of one to six characters, where the first character is a letter. The type of the name (i.e., the letter it begins with) does not matter, because no value will be assigned to the subroutine name itself. The name is chosen by the programmer. It must not be the same as any other name in the referencing program or in the subroutine. It is advisable to choose a descriptive name for the subroutine.

Actual Arguments and Dummy Arguments

We referred to variables A, B, and C in the CALL statement in Example 9.1 as "actual arguments," and the variables in the subroutine list as "dummy arguments." The symbolic name of a dummy argument does not have to be the same as the actual argument. Moreover, the actual arguments can be constants, variables, array names, array elements, and expressions. For example, all of the following CALL statements are syntactically correct.

    CALL MYNAME(A+B, M, 3)
    CALL SSS(X, Y*N, C, 3.0*Z)
    CALL BIG(50.60, A(1), B, U−V, S)

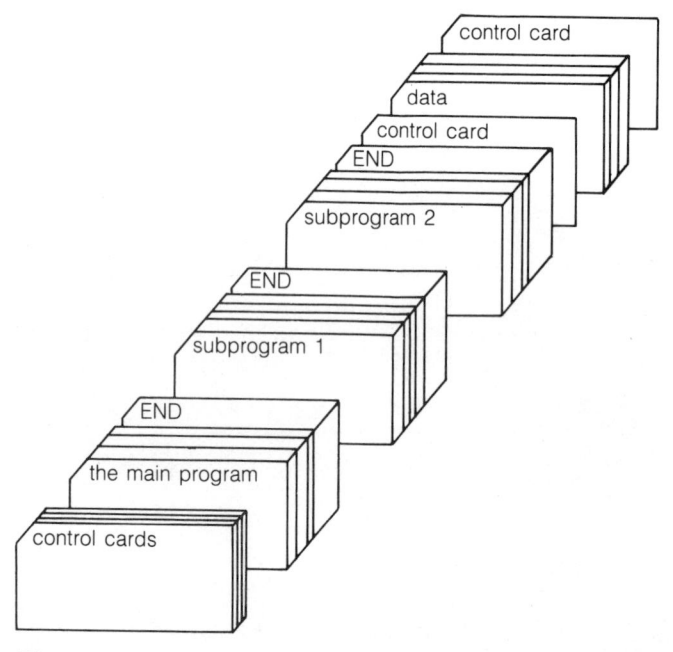

Figure 9.1 The deck setup for subprograms.

However, *the dummy arguments can only be variables*. The CALL statement passes values of the actual arguments to the dummy arguments when the subroutine is executed. If an actual argument is an expression, it will be evaluated before association. Then, the dummy arguments are matched with the actual arguments on a one-to-one basis. Therefore, the number, order, and type of the dummy arguments must correspond with the actual arguments. For example, the program in Example 9.1 can also be written as

```
        PROGRAM
        READ(5,10)A,B,C
        CALL TOTAL(A,B,C,SUM,AVG)
        WRITE(6,20)SUM,AVG
10      FORMAT(3F5.1)
20      FORMAT(11X,2(F6.2,5X))
        END
C
        SUBROUTINE TOTAL(X,Y,Z,S,G)
        S = X + Y + Z
        G = S/3.0
        RETURN
        END
```

The following is another example of a subroutine:

EXAMPLE 9.2

Problem: Write a program which calculates the sum, average, and product of three real numbers. Use a subroutine.

Program:

```
        PROGRAM DEMO2
C THE MAIN PROGRAM
        REAL MULT
        READ(5,10)X,Y,Z
        CALL PROD(X,Y,Z,MULT,SUM,3,AVG)
        WRITE(6,20)MULT,SUM,AVG
10      FORMAT(3F5.2)
20      FORMAT(11X,F10.2,5X,F8.2,5X,F6.2)
        END
C ************************************
C SUBROUTINE
C ************************************
        SUBROUTINE PROD(U,V,W,P,SSS,N,MEAN)
        REAL MEAN
        P = U*V*W
        SSS = U + V + W
        MEAN = SSS/N
        RETURN
        END
```

Notes:

1. MULT and MEAN are defined as real type in the main program and subroutine subprogram, respectively.

2. A one-to-one correspondence exists between the actual arguments and the dummy arguments, as shown below:

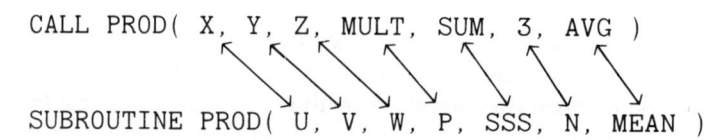

    CALL PROD(X, Y, Z, MULT, SUM, 3, AVG)

    SUBROUTINE PROD(U, V, W, P, SSS, N, MEAN)

    Therefore, the number, order, and mode of the dummy arguments must match those of the respective actual arguments.
3. One of the actual arguments is a constant (3), but the corresponding dummy argument is a variable (N).

Choosing a different argument name in a subprogram gives the programmer more flexibility. By this means, for example, the same subroutine can be used by several referencing programs (the main program and other subprograms). Also notice that when one of the actual arguments is a constant (3 in the previous example), or an expression (such as A + B), the corresponding dummy argument still is a variable (as it must be).

A subroutine does not have to have an argument. The following program is an example:

```
          PROGRAM SUB
          DO 10 I = 1,50
              READ(5,20)A,B
              IF(B .GT. A(C)CALL GREAT
              WRITE(6,30)A,B
10        CONTINUE
20        FORMAT(2F5.2)
30        FORMAT(11X,'A =',F6.2,'B=',F6.2)
          END
          SUBROUTINE GREAT
              WRITE(6,10)
              FORMAT(11X,'B IS GREATER THAN A')
              RETURN
              END
```

Completeness and Independence of Subroutines

A subroutine is a complete program. It has its own declaration statements (REAL, INTEGER, DIMENSION, CHARACTER, etc.), if necessary, and its own END statement. A subroutine will be executed independently also. The variables, instructions, and statement labels are all local to the subroutine. As mentioned earlier when the CALL statement is encountered, the values of the actual arguments will be passed to the dummy arguments, the instructions in the subroutine will be carried out, the results will be returned, and control will be given back to the referencing program.

A main program may have several subprograms. A subprogram in turn may also have one or several subprograms. The following is an example:

EXAMPLE 9.3

Problem: There are 50 students in a course, and each has taken three tests. Write a program which reads the test scores and prints the average of the two highest test scores (dropping the lowest score).

Program:

```
      PROGRAM DEMO3
C THE MAIN PROGRAM
      DO 10 I = 1,50
         READ(5,20)T1,T2,T3
         CALL AVRG(T1,T2,T3,AVG)
         WRITE(6,30)AVG
10    CONTINUE
20    FORMAT(3F5.2)
30    FORMAT(11X,'THE AVERAGE OF THE HIGHEST TWO ,
     *'TESTS IS =',F6.2)
      END
C
C SUBPROGRAM 1
      SUBROUTINE AVRG(A,B,C,D)
      REAL LOW
      CALL LOWEST(A,B,C,LOW)
      D = (A + B + C - LOW)/2.0
      RETURN
      END
C
C THE SUBPROGRAM OF SUBPROGRAM 1
      SUBROUTINE LOWEST(X,Y,Z,SMALL)
         SMALL = X
         IF(Y .LT. SMALL) SMALL = Y
         IF(Z .LT. SMALL) SMALL = Z
      RETURN
      END
```

Notes:

1. The main program has a subroutine named AVRG, and subroutine AVRG in turn has a subroutine named LOWEST.
2. The argument correspondences are as follows:

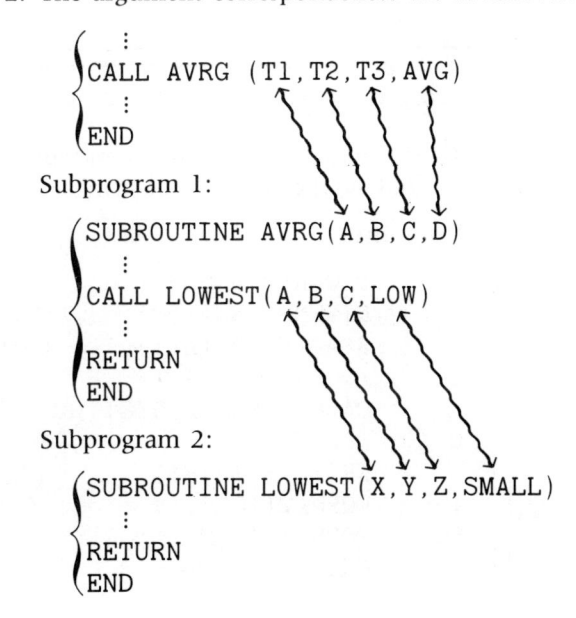

```
     ⋮
{ CALL AVRG (T1,T2,T3,AVG)
     ⋮
  END
```
Subprogram 1:
```
{ SUBROUTINE AVRG(A,B,C,D)
     ⋮
  CALL LOWEST(A,B,C,LOW)
     ⋮
  RETURN
  END
```
Subprogram 2:
```
{ SUBROUTINE LOWEST(X,Y,Z,SMALL)
     ⋮
  RETURN
  END
```

> 3. The CALL statement of the main program is in the DO loop. It will be executed, like any other statement in the loop, fifty times.

The RETURN Statement

Every subroutine subprogram ends physically with an END statement. The RETURN statement, however, terminates the execution of the subroutine subprogram and returns the control to the next executable statement after the CALL statement in the referencing program. In FORTRAN 77, the END statement may be used instead of the RETURN statement if it is the last statement in the subprogram. A subroutine has only one END statement, but may have more than one RETURN statement. The following is an example:

```
            READ(5,10)A
            CALL SIGN(A)
    10      FORMAT(F5.2)
            END
C
            SUBROUTINE SIGN(X)
            IF(X .LT. 0.0)GO TO 20
            WRITE(6,30)
    30      FORMAT(11X,'A IS POSITIVE')
            RETURN
    20      WRITE(6,40)
    40      FORMAT(11X,'A IS NEGATIVE')
            RETURN
            END
```

Subroutines and Arrays

An actual argument can be a variable, constant, individual array element, or array name. If the actual argument is an array name, the corresponding dummy argument must also be an array name. Therefore the array and its size must be declared in the subroutine as well as in the referencing program. The following is an example of the use of an array in a subroutine.

> ### EXAMPLE 9.4
>
> ■ **Problem:** Write a program which reads 50 values into the array A. Find the largest value using a subroutine. Print the largest value.
>
> **Program:**
>
> ```
>             PROGRAM DEMO4
>             DIMENSION A(50)
>             READ(5,10)(A(I),I = 1,50)
>     C
>             CALL LARGE(A,BIG)
>     C
>             WRITE(6,20)BIG
>     10      FORMAT(10F5.1)
>     20      FORMAT(1X,F6.1)
>             END
> ```

```
C
      SUBROUTINE LARGE(X,GREAT)
      DIMENSION X(50)
      GREAT = X(1)
      DO 100 I = 2,50
          IF(X(I) .GT. GREAT) GREAT = X(I)
100   CONTINUE
      RETURN
      END
```

Notes:

1. One of the actual arguments is an array name A, and the corresponding dummy argument X is also defined as an array in the subroutine.
2. The entire array is transferred to the subroutine. Then each individual element is compared with GREAT (the greatest value up to that point) in a loop in the subroutine.
3. The CALL statement is not in a loop in the main program, but the comparison is performed in a DO loop in the subroutine.
4. The following READ statement would also be valid (the entire-list method):

   `READ(5,10)A`

Obviously, if one of the actual arguments is a subscripted element, it will be treated as a single variable and there is no need to declare the array in the subprogram. For example, each of the array elements in the following statements will be considered as a single variable:

```
CALL THERM (A(1),B,X,N,Y(5))

CALL FIND(M,N,X,X(1),X(2),X(3))

I = 3
CALL SECOND (A(I),X(I),I)
```

As an alternative, the previous problem can be coded so that the individual elements are used in a loop in the main program. This is demonstrated below:

EXAMPLE 9.5

Problem: The same as Example 9.4: find and print the largest value of the elements of array A. Use a subroutine.

Program:

```
      PROGRAM DEMO5
      DIMENSION A(50)
      READ(5,10) (A(I),I = 1,50)
C
      BIG = A(1)
C
      DO 200 J = 2,50
          CALL COMPAR (A(J),BIG )
200   CONTINUE
```

358 Subprograms

```
C
      WRITE (6,20) BIG
10    FORMAT (10F5.1)
20    FORMAT (1X,F6.1)
      END
C
      SUBROUTINE COMPAR(Y,GREAT )
           IF (Y .GT. GREAT ) GREAT = Y
           RETURN
           END
```

Note: Each element of A is compared with BIG (the largest number up to that point) in a DO loop in the main program. An individual comparison is made within the subroutine.

The following example demonstrates the use of a multidimensional array in a subroutine.

EXAMPLE 9.6

Problem: Write a program which reads the elements of array PPP with 10 rows and 5 columns, calculate and print the average of the values. Find and print also the largest value in the array.

Program:
```
      PROGRAM DEM06
      DIMENSION PPP(10,5)
      REAL LARGE
      READ(5,100) PPP
100   FORMAT(10F5.1)
C
      CALL MEAN(PPP,AVG)
C
      CALL BIGEST(PPP,LARGE)
C
      WRITE(6,200) AVG,LARGE
200   FORMAT(11X,'THE MEAN IS =',F8.1,
     *'THE LARGEST NO. IS =',F8.1)
      END
C
      SUBROUTINE MEAN(X,A)
      DIMENSION X(10,5)
           SUM = 0.0
           DO 100 I = 1,10
                DO 100 J = 1,5
                     SUM = SUM + X(I,J)
100        CONTINUE
           A = SUM/50.0
           RETURN
           END
```

```
C
        SUBROUTINE BIGEST (Y,GREAT )
            REAL Y(10,5)
            GREAT = Y(1,1)
            DO 100 I = 1,10
                DO 100 J = 1,5
                    IF(Y(I,J) .GT. GREAT)GREAT = Y(I,J)
100         CONTINUE
            RETURN
            END
```

Notes:

1. The main program uses two subroutines; the first one calculates the mean of the elements of the array, and the second one finds the largest number in the array.
2. The variable LARGE is declared as a real variable in the main program. There is no need to declare MEAN, the name of the subroutine, as a real variable, because no value is assigned to it.
3. Statement label 100 appears in the main program and in the subprograms. This does not create a problem, because statement labels, as well as variables, are local to the program unit.

The next example shows how only part of an array can be processed in a subprogram.

EXAMPLE 9.7

Problem: An array B has 500 elements. Read N elements of B; calculate and print their mean. N is to be read in from a header record. Use a subroutine to calculate the mean.

Program:

```
        PROGRAM DEMO
        DIMENSION B(500)
        READ(5,10) N
        READ(5,20)(B(I),I = 1,N)
C
        CALL MEAN(B,AVG,N)
C
        WRITE(6,30)AVG
10      FORMAT(I3)
20      FORMAT(F5.1)
30      FORMAT(11X,'AVERAGE IS =',F6.1)
        END
C
        SUBROUTINE MEAN(X,A,M)
            DIMENSION X(500)
            SUM = 0.0
            DO 100 I = 1,M
                SUM = SUM + X(I)
```

```
100         CONTINUE
            A = SUM/M
            RETURN
            END
```

Note: The array in the subroutine is defined to have the same size as in the main program. However, only N elements of the array are processed in the main program and in the subprogram. N is passed to the subroutine as an argument.

The array size in a subprogram may be different than the array size in the referencing program. This is explained in the following section.

Adjustable Array Size

The array size in the DIMENSION statement of a subprogram can be a variable. For example:

```
DIMENSION A(N)
```

in a subprogram is valid as long as the array size N is passed to the subprogram. This is called an adjustable array size. The following example demonstrates how an adjustable array size is used.

EXAMPLE 9.8

Problem: Write a program which reads the values of an array B with 100 elements. Calculate the sum of the first N elements in a subroutine. N is to be read from a header record.

Program:

```
        PROGRAM DEMO8
        DIMENSION B(100)
        READ(5,10)(B(I),I = 1,100)
        READ(5,20) N
C
        CALL TOT(B,N,SUM)
C
        WRITE(6,30)SUM
10      FORMAT(F5.1)
20      FORMAT(I2)
30      FORMAT(11X,'THE SUM IS =',F8.1)
        END
C
        SUBROUTINE TOT(X,M,SS)
            DIMENSION X(M)
            SS = 0.0
            DO 100 I = 1,M
                SS = SS + X(I)
            CONTINUE
            RETURN
            END
```

> **Notes:**
> 1. The array size in the DIMENSION statement of the subroutine is a variable M.
> 2. M is one of the dummy arguments, and its value is passed to the subroutine by the actual argument N.
> 3. M is also used to control the DO loop for the array manipulation.

It should be emphasized that the adjustable array size applies only to a subprogram; also, the variable representing the array size must be included in arguments lists, and its value cannot be larger than the original array size.

An adjustable array size enables a programmer to write a more flexible and efficient subprogram. However, for the novice programmer, it is not advisable to use an adjustable array size for arrays with more than one dimension, because the internal storage order of the arrays may create a problem. The problem could arise because the values of array elements are passed to the subprogram in the order in which they are stored. Thus, when the sizes are different in different dimensions, an unexpected pattern may result. For example, if the array A in the main program is defined as

```
DIMENSION A(3,4)
```

and the corresponding array in the subprogram is defined with a different size:

```
DIMENSION AA(2,3)
```

then the correspondence between the elements is as follows:

```
A(1,1)   A(2,1)   A(3,1)   A(1,2)
  ↕        ↕        ↕        ↕
AA(1,1)  AA(2,1)  AA(1,2)  AA(2,2)

A(2,2)   A(3,2)   A(1,3)   A(2,3)  ...
  ↕        ↕
AA(1,3)  AA(2,3)
```

Notice the pattern of association [e.g., A(3,1) and AA(1,2)]. When the programmer does not anticipate this pattern, incorrect results occur. Therefore, if at all possible, avoid using adjustable array size for multidimensional arrays in a subprogram.

☐ Summary of the Important Points about Subroutines

1. A subroutine is an independent program. Its variables, statements, and statement labels are all local to the program unit. The communication between the variables of a referencing program and the subroutine is through the actual arguments and the dummy arguments.
2. When the CALL statement is encountered in a program, the values of the actual arguments will be passed to the dummy argument. The subroutine will then be executed, and control will be returned to the referencing program when the RETURN statement is encountered.
3. A subroutine is a complete program. It must have its own specification statements, if necessary, and its own END statement. A subroutine may use any other subprogram or have its own subprogram. A subroutine, however, cannot directly or indirectly (through another program) call itself.

4. The first statement of a subroutine subprogram must be the SUBROUTINE statement, and the last statement must be the END statement.
5. An actual argument (in the referencing program) can be a constant, variable, array name, or array element. The dummy arguments, however, must be variables.
6. The symbolic names of the dummy arguments may be different than the symbolic names of the actual arguments. However, a one-to-one correspondence exists between the actual arguments and the dummy arguments. Therefore, the number, order, and mode of the dummy arguments must match those of the actual arguments.
7. The name of a subroutine must be a FORTRAN name (up to six characters, the first one a letter), but the type of the name does not matter because no value will be assigned to it.
8. A subroutine may have several RETURN statements. In FORTRAN 77, the END statement may be used as a RETURN statement if it is the last statement.
9. If one of the actual arguments is an array name, the corresponding dummy argument must also be an array name as well; thus, the array must be defined in the subroutine as well as the referencing program.
10. The size in the DIMENSION statement of a subroutine may be a variable as long as the variable is included in the argument lists, and its value does not exceed the size in the referencing program. Using different array sizes in a referencing program and a subprogram for arrays with more than one dimension is not recommended.
11. If one of the actual arguments is character type, the corresponding dummy argument must be also character type (it must be declared with the CHARACTER statement if necessary), and the character length in the subroutine must not be greater than its corresponding length in the referencing program.

SOLVED PROBLEMS

9.1 Find all the errors in the following statements:
   a. CALL X = FUN1(A, B, C, X + Y)
   b. CALL AVERAGE (X, Y, X(INDEX))
   c. CALL SUB2(SQRT(A), B(I), C, 3)
   d. CALL SONY (ABS(A − B), 5, X(50))
   e. SUBROUTINE AYE(A, B, 3)
   f. SUBROUTINE BCD(A(1), X + Y, S)
   g. SUBROUTINE COMPUTER(A, B)
   h. SUBROUTINE REAL MEAN(X, Y, S)

☐ **Answers**
   a. The CALL statement cannot have an assignment statement.
   b. The subroutine's symbolic name is too long.
   c. This is syntactically correct.
   d. This is also syntactically correct.
   e. A dummy argument cannot be a constant.

    f. A dummy argument cannot be an array element or an expression.
    g. The subroutine's name is too long.
    h. The REAL statement cannot be used in the SUBROUTINE statement.

9.2 Are the general forms of the following CALL-SUBROUTINE structures correct?

    a.
```
       ⋮
       CALL SAL(WAGE, RATE)
       ⋮
       END
       SUBROUTINE SAL(X, Y, PAX)
       ⋮
       END
```

    b.
```
       ⋮
       CALL SSS(P, Q, R, KAY)
       ⋮
       END
       SUBROUTINE SSS(X, Y, Z, V)
       ⋮
```

    c.
```
       ⋮
       CALL SPA(X, 5,KA)
       ⋮
       END
       SUBROUTINE SPA(A,B,K)
       INTEGER B
       DATA A,K/2*0/
       ⋮
```

    d.
```
       REAL NAY(50), MON(50)
       ⋮
       CALL JANE(NAY, MON, 5)
       ⋮
       END
       SUBROUTINE JANE(I, J, P)
       DIMENSION I(50), J(50)
```

    e.
```
       ⋮
       CALL Q(X,Y)
       ⋮
       SUBROUTINE Q(A,B)
          B = 30.5
          A = 5.0*B + 2.0
          B = A*A + 3.0*A
          RETURN
          END
```

f.

```
      ⋮
   CALL SUB1(A, K, 5)
      ⋮
   END
   SUBROUTINE SUB1(X, I, J)
      J = 3*I
      X = X + 1
      RETURN
      END
```

g.

```
   DIMENSION XXX(100), AAA(100)
      ⋮
   CALL CHANGE(XXX, AAA)
      ⋮
   END
   SUBROUTINE CHANGE(U, P)
      U = P
      RETURN
      END
```

h.

```
   DIMENSION UP(200), DOWN(200)
      ⋮
   CALL TOT(UP, DOWN)
      ⋮
   END
   SUBROUTINE TOT(A, B)
   DIMENSION A(500), B(500)
      ⋮
```

i.

```
   DIMENSION T(100)
      ⋮
   N = 65
      ⋮
   CALL STEE(T,S)
      ⋮
   END
   SUBROUTINE STEE(A, B)
   DIMENSION A(N)
      ⋮
```

j.

```
      ⋮
   CALL SUM(3.0, 2.0, S)
   PRINT*,S
   END
   SUBROUTINE SUM(A,B,C)
   C = A + B
   STOP
   RETURN
   END
```

k.

```
            DIMENSION A(100)
              :
            SUM = 0.0
            CALL SSS(A, SUM)
              :
            END
            SUBROUTINE SSS(X, SS )
            DIMENSION X(100)
             SS = 0.0
            DO 10 I = 1,100
10              SS = SS + X( I)
            RETURN
            END
```

l.

```
            DIMENSION A(100)
              :
            SUM = 0.0
            DO 10 I = 1,100
                CALL TTT( A(I), SUM )
10          CONTINUE
              :
            END
            SUBROUTINE TTT(AA,S)
            DIMENSION AA(100)
            S = S + AA
            RETURN
            END
```

☐ **Answers**

- **a.** Incorrect; there are two actual arguments but three dummy arguments.
- **b.** Incorrect; the mode of the dummy argument V does not match with the mode of actual argument KAY (unless V is defined as an integer).
- **c.** This is correct; but it is not recommended to use the data statement in a subroutine because the data statement is effective only once (when the program starts execution). After the execution of the return statement the values become undefined, and the data statement is not effective the second time.
- **d.** Incorrect; the modes of the actual arguments do not match with those of the dummy arguments.
- **e.** This is correct; note that the values of all arguments (X and Y) are changed by the subroutine.
- **f.** Incorrect; one of the actual arguments is a constant (5), and therefore its value cannot be changed in the subprogram by its corresponding dummy argument J.
- **g.** Incorrect; U and P must be declared as arrays in the subroutine and then treated as arrays rather than as single variables.
- **h.** Incorrect; the array size in the subroutine must not be

> greater than the corresponding array size in the referencing program.
>
> **i.** Incorrect; the integer variable N representing the array size must be among the arguments. (Unless a COMMON statement is used. The COMMON statement is explained later in this chapter.)
>
> **j.** This is syntactically correct. But the STOP statement in the subroutine causes the execution of the entire program to stop. Thus, the value of S will not be printed.
>
> **k.** This is correct; but either SUM = 0.0 or SS = 0.0 can be omitted. Since the subroutine can be called elsewhere, the SS = 0.0 might be needed; thus SUM = 0.0 can be deleted.
>
> **l.** Incorrect; A(I) in the referencing program will be treated as a single variable. Therefore, not only is the DIMENSION statement in the subroutine unnecessary, but also it creates an error because of the statement S = S + AA.

FUNCTIONS

☐ Introduction

A function is a subprogram which can be invoked by appearance of its name in a statement. For example, if the function MIN is defined to return the smallest integer among three variables, then the statement

```
LES = MIN(I,J,K)
```

invokes the function and assigns the smallest of I,J,K to LES.

There are two main differences between a function and a subroutine:

1. A function is invoked by the appearance of its name in an expression or statement, whereas a subroutine is invoked by the CALL statement. The function's name can appear in an expression, an assignment statement, an IF statement, or an output statement. For example, if the function MIN is defined to return the smallest of three integer variables, then any of the following statements is correct:

   ```
   LOW = MIN(I,J,M)
   MAY = 3*MIN(I,J,M) + 5*M
   WRITE(6,10) MIN(I,J,M)
   IF (MIN(I,J,M) .GT. K) K = MIN(I,J,M)
   PRINT*,I,J,M,MIN(I,J,M)
   CALL XXX(MIN(I,J,K),B,C)
   PRINT*, 'Y=',2*MIN(I,J,K)
   ```

2. The name of the function itself is considered a variable, and normally the result of the function is assigned to it. Therefore, the first letter of the name of the function plays an important role in determining the type of the function (unless it is defined by a type statement).

There are three types of functions in FORTRAN:

1. Library functions
2. Function statement
3. Function subprograms

☐ Library Functions

Library functions are those functions which are defined by the system and are automatically available to all users. To use any of these functions, you need only to know the name of the function, the purpose, and the mode of the arguments used in that function. For example, SQRT is a library function which calculates the square root of a positive real number. The following program demonstrates how it is used in a program.

EXAMPLE 9.9

■ Problem: Write a program which reads the value of A and prints the square root of A.

Program:

```
PROGRAM LIB1
READ*,A
B = SQRT(A)
PRINT*,B
END
```

The library functions are also called *compiler-defined, intrinsic,* or *built-in functions.* Table 9.1 presents some of the most widely used library functions. Take a look at the table first and then review the following notes and examples.

EXAMPLE A

The standard deviation is equal to the square root of the variance of a set of data. If the variance is 395.5, write a program segment which prints the standard deviation.

The program segment is:

```
STD = SQRT(395.5)
PRINT*,STD
END
```

EXAMPLE B

If the sum of N values is equal to SUM, write a program segment which calculates the average of the values. Avoid mode-mixing operations.

The program segment is:

```
AVG = SUM/FLOAT(N)
```

EXAMPLE C

There are four test scores for each student. The instructor has decided to drop the lowest score. Write the program segment which calculates the average of the scores (dropping the lowest one).

The program segment is:

```
SMALL = AMIN1(TEST1,TEST2,TEST3,TEST4)
AVG = (TEST1 + TEST2 + TEST3 + TEST4 - SMALL)/
*FLOAT(3)
```

EXAMPLE D

Write a program which reads 100 values into an array and prints the absolute value of each.

Table 9.1 Some of the Library Functions

NAME AND FORM	MODE OF ARGUMENT	MODE OF RESULT	PURPOSE	EXAMPLE
ABS(a)	Real	Real	Returns the absolute value of an argument	X = ABS(-5.5)
IABS(i)	Integer	Integer		I = IABS(NET)
AINT(a)	Real	Real	Returns the value of an argument after truncation	X = AINT(Y) Z = AINT(S/Q)
ALOG(a)	Real	Real	Returns the natural logarithm of an argument	A = ALOG(B)
ALOG10(a)	Real	Real	Returns the base-10 logarithm of an argument	X = ALOG10(B)
MAXO(i,j,k,...)	Integer	Integer	Returns the value of the largest argument	L = MAX(M,N,K,I)
AMAX1(a,b,c,...)	Real	Real		C = AMAX1(6.9,95.2)
MINO(i,j,k,...)	Integer	Integer	Returns the value of the smallest argument	N = MINO(I,J,L)
AMIN1(a,b,c,...)	Real	Real		X = AMIN1(39.1,5.8)
DIM(a,b)	Real	Real	Returns a positive difference of two arguments ($a - b$); if a < b, DIM(a,b) = 0.0	D = DIM(Y,X)
IDIM(i,j)	Integer	Integer		IDEF = IDIM(K,L)
EXP(a)	Real	Real	Returns the exponential of an argument (e^a)	X = EXP(5.3) Y = EXP(A)
FLOAT(i)	Integer	Real	Returns the real equivalent from an integer argument	X = FLOAT(N) AVG = SUM/FLOAT(I)
IFIX(a) or INT(a)	Real	Integer	Returns the integer part from a real argument	L = IFIX(R) M = INT(X)
ICHAR(a)	Character	Integer	Returns an integer value from a character argument	K = ICHAR(A)
CHAR(i)	Integer	Character	Returns the character value from an integer argument	LETTER = CHAR(35)
LEN(a)	Character	Integer	Returns the length of a character string	N = LEN(NAME)
AMOD(a,b)	Real	Real	Returns the remainder of a divided by b	X = AMOD(X,Y)
MOD(i,j)	Integer	Integer		I = MOD(J,L)
NINT(a)	Real	Integer	Returns the nearest integer	M = NINT(X)
SQRT(a)	Real	Real	Returns the square root of an argument	S = SQRT(VAR)
DATE()[a]	—	Character	Returns the current date	D = DATE()
TIME()[a]	—	Character	Returns the current reading of the system clock	T = TIME()
CLOCK()[a]				
RANF()[a]	—	Real	Returns a random number between 0.0 and 1.0	X = 100.0*RANF() PRINT*,RANF()

[a] See Note 6 on page 369.

Notes for Table 9.1

1. *a,b,c,...* represent real arguments, and *i,j,k,...* represent integer arguments. Any of the arguments may be a variable, constant, array element, or another function if appropriate. However, all arguments in a list must be of the same type and must match with the name of the function (unless there is type conversion).
2. In FORTRAN 77, certain library functions are generic. A generic name is more flexible and may be used instead of a specific name. When using a generic name, the type of argument determines the type of result (unless there is type conversion). For example, MAX or MIN is a generic name which returns the largest or the smallest value of several arguments.
3. ICHAR(*a*) converts a character-type argument to an integer value equal to the collating sequence of the character in that system. The collating sequence of the first character (not the first letter) may be 0, the second character, 1, the third, 2, and so on, depending on the system. Normally, the integer value is between 0 and 63.
4. CHAR(*i*) converts the integer i to a character value; the returned character depends on the collating sequence used in that system (see note 3).
5. DIM(*a, b*) and IDIM(*i,j*) return the positive difference of two arguments; if the first argument is smaller than the second one, zero will be returned.
6. The following functions are not included in the standard, and the form depends on the system that you are using. Check the FORTRAN manual for your system.
    a. DATE() returns the current date in the form of year/month/day. The format might be different at a particular installation. The returned value is character type; DATE must be declared by

    CHARACTER * 10 DATE.

    b. TIME() or CLOCK() function returns the current reading of the system's clock in the form of

    hour.minute.second.

    The returned value is character type and must be declared as character (the form also depends on the system).
    c. In most systems, RANF() returns a random number between 0.0 and 1.0. But again the form of the function and the range of the number depend on the system being used.

The program:

```
      PROGRAM LIB4
      DIMENSION A(100)
      DO 50 I = 1,100
         READ*,A(I)
         PRINT*,ABS(A(I))
50    CONTINUE
      END
```

The array is not needed in this example, but is used to show the technique.

EXAMPLE E

Write a program segment which calculates the square root of the absolute value of A − B.

The program segment:

```
      X = SQRT(ABS(A - B))
```

370 *Subprograms*

EXAMPLE F

Write a small program which prints the current date.
The program:

```
PROGRAM DAY
CHARACTER DATE*10
PRINT*,DATE( )
END
```

Note that the DATE function depends on the system that you are using.

SOLVED PROBLEMS

9.3 What is the value of X or M in each of the following statements?

a. ```
N = 10
X = FLOAT(N)
```

b. `M = INT(6.329)`

c. `M = INT(6.999)`

d. `M = NINT(6.7)`

e. `X = AINT(6.7)`

f. `X = AMAX1(9.7,59.0,-5.0)`

g. `X = INT (195.3)`

h. `X = ABS(5.0 - 7.0)`

i. `M = MOD(9,5)`

j. `X = DIM(8.0,3.0)`

k. `X = DIM(3.0,8.0)`

l. `M = LEN('XYZ')`

m. `M = ICHAR('B')`
   [assume ICHAR('A') is equal to 32]

n. ```
NAME = 'D'
M = ICHAR(NAME)
```

o. `M = CHAR(33)`

p. ```
CHARACTER*10 M, DATE
M = DATE ( )
PRINT*, M
END
```

You should find out the DATE function for your system to run this program.

☐ **Answers**

a. X = 10.0
b. M = 6
```

c. M = 6
d. M = 7
e. X = 6.0
f. X = 59.0
g. X = 195
h. X = 2.0
i. M = 4
j. X = 5.0
k. X = 0.0
l. M = 3
m. M = 33
n. M = 35
o. M = 'B'   (M must be defined as character type.)
p. It prints the current date (depending on the system).

☐ Function Statement

A function statement is a single statement defined by the programmer in order to evaluate an expression. A function statement is similar to an arithmetic or character assignment statement and should be defined at the beginning of the program unit. Once the function is defined, it can be utilized many times in a program. A function statement is referenced in the same way as a library function. The following is an example:

EXAMPLE 9.10

■ **Problem:** The optimal production quantity can be calculated by the formula

$$Q = \sqrt{2DS/C}$$

where

D = product demand for a specified time interval,
C = carrying cost,
S = setup cost.

Write a program which reads D, C, and S for a product. Calculate and print the optimal production quantity for the product. Assume there are five products.

Program:

```
      PROGRAM ARFUN1
C DEFINING THE FUNCTION OPTQ
C
      OPTQ(X,Y,Z) = SQRT(2*X*Z/Y)
C
      DO 10 I = 1,5
         READ(5,20)D,C,S
         Q = OPTQ(D,C,S)
         WRITE(6,30)D,C,S,Q
10    CONTINUE
20    FORMAT(3F5.2)
```

```
30      FORMAT(6X,'D=',F6.2,5X,'C=',F6.2, 5X, 'S=',
       *F6.2, 10X,'Q=',F6.2)
        END
```

Note: D, C, and S in the statement Q = OPTQ(D,C,S) are actual arguments; X, Y, and Z in the function definition OPTQ(X,Y,Z) = SQRT(2 * X * Z/Y) are dummy arguments.

When the name of a function appears in a statement, the values of actual arguments will be passed to the dummy arguments, the expression will be evaluated, and the resulting value (assigned to the name of the function) will be returned.

An important point about the function statement is that it is a part of the program unit. It is not treated separately (like a subroutine). A variable in the function expression can be either a dummy argument or a reference to a variable in the program. For example, if a function is defined as

```
FFF(X,Y,Z) = A*X + B*Y + C*Z
```

the values of X, Y, and Z will be passed to the function through the actual arguments, and the values of A, B, and C will be taken as they were before the execution of the statement containing the name of the function. In other words, the current values of variables A, B, and C will be used in evaluating the function. The following is another example:

```
        PROGRAM ARFUN2
        GAMA(P,Q) = I*P + I*Q
        DO 100 I = 1,10
           READ*,A,B
           F = GAMA(A,B)
           PRINT*,A,B,I,F
100     CONTINUE
        END
```

Pay attention to the following points when using a function statement.

1. The statement function applies only to the program unit containing the definition of the function.
2. The function definition (the function statement itself) is nonexecutable, and it must be placed after the specification statements and before any executable statement.
3. The function name must be a FORTRAN name, consist of one to six characters, the first one being a letter; the mode of the name is determined by the first letter, unless it is specified by a type statement. The name is a variable and contains the value of the expression after execution.
4. The actual arguments can be variables, array elements, constants, or simple arithmetic expressions. The dummy arguments must be variables.
5. The number, mode, and the order of dummy arguments must match with those of the actual arguments.
6. The expression defining the function may include variables other than the dummy arguments. In this case, the current values of those variables will be used when the function is executed.
7. The parentheses with a function are required, but the argument list may be empty.

8. The name of the function cannot be any other name in the program unit.
9. A function must be defined before it is referenced.

SOLVED PROBLEMS

9.4 Identify the errors in the following statements.

a. FUNC1(A,B) = 2.0 * A ** 3 + 5.0 * B ** 3
b. FUNC2(X,2) = 5.0 * X + 2.0
c. QUE(E,F(1)) = E * F(1) + E + F(1)
d. TOTALPAY(W,H) = W * H
e. EXPONE(NT,P) = D * NT ** M + Q * P + 3.0
f. NET(M,N) = M − N
g. EXPO(P,Q) = P + Q + EXPO(A,B)
h. F(A,B) = 5.0 + SQRT(ABS(A * B + A + B))
i. ADDXY(A,B) = A//B

□ **Answers**

a. This is correct.
b. The dummy argument cannot be a constant.
c. The dummy argument cannot be an array element.
d. The name of the function is too long.
e. This is correct.
f. This is also correct.
g. A function cannot invoke itself.
h. This is correct.
i. This is correct; note that it is a character function, and the function name as well as A and B must have been defined as characters.

9.5 Identify any errors in the general form of the following function statements:

a.
```
FACT(M,N) = 3*M + 5*N
    ⋮
V = FACT(X,Y)
```

b.
```
EXPLOT(E,G) = X*G + 2*X*Y + E
    ⋮
D = EXPLOT(A,B,C)
    ⋮
```

c.
```
PAY = A*B
    ⋮
WAGE = PAY
    ⋮
```

d.

```
FUN1(X) = 2.0*X + A*B
FUN2(Y) = 5.0*A + FUN1(B)*Y
   ⋮
Z = FUN2(D)
```

e.

```
PIE(C,D) = 2*C + D + RO(5)
RO(A) = A + A**2
   ⋮
X = PIE(X,Y)
```

f.

```
QUET(FIRST,SECOND) = FIRST + SECOND*FIRST
QUET(A,B) = A*B + A + B
   ⋮
```

g.

```
FUNCT1(A,B) = A + B
   ⋮
X = FUNCT1(P,Q)
PRINT*,FUNCT1
   ⋮
```

☐ **Answers**

a. The modes of the dummy arguments do not match the modes of the actual arguments.
b. The number of dummy arguments is less than the number of actual arguments.
c. A function must have an argument or at least parentheses. Therefore, if PAY is a function, it must be corrected accordingly.
d. This is correct. Note that FUN1 is defined before FUN2.
e. The function RO must be defined before PIE.
f. A function with the same name cannot be defined twice.
g. The name of the function cannot appear as a single variable [PRINT *,FUNCT1(P,Q) would be correct].

■ **9.6** What is the value of V in the following program segments?

a.

```
GAMMA(X,Y) = 2.0*X + X*Y
   ⋮
A = 2.0
B = 3.0
V = GAMMA(A,B)
```

b.

```
BETA(G,H) = G*H + A
   ⋮
A = 5.0
P = 3.0
V = BETA(P,2.0)
```

c.

```
FUN2(X) = X*X + 1.0
FUN1(E,F) = E + F + FUN2(F)
   ⋮
V = FUN1(5.0,1.0)
```

d.

```
OVER(X) = 1.5*X
   ⋮
H = 42.0
IF(H .GT. 40.0) THEN
    V = OVER(H)
ELSE
    V = 0.0
ENDIF
```

☐ **Answers**

    a. V = 10.0
    b. V = 11.0
    c. V = 8.0
    d. V = 63.0

☐ **Function Subprograms**

A function subprogram performs a set of instructions independently when its name appears in an expression. A function subprogram returns a single value through its name, though it may return several other values through its arguments. Function subprograms are utilized the same way as subroutines; however, there are three differences between a function subprogram and a subroutine:

1. The function is invoked when its name appears in an expression, whereas a subroutine is invoked by the CALL statement.
2. A value is assigned to the function's symbolic name, but no value is associated with the subroutine's symbolic name.
3. The word FUNCTION is used as the first statement of the subprogram instead of the word SUBROUTINE.

Let's look at an example using a function subprogram before explaining it further.

EXAMPLE 9.11

■ **Problem:** Calculate and print the sum, and average of three variables.

Program: Style I

```
      PROGRAM FUNSUB
C STYLE I
      READ(5,10)A,B,C
      AVG = SUM(A,B,C)/3.0
      WRITE(6,20)SUM,AVG
20    FORMAT(11X,'SUM=',F7.2,'AVERAGE=',F6.1)
10    FORMAT(3F5.1)
      END
```

```
      C
            FUNCTION SUM(X,Y,Z)
            SUM = X + Y + Z
            RETURN
            END
```
Program: Style II
```
            PROGRAM FUNSUB
      C STYLE II
            READ(5,10)A,B,C
            TOT = SUM(A,B,C,AVG)
            WRITE(6,20)TOT,AVG
      10    FORMAT(3F5.1)
      20    FORMAT(11X,'SUM=',F7.2,AVERAGE=',F6.1)
            END
      C
            FUNCTION SUM(X,Y,Z,D)
            SUM = X + Y + Z
            D = SUM/3.0
            RETURN
            END
```

Notes:

1. Only the sum of three variables A,B,C is calculated by a function in style I, but the sum and the average are calculated by a function in style II. In both styles, the sum of the numbers is assigned to the function's name.
2. The function subprogram is an independent program which starts with the word FUNCTION and ends with an END statement.
3. The function is invoked when its name appears in an expression (the first style), or in an assignment statement (the second style).
4. The variables A, B, ... in the function list of the main program are the actual arguments, and X, Y, ... in the function list of the subprogram are the dummy arguments.

The general form of using a FUNCTION subprogram is as follows:

The referencing program:

$\vdots$

A = Fname (a1,a2,a3,...)
       ↑   ↑
       1   2

$\vdots$

END

The subprogram:

FUNCTION Fname (d1,d2,d3,...)
    ↑   ↑   ↑
    3   1   4

$\vdots$

Fname = ...
 ↑
 5 $\vdots$

```
    RETURN
      ↑
      6
    END
```

It has six important components:

1. The name of the function in a statement in the referencing program. This name must be the same as in the FUNCTION statement of the subprogram.
2. The actual-argument list in parentheses following the name of the function.
3. The word FUNCTION, which must be the first statement in the subprogram.
4. The dummy-argument list in parentheses, following the name of the function.
5. The assignment statement, which assigns a value to the function name.
6. The RETURN statement, which returns control to the referencing program.

The number, mode, and order of the dummy arguments must match those of the actual arguments.

Function Name

The function's symbolic name is a FORTRAN name; it must not exceed six characters, the first one being a letter. The first letter determines the function type unless it is defined by a type statement. For example,

```
    INTEGER A,B,GEE,C(100)
       ⋮
    IT = GEE(X,Y,A,B)
       ⋮
    FUNCTION GEE(U,P,Q,R)
    INTEGER GEE,Q,R
       ⋮
```

defines the function's name as integer. Notice that the mode must be defined both in the referencing program and in the subprogram. Furthermore, the type statement can include the function statement. For example,

```
    INTEGER FUNCTION GEE(U,P,Q,R)
```

is equivalent to

```
    FUNCTION GEE(U,P,Q,R)
    INTEGER GEE
```

and

```
    REAL FUNCTION MEAN(X,Y)
```

is equivalent to

```
    FUNCTION MEAN(X,Y)
    REAL MEAN
```

A function always carries a single value through its name. Nonetheless, a function can return several other values through its arguments, as demonstrated by Example 9.17.

The function's name cannot be the name of any other function, variable, subroutine, or array. In other words, a function's name must be a unique name in the program unit. For example, the program segment

```
DIMENSION PAY(100)
      :
X = PAY(HOURSE,WAGE)
      :
END
FUNCTION PAY(A,B)
      :
```

is not correct, because the function's name cannot be an array name.

Arrays and Functions

The actual arguments of the function can be constants, variables, expressions, array names, or array elements. However, as in subroutines, if one of the actual arguments is an array name, the corresponding dummy argument must be also an array name. The following is an example:

EXAMPLE 9.12

Problem: Write a program which calculates the sum of the elements of an integer array NORM with 20 rows and 10 columns. Use a function to calculate the sum.

Program:

```
      PROGRAM FUNARR
      DIMENSION NORM(20,10)
      INTEGER SUM
C
      READ(5,10) NORM
C
      III = SUM(NORM)
      WRITE(6,20) III
10    FORMAT(20I3)
20    FORMAT(11X,'THE SUM IS=',I10)
      END
C *************************************
      INTEGER FUNCTION SUM(X)
         INTEGER X(20,10)
         SUM = 0
         DO 100 I = 1,20
            DO 100 J = 1,10
               SUM = SUM + X(I,J)
100      CONTINUE
         RETURN
         END
```

Notes:

1. The name of the function is defined to be integer both in the main program and in the function subprogram.
2. The entire array NORM is transferred to the subprogram. It is also defined as an integer array in the subprogram.

Again, as explained for subroutines, if an array element is used as an actual argument, it is treated like a single variable and there is no need to define the array in the subprogram. The following example demonstrates two different styles for coding a problem. In the first style, the individual array element is used as an actual argument in a DO loop; and in the second style, the entire array is used as an actual argument in the referencing program. Comparing the two styles also answers the question whether a function can be used in a DO loop or as a "stand-alone" statement. Look at the example and compare the styles carefully.

EXAMPLE 9.13

Problem: There are five salesmen in a department store. Their amount of sales for a four-week period is available in tabular form. The data on one occasion are:

Salesman	Week 1	Week 2	Week 3	Week 4
1	6295.50	5629.60	8760.30	2754.50
2	9678.40	3948.30	2947.20	2574.70
3	3958.60	3927.50	3857.40	3867.90
4	7885.50	7369.30	6826.90	3730.50
5	3968.70	3869.40	8990.80	8905.60

Write a program which reads the data into an array. Use a function subprogram to calculate the total sales for each salesman and the grand total. Print the information.

Program: Style I

```
      PROGRAM SALES1
C STYLE I, USING A FUNCTION IN A DO LOOP
      DIMENSION SALES(5,4),TOT(5)
C
      DO 40 I = 1,5
          READ(5,10)(SALES(I,J), J = 1,4)
40    CONTINUE
C
      GTOTAL = 0.0
C CALCULATING EACH SALESMAN'S TOTAL IN A LOOP
      DO 50 I = 1,5
          TOT(I) = SUM(SALES(I,1),SALES(I,2),
     *    SALES(I,3),SALES(I,4))
C
          GTOTAL = GTOTAL + TOT(I)
C
50    CONTINUE
C WRITING THE INFORMATION
      DO 60 I = 1,5
          WRITE(6,20)(SALES(I,J),J = 1,4),TOT(I)
60    CONTINUE
C
      WRITE(6,30)GTOTAL
```

380 *Subprograms*

```
10      FORMAT(4(F7.2,2X))
20      FORMAT(11X,5(F10.2,5X))
30      FORMAT(22X,'TOTAL ='F10.2)
        END
C
        FUNCTION SUM(S1,S2,S3,S4)
            SUM = S1 + S2 + S3 + S4
            RETURN
            END
```

Program: Style II

```
        PROGRAM SALES2
C STYLE II USING A FUNCTION AS A STAND-ALONE
  STATEMENT
        DIMENSION SALES(5,4),TOT(5)
C
        DO 40 I = 1,5
            READ(5,10)(SALES(I,J),J = 1,4)
40      CONTINUE
C
        SSS = GTOTAL(SALES,TOT)
C
C WRITING THE INFORMATION
        DO 60 I = 1,5
            WRITE(6,20)(SALES(I,J),J = 1,4),TOT(I)
60      CONTINUE
        WRITE(6,30)SSS
10      FORMAT(4(F7.2,2X))
20      FORMAT(11X,5(F10.2,5X))
30      FORMAT(11X,'TOTAL =',F10.2)
        END
C
        FUNCTION GTOTAL(S,T)
            DIMENSION S(5,4),T(5)
            GTOTAL = 0.0
            DO 50 I = 1,5
                T(I) = S(I,1) + S(I,2) + S(I,3) +
       *               S(I,4)
                GTOTAL = GTOTAL + T(I)
            CONTINUE
            RETURN
            RUN
```

Notes:

1. In style I: The total sales for each salesman is calculated in the main program in a DO loop by the function SUM. Each time that the function is executed, the total sales for salesman I is assigned to variable SUM, and in turn will be assigned to the element TOT(I) of the array TOT in the main program. Thus, the function returns only one value (SUM) through its name at a time. The grand total is calculated in the main program. Note that the function is invoked in a DO loop in the main program.

2. In style II: The total sales for each salesman is calculated and assigned to array T in a DO loop in the function. The grand

> total is also calculated in the function, though it could be calculated in the main program. The function passes the values of array T to the array TOT (to the main program) through the arguments list, and its name carries the value of GTOTAL. Note that the function is not invoked in a DO loop in the main program.

The adjustable array size discussed for subroutines is also permitted in function subprograms, provided that (1) integer variables representing the sizes are included in the arguments lists, and (2) the sizes in the function do not exceed the sizes in the referencing program. The following is an example of a function with an adjustable array size:

```
          FUNCTION BIG(A,M)
          DIMENSION A(M)
          BIG = A(1)
          DO 200 I = 2,M
              IF(A(I) .GT. BIG) BIG = A(I)
 800      CONTINUE
          RETURN
          END
```

The Difference Between an Array and a Function

An array's name and function's name can appear in a statement in the same way. For example, in the statement:

P = AAA(I,J,K)   or   CUTE = 2.0 * AAA(X,Y,Z) + 1

AAA can be an array's symbolic name or a function's name. The compiler recognizes either by encountering a symbolic name followed by a left parenthesis. The difference, however, is that an array is defined at the beginning of a program unit, whereas a function is defined outside the program unit. Therefore, if you sometimes forget to define an array at the beginning of a program, don't be surprised if you see an error message under the array name saying

THE FUNCTION NOT DEFINED

This may refer to merely an undefined array.

Summary of the Important Points about Function Subprograms

1. A function subprogram is a complete program unit. It starts with the FUNCTION statement and ends with an END statement. A function may use another subprogram; however, it cannot directly or indirectly invoke itself.
2. A function subprogram is an independent program. Its variables, statements, and statement labels are all local to it. The subprogram communicates with the referencing program through its name and its arguments.
3. The name of a function carries a value; it must be a FORTRAN name, composed of one to six characters, the first being a letter. The first letter determines its type unless that is specified by the type statement. The name cannot be any other name in the program.
4. A function is invoked when its name appears in a statement or in an expression.
5. The number, the order, and the type of the dummy arguments of a

function must match with those of the actual arguments in the referencing program.
6. A function must have at least one RETURN statement. However, in FORTRAN 77, the END statement can also be used as a RETURN statement if it is the last statement.
7. Since the function name itself is a variable, its value must be defined. Normally a value is assigned to it in the subprogram.
8. If one of the actual arguments is an array name, the corresponding dummy argument must be an array name also. The arrays must be declared in the function subprogram as well as the referencing program. The array size, however, may be an integer variable (or even an integer expression) as long as its value is passed to the subprogram.

SOLVED PROBLEMS

9.7 Identify all the errors in the following statements:

    a. FUNCTION COMPUTE(A,B)

    b. FUNCTION FUNCT1(X + Y,B(1),5)

    c. INTEGER FUNCTION NORM(X,I,J,Y)

    d. FUNCTION REAL N(A,B,C,I,J)

    e. INTEGER FUNCTION PA(A,B,I)

    f. FUNCTION (A,B,N)

    g. FUNCTION IDEE(X,Y,N,M)
        REAL IDEE, N

    h. FUNCTION AYE(X,Y)
        DIMENSION AYE(100)

☐ **Answers**

    a. The name of the function is more than six characters.
    b. The dummy argument cannot be an expression, an array element, or a constant.
    c. This is correct.
    d. The word REAL must be before the word FUNCTION.
    e. This is correct.
    f. The name of the function is missing.
    g. This is correct.
    h. The name of the function cannot be an array.

9.8 Identify all the errors in the following program segments:

    a.

```
          ⋮
     Y = POP(A,B,C,I,M)
          ⋮
     FUNCTION POP(X,Y,Z,V,U)
          ⋮
```

b.

```
      INTEGER RRR
         ⋮
      JAY = RRR(SQRT(A)+B, M)
         ⋮
      FUNCTION RRR(X,I)
      INTEGER RRR
```

c.

```
         ⋮
      SUM = SUM + SMALL(A,B,C)
         ⋮
      FUNCTION SMALL(I,J,K)
```

d.

```
         ⋮
      X = S(A,B,C,SUM)
         ⋮
      FUNCTION S(X,Y,Z,P)
      P = X + Y + Z
      RETURN
      END
```

e.

```
      DIMENSION X(100)
         ⋮
      TOT = SUM(X,SSS)
      PRINT*,SSS
         ⋮
      FUNCTION SUM(Y,S)
      DIMENSION Y(100)
      S = 0.0
      DO 10 I = 1,100
   10 S = S + Y(I)
      SUM = 0.0
      RETURN
      END
```

f.

```
      REAL MEAN, N
         ⋮
      X = MEAN(X,Y)
         ⋮
      FUNCTION MEAN(A,B)
      INTEGER MEAN
```

g.

```
      REAL NNN(50),M(50)
         ⋮
      P = NNN(M,N)
         ⋮
      FUNCTION NNN(I,J)
      REAL NNN(50),I(50)
         ⋮
```

h.
```
        DIMENSION A(100),B(100,50)
            ⋮
        X = OLD(A,B,M,N)
            ⋮
        FUNCTION OLD(X,Y,I,J)
        DIMENSION X(I),Y(I,J)
```

i.
```
            ⋮
        P = FUN(A,B)
        PRINT*,FUN
            ⋮
        FUNCTION FUN(X,Y)
```

j.
```
        PROGRAM FUN1
            ⋮
        P = FUN1(X,Y)
            ⋮
        FUNCTION FUN1(A,B)
        A = 2.0*B
        RETURN
        END
```

☐ **Answers**

a. The modes of the dummy arguments V and U do not match the corresponding actual arguments.
b. This is correct.
c. The modes of the dummy arguments do not match those of the actual arguments.
d. The name of the function must be assigned a value.
e. This is correct; note that the name of the function is assigned a value.
f. The mode of the function must be the same in both the referencing program and the subprogram.
g. The function's name cannot be an array.
h. This is correct; note that the dimension sizes are passed to the subprogram.
i. The name of the function cannot be used without its arguments.
j. The name of the function must be a unique name (not the same as the program name), and a value should be assigned to it.

☐ **Comparing Functions with Subroutines**

As discussed before, both the function and the subroutine are useful tools for breaking a complex task into smaller units and then treating each smaller unit separately, thus making a complicated program manageable. They also provide a technique to avoid repeating a task in a program several times, hence saving storage space and programming time.

Most often, a function is used to return a single value, and a subroutine is utilized to return several values. If there is only one expression to be evaluated, a function statement is the best choice. But if there is more than one instruction to be performed, then either a subroutine or a function subprogram can be used.

Any task which can be done by a function can also be done by a subroutine. However, there can be some tasks which suit a subroutine more. The following is an example:

```
        READ(5,10)A,B
        IF(A - B)100,200,200
100     CALL PRTNEG
200     WRITE(6,30)A,B
        FORMAT(11X,'A=',F5.1,'B=',F5.1)
        END
```

Table 9.2 Summary of the General Form of Subprograms

SUBPROGRAM	GENERAL FORM	NOTES
Subroutine	$\vdots$ CALL XXX($a_1, a_2, \ldots$) $\vdots$ END SUBROUTINE XXX($d_1, d_2, \ldots$) $\vdots$ RETURN END	1. The subroutine is invoked by the CALL statement. 2. No value is associated with the subroutine's name; the results are returned through the arguments. 3. The subroutine is defined by the programmer outside the program unit. 4. A RETURN statement transfers control to the calling program. However, the last statement may be a STOP, if the execution of the entire program (main program and subprograms) is to be stopped. The END statement can also be used as a RETURN statement.
Library function	$\vdots$ Y = YYY($a, b, \ldots$) $\vdots$ END	1. The function is invoked by the appearance of its name in a statement. 2. Only one value is returned; the value is associated with the name of the function. 3. It is a compiler-defined function.
Function statement	ZZZ($d_1, d_2, \ldots$) = expression $\vdots$ Z = ZZZ($a_1, a_2, \ldots$) $\vdots$ END	1. The function is invoked by the appearance of its name in a statement. 2. Only one value is returned; the value is associated with the name of the function. 3. The function is defined inside the program unit, at the beginning of the program, by the user, and is limited to one statement.
Function subprogram	$\vdots$ F = FFF($a_1, a_2, \ldots$) $\vdots$ END FUNCTION FFF($d_1, d_2, \ldots$) $\vdots$ RETURN END	1. The function is invoked by the appearance of its name in a statement. 2. The function name returns a value, and other results may be returned through its arguments. The first letter of the function's name, or a type statement, defines the mode of the function. 3. The function is defined by the programmer outside the program unit. 4. A function must have at least one argument (or at least the parentheses). The END statement can be used as a RETURN statement.

```
      C
            SUBROUTINE PRTNEG
            WRITE(6,10)
            FORMAT(11X,'A - B IS NEGATIVE')
            STOP
            END
```

Similarly, there are places where a function is more convenient. A program also can use several functions or subroutines.

Table 9.2 (page 385) summarizes the forms of the subprograms and their differences.

COMMON BLOCK

☐ Unlabeled COMMON Block

A COMMON statement with a variable list at the beginning of a program causes the variables to be accessible to the other program units, thus eliminating the need for passing the variables through the argument list. Let us look at the following example before explaining the COMMON statement further.

EXAMPLE 9.14

Problem: Write a program which calculates and prints the average of three variables. Use a subroutine to calculate the average.

Program I: No COMMON Statement

```
            PROGRAM NOCOMM
            READ*,A,B,C
            CALL TOTAL(A,B,C,AVG)
            PRINT*,AVG
            END
      C
            SUBROUTINE TOTAL(X,Y,Z,D)
            D = (X + Y + Z)/3.0
            RETURN
            END
```

Program II: Using COMMON Statements

```
            PROGRAM WCOMM
            COMMON A,B,C
            READ*,A,B,C
            CALL TOTAL(AVG)
            PRINT*,AVG
            END
      C
            SUBROUTINE TOTAL(D)
              COMMON X,Y,Z
              D = (X + Y + Z)/3.0
              RETURN
              END
```

> **Notes:**
>
> 1. The COMMON statement causes the variables A, B, and C in the main program to be accessible to the subroutine. Those variables are referred to as X, Y, Z in the subroutine. The one-to-one correspondence between the two groups is as follows:
>
>   the main program:
>
>   COMMON A,B,C
>
>   the subprogram:
>
>   COMMON X,Y,Z
>
> 2. The COMMON statement must appear both in the main program and in the subprogram.
> 3. By defining the variables in the COMMON statement, there is no need to pass the values through the arguments.
> 4. The variable AVG (and D) could have been included in the COMMON list. In that case, the program would have looked like this:
>
> ```
> COMMON A,B,C,AVG
>     ⋮
> CALL TOTAL
>     ⋮
> END
> SUBROUTINE TOTAL
> COMMON X,Y,Z,D
>     ⋮
> ```

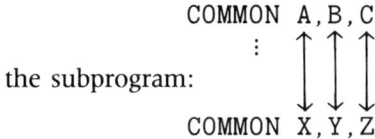

Internally, the COMMON statements cause the block of variables listed in the referencing program to share the same memory units with the block of variables listed in the subprogram. However, the same datum can be known by one symbolic name in a program unit, and by a different name in another program unit (Figure 9.2).

The variables in the COMMON statement of a subprogram must match those in the referencing program according to their order and type. The number of variables in an unlabeled COMMON statement, however, may be different in two program units. The following is an example:

Main program:

```
COMMON A,B,C,D,J,L
    ⋮
```

Subprogram 1:

```
COMMON A,B,C,D
    ⋮
```

Subprogram 2:

```
COMMON A,B,C,D,J
```

In some cases, we need to include additional variables in a COMMON statement to insure proper correspondence. The following subroutine is an example:

388 *Subprograms*

```
Memory names in          Memory locations          Memory names in
the main program              (data)                the subprogram

     A  ----------→   ┌─────────┐   ←----------  X
     B  ----------→   ├─────────┤   ←----------  Y
     C  ----------→   ├─────────┤   ←----------  Z
     AVG ---------→   ├─────────┤   ←----------  D
                      │    ⋮    │
                      └─────────┘
```

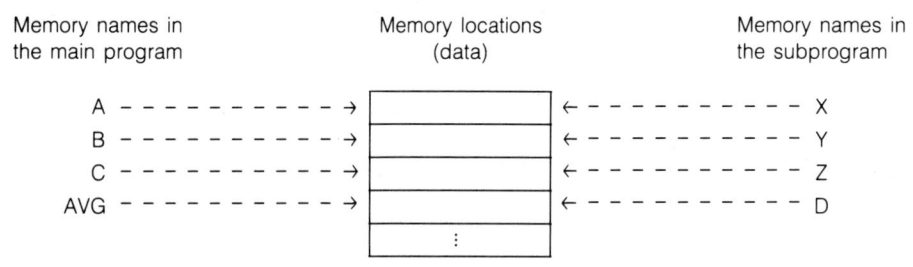

Figure 9.2 A COMMON block.

```
SUBROUTINE FIX
COMMON E,F,G,H,I,K
K = 2*I
RETURN
END
```

Notice that the variables E,F,G,H are listed only for proper access to K and I.

The COMMON statement can be used with function subprograms as well as with subroutines. The following is an example:

```
COMMON A,B,KAY
   ⋮
X = FUNCT1(Y)
   ⋮
END
FUNCTION FUNCT1(X)
COMMON E,F,MAY
   ⋮
FUNCT1 = . . .
RETURN
END
```

If an array name appears in a COMMON statement, all the array elements will be reserved in the COMMON block. For example, the statements

```
DIMENSION A(100)
COMMON A,B,C
```

cause 102 memory locations to be reserved in a COMMON block. Furthermore, when an array and its size appear in a COMMON statement, certain memory locations automatically will be reserved for the array elements, thus eliminating the need for a DIMENSION statement. For example,

```
COMMON A(100),B,C
```

is equivalent to

```
DIMENSION A(100)
COMMON A,B,C
```

The following are additional examples:

```
COMMON XXX(1000), YYY(50), C(10)
```

is equivalent to

```
DIMENSION XXX(1000), YYY(50), C(10)
COMMON XXX,YYY,C
```

and

```
COMMON ICON(500,3),KOUNT(100,10),X,Y,Z(500)
```

is equivalent to

```
DIMENSION ICON(500,3),KOUNT(100,10),Z(500)
COMMON ICON,KOUNT,X,Y,Z
```

When arrays are included in the variable list of a COMMON statement, pay attention to the fact that the memory locations will be reserved in the order in which they *appear* in the COMMON statement. For example, suppose the following statements are included in a program:

Main program:

```
COMMON X(100)Y,Z
  :
CALL STDENT
  :
CALL QUE
  :
END
```

Subprogram 1:

```
SUBROUTINE STDENT
COMMON B(50),A,D
  :
```

Subprogram 2:

```
SUBROUTINE QUE
COMMON EEE(102)
```

Then the one-to-one correspondence of the variables is as in Figure 9.3. Note that A and D refer to X(51) and X(52), respectively; EEE(101) and EEE(102) refer to Y and Z, respectively. Furthermore, the array elements are reserved in the order in which they are stored. For example, consider the following statements:

Main program:

```
COMMON A(3,4)
  :
```

Subprogram:

```
COMMON AA(2,3),BB(2,3)
  :
```

The one-to-one correspondence of array elements is as shown in Figure 9.4.

Variables in the main program	Memory locations	Variables in subroutine 1	Variables in subroutine 2
X(1)		B(1)	EEE(1)
X(2)		B(2)	EEE(2)
X(3)		B(3)	EEE(3)
:	:	:	:
X(50)		B(50)	
X(51)		D	
X(52)		A	
:	:		:
X(100)			EEE(100)
Y			EEE(101)
Z			EEE(102)

Figure 9.3 A COMMON block for two subprograms.

390 *Subprograms*

Variables in the main program	Memory locations	Variables in the main program
A(1,1) →		← AA(1,1)
A(2,1) →		← AA(2,1)
A(3,1) →		← AA(1,2)
A(1,2) →		← AA(2,2)
A(2,2) →		← AA(1,3)
A(3,2) →		← AA(2,3)
A(1,3) →		← BB(1,1)
A(2,3) →		← BB(2,1)
A(3,3) →		← BB(1,2)
A(1,4) →		← BB(2,2)
A(2,4) →		← BB(1,3)
A(3,4) →		← BB(2,3)

Figure 9.4 Variable correspondence in a COMMON block.

One more note: it is not allowed to define the initial values of the variables in an unlabeled COMMON block with a DATA statement. Either an assignment statement must be used for this purpose or a BLOCK DATA subprogram, explained in Chapter 11.

☐ Labeled COMMON Block

As discussed before, if two subprograms use different groups of variables, additional variables must be included in a COMMON statement to insure proper correspondence even though some of the variables are not used in a program unit. This may create inefficiency in writing programs. Fortunately, FORTRAN allows us to group and label the variables, and use only a desired group in a subprogram. The group label is a FORTRAN variable name; it must be placed within a pair of slashes and precede the group. Each group is called a block. For example, the following COMMON statement declares three common blocks A,B,C:

```
COMMON /A/X,Y,Z,V /B/T,U /C/P,Q,R,S
```

Each subprogram may use one or more blocks. The name of each block (the label) must be the same in all program units; but the variables' symbolic names may be different. The following example demonstrates using labeled common blocks:

EXAMPLE 9.15

■ **Problem:** Write a program which reads the values of A, B, I, J, and K. Calculate the average of A, B, C in a subroutine and the product of I, J, K in another subroutine. Use COMMON statements.

Program:

```
PROGRAM LABCOM
COMMON /FIRST/ A,B,C,AVG /SECOND/ I,J,K,M
READ*,A,B,C,I,J,K
CALL SUM
CALL MULT
PRINT*,AVG,M
END
```

```
      C
            SUBROUTINE SUM
                  COMMON /FIRST/ X,Y,Z,D
                  D = (X + Y + Z)/3.0
                  RETURN
                  END
      C
            SUBROUTINE MULT
                  COMMON /SECOND/ I,J,K,N
                  N = I*J*K
                  RETURN
                  END
```

Note: The block name must be the same in the main program and in the subprogram, but the variables' symbolic names may be different.

Thus, the general form of a COMMON statement is:

```
                    ┌──────Name of COMMON block (label)
                    ↓
COMMON/label/ list , /label/ list . . .
                  ↑ ↑
                  │ └─Comma (optional)
                  └──List of variable names or array names
```

The unlabeled COMMON statement discussed in the previous section is a special case of the general COMMON block when the label is "blank." Thus:

`COMMON // A,B,C`

is equivalent to:

`COMMON A,B,C`

Generally, when several variables are used in several subprograms, the COMMON statement leads to efficiency. Pay attention to the following points when using a COMMON statement:

1. Both labeled and unlabeled (blank) COMMON blocks can be used in a COMMON statement. However, if the first specification is for a blank COMMON block, the slashes may be omitted. For example, the statement

   `COMMON X,Y,Z /B/P,Q,R,/A/V,W`

   is equivalent to

   `COMMON // X,Y,Z /B/P,Q,R,/A/V,W`

2. Function names cannot be included in a COMMON block.
3. Dummy-argument names cannot be included in a COMMON block in a subprogram.
4. Variables and arrays are stored in the order in which they appear in a COMMON block.
5. The declarations of COMMON blocks are cumulative within a program unit. That is, each block in a list is considered as a continuation of the previous specifications. For example,

   `COMMON A,B,C(100)`
   `COMMON X(30),Y,Z`

   is equivalent to

```
COMMON A,B,C(100),X(30),Y,Z
```

and

```
COMMON /A/X,Y,Z /B/U,P,Q /A/V,W
```

is equivalent to

```
COMMON /A/X,Y,Z,V,W /B/U,P,Q
```

6. If one of the variables in a block is character type, all of the variables must be character type.
7. The size of a labeled COMMON block must be the same in all program units, but the unlabeled (or blank) common block may have different sizes in different program units.
8. The initial values of variables in an unlabeled COMMON block can be defined by a BLOCK DATA subprogram. BLOCK DATA subprogram is discussed in Chapter 11.

SOLVED PROBLEMS

9.9 Identify the errors in the following statements:

a. `COMMON /A/,X,Y,I,J,/B/ A,B,C`

b. `COMMON (A,X,Y,Z)`

c. `COMMON B/P,Q,R(30),`

d. `COMMON //A,B,C(10) // X,Y,Z`

e. ```
   COMMON P,Q,R
   COMMON /COOL/ A,B(10) /ROL/ X,
   *Y(100),A,B(5)
   ```

f. ```
   INTEGER P,Q,R(100)
   REAL JIM,IM,MAD(70)
   COMMON /X/ P,Q,IM,MAD/Y/
   *Q,R,JIM,IM
   ```

g. ```
   DIMENSION P(100)
   COMMON A,B,C(50) /D/ P,Q(50)
   ```

h. ```
   CHARACTER*12 NAME,ADR
   COMMON /A/ NAME, ADR, X,Y
   ```

□ **Answers**

a. There should not be a comma before X.
b. The variables should not be in parentheses.
c. There should be a slash before B and no comma after R(30).
d. Since this is a blank common block, all the slashes can be omitted.
e. This is correct.
f. This is also correct.
g. This is also correct.
h. If one of the variables in the common block is character type, all variables must be character type.

9.10 Identify all the errors in the following program segments. Also, specify the correspondence of variables for the correct ones.

a.
```
COMMON R,S,T(50)
   ⋮
CALL SQ(R,S,T)
   ⋮
SUBROUTINE SQ(A,B,C)
COMMON A,B,C(50)
```

b.
```
COMMON X,Y,Z
DATA X,Y,Z /3*0.0/
   ⋮
CALL QUE(A)
   ⋮
SUBROUTINE QUE(B)
COMMON G,H,I
   ⋮
```

c.
```
COMMON A(10), B(10), C(10), D
   ⋮
POT = COOL(X)
   ⋮
FUNCTION COOL(Y)
COMMON E(5),F(15),G(6),H(5)
```

d.
```
COMMON P,Q,R,S,T
   ⋮
CAL = POP(X)
   ⋮
FUNCTION POP(Y)
COMMON A,B,C,D
Y = 0.0
D = A + B + C
A = B + C
B = 0.0
POP = 0.0
RETURN
END
```

e.
```
COMMON FUN1, A,B
   ⋮
X = FUN1
   ⋮
FUNCTION FUN1
COMMON FUN1,X,Y
FUN1 = X + Y
RETURN
END
```

f.
```
      COMMON A,D(5,2)
         :
      CALL TOT(X)
         :
      SUBROUTINE TOT(S)
      COMMON E(11)
         :
```
g.
```
      REAL A,B,C,I,J,K
      COMMON A,B,C,I,J,K
         :
      CALL GEE
         :
      SUBROUTINE GEE
      COMMON X,Y,Z,U,P,Q
```
h.
```
      CHARACTER*20 A,B
      COMMON A,B
      DATA A,B/'FIRST','SECOND'/
         :
      CALL RITE
         :
      SUBROUTINE RITE
      CHARACTER*20 X,Y
      COMMON X,Y
      PRINT*,X,Y
      RETURN
      END
```

□ **Answers**

a. Since the variables R, S, and T are in the COMMON block, there is no need to list them in the argument lists. In fact, the dummy arguments cannot be included in the COMMON statement.

b. The mode of the variable Z in the main program does not match that of I in the subprogram. Also, variables X, Y, and Z cannot be initialized with the DATA statement. Assignment statements must be used for this purpose (or a BLOCK DATA subprogram).

c. This is correct, and the variable correspondences are

```
A(1)...A(5) A(6)...A(10) B(1)...B(10)
  ↕           ↕     ↕       ↕       ↕
E(1)...E(5) F(1)...F(5) F(6)...F(15)

C(1)...C(6) C(7)...C(10)  D
  ↕           ↕     ↕       ↕    ↕
G(1)...G(6) H(1)...H(4)  H(5)
```

d. This is also correct, and the variable correspondences are

```
P    Q    R    S
↕    ↕    ↕    ↕
A    B    C    D
```

e. First, a function's name cannot be included in the COM-

MON block. Second, a function must have at least one argument or a pair of parentheses.

f. This is correct, and the variable correspondences are as follows:

```
A    D(1,1) D(2,1) D(3,1) D(4,1) D(5,1)
↕      ↕      ↕      ↕      ↕      ↕
E(1)  E(2)  E(3)  E(4)  E(5)  E(6)
```

```
D(1,2) D(2,2) D(3,2) D(4,2) D(5,2)
  ↕      ↕      ↕      ↕      ↕
E(7)   E(8)   E(9)  E(10)  E(11)
```

g. This is correct also; the variable correspondences are as follows (note that I, J, K are real type):

```
A   B   C   I   J   K
↕   ↕   ↕   ↕   ↕   ↕
X   Y   Z   U   P   Q
```

h. The variables in a COMMON block cannot be initialized with the DATA statement.

9.11 Suppose the variables A, B, C, D, E, F of a main program have to be accessible to two subprograms. Subprogram 1 uses A, B, C (referred to as X,Y,Z in the subprogram), and subprogram 2 uses D,E,F (referred to as S,T,U in the subprogram). Write the COMMON statements for the main program and the subprograms.

☐ **Answer**

Method I

```
COMMON A,B,C,D,E,F
   ⋮
CALL SUB1
   ⋮
CALL SUB2
   ⋮
SUBROUTINE SUB1
COMMON X,Y,Z
   ⋮
SUBROUTINE SUB2
COMMON P,Q,R,S,T,U
   ⋮
```

Method II

```
COMMON /S1/A,B,C /S2/ D,E,F
   ⋮
CALL SUB1
   ⋮
CALL SUB2
   ⋮
SUBROUTINE SUB1
COMMON /S1/ X,Y,Z
   ⋮
SUBROUTINE SUB2
COMMON /S2/ S,T,U
```

ORDER OF STATEMENTS

It will be helpful at this time to summarize the order of various statements within a program unit. The order is shown in Figure 9.5.

Figure 9.5 Order of statements.

I. Type specification statements

```
INTEGER
REAL
CHARACTER
   :
```

II. Specification statements

```
DIMENSION
COMMON
   :
```

III. Arithmetic function statements

IV. DATA statements
The DATA statement can appear anywhere in a program unit after the specification statements. It is a nonexecutable statement and it is effective only once (when the program begins execution).

V. Executable statements
Assignment

```
DO
IF
CALL
READ
WRITE
   :
```

VI. END statement

Each group of statements must be ordered as shown. However, statements within each group can be ordered as desired. COMMENT and FORMAT statements can appear anywhere within a program unit. A COMMENT statement after the END statement is considered as a part of the next program unit.

EXERCISES

☐ **Subroutines**

9.1 Find all the errors in the following statements:

a. CALL A = CALPAY(X(11), Y(10), YOU)

b. CALL SUBMARINE(BOAT,SHIP)

c. CALL PION, X(100), P,Q, 5

d. CALL REAL MUCH(C, D(2))

e. SUBROUTINE ABCCO(A(1),50)

f. SUBROUTINE TABOO(A + B, Q(100))

g. `SUBROUTINE INTEGER ASK(B)`

h. `SUBROUTINE PAY(A,B)`
`DIMENSION PAY(50), A(50), B(50)`

9.2 Find all the errors in the following general form of the CALL-SUBROUTINE structures:

a.
```
    ⋮
CALL MOY(A, B, C, I, J)
    ⋮
END
SUBROUTINE MOY(X, Y, Z, P, Q)
    ⋮
```

b.
```
    ⋮
CALL BO(XXX, AAA, III, JJJ)
    ⋮
END
SUBROUTINE BO(A, B, I)
```

c.
```
    ⋮
DIMENSION PP(100), QQ(200)
    ⋮
CALL CLEAR(PP, QQ(200))
    ⋮
END
SUBROUTINE CLEAR(X, Y(200))
    ⋮
```

d.
```
    ⋮
DIMENSION A(100), B(100)
    ⋮
CALL PLAY(A, B, S, U)
    ⋮
END
SUBROUTINE PLAY(X, Y, P, Q)
DIMENSION X(N), Y(N)
    ⋮
```

e.
```
    ⋮
CALL AXE(A, B, I, X, K, Y)
    ⋮
END
SUBRPITINE AXE(P, Q, R, S, T, L)
    ⋮
```

f.
```
CALL AVERG(A, B, C, I)
    ⋮
END
SUBROUTINE REAL AVRG(X,Y,Z,K)
```

Library Functions

9.3 Write a statement or program segment to:

a. Calculate $A = \sqrt{X + Y}$.
b. Calculate: AVG = SUM/N (avoid mode mixing).
c. Calculate X_1 and X_2 with the formula
$$\frac{-b \pm \sqrt{b^2 - 4aC}}{2a}$$
d. Calculate $\log_{10} X$.
e. Calculate $\log_e A$.
f. Calculate the integer part of SCORE/10.
g. Calculate the absolute value of $P - Q$.
h. Calculate $\sqrt[4]{X}$ (this can be written as $\sqrt{\sqrt{X}}$).
i. Find the largest of the variables X, Y, Z, P, Q, R, and S.
j. Find the remainder of 95 divided by 13.
k. Find the length of NAME = 'JOHN'.
l. Print the current date.
m. Print the time.

(Make sure to find out the appropriate function for your system for exercises **l.** and **m.**)

Function Statements

9.4 Identify all the errors in the following function statements:

a. FF(A, B(2)) = 5 * A + 6 * B(2)
b. FORMULA(X,Y) = SQRT (X ** 2 − 4 * Y * C)
c. GEE(P, 5) = 5 * P + P ** 5
d. RO(K, G + A) = 2 * K + 5 * (G + A)
e. GAMA(A, B) = B ** 2 + A + X + GAMA (X, Y)

9.5 Identify all the errors in the following general form of the function statements:

a.
```
FUN(A, B, I) = I*A + B
    ⋮
X = F(X, Y, 2)
    ⋮
END
```

b.
```
BETA(F, G, A, B) = F*G + A*B
    ⋮
Y = BETA(X, Y)
    ⋮
END
```

c.
```
F (Z, B) = A*B + A + B + GE (A, B)
GE (X, Y) = 3.0*X**2 + Y
    ⋮
   P = F (X, Y)
    ⋮
END
```

d.

```
PAGE (FIX, Z) = FIX**5 + FIX + Z
   ⋮
EX = PAGE (I, 6)
   ⋮
END
```

e.

```
PA (X, Y) = 2.0*X**3 + Y
   ⋮
X = PA (A, B)

PRINT*, PA
   ⋮
END
```

9.6 What is the value of X in each of the following program segments?

a.

```
F (XX, YY) = XX + YY
   ⋮
 X = F (5.0, 6.0)
   ⋮
END
```

b.

```
G (A, B) = A*B + C
   ⋮
Y = 2.0
C = 5.0
X = G (Y, 3.0)
   ⋮
END
```

c.

```
FUNCT (I, P, Q) = P**I + Q
   ⋮
J = 2
AKE = 3.0
B = 2.0
X = FUNCT (J, B, AKE)
   ⋮
END
```

d.

```
FUN1(A) = A + 3.0 + Y
FUN2(B) = B + FUN1(B)
   ⋮
Y = 2.0
C = 3.0
X = FUN2(C)
   ⋮
END
```

Function Subprograms

9.7 Identify all the errors in the following statements:

a.
```
FUNCTION EX(A, B(2))
```
b.
```
FUNCTION IKE(Y, 5)
```
c.
```
FUNCTION REAL KEY(K)
```
d.
```
FUNCTION (A, B, C)
```
e.
```
FUNCTION DIMENSION A(X)
```
f.
```
FUNCTION AAA(X,Y)
DIMENSION AAA(100), X(100), Y(100)
```
g.
```
FUNCTION XYZ
```

9.8 Identify all the errors in the following function subprograms:

a.
```
         ⋮
      P = POL (A, B, C, I, J)
         ⋮
      END
      FUNCTION POL (X, Y, Z, P, Q)
         ⋮
```
b.
```
         ⋮
      IQUE = MVE(PAL, CAL, JAL)
         ⋮
      END
      FUNCTION IQUE(AM, B, I)
```
c.
```
      REAL ME, I
         ⋮
      PAY = ME(X, Y, Z, I, J)
         ⋮
      END
      FUNCTION ME(A, B, C, D, J)
      REAL J
```
d.
```
         ⋮
      KAY = MAY (I + J, K)
         ⋮
      END
```

```
      FUNCTION MAY (L, M)
      L = L**3
      RETURN
      END
```

e.

```
      DIMENSION AXE(100), BOX(100), COX(100)
         :
      DO 10 J = 1, 100
      BOX (J) = AXE (J, K, COX)
10    CONTINUE
         :
      END
      FUNCTION AXE(M, N, C)
      DIMENSION C(100), AXE(100)
         :
```

f.

```
      REAL MALE
         :
      PALE = MALE (A, B, C)
         :
      FUNCTION REAL MALE (X, Y, Z)
         :
```

g.

```
         :
      X = SOON (A, B, C)
      PRINT*, SOON
         :
      END
      FUNCTION SOON (X, Y, Z)
         :
```

☐ COMMON Statement

9.9 Identify all the errors in the following program segments:

a.

```
COMMON X, Y, I, M, Q(100)
   :
CALL PLAY(SUM)
   :
END
SUBROUTINES PLAY (U)
COMMON A, B, J, X, M(100)
   :
```

b.

```
COMMON BIO(100), COLA
   :
X = COLA (X, Y)
   :
END
FUNCTION COLA(A,B)
COMMON B(100), COLA
   :
```

c.

```
COMMON AA, BB, CC, SUM
    ⋮
CALL PLU(AA, BB, SUM)
    ⋮
END
SUBROUTINE PLU(X, Y, Z)
COMMON X, Y, Z, D
```

d.

```
COMMON /FIRST/ A(100), B, C /SECOND/ X, Y, Z
    ⋮
CALL GAMA
    ⋮
END
SUBROUTINE GAMA
COMMON /FIRST/ M(100), N, L /SECOND/ X(100), D, E
    ⋮
```

e.

```
CHARACTER*12 M, N
COMMON /X/ M, J, K /Y/ A, B, C, N
    ⋮
CALL SORT
    ⋮
END
SUBROUTINE SORT
CHARACTER*12 I, L
COMMON /S/ I, J, K, /Y/ X, Y, Z, L
    ⋮
```

f.

```
COMMON /ONE/ A, B, C, D, E /TWO/ X, Y, Z
    ⋮
CALL POL
    ⋮
END
SUBROUTINE POL
COMMON /ONE/ F, G /TWO/ U, P, Q, R
    ⋮
```

g.

```
COMMON /B1/ X, Y, Z /B2/ MAY, KAY
    ⋮
J = MAY(A, B)
    ⋮
END
FUNCTION MAY(P, Q)
COMMON /B2/ MAY, KAY,Z
    ⋮
```

9.10 Identify the variable correspondences in each of the following program segments.

a.

```
COMMON A, B, C, I, J
    ⋮
```

```
    CALL ROOT
       ⋮
    SUBROUTINE ROOT
    COMMON X, Y, Z, K, L
```

b.
```
    REAL MUCH, LUNCH
    COMMON X, Y, MUCH, LUNCH
       ⋮
    CALL POP(JOY)
       ⋮
    SUBROUTINE POP(JAY)
    REAL LOOK, KOOK
    COMMON A, B, LOOK, KOOK
       ⋮
```

c.
```
    COMMON A(100), B(20), X, Z(100)
       ⋮
    F = FUN(G)
       ⋮
    FUNCTION FUN(BB)
    COMMON X(160), Y(60), YY
       ⋮
```

d.
```
    DIMENSION A(5, 2), B(7, 3), D(10)
    COMMON A, B, D
       ⋮
    CALL RUDE
       ⋮
    SUBROUTINE RUDE
    COMMON X(10), Y(21), Z(5,2)
       ⋮
```

e.
```
    COMMON/A/GEE(10), BEE(10), C /B/ DEE(10), EF(10)
       ⋮
    CARE = DARE(S)
       ⋮
    FUNCTION DARE (X)
    COMMON/A/GEE(6), B(10), CC(5) /B/ QUE(5, 3), TEE(5)
```

9.11 Identify the errors in the following statements:

a.
```
    COMMON A1, B, C, D, E/F, G, H/
```
b.
```
    COMMON /XY/, V(10), B(920)//C, D, E
```
c.
```
    COMMON REAL M, G, L, K /A/ X, Y
```
d.
```
    COMMON /A/ (AA, BB, CC), D, E
```

e.

```
COMMON //U, P(10), G // MU, CO,
```

f.

```
CHARACTER*12 A, B
INTEGER C, D
COMMON /ONE/ A, B, C, D, X
```

9.12 Suppose arrays A(100), B(100), and C(100) of a main program are to be in a COMMON block. Subprogram 1 uses only A, subprogram 2 uses only B, and subprogram 3 uses only C. Write the COMMON statements for the main program and the subprograms.

■ PROGRAMMING EXERCISES

Write a complete program for each of the following cases:

9.13 Write a program which reads the employee name, ID number, hourly rate, and number of hours worked during a week for 100 employees. Calculate the gross pay (time over 40 hours is paid at 1.5 times the regular rate), the deductions, and the net pay for each employee. The deduction is 15% of the gross pay. Use a subroutine for calculating the pay. Print the information.

9.14 Write a subroutine (call it EXIT) which prints THE END OF THE REPORT and stops execution when invoked.

9.15 Write a program which reads N elements of an array ZIP. The array can have up to 50 elements. Calculate the sum of N elements in a subroutine called SSS. Print the array and the sum.

9.16 One-dimensional arrays A,B,C each can contain up to 100 elements. Write a program which has a subroutine to read the first N values of A. Use the same subroutine to read the first M values of B. The integers M and N are to be read from two header records. Calculate an array C = A − B in another subroutine. Assume N is larger than M and the remaining values of B, up to N, are zeroes. Write the values of the arrays with a third subroutine.

9.17 Write a program which reads a salesman's ID number and three sales amounts, for several salesmen in an array. Prepare a sales report as follows:

  a. Calculate the total sales for each salesman in a subroutine.
  b. Find the largest sales amount of each salesman in another subroutine.
  c. Calculate the total sales of all salesmen and the average sales in another subroutine.
  d. Print the individual sales amounts, the total sales, and the largest sales amount for each salesman in a report form.
  e. Print the total sales at the end of the report.
  f. Print each individual sale which is above the average (of all sales).

9.18 Assume 100 integers are stored in an array SET. Each is between 1 and 10. Write a subroutine which counts the number of times each number occurs in the set.

9.19 Write a subroutine which finds the smallest among a set of data stored in an array AMT, which has 200 elements.

9.20 Write a subroutine which finds the largest among a set of data stored in an array BET. Assume BET has a maximum of 500 elements but only N elements ($N < 500$) are read and stored.

9.21 Assume the array ART contains N values ($N < 100$). Write a subroutine which inserts the value A into the Kth element of ART while maintaining the existing data in the same order.

9.22 Write a subroutine which counts the number of commas in a sentence. The sentence is less than 80 characters long and is stored in an array A(80).

9.23 Write a subroutine (call it CHANGE) which interchanges the values of two variables: X and Y.

9.24 Write a program which reads an integer M and prints all the divisors of M. (Hint: use a library function to find whether or not the remainder of M when divided by 1, 2, 3, ..., M is equal to zero.)

9.25 A prime number is a number which does not have any divisor except 1 and itself. For example, 1, 3, 5, 7, 19, 39 are prime numbers, but 21 is not (because it is divisible by 7 and 3). To find out if a number N is a prime number, we must check to see whether it is divisible by 1, 2, 3, ... up to $\sqrt{N}$. Write a program which reads a number N and prints a message saying whether or not it is a prime number. Use a subroutine for the checking.

9.26 Suppose X, Y, Z can be calculated by the following formulas:

$X = \log_{10} A + \log_e A$
$Y = 3A^2 + 4A - 3$
$Z = Ae^{3A+2}$

Write a program which reads A and prints the values of X, Y, Z. Use a FUNCTION statement for each of the formulas.

9.27 The area of a triangle can be calculated by:

Area $= \sqrt{s(s-a)(s-b)(s-c)}$

where

$s = \frac{1}{2}(a + b + c)$

and a, b, c are the sides. Write a program which reads the sides of a triangle and calculates the area. Use two FUNCTION statements.

9.28 Write a program to calculate the kinetic energy T:

$T = \frac{1}{2}MV^2$

where M (the mass) and V (the velocity) are to be read as input. Use a FUNCTION statement.

9.29 Write a function subprogram (call it FACT) which calculates the factorial of a given number N. (Factorial N is $N! = 1 \times 2 \times 3 \times 4 \times 5 \times 6 \times \cdots \times N$.)

9.30 The number of combinations (C) of N items taken M at a time can be calculated by

$$C = \frac{N!}{M!(N-M)!}.$$

Write a program which reads M and N, and calculates and prints C. Use a function to calculate the factorial.

9.31 Write a program which reads a positive number N and calculates

$$X = \frac{1}{1^2} + \frac{3}{2^2} + \frac{5}{3^2} + \cdots + \frac{2N-1}{N^2}$$

in a function called EX.

9.32 Write a program which reads a real number X and an integer N, and then calculates

$$\text{EXPOX} = 1 + X + \frac{X^2}{2!} + \frac{X^3}{3!} + \cdots + \frac{X^N}{N!}$$

in a function. Print X, N, and EXPOX.

9.33 Write a program which reads a real number X, and an integer N. Calculate:

$$\text{SINX} = X - \frac{X^3}{3!} + \frac{X^5}{5!} - \frac{X^7}{7!} + \cdots \pm \begin{cases} \dfrac{X^N}{N!}, & N \text{ odd,} \\ \dfrac{X^{N-1}}{(N-1)!}, & N \text{ even.} \end{cases}$$

in a function. Print X, N, and SINX.

9.34 Design and write a program for printing a weekly salary report as follows:

Input: Name, ID number, Hours worked during the week, Department code (1 to 5), Wage code (1 to 3) for each employee.

Processing:

1. Calculate the pay. Time over 40 hours is paid at one and a half times the regular rate, and the wage-rate table is as follows:

Department Code	Wage Code 1	Wage Code 2	Wage Code 3
1	5.85	6.45	5.95
2	6.55	7.30	6.90
3	7.60	8.70	8.05
4	8.90	9.95	9.10
5	10.50	11.50	10.50

2. Calculate the deductions. The deduction is calculated as follows:

Pay between   0   and $100: 2%
Pay between $100   and $200: 5%
Pay over $200:   9%

Write the input information as well as the gross pay, deduction, and net pay for each employee in report form. Print also the total pay, total deductions, and total net pay at the end of the report. Use arrays for storing the information. There are less than 100 employees, and the first data record is a header record.

Note: You must break down the problem into modules such as:

· Reading the data
· Calculating the pay
· Calculating the deductions
· Writing the information

Then use one or more subprograms for each module.

9.35 Consumer Electric Company would like to have a report about the customers' power use for a given period. The requirements and information are as follows:

Input:

1. Customer name	12 characters
2. Customer account number	9 digits
3. Date of the last bill (as 12/10/79)	8 characters
4. Last-month balance	7 columns
5. Payment	7 columns
6. Last meter reading (in kwh)	7 columns
7. Current meter reading (in kwh)	7 columns
8. Code for industrial, residential, or farm: I, R, F	1 character

Process requirements:

1. The usage is calculated as

   U = current meter reading − last meter reading.

2. Residential customers are charged a flat fee of $6.50 plus a rate based on the following table:

    First 250 kwh at $.0589 per kwh
    More than 250 to 400 kwh at $.0659 per kwh
    More than 400 kwh at $.0789 per kwh

3. Industrial customers are charged a flat fee of $19.50 plus the following rate:

    First 500 kwh at $.0618 per kwh
    More than 400 kwh at $.0700 per kwh

4. The charge to farm customers is calculated by the formula

$$\text{AMOUNT} = U \times .0685 + \frac{U + 55}{1000} + \frac{U + 25}{U}$$

   where U is the amount used.

5. The amount due is calculated as:

   Current charges + last month's balance − payment.

Write a complete program which reads the input information and calculates:

1. the amount used by each customer,
2. the charge to each residential user,
3. the charge to each industrial user,
4. the total charge to each farm user,
5. the amount due from each user,
6. the total power used and the total charge to all users.

Print the input information and the charges in report form. There are less than 500 customers. The first data line is a header record showing the number of customers. Also print the date at the top of the report. Again you must break down the problem into several modules and submodules, and use one or more subprograms for each module.

SUMMARY OF CHAPTER 9

You have learned:

1. Subprograms allow us to assign a name to one or more instructions and to use the name whenever we wish to execute these instructions. The use of subprograms facilitates the modular approach in writing programs, and also avoids the necessity of repeating a sequence of instructions several times in a program. Subprograms in FORTRAN are of two types:

   a. Subroutines
   b. Functions

2. A subroutine is a complete and independent program unit which can be executed when its name is invoked in a CALL statement. Communication between the variables in the referencing program and the subroutine is through its argument list or through a COMMON statement. The RETURN statement of a subroutine transfers control back to the referencing program.

3. A function is a subprogram which can be invoked by the appearance of its name in a statement. There are three kinds of functions:

   a. A *library function* is defined by the system and is available to all users. The following statements are examples of using library functions:

   ```
   Q = SQRT(ABS(A - B))
   PRINT*, AMIN1(X, Y, Z)
   IY = NINT(Y)
   ```

   b. A *function statement* is a single statement defined by the programmer at the beginning of a program unit in order to evaluate an expression. It is similar to an arithmetic or character assignment statement.

   c. A *function subprogram* is a complete and independent program unit which can be invoked by the appearance of its name in a statement. The difference between a function and a subroutine is: (1) a function is invoked by the appearance of its name in a statement, but a subroutine is invoked by a CALL statement, and (2) a value is assigned to the function's symbolic name, whereas no value is associated with the subroutine's symbolic name.

4. COMMON blocks allow several program units to share the same memory space. The data can be known by one symbolic name in one program unit, and by a different name in another program unit. Furthermore, the group of variables in a COMMON statement can be labeled or unlabeled. The following are some examples of COMMON statements:

   ```
   COMMON X, Y, Z, D

   COMMON /FIRST/ Q,R,X /SECOND/ V,W,P,A,B

   COMMON A,B,C,P,Y /ONE/ EEE,NET,PAY,X /TWO/ M,N,RHO,AVG
   ```

5. The order of different kinds of statements in a program must be as follows:

   a. Type specification statements: CHARACTER, REAL, INTEGER, ...
   b. Specification statements: DIMENSION, COMMON, ...
   c. Arithmetic function statements.
   d. Executable statements.

SELF-TEST REVIEW

9.1 Which of the following statements *are* syntactically correct?
  a. CALL X = Y(A,B)
  b. CALL CUTE P,G,Q
  c. CALL ALOY(X,I,Y,J), JOY(A,B,N)
  d. CALL PAN(X(10),Y(20),KZ(I))
  e. CALL REGULATOR(A,B(1),10)
  f. CALL PAIL
  g. SUBROUTINE SONY(X,Y(2),I,J)
  h. SUBROUTINE TGER(POL,QUE,I)
  i. SUBROUTINE PAGE(X(N),A,B)

9.2 Which of the following pair of CALL-SUBROUTINE statements *is* correct?
  a.
```
CALL POP(X,I,Y(2), J)
    ⋮
SUBROUTINE POP(A,L,B,K)
```
  b.
```
CALL MUCH(SAY,PAY,JAY)
    ⋮
SUBROUTIN MUCH(NAY,KAY,RAY)
```
  c.
```
CALL PUNCH(A,B,C,I,J,K)
    ⋮
SUBROUTINE PUNCH(X,Y,L,M)
```

9.3 Write a statement, using a library function, to:
  a. calculate AVG = SUM/N
  b. find the smallest of three scores: T1, T2, T3
  c. find the remainder of 39 divided by 5
  d. find the nearest integer of 98.7

9.4 The area of a triangle can be calculated by

$$\text{Area} = \sqrt{s(s-a)(s-b)(s-c)}$$

where

$$s = \tfrac{1}{2}(a + b + c)$$

and a, b, c are the sides.

Write the program segment with two arithmetic function statements which calculates the area of a triangle.

9.5 Which of the following statements are correct:
  a. FUNCTION EXPLOREE(X,Y,Z,I)
  b. FUNCTION REAL NET(A,B,C)
  c. FUNCTION POT(Z(1),5,X)
  d. INTEGER FUNCTION PLAY(X,I,Y,K)
  e. REAL FUNCTION(A,B,C,I,J)

9.6 Which of the following function pairs is correct:

a.

```
IQ = MM(A,B,8,J,Y)
    ⋮
FUNCTION MM(X,Y,Z,I,A)
INTEGER Z
    ⋮
```

b.

```
B = BATTLE(NAIL,PAIL,MAIL(10))
    ⋮
FUNCTION BATTLE(M,X,Y)
    ⋮
```

c.

```
POP = (X + Y(A,B))/Y(2)
    ⋮
FUNCTION Y(D,E)
    ⋮
```

d.

```
M = MARY(A(10),J(I),Q)
    ⋮
FUNCTION MARY(A,J,P)
    ⋮
```

9.7 Specify the variables correspondence between the COMMON statements in each of the following pairs:

a.

```
COMMON MUNCHO,PURE,KORE(10)
    ⋮
CALL SOUP
    ⋮
SUBROUTINE SOUP
COMMON I,X,IY(10)
    ⋮
```

b.

```
COMMON X(5,2),Y(3,2),Z(4),A
    ⋮
ZIP = FUNZIP(T)
    ⋮
FUNCTION FUNZIP(U)
COMMON PAT(16),ZAP(5)
```

9.8 Write a program which prints a customer-order report for a department store. Assume there are less than 100 customers, and each customer orders only one item. The first record shows the number of customers with orders. Use arrays for storing the information. The information and requirements are as follows:

Input: The customer's name (12 characters), the quantity ordered (less than 99), the unit price (less than $99.99), and a code for in- and out-of-state orders (I for in-state, O for out-of-state).

Processing:

1. The amount ordered (price times the quantity); use a subroutine.
2. The tax (4% for the in-state customers only); use a function.
3. The total amount for each customer.

Output: The input information, amounts, tax, and total for each customer, in report form. Also print the totals at the end of the report.

☐ Answers

9.1 d., f., h.

9.2 a.

9.3
  a. AVG = SUM/FLOAT(N)
  b. S = AMIN1(T1,T2,T3)
  c. IR = MOD(39,5)
  d. N = NINT(98.7)

9.4

```
S(A,B,C) = (A + B + C)/2
AREA(A,B,C) = SQRT(S*(S(A,B,C) - A)*
    *(S(A,B,C) - B)*(S(A,B,C) - C)
           ⋮
    X = AREA(X,Y,Z)
```

9.5 Only **d.**

9.6 Only **d.**

9.7 a.

```
MUNCHO   PURE   KORE(1)   KORE(2)  ...  KORE(10)
   ↕       ↕       ↕         ↕             ↕
   I       X      IY(1)     IY(2)   ...  IY(10)
```

b.

```
X(1,1)  X(2,1)  X(3,1)  X(4,1)  X(5,1)
   ↕       ↕       ↕       ↕       ↕
PAT(1)  PAT(2)  PAT(3)  PAT(4)  PAT(5)

X(1,2)  X(2,2)  X(3,2)  X(4,2)  X(5,2)
   ↕       ↕       ↕       ↕       ↕
PAT(6)  PAT(7)  PAT(8)  PAT(9)  PAT(10)

Y(1,1)  Y(2,1)  Y(3,1)  Y(1,2)  Y(2,2)  Y(3,2)
   ↕       ↕       ↕       ↕       ↕       ↕
PAT(11) PAT(12) PAT(13) PAT(14) PAT(15) PAT(16)

Z(1)    Z(2)    Z(3)    Z(4)    A
   ↕       ↕       ↕       ↕       ↕
ZAP(1)  ZAP(2)  ZAP(3)  ZAP(4)  ZAP(5)
```

9.8 The following program lacks internal documentation:

```
      CHARACTER NAME(100)*12, INOUT(100)*1
      COMMON QUANT(100), PRICE(100), AMOUNT(100),
     *TAX(100), ACHRGE(100)
C
      READ(5,10)N
      DO 100 I = 1,N
           READ(5,20) NAME(I), QUANT(I), PRICE(I),
     *     INOUT(I)
      CONTINUE
```

```
C
      CALL CHARGE(N, INOUT, TOTAMT, TOTTAX, TOTAL)
C
C WRITING THE HEADING
      WRITE(6,30)
C     DO 110 J = 1,N
          WRITE(6,40)NAME (J),QUANT(J),PRICE(J),
     *           INOUT(J),AMOUNT(J),TAX(J),
     *           TCHRGE(J)
110   CONTINUE
C
      WRITE(6,50) TOTAMT, TOTTAX, TOTAL
C
10    FORMAT (I2)
20    FORMAT (A12,1X,F2.0,1X,F5.2,1X,A1)
30    FORMAT (9X, 'NAME', 9X, 'QUANTITY', 3X,
     *'PRICE', 5X, 'STATE', 3X, 'AMOUNT', 5X,
     *'TAX', 5X, 'TOTAL CHARGES')
40    FORMAT (6X, A12, 6X, F3.0, 6X, F6.2, 6X, A1,
     *4X, F8.2, 3X, F5.2, 8X, F8.2)
50    FORMAT (36X, 'TOTALS', 6X, F10.2, 2X, F6.2,
     *2X, F8.2)
      END
C ****************************************************
      SUBROUTINE CHARGE (M,INO, TAMT, TTAX, TOT)
      CHARACTER*1 INO(100)
      COMMON QU(100), PR(100), AMT(100), TX(100),
     *TCHARG (100)
      TAMT = 0.0
      TTAX = 0.0
      TOT = 0.0
      DO 100 I = 1,M
          AMT(I) = QU(I)*PR(I)
          TX(I) = TAXFUN(INO(I), AMT(I))
          TCHARG (I) = AMT(I) + TX(I)
          TAMT = TAMT + AMT(I)
          TTAX = TTAX + TX(I)
          TOT = TOT + TCHARG (I)
      CONTINUE
      RETURN
      END
C ****************************************************
      FUNCTION TAXFUN(KODE,A)
      CHARACTER*1 KODE
      IF (KODE .EQ. 'I') THEN
                    TAXFUN = .04*A
      ELSE
                    TAXFUN = 0.0
      END IF
      RETURN
      END
```

Chapter 10

Problem Solving and Programming Techniques

INTRODUCTION

    Algorithms
    Program Efficiency

DATA PROCESSING

    Sorting
    Searching
      Binary Search
    Merging

STATISTICS

    Median
    Summation Notation
    Regression Analysis

MATHEMATICS

    Binary Numbers
    Finding Roots of an Equation—Bisection Method
    Simultaneous Linear Equations

GRAPHING TECHNIQUES

    Graphing
    Plotting a Histogram

SIMULATION

    Random Numbers
    Generating Random Numbers
    Monte Carlo Simulation

MANAGEMENT INFORMATION AND DECISION SUPPORT SYSTEMS

INTRODUCTION

☐ Algorithms

As discussed in Chapter 3, a computer program is merely a tool for solving problems. Before coding, it is necessary to make a thorough analysis of the problem and develop an appropriate solution procedure—an algorithm. In the previous chapters, most of the solution procedures have been rather straightforward. However, this is not always the case. Developing an algorithm is itself quite an art. Throughout this chapter you will be seeing examples of algorithms which are not so simple.

We define an algorithm as a detailed solution procedure suitable for computer programming. A well-developed algorithm must be:

1. sufficiently detailed to allow programming,
2. concise,
3. unambiguous,
4. flexible,
5. effective in solving the specified problem,
6. efficient in terms of resource usage.

These characteristics are illustrated by the following example. A possible algorithm for adding a series of data is as follows:

1. Start, identify a variable, SUM, which will accumulate the answer, set SUM equal to zero.
2. Read the number of data, N.
3. Repeat the following steps, N times:
   3.1 Read an item of data.
   3.2 Cumulate the data in SUM.
   3.3 Print the data item.
   3.4 End of the loop.
4. Print the cumulated data (the sum).

This algorithm is detailed enough to be coded as a program. It is clear enough to be unambiguous. It is flexible in that it can handle a variety of input data rather than a specific set of data. It is effective in that it provides an answer to the problem in a small number of steps. The efficiency of the algorithm, however, depends on the total resources used for developing and implementing the algorithm.

☐ Program Efficiency

A problem can be solved with various algorithms. Obviously an efficient algorithm will result in an efficient program if the algorithm is implemented correctly. But the efficiency of a program in turn depends upon:

1. the amount of computer time required to compile and execute the program,
2. the amount of storage required to run the program,
3. the human effort in writing, documenting, and maintaining the program,
4. the desired accuracy of solution,
5. the required generality of the algorithm (e.g., must negative numbers be ignored?).

Unfortunately, there is always a tradeoff among these factors. For example, there is a tradeoff between time and memory requirements. One program may require less running time but more memory space than another. A program that requires less storage area and less running time may require greater human effort to develop, test, and maintain than another.

Sometimes incorporating certain mathematical techniques in an algorithm results in a more efficient program. For example, consider the function:

$$Y = 3X^2 + 2X + 5$$

If this function is written as

```
Y = 3*X*X + 2*X + 5
```

it requires three multiplications and two additions. However, if the same function is written as

```
Y = (3*X + 2)*X + 5
```

it requires only two multiplications and two additions. (This particular kind of reduction in mathematical steps is called Horner's method.) This change not only causes a more efficient operation but may also increase accuracy. Thus, it is sometimes possible to apply standard techniques to a given algorithm to get a more efficient process, improved speed, and greater accuracy.

Good programming style directly affects the efficiency of a program. The following are some recommendations:

Language Dependence

Consider the features of the language being used. For example, in FORTRAN, mixed modes of operation reduce the efficiency of a program and should be avoided. Thus, you can write

```
AVG = SUM/2.0
```

instead of

```
AVG = SUM/2
```

as discussed in Chapter 2. Similarly, in FORTRAN, using a DATA statement to set the initial values for variables is more efficient than using several assignment statements.

Arrays Use

If possible, it is best not to use arrays in a program. Arrays require a large amount of memory space and extra computer time. For example, if the objective is only to calculate the sum of 1000 data, the program segment

```
        SUM = 0.0
        DO 10 I = 1,1000
            READ*,X
            SUM = SUM + X
 10     CONTINUE
```

is more efficient than

```
        DIMENSION A(1000)
        SUM = 0.0
        DO 10 I = 1,1000
            READ*,A(I)
            SUM = SUM + A(I)
 10     CONTINUE
```

If you find it necessary to use an array, keep its size as small as possible.

Algorithm Design

Choose your algorithm with program efficiency in mind. For example, suppose we want to interchange the values of memories A and B. Of course, the program segment

```
A = B
B = A
```

will not work (because by placing the value of B in A, the original content of A will be erased). A possible method is to copy A and B into temporary memories C and D, and then transfer them back into the original memories A and B in reverse order. Thus the program segment would be

```
C = A
D = B
A = D
B = C
```

However, this method requires two additional memory locations and four assignments. Another method is simply to save the content of A into a temporary memory TEMP, before assigning B to A, and then assign the value of TEMP to B:

```
TEMP = A
A = B
B = TEMP
```

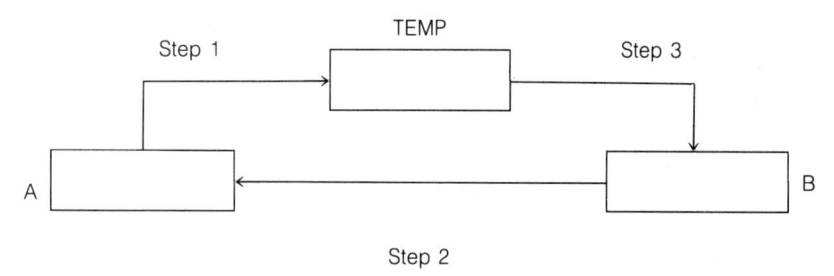

This method requires only one additional memory location and three assignments. The value of this method becomes more apparent when we try to do the same kind of interchanging in the following example with an array which has 1000 elements. The first method requires 1000 additional locations, but the second method requires only one.

EXAMPLE 10.1

■ **Problem:** Assume data are stored in arrays A and B with 1000 elements each. Develop a procedure and write the program segment which interchanges the values of the elements of these two arrays.

Solution Plan:

1. Store the content of the first element of A in a temporary memory TEMP.
2. Assign the first element of B to the first element of A.
3. Assign TEMP to the first element of B.
4. Repeat steps 1 through 3 for the second, third, . . . , thousandth elements of the arrays.

Program Segment:
```
      DO 10 I = 1,1000
         TEMP = A(I)
         A(I) = B(I)
         B(I) = TEMP
10    CONTINUE
```

Techniques for searching for an element in an array provide another example of how an algorithm affects the efficiency of a program. There are several methods which can be used to find a given value in a list of values. One is to search the list from beginning to end, one by one, until the desired value is found. This is called the straight search method. Another method, called binary search, is to check the item in the middle of the list. If the item is not the one sought, find out whether it is in the first or second half of the list, and check the middle of the appropriate half. Repeating this process until the item is found results in a shorter search time. Of course, the binary search works only if the list is already ordered. To compare the efficiency of these two methods, try to apply them to finding someone's name in a telephone directory. Obviously, finding a name in a directory by straight search is cumbersome, especially when the directory is a large one. Search methods will be discussed later in this chapter.

Mathematical Techniques

Use mathematical techniques which lead to efficient operation. For example, using Horner's rule to evaluate a polynomial minimizes the number of multiplications. Here is another example of this technique. Consider the equation

$$Y = 5X^3 - 6X^2 + 10X - 3.$$

If it is coded as

```
Y = 5*X*X*X - 6*X*X + 10*X - 3
```

it requires six multiplications and three additions. But if it is coded as

```
Y = (5*X*X - 6*X + 10)*X - 3
```

it requires four multiplications and three additions. Furthermore, if it is coded as

```
Y = ((5*X - 6)*X + 10)*X - 3
```

it requires only three multiplications and three additions. The following simple example also demonstrates how mathematical procedures are important in developing an algorithm.

EXAMPLE 10.2

Problem: Develop a solution procedure and write a program which determines whether or not a given integer greater than one is a prime number.

Solution Procedure: A prime number is a number which does not have any positive integer divisor except one and itself. For example, 17 is a prime number because none of the numbers 2, 3, 4, ...,

15, 16 are its divisor. It can be proven mathematically that N is prime if none of the numbers from 2 to the closest integer number to $\sqrt{N}$ divides N exactly. Thus, if any remainder of N divided by 2, 3, 4, ..., $\sqrt{N}$ is zero, the number is not prime.

Program Plan:

1. Read the number N.
2. Calculate the integer of the square root of N (call it M).
3. Generate sequence numbers 2, 3, 4, ..., M
4. Divide N by each of the generated sequence numbers.
5. Find the remainder.
6. If any of the remainders is equal to zero, then the number is not prime; otherwise, the number is prime.

Program:

```
      PROGRAM PRIME
      READ*,N
C
      M = INT(SQRT(N))
      DO 10 I = 2,M
C
         L = MOD(N,I)
         IF(L .EQ. 0)THEN
            PRINT*,'THE NUMBER IS NOT PRIME'
            STOP
         ENDIF
10    CONTINUE
      PRINT*,'THE NUMBER IS PRIME'
      END
```

Notes:

1. The efficiency of the program also depends on the method used to compute the square roots.
2. The efficiency of the algorithm could be improved substantially if divisibility by 2 were checked separately and then only odd numbers were considered in the loop (with DO 10 I = 3,M,2).

Human Factors

Human factors are an extremely important consideration in program efficiency because the computer's time has become far less expensive than people's time, given the cost of developing, debugging, testing, documenting, and maintaining a program. The doctrine of structured programming emphasizes this point. Using structured programming techniques makes a program easier to understand, debug, maintain, and update. Structured programming techniques are discussed further in Chapter 12.

We defined an algorithm as a detailed solution procedure suitable for computer programming. This term, however, also has a much deeper and more precise mathematical definition. We avoid discussion of the theory of algorithms, and in this chapter use the phrase *program plan*, or simply *plan*, interchangeably with *algorithm*.

Further examples of algorithms and programs for specific applications in data processing, statistics, mathematics, graphing, and simulation are presented in the following sections. The examples are deliberately made simple. Understanding them will help you in developing your own algorithms for more complicated problems. Of course, knowledge of the subject matter of an application is necessary in order to understand the algorithm for that application. Explaining the details of each subject area is beyond the scope of this text. Therefore, if you are not already familiar with the subject matter of an example, skip that example.

DATA PROCESSING

One of the differences between data processing and scientific calculations is that in data processing several files are normally involved. For example, a payroll system may involve a master file (containing permanent information such as names and Social Security numbers), a transaction file (containing temporary information such as the number of hours worked by employees), and other files for wage tables and tax tables. Often, these files must be sorted, updated, merged, and processed. In this section, simple examples of certain techniques for sorting, updating, and merging files are presented.

☐ Sorting

Sorting is the process of arranging unordered items (or records) in a list in a predetermined order. The main advantage of a sorted list is rapid retrieval of a particular item. For example, finding the name of a person in an alphabetized telephone directory is much faster and more efficient than finding it in an unsorted directory.

If records of a file contain more than one item, we can sort the file based on only one of the items. However, for demonstrating the process, in this section we consider records which consist of only one item.

There are several techniques of sorting. One of the most common and straightforward method is the *bubble sort*. The basic idea behind it is to go through the list on the first round and compare each pair of contiguous elements, one pair at a time, and interchange the two values if the second value belongs before the first one. (See Figure 10.1.) Thus, if items are to be sorted in ascending order of size, the first-round comparison forces the largest item in the list into the end of the list. This can be accomplished by a loop such as

```
        DO 20 K = 1, N - 1
            if A(K) > A(K + 1), interchange A(K) and A(K + 1)
 20     CONTINUE
```

[$N - 1$ is used because $A(K + 1)$ will take the last element into account.] Then the comparison must be repeated for the second round. However, there is no need to go to the last item in the list for the second round, because the largest number is already there. This process is repeated for the third round, fourth round, ..., and $(N - 1)$th round (where N is the number of items in the list). The following example illustrates the procedure and the program.

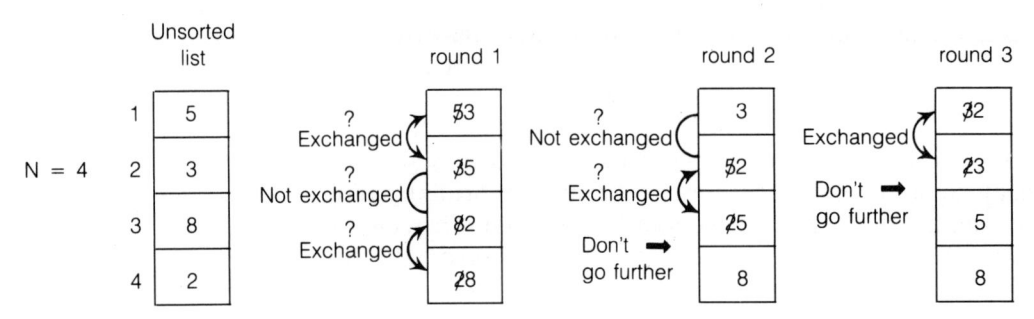

420 Problem Solving and Programming Techniques

Figure 10.1 Bubble-sort procedure.

EXAMPLE 10.3

■ **Problem:** Develop the bubble-sort plan and write a program which sorts the elements of an array A in ascending order.

Procedure Plan:

1. Read the number of data, N.
2. Read the data into the array A.
3. Print the unsorted data.
4. For $I = 1$ to $N - 1$, repeat the following steps, up to step 5 ($N - 1$ times):
   4.1 Find the last necessary item to be searched: LAST = $N - I$.
   4.2 For $K = 1$ up to $K = $ LAST repeat:
       4.2.1 Compare $A(K)$ and $A(K + 1)$; if $A(K) > A(K + 1)$, then interchange the values of $A(K)$ and $A(K + 1)$.
   4.3 Next K.
5. End of the loop.
6. Print the sorted data.

Program:

```
      PROGRAM SORT1
      DIMENSION A(100)
      READ*, N
      READ*, (A(I), I = 1,N)
      PRINT*, 'UNSORTED LIST: ', (A(I), I = I, N)
C
      DO 20 I = 1, N - 1
         LAST = N - I
         DO 20 K = 1, LAST
            IF (A(K) .GT. A(K + 1)) CALL
     *      CHANGE (A(K), A(K + 1))
 20   CONTINUE
      PRINT*,'SORTED LIST: ', (A(I), I = 1,N)
      END
      SUBROUTINE CHANGE(X,Y)
      TEMP = X
      X = Y
      Y = TEMP
      RETURN
      END
```

The program in the above example sorts the list in increasing order. To sort a list in a decreasing order, the IF statement may be changed to

```
IF (A(K) .LT. A(K + 1)) CALL CHANGE (A(K), A(K + 1))
```

It is obvious how numbers can be compared with each other to determine the ordering. But how about character data? As discussed in Chapter 6, character data are represented in the computer by numerical quantities in the sequence of the alphabetical order. Thus, for sorting an alphabetical list, the same procedure as the numerical list can be used. The following program is an example:

EXAMPLE 10.4

Problem: Write a program which reads 100 names, sorts the names, and prints the sorted list.

Program:
```
      PROGRAM SORT2
      CHARACTER*20 NAME(100)
      READ*, (NAME(I), I = 1, 100)
      DO 10 I = 1, 99
          LAST = 100 - I
          DO 10 K = 1, LAST
              IF(NAME(K) .GT. NAME(K + 1))CALL
     *        CHANGE(NAME(K),NAME(K + 1))
10    CONTINUE
      PRINT*, 'SORTED NAMES: ', (NAME(I), I = 1,
     *100)
      END
      SUBROUTINE CHANGE (X,Y)
      CHARACTER*20 X,Y,TEMP
          TEMP = X
          X = Y
          Y = TEMP
      RETURN
      END
```

The bubble sort procedure is easy to understand, but generally it is not efficient, although many improvements are possible. For example, one of the disadvantages of the program in Example 10.3 or 10.4 is that the computer must go through the list $N - 1$ rounds and check the contiguous pairs in each round even if the list is already sorted or partially sorted. The efficiency of the procedure can be improved by placing a statement to terminate the loop when no interchange takes place in a particular round.

Many other sorting procedures have been proposed. The following are two examples:

1. Start by sorting only two items. Then place the third item in the proper position, then the fourth item, and so on.
2. Start by sorting the first pair, then the second pair, then the third pair, and so on. After sorting the pairs, merge pairs of pairs to form sorted quadruplets (sequences of four). Then merge pairs of quadruplets to form sorted sequences of eight. This process continues until the entire

list is sorted. This method is called the merge-sort method and is especially efficient for large numbers of data.

☐ Searching

Searching is necessary for locating and retrieving information. It is also important for deleting or updating an item. (Of course, if the location (or the address) of the desired information is known, searching is not necessary and the information can be retrieved easily.) The following example demonstrates the simple *sequential search method*.

EXAMPLE 10.5

■ **Problem:** Assume 500 ID numbers are stored in the array ID. Write a program segment which reads an ID number and searches the array for it. Print the location of the number. If there is no ID number equal to that number, print a message.

Program Segment:

```
             ⋮
      READ*,NO
      DO 10 I = 1,500
          IF (ID(I) .EQ. NO) THEN
              PRINT*,'LOCATION',I
              GO TO 30
          END IF
10    CONTINUE
      PRINT*,'NOT FOUND'
30    ...
```

Note: The items in the array are compared with NO one by one (Figure 10.2a); this requires 500 comparisons if the item is in the last location, and only one comparison if the item is in the first location (an average of 250 comparisons).

Binary Search

A more efficient searching method is called *binary search*. It is applicable to *a list which is already sorted*. The method starts by examining the item in the middle (or near the middle) of the list. If the item is not the one sought, then it is determined whether the item is in the upper half of the list or in the lower half of the list. The item in the middle of the appropriate half is examined next. If it is not the one sought, it is determined whether the item is in the upper quarter or the lower quarter, and the search continues in the appropriate quarter. This process is repeated until the desired item is located (Figure 10.2b). The following example illustrates this method further.

EXAMPLE 10.6

■ **Problem:** Show the programming plan and write a program segment which reads an item (NO) and finds the location of the item in an array ID. Assume the information is stored in N elements of ID.

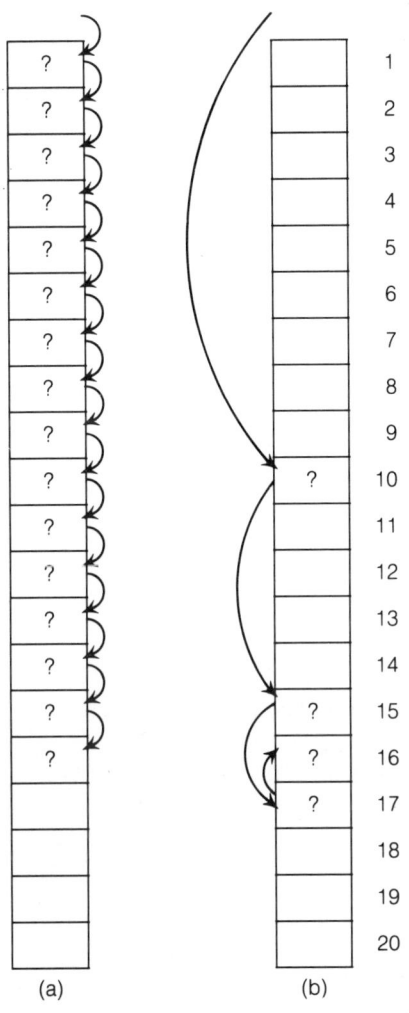

Figure 10.2 Searching for an item: (a) sequential search, (b) binary search.

Plan:

1. Make sure the array ID is sorted in ascending order.
2. Read the item, NO.
3. Set the first item and the last item in the list to be searched as:

   FIRST = 1
   LAST = N

4. Find K, the middle of the list (the location of the item to be examined) as

   K = integer of ((FIRST + LAST)/2)

5. If ID(K) = NO, then Location = K, terminate the search. Otherwise:

   5.1 If ID(K) > NO, then reset LAST as
       LAST = K
       (consider K as the last item of the list to be searched).

   otherwise, reset FIRST as
   FIRST = K
   (consider K as the first item in the list to be searched).

6. Go to step 4.

Program Segment:

```
         INTEGER FIRST
            ⋮
C        ARRAY ID MUST BE SORTED IN ASCENDING ORDER
            ⋮
         READ*,NO
         FIRST = 1
         LAST = N
10       K = INT((FIRST + LAST)/2)
         IF (ID(K) .EQ. NO) THEN
              LOC = K
              GOTO 50
         ELSE IF (ID(K) .GT. NO) THEN
                    LAST = K
         ELSE
                    FIRST = K
         END IF
         GOTO 10
50       ...
```

Note: The program has several flaws and can be improved as follows:

1. The algorithm and program should be modified to print a message if NO is not in the list. Otherwise, it creates an infinite loop.
2. The program also should be modified for the cases when N > 1 and ID(N) = NO (i.e., NO is the last item in the list).

☐ Merging

Merging is the process of combining two sorted lists into a single sorted list. We could simply combine the two lists and then sort them. This process, however, is not efficient. The following example demonstrates a more efficient method for merging two arrays.

EXAMPLE 10.7

■ **Problem:** Develop a solution procedure and a program plan which merges arrays A and B, with NA and NB elements, into array C. Assume arrays A and B are already sorted in increasing order.

Solution Procedure: Compare the first elements of A and B, and move the smaller of the two into C. Continue the comparison and moving for the first of the remaining elements of A and B until all the elements of one array are exhausted; then move the remainder of the other to C.

Plan:

1. Set the size of C to NC given by

   NC = NA + NB

2. Set INA = 1   (INA will represent the subscript of A.)
3.   INB = 1   (INB will represent the subscript of B.)
4.   INC = 1   (INC will represent the subscript of C.)

> 5. If A(INA) < B(INB), then
>   5.1 Do the following:
>   C(INC) = A(INA)
>   INC = INC + 1   (increasing the subscript of C)
>   INA = INA + 1   (increasing the subscript of A)
>   5.2 If INA > NA, then move the remaining elements of B to C, and go to step 6; else go to step 5.
>
>   Otherwise do the following:
>   5.3 C(INC) = B(INB)
>   INC = INC + 1
>   INB = INB + 1
>   5.4 If INB > NB, move the remaining elements of A to C, and go to step 6; else go to step 5.
> 6. Continue the program.
>
> **Program Segment:**
>
> ```
>             ⋮
>          NC = NA + NB
>          INA = 1
>          INB = 1
>          INC = 1
> 50       IF (A(INA) .LT. B(INB)) THEN
>             C(INC) = A(INA)
>             INC = INC + 1
>             INA = INA + 1
>             IF (INA .GT. NA) GO TO 60
>          ELSE
>             C(INC) = B(INB)
>             INC = INC + 1
>             INB = INB + 1
>             IF (INB .GT. NB) GO TO 70
>          END IF
>          GO TO 50
> C
> 60       DO 10 L = INC,NC
>             C(L) = B(INB)
> 10       INB = INB + 1
>          GO TO 80
> 70       DO 20 L = INC, NC
>             C(L) = A(INA)
> 20       INA = INA + 1
> 80       ...
> ```

■
STATISTICS

Statistics is concerned with collecting, summarizing, and analyzing data, and making inferences. Statistical analysis involves large amounts of data, and is well suited for computer applications. Calculations of the mean, variance, and standard deviation (discussed in the previous chapters) are examples of the applications of programs in statistics. Finding the median and estimating regression coefficients are presented in this section as further examples.

Median

The median is a useful measure of the typical size of observations or data. The median is the middle observation, such that half of the observations are greater than the median, and half of them are less. For example, if the seven results of a test are 99, 95, 92, 90, 85, 75, 70, the median is 90. If there are six results, 95, 92, 90, 85, 80, 70, then the median is taken as (90 + 85)/2.

The median can be defined as follows:

1. The data must be sorted.
2. If n (the number of observations) is odd, the median is the kth observation, where $k = (n + 1)/2$. If n is even, the median is the arithmetic average of the two observations in the middle, that is, $(X_k + X_{k+1})/2$, where $k = n/2$.

The following example demonstrates how to find the median in a program.

EXAMPLE 10.8

■ **Problem:** Write the program segment which calculates the median of N test scores stored in an array A. Assume A is sorted.

Program Segment:

```
REAL MEDIAN
      ⋮
IF(MOD(N,2) .EQ. 0) THEN
        K = N/2
        MEDIAN = (A(K) + A(K + 1))/2
ELSE
        K = (N + 1)/2
        MEDIAN = A(K)
ENDIF
```

Summation Notation

Before explaining the regression analysis, let us explain the summation symbol. An oversize Greek letter sigma (Σ) is used often in textbooks to denote adding a series of variable. For example, $\sum_{i=1}^{n} X_i$ is defined as:

$$\sum_{i=1}^{n} X_i = X_1 + X_2 + X_3 + \cdots + X_n.$$

The following are further examples:

$$\sum_{i=1}^{n} X_i^2 = X_1^2 + X_2^2 + X_3^2 + X_4^2 + \cdots + X_n^2,$$

$$\sum_{i=1}^{5} (Y_i - 3)^2 = (Y_1 - 3)^2 + (Y_2 - 3)^2 + (Y_3 - 3)^2 + (Y_4 - 3)^2 + (Y_5 - 3)^2,$$

$$\sum_{i=1}^{5} X_i Y_i = X_1 Y_1 + X_2 Y_2 + X_3 Y_3 + X_4 Y_4 + X_5 Y_5.$$

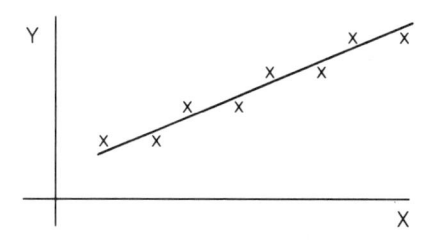

Figure 10.3 A regression line.

The summation notation can be translated into a FORTRAN program segment easily by a DO loop. For example

$$S = \sum_{i=1}^{n} A_i = A_1 + A_2 + A_3 + \cdots A_n$$

can be performed by

```
        S = 0.0
        DO 10 I = 1,N
            S = S + A(I)
10      CONTINUE
```

☐ Regression Analysis

The purpose of regression analysis is to predict the value of a variable by using known values of other variables. The prediction is made through a regression equation (represented graphically by a curve or line). Fitting a regression equation to a set of data corresponds to finding the best curve through a set of points (Figure 10.3). The general form of a regression equation with only one independent variable is $Y = b + mX$.

A common technique for finding the values of m and b is known as the method of least squares. The method is based on minimizing the sum of the squares of the distances between the data point and the value of Y in the equation. The following formulas are the results of the least-squares method for estimating m and b:

$$m = \frac{\sum_{i=1}^{n} X_i Y_i - N \cdot XM \cdot YM}{\sum_{i=1}^{n} X_i^2 - N \cdot XM}$$

and

$$b = YM - m \cdot XM$$

where

  XM = mean (average) of the X data,
  YM = mean (average) of the Y data,
  N = number of data.

We use these formulas to calculate the regression coefficients in the following example.

EXAMPLE 10.9

■ **Problem:** Write a program which reads N pairs of data and stores them in arrays X and Y. Calculate the estimated regression coefficients b and m using the least-squares method. There are less than 50 data for each array. The first record is a header record.

Program:

```
      PROGRAM CURVE
      REAL M
      DIMENSION X(50), Y(50)
      DATA SUMXY, SUMX, SUMY, SUMXX /4*0.0/
C
      READ*, N
C
      READ*,(X(I),Y(I), I = 1,N)
C
      DO 10 I = 1,N
          SUMXY = SUMXY + X(I)*Y(I)
          SUMX = SUMX + X(I)
          SUMY = SUMY + Y(I)
          SUMXX = SUMXX + X(I)*X(I)
10    CONTINUE
C
      XM = SUMX/N
      YM = SUMY/N
      M = (SUMXY - N*XM*YM)/(SUMXX - N*XM)
      B = YM - M*XM
      DO 20 I = 1,N
          PRINT*,X(I), Y(I)
20    CONTINUE
      PRINT*,'B = ',B,'M = ',M
      PRINT*, 'Y = ',B,' + ',M,'X'
      END
```

■
MATHEMATICS

Computers are often used to carry out numerical calculations. Common computer applications in mathematics include finding the area under a graph of a function, finding approximations to solutions of differential equations, and solving a system of algebraic equations. In this section, three simple examples are presented. These examples in no way exhaust the capabilities of computers in solving mathematical problems. To understand any of them, you must have some background in mathematics. It is assumed in this section that the reader is familiar with binary numbers, finding the roots of an equation, and simultaneous linear equations.

☐ Binary Numbers

We are accustomed to dealing with decimal numbers (base ten), where each digit (0 through 9) is taken as multiplied by a power of ten. For example,

$$297 = 7 + 9 \times 10 + 2 \times 10^2$$

When using the binary system (base two), each digit is written with binary digits 0 and 1, and is taken as multiplied by a power of two. For example, 1101 in binary is equivalent to:

```
 1   1   0   1
             └→1 × 1 = 1
         →0 × 2 = 0
     →1 × 2² = 4
 →1 × 2³ = 8
1 + 0 + 4 + 8
```

or 13 in the decimal system.

A simple method for converting a decimal number to its binary equivalent is by successive divisions by 2, recording the remainder from each division. That is, on dividing the number by 2, the remainder constitutes the first digit (from the right) of the binary number; on dividing the result (quotient) by 2, the remainder then constitutes the second-to-last digit of the binary number, and so on. We continue this procedure until the result is zero. The following program uses this technique.

EXAMPLE 10.10

■ **Problem:** Write a program which reads an integer N and converts it to its binary equivalent.

Program:

```
      PROGRAM BINARY
      INTEGER RESULT
      DIMENSION IREM(10)
C
      DATA IREM/10*0/
C
      READ*,N
      M = N
      DO 10 I = 1,10
          RESULT = INT(M/2)
          IREM(I) = MOD(M,2)
C
          IF (RESULT .EQ. 0)GO TO 50
          M = RESULT
10    CONTINUE
50    PRINT*,'THE NO. IN DECIMAL:',N
      PRINT*,'THE NO. IN BINARY:',(IREM(K),K = 10,
     *1 , -1)
      END
```

Notes:

1. It is assumed that the binary number has less than 10 digits [IREM(10)].
2. The first remainder is stored in IREM(1), the second remainder in IREM(2), and so on. But IREM(10) is printed first, IREM(9) second, and so on in descending order.

☐ Finding Roots of an Equation—Bisection Method

The bisection method is a method for finding an approximation to the root of an equation. An example demonstrates the method. Suppose we would like

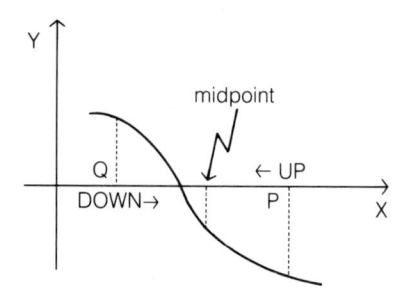

Figure 10.4 Bisection method of finding the root of an equation.

to find a root of the equation

$$f(X) = 2X^3 - 6X^2 + X + 5 = 0.$$

The problem involves two steps:

1. finding vicinities of the solution, and
2. searching for an approximation to the root.

If we find two values P and Q at which the values of the function have opposite signs, then the root is between P and Q (assuming the function is continuous between P and Q).

For example, for $P = 1$ and $Q = 2$ the values of the function are 2 and -1, respectively. Thus, one of the roots is between 1 and 2. The search starts by evaluating the function at the midpoint between P and Q, say X_0 [Note: $X_0 = (p + q)/2$]. If $f(X_0) = 0$, or close enough to zero according to a predetermined approximation, then the root is X_0 (see Figure 10.4). However, if the value of the function is not close enough to zero, we know the sign of the function will change either between P and X_0 or between X_0 and Q. The function will then be evaluated at a new midpoint between P and X_0 or between X_0 and Q, depending which one alters the sign of the equation. The process of finding a new midpoint will be repeated until the desired approximation is reached. The following example illustrates the technique.

EXAMPLE 10.11

■ **Problem:** Write a program which finds one of the roots of an equation of the form

$$AX^3 + BX^2 + CX + D = 0$$

where the coefficients A, B, C, D and the vicinities to the root, P and Q, are given as input. Find the root with an approximation of .0001. Use the bisection method.

Plan:

1. Read the coefficients A, B, C, D.
2. Read P and Q.
3. Assign P and Q to the variables UP and DOWN.
4. Calculate X, the midpoint of UP and DOWN: $X = (UP + DOWN)/2$.
5. Calculate $Y = AX^3 + BX^2 + CX + D$.
6. If the absolute value of UP − DOWN is less than .0001, X is the root; go to step 8;
   Otherwise:

              if $Y < 0$, reset DOWN $= X$,
              if $Y > 0$, reset UP $= X$.
        7. Go to step 4.
        8. Print the root, X.
Program:

```
        PROGRAM BISEC
        READ*,A,B,C
        READ*,P,Q
C
        UP = P
        DOWN = Q
C
50      X = (UP + DOWN)/2.0
        Y = A*X**3 + B*X**2 + C*X + D
        IF(ABS(UP - DOWN) .LT. .0001) GO TO 99
        IF(Y .LT. 0.0) THEN
            DOWN = X
        ELSE
            UP = X
        END IF
        GOTO 50
99      PRINT*, 'THE ROOT IS=', X
        END
```

☐ Simultaneous Linear Equations

A common application of computer programs in mathematics is solving a system of simultaneous equations of the form

$$a_{11}X_1 + a_{12}X_2 + \cdots + a_{1n}X_n = b_1,$$
$$a_{21}X_1 + a_{22}X_2 + \cdots + a_{2n}X_n = b_2,$$
$$\vdots$$
$$a_{n1}X_1 + a_{n2}X_2 + \cdots + a_{nn}X_n = b_n.$$

Several methods have been proposed for solving systems of this type. One of the most frequently used is known as the Gauss elimination method. This method replaces certain equations of the system with combinations of other equations, to produce a triangular system of the form

$$a_{11}X_1 + a_{12}X_2 + \cdots + a_{1n}X_n = b_1,$$
$$a_{22}X_2 + \cdots + a_{2n}X_n = b_2,$$
$$\vdots$$
$$a_{nn}X_n = b_n.$$

(Note that the a_{ij} and b_i in the new system need not be the same as in the original system.) Once in the triangular form, the system can be easily solved by back substitution. That is, the last equation yields

$$X_n = b_n/a_{nn}.$$

X_n can then be substituted in the equation before the last, which can then be solved for X_{n-1}, and so on.

The elimination of variables is accomplished by *elementary row operations*—that is, multiplying one row by a ratio, and adding it to another. Mathematically it can be shown that the elementary row operations, though they change the coefficients of variables, will not change the solution to the system.

The solution procedure goes through several stages. First, the coefficient of $a_{11}, a_{22}, ..., a_{nn}$ (called the pivots) in the triangular form must not be equal to zero (except in special cases). Therefore, the first stage is to search for a nonzero coefficient in column i (but below row i), and interchange it with the ith equation if the coefficient of the ith variable in the ith equation is zero. Furthermore, because the roundoff errors can be a serious problem if the pivots are small, it is customary to search column i for the coefficient of largest magnitude rather than merely a nonzero coefficient. (This also guarantees a nonzero coefficient.)

The second stage of the solution procedure is to create a triangular form. First, X_1 is to be eliminated from all equations except the first one. Next, X_2 is to be eliminated from all the equations after the second one, and so forth. In order to eliminate X_1 from the second equation, a_{21}/a_{11} times all the elements in the first equation should be subtracted from the second equation. To eliminate X_1 from the third equation, a_{31}/a_{11} times the first equation should be subtracted from the third one, and so on. This process must be repeated for X_2, X_3, ..., X_{n-1}. Finally, in the third stage, the roots are calculated by back substitution. The following example further illustrates the method.

EXAMPLE 10.12

Problem: Prepare a plan and a subroutine which solves a set of N linear equations. Assume N is 2 or greater.

Plan:

1. Start with the name of the subroutine and define the arrays.
2. For $I = 1$ to $I = N - 1$ repeat the following steps:
   2.1 Find and interchange the largest coefficient as follows:
      2.1.1 Find the row with the largest absolute value in column I, call it row L, and call the coefficient the *pivot*
      2.1.2 If the pivot is equal to zero, let MESSAGE = 1 and return.
      2.1.3 If necessary (if $I \neq L$), interchange the Lth row with the Ith row.
   2.2 Do the row elimination as follows:
      2.2.1 For $J = I + 1$ to $J = N$ repeat the following steps:
         a'. Find RATIO = a_{ji}/a_{ii}.
         b. Calculate the new coefficients by multiplying RATIO by each coefficient in row i, and subtract them from the corresponding coefficients in row j.
3. Find the value of each variable by back substitution as follows:
   3.1 Find the solution for the last variable.
   3.2 For $I = N - 1$ to 1 repeat the following steps:
      3.2.1 Calculate SUM, the sum of the products of the coefficients with the values of the corresponding variables, i.e., the sum of $a_{im} \cdot X_m$, where m runs from $I + 1$ to N.
      3.2.2 Calculate the solution for each X as $X_i = (b_i -$ SUM$)/a_{ii}$.
4. End of the subroutine.

Subroutine:

```fortran
      SUBROUTINE SOLVE(A,B,X,N,MSGE)
      DIMENSION A(10,10),B(10),X(10)
C
      MSGE = 0
C
      DO 10 I = 1, N - 1
C FINDING THE LARGEST COEFFICIENT
C
         L = I
         PIV = ABS(A(I,I))
         DO 20 J = I + 1, N
            BB = ABS(A(J,I))
            IF(BB .GT. PIV) THEN
                        L = J
                        PIV = BB
            END IF
20       CONTINUE
         IF(PIV .EQ. 0.0) THEN
                     MSGE = 1
                     RETURN
         ENDIF
C INTERCHANGING THE ROWS IF NECESSARY
         IF(L .EQ. I) GOTO 90
         DO 30 J = 1,N
            TEMP = A(L,J)
            A(L,J) = A(I,J)
            A(I,J) = TEMP
30       CONTINUE
         TEMP = B(L)
         B(L) = B(I)
         B(I) = TEMP
C CREATING THE TRIANGULAR FORM
90       DO 10 J = I + 1,N
            RATIO = A(J,I)/A(I,I)
            DO 40 K = I,N
               A(J,K) = A(J,K) - RATIO*A(I,K)
40          CONTINUE
            B(J) = B(J) - RATIO*B(I)
10    CONTINUE
C THE BACK SUBSTITUTION STEP
      X(N) = B(N)/A(N,N)
      DO 200 I = N - 1, 1, -1
C
         SUM = 0.0
         DO 210 M = I + 1, N
C
            SUM = SUM + A(I,M)*X(M)
210      CONTINUE
         X(I) = (B(I) - SUM)/A(I,I)
200   CONTINUE
      RETURN
      END
```

GRAPHING TECHNIQUES

Computers can be utilized to print a plot, graph, chart, diagram, or picture. This is accomplished by printing a sufficient number of appropriate characters in appropriate places. Two simple examples are presented here to demonstrate some of the techniques.

☐ Graphing

Suppose we would like to plot the graph of the function

$$Y = f(X).$$

The simplest method is to (1) assign a value to X, (2) calculate the value of Y, (3) print a symbol at the distance Y, and (4) repeat this process for different values of X. The printed symbol can be any character, such as an asterisk (*), a plus sign (+), or a dot.

The following program demonstrates the method in the simplest form. For simplicity we consider the horizontal axis as the Y axis, and the vertical axis as the X axis.

EXAMPLE 10.13

■ **Problem:** Write a program which plots the graph of the function

$$Y = 50.0/2**(X**2)$$

for values of X between -2 and 2 with an increment of $.1$. Explain the plan before writing the program.

Plan:

1. Consider the horizontal axis as the Y axis and the vertical axis as the X axis.
2. Define a character array LINE(50) for printing a horizontal line of the graph.
3. Set the LINE equal to blanks.
4. Write the appropriate heading.
5. Repeat the following step, for values of x from -2 to $+2$, at $.1$ intervals.
   5.1 Calculate the value of Y for a given X; denote the integer part of Y as IY.
   5.2 Place an * in the IYth position of the array LINE.
   5.3 Print X, Y, and LINE.
   5.4 Erase the * in LINE.
   5.5 Increase the value of X by the appropriate increment.
6. Write the footing.
7. End.

Program:

```
      PROGRAM GRAPH
      CHARACTER*1 LINE(50)
      DATA LINE/50*' '/
C WRITING A HEADING FOR X AND Y
      WRITE(6,10)
```

```
10      FORMAT(3X, 'X',5X,'Y')
        DO 40 X = -2.0,2.0,.1
        Y = 50.0/2**(X**2)
        IY = INT(Y)
            LINE(IY) = '*'
            WRITE(6,20) X,Y,LINE
            LINE(IY) = ' '
40      CONTINUE
C
        WRITE(6,30)
20      FORMAT(2X,F4.1,1X,F5.2,1X, 'I',1X,50A)
30      FORMAT(12X,50('-'))
        END
```

This program printed the graph shown in Figure 10.5.

Plotting a Histogram

A *histogram* or *bar chart* is a graph illustrating the number of occurrences of given values. For example, Figure 10.6 shows a histogram of the number of A's, B's, C's, D's, and E's in a course.

There are several methods for plotting this type of graph. The simplest is: (1) consider the entire graph as a two-dimensional array, (2) place symbols (such as asterisks) in the appropriate places in the array, and (3) print the entire array. The size of the array can be determined by the maximum number of rows and columns necessary for a particular histogram. One example for this method was given in Chapter 8. Here is another.

EXAMPLE 10.14

Problem: Suppose the numbers of A's, B's, C's, D's, and E's in a test are as follows:

3 A's
5 B's
7 C's
4 D's
2 E's

Write a program which reads the data and plots the histogram.

Program:

```
        PROGRAM HISTO
        CHARACTER*1 GRAF(10,5),S
C CLEARING THE GRAF
        DATA GRAF/50*' '/, S/'*'/
C FILLING THE GRAF WITH * FROM TOP TO BOTTOM
        DO 10 I = 1,5
            READ*,NUMBER
            IF(NUMBER .EQ. 0) GO TO 10
                DO 10 J = 1,NUMBER
                    GRAF(J,I) = S
10      CONTINUE
```

436 *Problem Solving and Programming Techniques*

```
  X      Y
-2.0    3.13  I  *
-1.9    4.09  I   *
-1.8    5.29  I    *
-1.7    6.75  I     *
-1.6    8.48  I      *
-1.5   10.51  I       *
-1.4   12.85  I        *
-1.3   15.50  I         *
-1.2   18.43  I          *
-1.1   21.61  I           *
-1.0   25.00  I             *
 -.9   28.52  I              *
 -.8   32.09  I               *
 -.7   35.60  I                *
 -.6   38.96  I                 *
 -.5   42.04  I                  *
 -.4   44.75  I                   *
 -.3   46.98  I                    *
 -.2   48.63  I                     *
 -.1   49.65  I                      *
  .0   50.00  I                       *
  .1   49.65  I                      *
  .2   48.63  I                     *
  .3   46.98  I                    *
  .4   44.75  I                   *
  .5   42.04  I                  *
  .6   38.96  I                 *
  .7   35.60  I                *
  .8   32.09  I               *
  .9   28.52  I              *
 1.0   25.00  I             *
 1.1   21.61  I           *
 1.2   18.43  I          *
 1.3   15.50  I         *
 1.4   12.85  I        *
 1.5   10.51  I       *
 1.6    8.48  I      *
 1.7    6.75  I     *
 1.8    5.29  I    *
 1.9    4.09  I   *
 2.0    3.12  I  *
```

Figure 10.5 Graph of the function 50.0/2**(X**2).

```
            *
            *   *
        *   *   *
        *   *   *
        *   *   *
        *   *   *
        *   *   *   *
    *   *   *   *   *
    *   *   *   *   *
    ─────────────────
    A   B   C   D   E
```

Figure 10.6 A histogram.

```
C  PRINTING THE HISTOGRAM, THE GRAF IS TO BE
C  PRINTED UPSIDE DOWN
       DO 20 M = 10,1,-1
C
           WRITE(6,50)(GRAF(M,J),J = 1,5)
20     CONTINUE
       WRITE(6,60)
50     FORMAT(11X,5(3X,A1))
60     FORMAT(14X,'A',3X,'B',3X,'C',3X,'D',3X,'E')
       STOP
       END
```

Note: First, the array GRAF is filled with necessary numbers of asterisks in the appropriate places from the top of the array to bottom (Figure 10.7); then this upside-down graph is printed from bottom to top; i.e., the last row is printed first, the row before the last row second, and so on.

SIMULATION

Simulation is a process of imitating a system by using a model and then studying its behavior. One can use simulation models to study both simple and complex systems. The basic principle behind computer simulation is to have a program imitate the behavior of a process as closely as possible. For example, a program can be developed to imitate the number of automobiles arriving at a specific intersection, or the number of a particular item sold in a department store.

Unfortunately, real-life situations are usually too complicated to be exactly duplicated by a computer program. However, if the model is somewhat representative of an actual situation, the result of the simulation can give useful information about the behavior of the real system. Of course, the closer the model is to the real system, the more accurate the results will be.

Simulating a process has several advantages. One advantage is the ability to predict the behavior of a system. Another is the ability to experiment with different strategies in order to find the best one.

☐ Random Numbers

Random numbers are quantitative presentations to chance. To be considered a random number, it must be chosen from a range in such a way that all numbers in the range have an equal chance of being chosen.

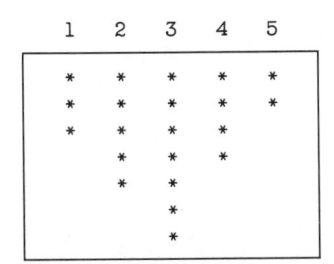

Figure 10.7 Filling the array A with asterisks.

Normally any process in a real-world situation involves some kind of randomness. Therefore, the generation of random numbers by the computer is the main tool of computer simulation. For example, a random number generated by the computer can represent the draw of a single playing card from a deck, or the result of casting dice. Even if the process does not follow a uniform distribution, random numbers generated by the computer can be used to produce nonuniformly distributed random numbers governed by an appropriate distribution function (for example, the distribution of customers arriving at a department store).

☐ Generating Random Numbers

Random numbers can be generated by a library function in almost any computer system. Look at the FORTRAN manual for the system that you are using to find the name of the function. It is likely that you will find the name and the form of the function as

```
RANF ( )
```

It is also likely that this function generates a random number between 0.0 and 1.0 whenever the function is invoked. For example,

```
      DO 10 I = 1,10
      X = RANF()
10    PRINT*,X
      END
```

executed by a CDC system generated the following numbers:

```
.9505127
.2976203
.0062619
.3056509
.3826622
.8318579
.0986253
.6204460
.9903771
.6938844
```

Notice that the numbers are between 0.0 and 1.0. If it is desired to generate random numbers between A and B, then the following formula can be used:

$$X = (B - A)*RANF( ) + A$$

For example, by using the statements

```
      DO 10 J = 1,10
         X = 100.0*RANF()
10    PRINT*,X
      END
```

the following numbers (between 0.0 and 100.0) were obtained:

```
58.01136
95.05127
78.63714
29.76202
45.36999
00.62619
27.57364
```

```
30.56509
68.91007
38.26622
```

The following statement produces integer random numbers between M and N:

```
L = INT((N - M + 1)*RANF()) + M
```

For example,

```
L = INT(6*RANF()) + 1
```

generates a random number between 1 and 6. The following program simulates numbers represented by rolling a regular die 10 times:

```
DO 10 J = 1,10
L = INT(6*RANF()) + 1
    PRINT*,L
CONTINUE
END
```

When this program was executed by a CDC system, the results were

```
4
6
5
2
3
1
2
2
5
3
```

In a program, each random number generated by the computer has an equal chance of occurring. Nevertheless, the stream of numbers generated is predetermined by a mathematical formula. Therefore, whenever the generator restarts, it produces the same sequence of numbers. That is, if you reexecute a program, you will see the same sequence of random numbers. In order to avoid this occurrence, you may initialize the seed of the random-number generator (the starting point). For example, RANSET is a subroutine in the CDC system which initializes the seed of RANF(). This subroutine is used in the following program to generate ten random numbers:

```
S = SECOND()
CALL RANSET(S)
DO 10 I = 1,10
    X = RANF()
PRINT*,X
END
```

Note that the seed is the systems clock time in seconds invoked by a library function: SECOND(). But note again that the functions SECOND, RANSET, and even RANF are not standard FORTRAN 77 functions and depend on the system.

☐ Monte Carlo Simulation

There are several forms of simulation; among them, the *Monte Carlo* simulation is especially suitable to be implemented by a program. The Monte Carlo simulation technique uses random numbers and a known probability distribution of the variables. It also incorporates rules, policies, and any other factors in the system to obtain simulated values. When simulation is repeated a large number of times, the average of the simulated values will be close to the mathematical expectation for the model. Of course, how close this is to the results in a real situation depends on how closely the model resembles the real system. The following is an example of a simulation program.

EXAMPLE 10.15

■ **Problem:** Suppose the newest casino in Las Vegas has invented a die with 10 sides. Any time the die is rolled, each number between 1 and 10 can occur with equal probabilities. You have heard that the payoff is 6 for 1; that is, any time that you bet $1, you will win $6 if you are lucky and the number of your choice comes up. Suppose that you take $500 with you and play the game 500 times; since you believe your lucky number is 7, you bet always on 7 and will not change your strategy. Now, before going to Las Vegas, write a program which simulates the game. The program should show your expectation gains and losses and your net gain or loss after 500 trials.

Program:

```
        PROGRAM DICE1
        DATA IWIN, LOSS, LUCK /0,0,7/
        DO 10 I = 1,500
            CALL ROLL(L)
            IF (L .EQ. LUCK) THEN
                IWIN = IWIN + 6
                LOSS = LOSS + 1
            ELSE
                LOSS = LOSS + 1
            ENDIF
10      CONTINUE
        NET = IWIN - LOSS
        PRINT*,'WINS = $',IWIN,'     LOSS = S',LOSS
        PRINT*,'          NET = $',NET
        END
        SUBROUTINE ROLL(L)
        L = INT(10.0*RANF()) + 1
        RETURN
        END
```

Notes:

1. LOSS = LOSS + 1 is always executed even when the player wins (the policy is that the house keeps the $1 wager).
2. The subroutine ROLL works as follows: when the random number is

   between 0.0 and .099999, the roll is 1;
   between .1 and .199999, the roll is 2;
   between .2 and .299999, the roll is 3;
   between .3 and .399999, the roll is 4;
   between .4 and .499999, the roll is 5;
   between .5 and .599999, the roll is 6;

between .6 and .699999, the roll is 7;
between .7 and .799999, the roll is 8;
between .8 and .899999, the roll is 9;
between .9 and .999999, the roll is 10.

This is a uniform distribution (notice the equal intervals).
3. When the program was executed by a CDC system, the results were

    WINS = $300   LOSS = $500
             NET = $-200

As mentioned before, random numbers generated by the computer can be used to generate random numbers from an appropriate distribution function. To show the method, pay close attention to the subroutine of the following example (although the function is still uniform).

EXAMPLE 10.16

Problem: Suppose gambling is legal in your town and you would like to play the dice in the local casino. The game is simple; when a regular die is rolled and you bet $1 on a number, you win $4 if the number of your choice comes up. Write a program which simulates the game for 500 plays. Pick 5 as your lucky number.

Program:

```
      PROGRAM DICE2
      DATA IWIN,LOSS,LUCK /2*0,5/
      DO 10 I = 1,500
           CALL ROLL(L)
           IF(L .EQ. LUCK) IWIN = IWIN + 4
           LOSS = LOSS + 1
      CONTINUE
      NET = IWIN - LOSS
      PRINT*, 'WINS=$',IWIN, '    LOSSES = $',
     *LOSS
      PRINT*, 'NET =$',NET
      END
      SUBROUTINE ROLL(M)
      X = RANF()
      L = 1
      IF(X .GT. 0.166665) L = 2
      IF(X .GT. 0.333332) L = 3
      IF(X .GT. 0.499998) L = 4
      IF(X .GT. 0.666665) L = 5
      IF(X .GT. .8333332) L = 6
      RETURN
      END
```

Notes:

1. The subroutine works as follows: The probability of each number from 1 to 6 is the same ($\frac{1}{6}$ = .166666); thus, if the generated random number is:

between 0.0 and 0.166665, then L = 1,
between 0.166666 and 0.333332, then L = 2,
between 0.333333 and 0.499998, then L = 3,
between 0.499999 and 0.666665, then L = 4,
between 0.666666 and 0.833332, then L = 5,
between 0.833333 and 1.000000, then L = 6.

2. The following subroutine can also perform the same function:

```
      SUBROUTINE ROLL(L)
      DIMENSION PROB(6)
      DATA PROB/0.166666,0.333333,0.499999,
     *0.666666,0.833333,1.0/
      X = RANF()
      DO 10 I = 1,6
          IF(X .LT. PROB(I)) GO TO 20
10    CONTINUE
20    L = I
      RETURN
      END
```

3. Because the distribution is uniform (equal intervals), the following subroutine could also have been used:

```
      SUBROUTINE ROLL(M)
      L = INT(6*RANF()) + 1
      RETURN
      END
```

The method in the subroutine of the previous example is often used to produce a random number from a nonuniformly distributed process (whereas the statement

```
L = INT( (N - M + 1)*RANF() ) + M
```

is only for a uniformly distributed process). An example of a distribution which is not uniform is given next. This example will demonstrate the Monte Carlo simulation technique as a business application.

EXAMPLE 10.17

■ **Problem:** Farmer Bill's supermarket is known for its fresh milk. The store has the following sale policies:

1. The milk is purchased for $1.10/gallon, and sold for $2.10/gallon.
2. Milk is ordered by telephone at the end of each day based on that day's demand, and is received early the following morning.
3. Any milk not sold during a day is discarded.
4. Past experience has shown that the daily demand for milk varies between 50 and 350 gallons with the following probabilities:

Demand (gallons/day)	Probability
less than 50	.05
less than 100	.15
less than 150	.35
less than 200	.65
less than 250	.85
less than 300	.95
less than 350	1.0

Write a program which simulates the milk-selling process for 360 days; assume the first day's order is for 200 gallons of milk. Calculate the average daily demand, average number of gallons sold per day, average daily order, and average profit per day. Explain the plan before writing the program.

Plan:

1. Housekeeping
   1.1 Set initial order = 200.
   1.2 Set price = 2.10.
   1.3 Set cost = 1.10.
   1.4 Set the cumulator for the total profit, total sales, total order, and total demand equal to zero.
2. From day 1 to day 360 repeat the following steps, to step 2.7.
   2.1 Determine the demand X (from the demand function).
   2.2 Set sale = demand
       2.2.1 If the sale is more than the day's order (stock out), then reset sale = order.
   2.3 Calculate the profit for the day.
   2.4 Print the first 10 days' activity.
   2.5 Calculate the totals.
   2.6 Set the next day's order equal to that day's demand.
   2.7 End of the day's simulation.
3. Calculate the averages.
4. Print the averages.

Program:

```
      PROGRAM SALES
      INTEGER DAY
      DATA ORDER, PRICE, COST /200.0,2.10,1.10/,
     *TDEM, TSALES, TORDER, TPROF / 4*0.0/
      DO 10 DAY = 1,360
          CALL DEMAND(X)
          SALES = X
C RESET THE SALES EQUAL TO THE ORDER, IF DEMAND
C IS MORE THAN THE ORDER
          IF(SALES .GT. ORDER) SALES = ORDER
          PROFIT = SALES*PRICE - ORDER*COST
C WRITE THE 1ST 10 DAYS ACTIVITIES
          IF(DAY .LE. 10) PRINT*,DAY,X,SALES,
     *    ORDER,PROFIT
```

```
C CALCULATE TOTALS
      TDEM = TDEM + X
      TSALES = TSALES + SALES
      TORDER = TORDER + ORDER
      TPROF = TPROF + PROFIT
C SET THE NEXT DAYS ORDER EQUAL TO THAT DAYS DEMAND
      ORDER = X
10    CONTINUE
C CALCULATE THE AVERAGES
      ADEM = TDEM/360.0
      ASALE = TSALES/360.0
      AORDER = TORDER/360.0
      APROF = TPROF/360.0
      PRINT*,ADEM,ASALE,AORDER,APROF
C
      SUBROUTINE DEMAND(Y)
      DIMENSION GALLON(7), PROB(7)
      DATA GALLON / 50.0,100.0,150.0,200.0,250.0,
     *300.0,350.0/
      DATA PROB/.05,.15,.35,.65,.85,.95,1.0/
      X = RANF()
      DO 10 I = 1,7
         IF(X .LT. PROB(I)) GO TO 20
10    CONTINUE
20    Y = GALLON(I)
      RETURN
      END
```

Note: The following subroutine is equivalent to the above subroutine:

```
SUBROUTINE DEMAND(Y)
Y = 50.0
X = RANF()
IF(X .GT. 0.05) Y = 100.0
IF(X .GT. 0.15) Y = 150.0
IF(X .GT. 0.35) Y = 200.0
IF(X .GT. 0.65) Y = 250.0
IF(X .GT. 0.85) Y = 300.0
IF(X .GT. 0.95) Y = 350.0
RETURN
END
```

MANAGEMENT INFORMATION AND DECISION SUPPORT SYSTEMS

The concern of *management information systems* (MIS) is to provide information to managers so they can perform their function effectively and efficiently. FORTRAN is a useful tool for processing data and providing information to support management functions. Typical applications include labor analysis, sales analysis, sales forecasting, production control, optimization of resources, scheduling, and reporting.

A *decision support system* (DSS) goes beyond transaction processing, reporting and analysis. A DSS's main objective is to provide necessary information and feedback in order to improve the effectiveness of the management's problem-solving abilities (not to eliminate managers from the decision-making process). A DSS utilizes external and internal information, decision rules, and models, as well as the decision maker's insight and judgment in an interactive decision-making process. The flexibility and effectiveness of FORTRAN in model building and simulation, along with the development of interactive FORTRAN, make it a powerful tool for DSS projects. For example, an interactive program can be developed to simulate the results of different decisions and strategies with a given set of factors. This kind of program, then, allows the managers to play "what if" and "what would happen" for each decision. Consequently, the manager can experiment with several alternative decisions before actually selecting a course of action.

The following program is a very short and simple example of an interactive program.

EXAMPLE 10.18

■ **Problem:** Write an interactive program that informs a manager of the necessary number of items (N) to be sold in order to break even. Use the following formula:

$$N = \frac{\text{fixed costs}}{\text{price} - \text{variable cost}}$$

Program:

```
      PROGRAM VERSIM
      CHARACTER*3 ANSWER
      PRINT*, 'WOULD YOU LIKE TO KNOW THE BREAK
     *'EVEN POINT'
      READ(*,10) ANSWER
      IF (ANSWER .EQ. 'NO') GOTO 999
59    PRINT*, 'WHAT PRICE DO YOU HAVE IN MIND'
      READ*, PRICE
      PRINT*, 'WHAT IS THE TOTAL FIXED COST'
      READ*, FIXED
      PRINT*, 'WHAT IS THE VARIABLE COST PER
     *  UNIT'
      READ*, VAR
      N = FIXED / (PRICE - VAR)
      PRINT*, 'IN ORDER TO BREAK EVEN ', N, '
     *  ' UNITS MUST BE SOLD'
      PRINT*, 'WOULD YOU LIKE TO SEE THE ',
     *  'RESULT OF A PRICE CHANGE'
      READ(*,10)ANSWER
      IF (ANSWER .EQ. 'YES') GOTO 59
999   PRINT*, ' *** TAKE CARE *** '
10    FORMAT (A3)
      STOP
      END
```

> **Note:** The following is a sample of running the program:
>
> ```
> WOULD YOU LIKE TO KNOW THE BREAK EVEN POINT
> ? YES
> WHAT PRICE DO YOU HAVE IN MIND
> ? 1200
> WHAT IS THE TOTAL FIXED COST
> ? 2000000
> WHAT IS THE VARIABLE COST PER UNIT
> ? 30
> IN ORDER TO BREAK EVEN 1709 UNITS MUST BE
> *SOLD
> WOULD YOU LIKE TO SEE THE RESULT OF A PRICE
> *CHANGE
> ? NO
>    *** TAKE CARE ***
> ```
>
> Note that the underscored information was typed in by the user.

EXERCISES

Write a complete FORTRAN program for each of the following problems.

10.1 Write a program which prints the prime numbers from 3 to 499.

10.2 Develop an algorithm and write a program which converts

  a. Feet and inches to inches.
  b. Inches to feet and inches.
  c. Hour, minutes, and seconds to seconds.
  d. Seconds to hours, minutes, and seconds.

10.3

  a. Write a program which reads a Social Security number as one number with 9 digits (such as: 123456789), and prints it in the form: 123-45-6789.
  b. Write a program which reads a date as one number such as 091885 and prints it in the form: 09/18/85.

10.4 Write a program which reads values for M and N, and prints a message indicating whether or not M is a divisor of N.

10.5 Write a subroutine which rearranges the elements of an array A so that the value of the last element will be in the first element, the second to the last will be in the second element, and so on. Assume A has 100 elements.

10.6 Design a calendar for a year and write a program which prints the calendar. Input could be (1) the day of the week that January first falls on, and (2) a digit indicating whether the year is a leap year. Consider also the following information:

Months 01, 03, 05, 07, 08, 10, and 12 have 31 days.
Month 02 has 29 days in leap years only.
Month 02 has 28 days if the year is not a leap year.
Months 04, 06, 09, and 11 have 30 days.

☐ Sorting

10.7 Write a program which reads the names of the students in a course, and sorts them in alphabetical order. Print the unsorted list first and then the sorted list. There are less than 50 students in the course; the first data record is a header record.

10.8 Write a program to print a cross-reference telephone directory, which reads up to 100 names and telephone numbers, sorts the list first in alphabetical order, and prints the list. The list then should be sorted according to the phone numbers and printed. Use a header record. [Hint: When rearranging one list (the names), the other list (phone numbers) should be rearranged too. That is, whenever one pair of values are interchanged in one list, the corresponding pair in the other list must be interchanged as well.]

10.9 Write a program which reads the names of students and their Social Security numbers (9 digits, without the hyphens). Print their names and numbers:

  a. sorted according to the names.
  b. sorted according to the numbers.

(Pay attention to the hint in the previous problem.)

☐ Search

10.10 Write a program which reads the names of your friends and their phone numbers. Then the program should print the name after reading a phone number or print the phone number after reading a name.

  a. Use a sequential search.
  b. Use a binary search.

10.11 Modify the binary-search program in Example 10.6 so that if the desired number is not found after all the elements are searched, the program prints a message indicating that the number is not in the list.

☐ Merge

10.12 Write a program segment which combines arrays A and B to create a single sorted array (C). Assume the size of A is N, and the size of B is M. Read the values of N, M, A, and B at the beginning of the program. A and B each have less than 100 elements, and are not sorted.

10.13 Assume arrays A and B have 100 elements each. Write a program segment which creates an array C with entries arranged in the following order: A(1), B(1), A(2), B(2), ..., A(100), B(100).

10.14 Write a program which reads arrays X and Y, and merges them into an array Z. Assume:

  a. X and Y are sorted in decreasing order.
  b. X and Y are not sorted.

X and Y each have 250 elements, and Z must be sorted in decreasing order.

☐ Statistics

10.15 In statistics, the probability of M successes in N trials can be calculated by what is called the binomial probability distribution. The formula is

$$X = \frac{N!}{M!(N-M)!} p^M q^{M-N},$$

where

X = probability of M successes in N trials,
p = probability of success,
q = probability of failure = $1 - p$.

Write a program which generates a table of probabilities for $N = 10$, $M = 1, 2, 3, 4, 5$ and $p = .01, .02, .03, .04, .05, .06, .07, .08$. Use a subprogram for calculating the factorial of a number. The following table is an example of the output:

N = 10

P M	.01	.02	.03	.04	.05	.06	.07	.08
1	.0944							
2								
3								
4								
5								

10.16 The mean, median, variance, standard deviation, and range are useful measures of a group of observations. Write a program which reads a series of N observations into an array X and calculates these measures. Print the unsorted data, the sorted data, and the measures. Use the following formulas for calculations:

1. The mean:

$$\frac{\sum_{i=1}^{N} X_i}{N}$$

2. The variance:

$$\text{VAR} = \frac{\sum_{i=1}^{N}(X_i - \text{mean})^2}{N} = \frac{\sum_{i=1}^{N} X_i^2}{N} - (\text{mean})^2.$$

3. The median XMED (the data must be sorted):

if N is odd,

$\text{XMED} = X_k,$ where $k = (N+1)/2;$

if N is even,

$\text{XMED} = (X_k + X_{k+1})/2.0,$ where $k = N/2.$

4. The standard deviation:

$\text{STD} = \sqrt{\text{VAR}}.$

5. The range: (largest number) − (smallest number).

10.17 The following data shows the sales of products for a company:

Year	Sales (millions of dollars)
1	20
2	22
3	25
4	24
5	27
6	29
7	28
8	29
9	32
10	31

Write a program to estimate a straight-line relationship by the method of least squares. Estimate the expected sales for year 11.

10.18 The results for the amount X of a chemical reagent which was needed to obtain an amount Y precipitate of a substance from a given solution are:

X	Y
5.5	7.0
4.6	5.5
8.4	7.1
5.4	4.9
6.2	6.1
6.7	6.8
8.0	7.9
4.4	4.8
9.2	8.5
10.4	10.9
11.0	11.2

Write a program to estimate a linear relationship by the method of least squares. Estimate the expected amount of precipitate if $X = 7.9$.

10.19 The *correlation coefficient* shows how two sets of observations are related. Write a program that reads N pairs of observations. Calculate the correlation coefficient as follows:

1. Calculate the mean and standard deviation for each set of data separately; call them XMEANA, STDA, XMEANB, and STDB. See Exercise 10.16 for the formulas.
2. Calculate the correlation coefficient by the following formula:

$$R = \frac{\sum_{i=1}^{N} (A_i - \text{XMEANA})(B_i - \text{XMEANB})}{N \cdot \text{STDA} \cdot \text{STDB}}$$

R is always between -1 and $+1$.

Read N, the number of observations, from a header record.

10.20 The formulas for calculating the coefficients of a linear equation

$$y = a_1 + b_1 x$$

were discussed earlier in the text. But in some situations the data may not fit a linear equation. Assume the price-to-earnings relation for stocks follows the function

$$y = ax^b.$$

Write a program which reads N pairs of data for x (earnings) and y (prices); calculate the parameters a and b. [Hint: The formula can be changed to $\log y = \log a + b \log x$, which has the form $Y = A + BX$.]

☐ Mathematics

10.21 Write a program which finds the greatest common divisor of two positive numbers, A and B. (Hint: A simple method would be to choose the smaller of the two numbers, say A; test whether or not it is a divisor of the other; if not, check all the numbers less than A to see if they divide both A and B.)

10.22 Assume a computer cannot multiply, but can add and subtract. Develop an algorithm and write a program which multiplies two integer numbers, M and N.

10.23 Develop and write a program which reads a number in binary and prints its equivalent in decimal.

10.24 Develop and write a program which adds and subtracts two numbers in binary.

10.25 Develop and write a program which reads a number in base 8 and converts it to its equivalent in base 10 (decimal).

10.26 Develop and write a program which reads an integer number in decimal and converts it to its equivalent in base 8.

10.27 The equation

$$x^4 - 10x^3 + 35x^2 - 50x + 24 = 0$$

has four roots. The vicinities of the roots are $0 - 1.5$; $1.5 - 2.5$; $2.5 - 3.5$; and $3.5 - 5$. Write a program which finds the roots with an approximation of .00001.

10.28 In the year A.D. 1225, the following equation was solved:

$$x^3 + 2x^2 + 10x - 20 = 0$$

and it was found to have a root equal to

$$x = 1.3688$$

Write a program which calculates the root of the equation between 1 and 1.5; compare the result with the solution found in 1225.

10.29 Write a program which solves the following systems of equations:

a.
$$5x_1 + 6x_2 - 2x_3 + 8x_4 = 32$$
$$2x_1 + 3x_2 + 5x_3 - 2x_4 = 15$$
$$-3x_1 - 9x_2 + 8x_3 + 3x_4 = 8$$
$$9x_1 - 5x_2 + 3x_3 + 5x_4 = 6$$

b.
$$x_1 + \tfrac{1}{2}x_2 + \tfrac{1}{3}x_3 = 1$$
$$\tfrac{1}{2}x_1 + \tfrac{1}{3}x_2 + \tfrac{1}{4}x_3 = 0$$
$$\tfrac{1}{3}x_1 + \tfrac{1}{4}x_2 + \tfrac{1}{5}x_3 = 0$$

c.
$$9x + 6y + 6z + 6v = 31$$
$$6x + 4y + 5z + 4v = 20$$
$$4x + 10y + 9z + 3v = 28$$
$$5x + 11y + 8z + 2v = 50$$

10.30 The *inverse* of a square matrix A is a matrix B such that

$$AB = I,$$

where I is an identity matrix (i.e., a square matrix with 1's on its principal diagonal and 0's everywhere else). Write a program using the Gauss elimination method to produce the inverse of the matrix A. [Hint: Convert the array (A,I) into (I,A); A in the latter form is the inverse of the original matrix.]

☐ Graphing

10.31 Write a program which plots the graph of the following functions:

a. $Y = 3x^2 + 6x + 5$ for $-2.5 \leq x \leq +2.5$ with an interval of .1.
b. $Y = 50/(1 + x^2)$ for $-5 \leq x \leq +5$ with an interval of .2.
c. $Y = (x^2 - 1)(x^2 - 4)(x^2 - 9) + 25$ for $-2.5 \leq x \leq +2.5$ with an interval of .1.
d. $Y = 1 + x^2$ for $-10 \leq x \leq 10$ with an interval of .3.

10.32 Modify the program in Example 10.13 so that it prints the graph in the following form:

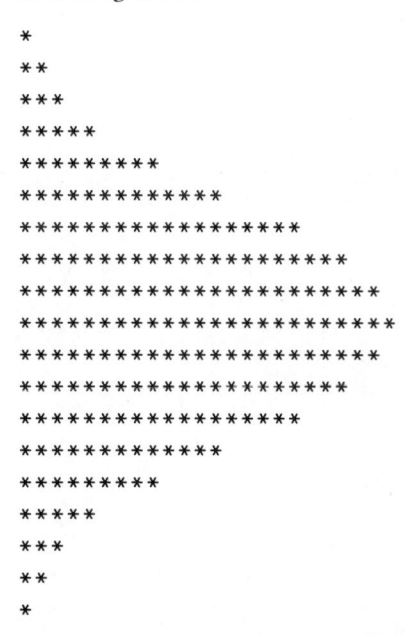

```
*
**
***
*****
********
************
****************
******************
********************
**********************
**********************
********************
******************
****************
************
********
*****
***
**
*
```

10.33 Write a program which reads the result of a test taken by 50 students. The scores are between 0 and 100. Calculate the number of A's (90 to 100), B's (80 to 90), C's (70 to 80), D's (60 to 70), and E's (below 60). Plot a histogram showing the number of each grade.

10.34 Write a program which reads a set of 100 real data. The data are between 1.0 and 10.0. Calculate the number of occurrences of data in an interval of length one; i.e., 1.0 to 2.0, 2.0 to 3.0, and so on. Print a histogram showing the number in each interval.

Simulation

10.35 Write a program which simulates the rolling of two dice 100 times. Print the number of occurrences of 7 and 11.

10.36 Write a program which simulates the rolling of three dice 100 times. Print the number of 7's and the number of 11's.

10.37 Computerize the "calls" in a game of Bingo. The process is as follows: 75 tokens, numbered 1 to 75, are placed in a container. The caller draws out one token at a time, and reads it. If the number is between 1 and 15 he also states that the number is under the B; between 16 and 30, under the I; between 31 and 45, under the N; between 46 and 60, under the G; between 61 and 75, under the O (Figure 10.8). Write a program which simulates the drawing and calling.

10.38 Suppose you are planning to take a trip to Las Vegas. Simulate the crap game before you try it. The rules are as follows:

  a. Two dice are rolled.
  b. If the roll (the total number of dots) shows 7 or 11 (called a natural), the player wins.
  c. If the roll is 2, 3, or 12 (called craps), the player loses.
  d. If the roll is other than those mentioned (i.e., 4, 5, 6, 8, 9, or 10), the result is not yet decided and the number becomes the player's *point*. In this case, the player must roll both dice again until he gets either his point or a 7. The player wins if his point appears, but loses if a 7 appears first. All rolls other than the point or 7 are meaningless.

Simulate the game 500 times. Find how many times you win.

Figure 10.8 A bingo card.

10.39 Some gamblers believe that doubling the bet after each loss works well. The system works like this: one starts with a small bet, say $1; if he loses, he bets $2 on the next game. If he loses again, he bets $4, then $8, and so on, until he wins a game. Suppose that you have $100 and would like to play the dice game. Write a program which simulates the play. Assume the payoff is one to four. Find the probability that you run out of money before you win anything. Repeat the simulation for the crap game and find the probability of losing all your money.

10.40 ABC Car Dealership has experienced the daily demand for cars as follows:

Demand/day	Probability
less than 1	.05
less than 2	.15
less than 3	.35
less than 4	.55
less than 5	.75
less than 6	.85
less than 7	.95
less than 8	1.0

Write a program which simulates the car-selling process for 500 days. Assume the cars can be ordered now and delivered to the customers later. Calculate the average daily demand.

10.41 Cute Bakery Shop bakes fine birthday cakes every morning. The production costs are $1.75/cake, and the cakes are sold for $4.50/cake. The leftover cakes are sold to a local supermarket at $1.30/cake. The baker bakes four cakes every day. But, historically, the daily demand for birthday cakes are as follows:

Demand/day	Relative frequency
0	.05
less than 1	.15
less than 2	.30
less than 3	.45
less than 4	.65
less than 5	.75
less than 6	.9
less than 7	1.0

a. Write a program which simulates the selling process for 360 days. Calculate the average daily demand, average number of cakes sold to customers, average number of cakes sold to the local supermarket, and average profit per day.

b. Repeat the simulation several times, each time using trial production quantities of 1, 2, 3, 4, 5, 6, 7, and 8 cakes per day. Calculate and print the average profit per day for each trial quantity. Which production quantity do you recommend?

10.42 Write a simulation program which simulates a waiting line for a service station. Customers arrive randomly for service, which takes exactly ten minutes. Calculate the average waiting time for each customer if they arrive irregularly at an average rate of six customers every 100 minutes. (The service time is not counted as part of the waiting time.)

```
        N
W  ┌──┬──┬──┬──┐
   │  │  │  │  │
   │1,1│1,2│1,3│
   ├──┼──┼──┼──┤
   │  │  │  │  │
   │2,1│2,2│2,3│
   ├──┼──┼──┼──┤
   │  │  │  │  │
   │3,1│3,2│3,3│
   ├──┼──┼──┼──┤
   │  │  │  │  │E
   └──┴──┴──┴──┘
 S
```

Figure 10.9 Street intersections.

10.43 A dog is lost in Manhattan. At each intersection he chooses a direction at random and proceeds to the next intersection, where he again chooses another direction at random, and so on. Assume the blocks are all square and there are only nine intersections, as shown in Figure 10.9. Write a program to simulate many of these random walks. Find the probability that the dog starting at a corner will emerge on the north side in less than 1,000 walks. Assume after reaching any exit, his walk is over. [Hint: The process can be simulated many times. Each time one of the intersections can be chosen as the starting point. Random numbers 1, 2, 3, and 4 (for south, west, east, and north) determine which direction is taken. The walk just finished is then counted as either a success or a failure, and another walk begins. After all trials, the ratio of number of successes to the total number of trials is an approximation to the probability.]

☐ Decision Support

10.44 Design, develop, and write an interactive program which helps a manager to find information about a product in order to determine one or several of the following factors:

   a. The price
   b. The variable cost. (For example, can more expensive materials be used?)
   c. The fixed cost. (For example, can a new piece of equipment be purchased?)
   d. Number of units which must be sold in order to break even.
   e. Total revenue if a certain number of units of the product are sold.
   f. Total profit if a certain number of units of the product are sold.

Use the following formula in the model:

$$\text{Revenue} = M * P$$
$$\text{Total profit} = \text{Revenue} - \text{costs}$$
$$= M * P - (F + V * M)$$
$$N = \frac{F}{P - V}$$

where

M = number of units sold
P = the price of the product
F = total fixed costs
V = variable costs per unit
N = the break even point

The program should ask questions and accept the information. For example, it can ask what decision is to be made.

a. Price
b. Costs
c. Number of units

etc.

Then, after the computer asks enough information about a specific area, it prints the decision variables and the information such as:
If:

```
the fixed cost is $XX
the variable cost is $XX
the estimated number of units which can be
sold is XX
```

then

```
the price should be more than $XX to make
any profit
```

or
If:

```
equipment is bought for $XX
the variable cost is $XX per unit
price is $XX
other information the same as before
```

then

```
the break even is XX units
```

☐ Operations Research

10.45 If you are familiar with the simplex method, develop a program to solve a linear-programming problem. Assume the problem is in the following form:

Maximize $Z = C_1X_1 + C_2X_2 + \cdots + C_kX_k$

subject to

$$a_{11}X_1 + a_{12}X_2 + \cdots + a_{1k}X_k \leq b_1$$
$$a_{21}X_1 + a_{22}X_2 + \cdots + a_{2k}X_k \leq b_2$$
$$\vdots$$
$$a_{m1}X_1 + a_{m2}X_2 + \cdots + a_{mk}X_k \leq b_m$$
$$X_1, X_2, \ldots, X_k \geq 0$$

The program starts by reading the input data as follows:

1. The number of variables, K.
2. The number of constraints, M.
3. Whether the problem is for maximization or minimization.
4. The coefficients of the objective function.
5. The coefficients of the constraints.
6. The coefficients of the right-hand side.

The output is the final tableau, the value of each variable, and the value of the objective function. You may find the following steps helpful in solving the problem:

$$\begin{array}{c} \text{I} \downarrow \\ \text{III} \left\{ \begin{bmatrix} -A_{1,1} & -A_{2,1} & \cdots & -A_{1,k} & | & -A_{1,k+1} & \cdots & -A_{1,n} & | & 0.0 \\ \hline A_{2,1} & A_{2,2} & \cdots & A_{2,k} & | & A_{2,k+1} & \cdots & A_{2,n} & | & A_{2,n1} \\ A_{3,1} & A_{3,2} & \cdots & A_{3,k} & | & A_{3,k+1} & \cdots & A_{3,n} & | & A_{3,n1} \\ \vdots & \vdots & & \vdots & | & \vdots & & \vdots & | & \vdots \\ A_{m1,1} & A_{m1,2} & \cdots & A_{m1,k} & | & A_{m1,k+1} & \cdots & A_{m1,n} & | & A_{m1,n1} \end{bmatrix} \right\} \text{II} \\ \text{IV} \uparrow \end{array}$$

Figure 10.10 Array A for the simplex algorithm.

Step I. Create the standard form of the linear programming problem. This is done by adding M slack variables to the model, thus having

$N = K + M$

variables in the system. Place all the coefficients in two-dimensional array A (Figure 10.10). A has $N1$ columns and $M1$ rows, where

$N = M + K$
$N1 = N + 1$
$M1 = M + 1$

The other elements of A can be grouped as follows:

- Part I, the first row, contains the negatives of the coefficients of the objective function, including the coefficients for the slack variables. The coefficients of the slack variables in the objective function are zero for a maximization problem (and a large number, such as -99999.99, for a minimization problem).
- Part II, the last column, contains the coefficients of the right-hand side.
- Part III contains the coefficients of the constraints.
- Part IV contains the coefficients of the slack variables of the constraints. These coefficients are all equal to zero except the ith one in the ith constraint, which has a value of one (i.e., this submatrix has 1's in its principal diagonal and 0's everywhere else). This can be accomplished by first setting all of them equal to zero in a DO loop:

```
      DO 10 I = 2, M1
         L = K + 1
         DO 10 J = L,N
            A(I,J) = 0.0
10    CONTINUE
```

and then

```
      DO 20 I = 2,M1
         L = K + I - 1
         A(I,L) = 1.0
20    CONTINUE
```

Step II. Complete the following tasks:

1. Define an array X with N elements to hold the values of the variables.
2. Keep track of the basic variables in an array XB, defined as an integer array with M elements. There are always M basic variables. For example, the initial basic variables are set by

```
            DO 100 I = 1,M
              XB(I) = K + I
10          CONTINUE
```

3. Find the basic feasible solution. All the variables in the basic feasible solution are equal to zero, except the variables in the basic solution. The statements

```
            DO 200 I = 1,N
              X(I) = 0.0
200         CONTINUE
```

can be used to set all the variables equal to zero. Then the statements

```
            DO 300 I = 1,M
              K = XB(I)
              L = I + 1
              X(K) = A(L,N1)
300         CONTINUE
```

determine the values of the basic variables.

* **Step III.** Choose the variables to be introduced into the basic solution. This can be done by finding the *pivot column*, i.e., the column which has the smallest negative value in the first row (the objective function). Call the subscript of this column IN. If there is no negative value in the first row, the solution is optimal; then go to step VIII. Use a subroutine for this step. Consider IN = 0 to indicate the optimal solution.

* **Step IV.** Choose the variable which must leave the basic solution by finding the *pivot row*. The pivot row can be found as follows:

1. For each row i compute the ratio b_i/a_{ij} in the pivot column for every a_{ij} greater than zero (note that b_i is $A_{i+1,n1}$ and a_{ij} is $A_{i+1,j}$ in A).
2. The row with the minimum value of this ratio is the pivot row. Call this row IROW. Use a subroutine for this purpose. Consider IROW = 0 to indicate no positive a_{ij} in column IN.

Step V. Find the variable which enters the basic solution:

XB(IROW - 1) = IN

* **Step VI.** Perform the necessary row operations to convert column IN into a unit column. This can be accomplished as follows:

1. Find the pivot, which is: A(IROW,IN)
2. Divide the pivot row by the pivot. This makes the value of the pivot equal to one.
3. Make all the elements in the pivot column equal to zero except the element in the pivot row, which must stay equal to one. This can be accomplished as follows:

   a. for each row i, except the pivot row, find the multiplier XR as

   XR = A(I,IN)

   b. Multiply each element in the pivot row by XR, and subtract the result from the corresponding element in row i.

---

* Answers at the end of the chapter.

Make sure to include all rows and columns of array A in the operations (including the first row and the last column). Use a subroutine for this purpose.

Step VII. Go to step III.

Step VIII. If the optimal solution is reached, print the information in the array A, values for the variables (as shown in step II.3), and the value of the objective function. Note that the value of the objective function is stored in element $A(1,N1)$.

SUMMARY OF CHAPTER 10

You have learned:

1. Developing an algorithm to solve a problem is itself an art. An algorithm was defined as the detailed solution steps necessary for computer programming.
2. The following hints help to write a more efficient program:
    a. Using structured programming techniques improves the efficiency of human effort in developing, debugging, testing, documenting, and maintaining a program.
    b. Using an effective and appropriate algorithm and mathematical techniques for a particular problem.
    c. Considering the features of the language being used.
    d. Not using arrays unless it is necessary.
3. Some of the techniques used for data processing are as follows:
    a. *Sorting* is the process of rearranging the items of a list into a required order. The bubble sort procedure starts by placing the largest (or the smallest) item at the bottom (or at the top) of the list by comparing contiguous items, one pair at a time, and interchanging them if necessary. The process is then repeated as many times as necessary until all the items are in the desired order. However, there are many more efficient techniques for sorting than the bubble sort.
    b. *Searching* is the process of finding an item in a list. Several techniques are available for searching.
    c. *Merging* is the process of combining two or more lists into one single sorted list.
4. The methods of finding the median and coefficients of a regression line by a computer program are examples of applications of computer in statistics.
5. The methods of converting decimal numbers to binary numbers, finding roots of an equation, and solving a set of linear equations are a few examples of the applications of computers in mathematics.
6. Computers can be instructed to print a graph, chart, or picture. The technique is to print characters in appropriate places. Plotting a function and a histogram were discussed as examples of plotting techniques.
7. In computer simulation, a program imitates a process in order to study its behavior. One of the advantages of computer simulation is the ability to experiment with the system. Generating random numbers is a necessary process for most simulation programs.

SELF-TEST REVIEW

10.1 Develop a solution procedure and write the program segment which reads an integer and determines whether it is odd or even.

10.2 A student has written the following program to simulate the profit of a product with a fixed investment ($50 per day) and fixed cost ($9). Can you rewrite the program, using structured programming or any other technique, to make it more efficient?

```
         DIMENSION DEMAND(1000),X(1000),PROF(1000)
         TOPROF = 0
         Y = 1
5        X(Y) = RANF()
         IF(X(Y) .LT. .5) GO TO 10
         IF(X(Y) .GT. .5) GO TO 20
10       DEMAND(Y) = 5
         GO TO 100
20       IF(X(Y) .LT. .8) GO TO 30
         IF(X(Y) .GT. .8) GO TO 40
30       DEMAND(Y) = 9
         GO TO 100
40       IF(X(Y) .LT. 1.0) GO TO 50
         IF(X(Y) .GT. 1.0) GO TO 60
50       DEMAND(Y) = 14
         GO TO 100
100      INVEST = 50
         PRICE = 9
         PROF(Y) = DEMAND (Y)*PRICE - INVEST
         TOPROF = TOPROF + PROF(Y)
         Y = Y + 1
         IF(Y .GE. 1000) GO TO 99
         GO TO 5
99       AVPROF = TOPROF/1000
         PRINT*,'AVG PROF =',PROF
60       STOP
         END
```

☐ **Answers**

10.1 Solution procedure: An even number is a number which is divisible by 2. Thus, if we divide the number by 2 and the remainder is equal to zero, the number is even; otherwise the number is odd.

Program plan:

    a. Read the number, call it N
    b. Find the integer of the number divided by 2
    c. Find the remainder
    d. If the remainder is zero, then
                      Print EVEN;
        otherwise
                        print ODD.

460 Problem Solving and Programming Techniques

Program:

```
      READ*,N
      M = INT(N/2)
      L = N - 2*M        (or L = MOD(N,2))
      IF(L .EQ. 0)THEN
               PRINT*,' EVEN'
      ELSE
               PRINT*,' ODD'
      ENDIF
      END
```

10.2

```
      PROGRAM PROFIT
      REAL INVEST
      DATA INVEST, PRICE , TOPROF/ 50.0 , 9.0 ,
     *0.0/
C SIMULATING THE PROCESS FOR 1000 DAYS
      DO 10 I = 1,1000
           CALL DEMAND(DEM)
C CALCULATING THE PROFIT, FIXED INVESTMENT EACH
C DAY
           PROF = DEM*PRICE - INVEST
           TOPROF = TOPROF + PROF
10    CONTINUE
C CALCULATING THE AVERAGE PROFIT
      AVGP = TOPROF/FLOAT(1000)
      PRINT*,'AVERAG PROFIT = ',AVGP
      END
C  ******************************************
      SUBROUTINE DEMAND(D)
          X = RANF()
          IF(X .LT. .5) THEN
                   D = 5.0
              ELSE IF(X .LT. .8) THEN
                   D = 9.0
              ELSE
                   D = 14.0
          END IF
      RETURN
      END
```

ANSWERS TO SELECTED EXERCISES

10.45

For step III:

```
      SUBROUTINE CEJAY(A,M1,N1,IN)
      DIMENSION A(M1,N1)
      N = N1 - 1
      IN = 0
```

Answers to Selected Exercises

```
              SMALL = 0.0
              DO 100 I = 1,N
                   IF (A(1,I) .GE. 0.0) GOTO 100
                   IF (A(1,I) .LT. SMALL) THEN
                        SMALL = A(1,I)
                        IN = I
                   ENDIF
100           CONTINUE
              RETURN
              END
```

For step IV:

```
              SUBROUTINE BEES(A,M1,N1,IN,IROW)
              DIMENSION A(M1,N1)
              IROW = 0.0
              SMALL = 9999.9
              DO 100 I = 2,M1
                   IF(A(I,IN) .LT. 0.0) GOTO 100
                   RATIO = A(I,N1)/A(I,IN)
                   IF(RATIO .LT. SMALL) THEN
                        SMALL = RATIO
                        IROW = I
                   ENDIF
100           CONTINUE
```

For step VI:

```
              SUBROUTINE OPERAT(A,M1,N1,IN,IROW)
              DIMENSION A(M1,N1)
              PIV = A(IROW,IN)
              DO 10 J = 1,N1
                   A(IROW,J) = A(IROW,J)/PIV
10            CONTINUE
              DO 200 I = 1,M1
                   XR = A(I,IN)
                   DO 200 J = 1,N1
                        IF(I .EQ. IROW) GOTO 200
                        A(I,J) = A(I,J) - A(IROW,J)*XR
200           CONTINUE
              RETURN
              END
```

Chapter 11

Additional Features of FORTRAN

MORE ABOUT DATA TYPE

　　Exponent Form of Real Data
　　Precision of Data in a Program
　　Double Precision
　　Logical Data
　　Complex Data

MORE ABOUT FORMAT

　　G Descriptor
　　P Descriptor—Scale Factor
　　BN and BZ Descriptors—Blank Interpretation

ADDITIONAL FEATURES AND STATEMENTS

　　EQUIVALENCE Statement
　　IMPLICIT Statement
　　BLOCK DATA Subprograms
　　General Form of READ and WRITE
　　Variable FORMAT
　　OPEN and CLOSE Statements
　　EXTERNAL and INTRINSIC Statements
　　PAUSE Statement

Chapter 6 provided a brief introduction to data types. In the first part of this chapter we continue our discussion of data types. In the second part, additional features of FORMAT are considered. Finally, in the third part we look at some additional FORTRAN statements.

■ MORE ABOUT DATA TYPE

In Chapter 6 we mentioned that the data types in FORTRAN are

- Real
- Integer
- Character
- Double precision
- Logical
- Complex

In this chapter we discuss double-precision, logical, and complex data.

Furthermore, exponent form and precision of data are two important topics in FORTRAN. They are also important to discuss in order to better understand double precision. Thus, these two topics are covered first.

☐ Exponent Form of Real Data

Often the letter E is used to indicate the order of magnitude of real data when the decimal point is displaced. For example, 57200000.0 may be written as 572.0E5, by moving the decimal point 5 places to the left. An obvious advantage of using this notation is that a large number can be written (and stored in the memory of a computer) in a shorter form. For example, the number

26759000000000000000000000000000000

can be written and stored as

.26759E35

In mathematical terms, 572.5E3 means "572.5 times 10 to the third power," or 572.5×10^3. Table 11.1 provides additional examples of numbers in E form. Notice that when the exponent is negative, the decimal point will be displaced n positions to the left.

Table 11.1 Presenting Data in E Form

	MATHEMATICAL NOTATION	
E FORM	EXPONENT FORM	COMMON FORM
1.0E0	1.0×10^0	1.0
5612867.E5	5612867.0×10^5	561286700000
.9999999E+15	$.9999999 \times 10^{15}$	999999900000000
−632.5213E+9	-632.5213×10^9	−632521300000
2.321367E12	2.321367×10^{12}	2321367000000
892523.5E−17	892523.5×10^{-17}	.000000000008925235
+5623.527E09	$+5623.527 \times 10^9$	5623527000000

Table 11.2 Format of Data in E Form

DATA	FORMAT	VALUE (READ)	MATHEMATICAL NOTATION EXPONENT FORM	COMMON FORM
6.50E5	E6.2	6.50E5	6.5×10^5	650000
56.20E+10	E9.2	56.20E+10	56.20×10^{10}	562000000000
0.95321E+12	E11.5	0.95321E+12	$.95321 \times 10^{12}$	953210000000
59232.1E−5	E10.1	59232.1E−5	59232.1×10^{-5}	.592321
−29321.0E−7	E11.1	−29321.0E−7	-29321.0×10^{-7}	.00293210
+529321E+8	E10.4	+52.9321E+8	$+52.9321 \times 10^8$	+5293210000

Input and Output of Data in E Form

E specification is used to input or output real type data in E form. The descriptor has the form

$En.m$

where n indicates the number of columns which are occupied by the number (including plus or minus signs, digits, the decimal point, and the letter E), and m indicates the number of digits to the right of the decimal point. For example, E10.2 accommodates the number

```
5625.63E+5
```

i.e., a total of 10 columns, with 2 decimal places. Table 11.2 presents additional examples of E specifications.

Note that if a decimal point is provided externally with the data, the m in $En.m$ is ignored. Also, the + is optional if the exponent is positive. For example, all the following forms are equivalent:

E5   E05   E+5   E+05

When E form is used for output, the width n must be large enough to contain the digits, minus sign (if any), the decimal point, and the letter E. If the width is not large enough to contain the value, asterisks will be printed. If the field is larger than the value, the value is printed right justified with blanks on the left. Data in integer mode cannot use the E form. The following is an example of using the E form in a program.

EXAMPLE 11.1

■ **Problem:** Write a program which reads the following data line, and prints the data and the sum of the numbers in E form.

Data:

```
6543.2E+12 3296.583
```

466 Additional Features of FORTRAN

> **Program:**
>  ```
>       PROGRAM EXPO
>       READ(5, 10)A, B
>       SUM = A + B
>       WRITE(6, 20)A, B, SUM
>  10   FORMAT (E10.1, 1X, F8.3)
>  20   FORMAT (1X,3E20.2)
>       END
>  ```

☐ Precision of Data in a Program

The digits in a real number, not including the decimal point or any leading or trailing zeros, are referred to as significant digits (that is, they are the digits of the whole part plus the digits in the fractional part). For example,

493.5632

has seven significant digits, and

98236.2

has six significant digits.

Because the size of a memory cell (called a memory word) which can hold data is limited, so are the magnitude and the number of significant digits of an individual datum. Table 11.3 presents examples of the maximum number of significant digits and magnitude of numbers for several computer systems. For example, the limitation for an IBM-370 is as follows:

Maximum significant digits: 7
Magnitude: 10^{-78} to 10^{75}

Thus, the following are invalid numbers for the IBM-370 example:

5619.3259E+5	Significant digits are more than seven
325.52E075	Exponent has more than two digits
00.982578987E4	Significant digits are more than seven
5232.25678E3	Significant digits are more than seven
2153.25E85	Exponent is larger than 75
621.52E−82	Exponent is smaller than −78

Table 11.3 Typical Ranges of Data

	IBM 360, 370, 43XX, 30XX	CDC CYBER 170	BURROUGHS 6700	UNIVAC 1100	YOUR SYSTEM
Word length, bits	32	60	48	36	
Range of magnitudes of data	10^{-78} to 10^{75}	10^{-293} to 10^{322}	10^{-47} to 10^{67}	10^{-38} to 10^{+38}	
Significant digits (single precision)	7	14	11	9	
Significant digits (double precision)	15	29	23	18	

If a number's significant digits are more than the allowable number n, the number will usually be rounded off internally. That is, all the digits after nth will be discarded. However, if the first discarded digit is 5 or greater, one will be added to the last retained digit. Here are some examples of numbers rounded off to seven significant digits:

Unrounded	Rounded
65.6328729512	65.63287
5.693283621	5.693284
3657823.2515	3657823.

However, some computer systems simply truncate the data.

In numerical applications, roundoff errors may create serious problems. Fortunately, the double precision feature of FORTRAN, discussed in the next section, can be used if additional precision is necessary.

SOLVED PROBLEMS

11.1 Express the following numbers in E notation with two decimal places, and write the FORMAT descriptor for each. Assume that the computer being used will not accept more than seven significant digits or a number larger than 10^{75}.

a. 623150000000
b. -59826100000
c. $+.00008262100$
d. $.621315 \times 10^{025}$
e. -56.2986×10^{-50}
f. 56283
g. 632.546×10^{89}
h. $.638956214 \times 10^{-5}$
i. $+.8932526432 \times 10^{-10}$
j. 562398275932
k. .000000008923

☐ **Answers**

a. 623.15E9   E8.2
b. -5982.61E7   E10.2
c. $+826.21$E-7 E10.2
d. 6213.15E21   E10.2
e. -5629.86E-52 E12.2
f. 562.83E2   E8.2
g. The number is too large.
h. 63895.62E-10 E12.2
i. $+89325.27$E-15 E13.2
j. 56239.83E7   E10.2
k. 89.23E-10 E9.2

Note that the numbers in **h**, **i**, and **j** are rounded off.

11.2 Are the following constants correct? Assume the computer being used will not accept more than seven significant digits or a number larger than 10^{75}.

a. 623.68E−072
b. 62.5000E−3
c. 56231.5281E2
d. 62.16231E−69
e. 5.932168E+71
f. .5621983E−07
g. 3.21933281E108

☐ **Answers**

a. Incorrect; there are more than two exponent digits.
b. This is correct.
c. Incorrect; the significant digits are more than seven.
d. This is correct.
e. This is correct.
f. This is correct.
g. Incorrect; the exponent is more than 75, and the significant digits are more than seven.

■ **11.3** What is the value of each variable read by the following statements? (Assume the character variables have been defined.)

a.
```
        READ(5,15)M,N,P,Q
15      FORMAT(A4,1X,I3,1X,E9.4,F6.3)
```
Data:

```
BITEϕ562ϕ962.50E+7529.31
```

b.
```
        READ(5,20)XX,KOOL,NORM,PAT
20      FORMAT(5X,E10.6,2I4,A4)
```
Data:

```
1E2593212825E-759623118E59E1
```

c.
```
        READ(5,50)A,B,C,K
50      FORMAT(F9.3,E13.3,I3,A4)
```
Data:

```
4962582ϕϕϕϕ5341754E+7ϕ93ϕEXPO
```

d.
```
        READ(5,60)X,Y
        FORMAT(2E15.3)
```
Data:

More about DATA Type   469

```
    ␣␣-623.532␣␣E2␣␣+81532␣␣E20
```

☐ **Answers**

a.

```
M = BITE
N = 562
P = 962.50E+7
Q = 529.31
```

Note: The 4 in E9.4 and the 3 in F6.3 are ignored because the numbers are entered with decimal points.

b.

```
XX = 3.212825E-7
KOOL = 5962
NORM = 3118
PAT = E59E
```

c.

```
A = 4962.582
B = 5341.754E+7
C = 93
K = EXPO
```

Note: The blank spaces are ignored in FORTRAN 77.

d.

```
X = -623.532E2
Y = +81.532E20
```

☐ **Double Precision**

Real data in their usual form as described in earlier chapters are referred to as *single precision*. The precision of data can be increased by using the *double-precision* feature of FORTRAN. Double-precision values are represented internally by two memory words. Thus, the maximum number of significant digits of double-precision data is approximately two times that of single-precision data.

Double-Precision Constants

A double-precision constant is written in the same way as a real constant in exponent form, except the letter E is replaced by the letter D. Following are examples of valid double-precision constants:

85.9362D3	Value: 85936.2
693.28D−2	Value: 6.9328
5628.9D05	Value: 562890000

Following are examples of invalid double precision constants:

6.9E5	Letter D should be used instead of E
6.9D	Exponent missing
5662.39	D and exponent missing

DOUBLE PRECISION Statement

In order to use double-precision data, the variable must be defined by a DOUBLE PRECISION Statement. The general form of the statement is

```
DOUBLE PRECISION var1, var2, ...
```

This statement is a specification statement, and overrides previously defined variables. For example,

```
DOUBLE PRECISION RHO , NUM , L
```

declares the variables RHO, NUM, and L as double precision, although NUM and L would otherwise have been integers.

The DOUBLE PRECISION statement can also be used to define arrays and function names as double-precision type. Following are some examples:

```
DOUBLE PRECISION A(100),DROP(50,5),BAY,NORM
DOUBLE PRECISION KAY(50,40),NUM
DOUBLE PRECISION FUNCTION FUN(A,B)
```

Double-Precision Input and Output

Double-precision data are inputted or outputted in the same way as real data. However, when the data are written in exponent form, a D specification must be used. The general form of the D specification is:

D$n.m$

This specification is like E specification except that the letter D replaces E. The following is an example of using double-precision data in a program.

EXAMPLE 11.2

■ **Problem:** Write a program which calculates and prints 22/7 with 14 significant digits (assuming seven significant digits for real data).

Program:

```
      PROGRAM PRECIS
      DOUBLE PRECISION X
      X = 22.0D0 / 7.0D0
      WRITE(6,10)X
10    FORMAT(1X,D20.14)
      END
```

■ SOLVED PROBLEMS

■ **11.4** What value is stored for each of the variables in the following cases? Assume the number of allowable significant digits of the computer is seven, and the variables are single precision, unless otherwise declared.

    a.   X = 56.50000000000

    b.   Y = 392.2358214

    c.   J = 65.98786293800

d.   DOUBLE PRECISION Y,J
            J = 65.98786293800D0
            AREA = 69532.532896

e.   DOUBLE PRECISION W
            W = 3.5682539687D25

f.   DOUBLE PRECISION P,Q
            Q = 3.0000000000000
            P = 5.0*Q + 3.0
            R = P + Q

g.   DOUBLE PRECISION X,Y
            READ(5,10)X,Y
      10   FORMAT(2D10.4)

Data:

```
693.2387D956238954D22
```

h.   DOUBLE PRECISION X,Y,Z
            X = 111.11111111D0
            Y = 222.22222222D0
            Z = X + Y
            WRITE(6,10)X,Y,Z
      10   FORMAT(1X,D20.8)
            END

What values are printed?

☐ **Answers**

a. X = 56.50000

b. Y = 392.2358

c. J = 65

d. J = 65.98786293800D0
       AREA = 69532.53

e. W = 3.5682539687D25

f. Q = 3.0000000000000
       P = 18.000000000000
       R = 21.00000

(R is not declared as double precision.)

g. X = 693.2387D9
       Y = 5623.8954D2

h. Z = 333.33333333D0

Printed data:

```
1111.11111111D-01
2222.22222222D-01
3333.33333333D-01
```

Note: Compare this result with that of **f**.

472 Additional Features of FORTRAN

☐ Logical Data

Logical-type data can be used for applications where the expected data are in the form of yes/no, true/false, positive/negative, on/off, or one/zero.

Logical Constants

In contrast to numerical constants, which can assume any numerical values, logical constants can take only two values:

    .TRUE.
    .FALSE.

The periods are part of the constants and must appear. For example, if X and Y are logical variables, then the statements

    X = .TRUE.
    Y = .FALSE.

are correct, but

    X = .T.
    Y = .F.

are not. Logical variables must be defined with the LOGICAL statement.

LOGICAL Statement

The LOGICAL statement can be used to define a logical variable, array, or function name. The general form of the statement is

    LOGICAL var1, var2, ...

Example:

    LOGICAL A, B, NORM, TEST

    LOGICAL POT(100), CORD, N

    LOGICAL FUNCTION X(A,B)

Logical Expressions

A *logical expression* is used to express a logical computation. The relational operators

    .GT.
    .GE.
    .LT.
    .LE.
    .EQ.
    .NE.

are used to construct a logical expression. The following are examples:

    (37.50 .LT. 5.0)

    (B .GT. 89.50)

    (C .EQ. A)

    (X + Y .LT. 2*X + Z)

    (I .GT. K - 10)

Note that the parentheses are not part of the expression but can be used if desired.

Evaluation of a logical expression produces a value of either .TRUE. or .FALSE.. The simplest use of logical expressions is in the logical IF statement discussed in Chapter 4.

Logical Assignment

A logical variable, logical expression, or logical constant can be assigned to a logical variable by an assignment statement. For example:

```
LOGICAL X, Y, Z, V, NET
A = 50.0
X = (A .LT. 45.0)
Y = 6.0 .GT. A
NET = .TRUE.
Z = NET
    ⋮
```

Again, the use of parentheses is optional. *Logical values and numerical values cannot be mixed.* For example, the logical assignment statement in:

```
LOGICAL X,A
A = .TRUE.
X = A + 5.0
```

is not valid. However, the following expression is valid:

```
X = B + 6.5 .GT. C - 5.5
```

The initial value of a logical variable can also be assigned by a DATA statement. For example:

```
      LOGICAL A, B, C
C
      DATA A, B, C/.TRUE., 2* .FALSE./ X, Y/2*0.0/
```

The following example illustrates how logical variables can be used in a program to turn a FLAG to "OFF" when the value of an integer variable is even (note that ON, OFF, and FLAG are symbolic names of variables in this example):

```
LOGICAL ON,OFF,FLAG
DATA ON,OFF/.TRUE.,.FALSE./
READ*,N
L = MOD (N,2)
FLAG = ON
IF(L .EQ. 0)FLAG = OFF
   ⋮
```

Logical Operators

A *compound logical expression* can be formed by using one or more logical expressions linked with the following *logical operators:*

.AND.
.NOT.
.OR.
.EQV.   (for "equivalent")
.NEQV. (for "not equivalent")

Here are some examples:

```
LOGICAL X,Y
I = 13
```

```
K = 24
X = (I .GT. K) .OR. (I .LT. L)
Y = (Q .LT. 3.5) .AND. (Q .GT. 0.0)
```

Two logical operators cannot be used in sequence except when the second one is .NOT.. For example,

```
A .LT. B .AND. .NOT. A .GT. C
```

is valid, but

```
X .LT. Y .AND. .OR. X .GT. Y
```

is not valid.

The outcome of a compound logical expression depends on the outcomes of the logical expressions used. Table 11.4 summarizes the outcomes of simple compound logical expressions in the forms:

(logexp1 .AND. logexp2),
(logexp1 .OR. logexp2),
(.NOT. logexp).

For example, if A = 1.0, and B = 2.0, then the value of logical variable X in the statement

```
X = A .GT. B .AND. A .GT. 0.0
```

is false, as can be seen from this diagram:

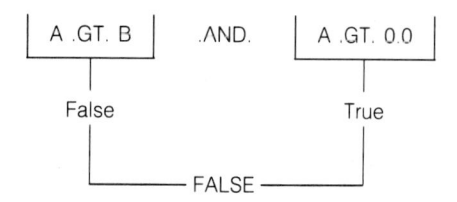

```
| A .GT. B |  .AND.  | A .GT. 0.0 |
   False              True
   |_____ FALSE _____|
```

The operator .EQV. checks for the *equivalence* of two logical expressions. The outcome of a simple compound expression with this operator is true when the outcomes of the first and second expression are the same, i.e., either both are true, or both are false. For example, the outcome of the expression

```
((3 .LT. 5) .EQV. (10 .LT. 11))
```

or

```
((30 .LT. 20) .EQV. (90 .GT. 100))
```

Table 11.4 The Outcomes of Simple Compound Logical Expressions

logexp1	logexp2	logexp1 .AND. logexp2	logexp1 .OR. logexp2
True	True	True	True
False	False	False	False
True	False	False	True
False	True	False	True

logexp	.NOT. logexp
True	False
False	True

Table 11.5 Examples of Compound Logical Expressions

THE EXPRESSION	FIRST PART	SECOND PART	THE OUTCOME
B .GT. A .AND. A .GT. 0.0	True	True	True
A .GT. B .AND. A .LT. 0.0	False	False	False
B .LT. A .OR. A .GT. 0.0	False	True	True
B .LT. A .OR. A .LT. 0.0	False	False	False
.NOT. A .LT. B	Not true		False
.NOT. A .GT. B	Not false		True
A .LT. B .AND. .NOT. A .LT. 0.0	True	Not false	True
B .GT. A .EQV. B .GT. 0.0	True	True	True
A .GT. B .EQV. B .GT. A	False	True	False
A .GT. B .NEQV. B .GT. A	False	True	True

is true, whereas the outcome of

 ((5 .LT. 6) .EQV. (10 .GT. 20))

is false. The operator .NEQV. checks for *inequivalence* of the outcome of two expressions. For example, the outcome of the expression

 (70 .LT. 60 .NEQV. 50 .GT. 40)

is true. Table 11.5 gives further examples of simple compound logical expressions and their outcomes, assuming A = 1.0 and B = 2.0.

If a logical expression contains two or more logical operators, the priority of execution from highest to lowest is as follows (unless changed by the use of parentheses):

first: .NOT.
second: .AND.
third: .OR.
fourth: .EQV. or .NEQV.

For example, assuming X, Y, and Z are logical variables, the expression

 (X .OR. Y .AND. Z)

is interpreted in the following order:

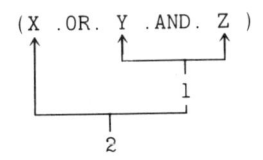

 (X .OR. Y .AND. Z)
         └───1───┘
   └─────2─────┘

Furthermore, if the logical expression contains operators with the same priority, the quantities are combined from left to right.

Logical expressions can become quite complicated. For example, the statement

 Y = (A .LT. B) .OR. .NOT. (A .GT. C) .AND. (A + B .GT. Q)

is valid. So complex an expression requires extra caution so as not to violate the rules for compound logical expressions.

IF Statement and Logical Expressions

Logical expressions and logical variables can be used with the IF statement. The following are some examples:

 IF (A .LT. B) GO TO 10

```
      IF ((A .LT. B) .AND. (A .LT. C)) DIS = .05
      IF ((Q .EQ. 55.0) .AND. (P .LT. 99.9)) THEN
                                 P = 75.0
      ELSE
                                 Q = 75.0
      ENDIF
      IF ((I .LT. 100) .AND. .NOT. (I .GT. 75)) GO TO 50
      IF (.NOT. (L1 .OR. L2)) GO TO 100
```

(assuming L1 and L2 are logical variables).

```
          IF ( A .EQ. B .AND. C .LT. D
         *.OR. X .EQ. Y .AND. V .LT. W)
         *     GO TO 30
          LOGICAL D, EEE, B, X
             ⋮
          EEE = C .GT. 99.9
             ⋮
          IF(EEE) GO TO 30
             ⋮
          IF(EEE)   THEN
                        Q = 55.5
          ELSE
                        P = 75.0
          ENDIF
             ⋮
          IF(B .AND. X)GO TO 100
             ⋮
          IF(D .AND. .NOT. B)PRINT*,X
             ⋮
          IF(D .EQV. B)X = .TRUE.
```

Relational operators (.GT., .LT., .EQ., etc.) are used only to compare numerical values, and cannot be used for logical values. For example,

```
LOGICAL X,Y
   ⋮
IF(X .EQ. Y)GO TO 10
```

is not valid. One way to compare logical variables is to use operators (.AND., .OR., .NOT.). For example, if FLAG and ON are logical variables, then the expression (FLAG .AND. ON) in

```
IF(FLAG .AND. ON)GO TO 10
```

is true only if FLAG and ON are both true. The logical expression in

```
IF(FLAG .AND. ON .OR. .NOT. (FLAG .OR. ON))GO TO 10
```

is true only if FLAG is equivalent to ON, i.e., either both are true, or both are false.

Of course, the operators .EQV. and .NEQV. can be used to compare logical variables. For example, the logical expression in the statement

```
IF(FLAG .EQV. ON)GO TO 10
```

is true if FLAG and ON are the same; either both are true or both are false.

The following example demonstrates using logical operators in an IF statement with numeric data.

EXAMPLE 11.3:

Problem: A data record is prepared for each of the 100 students in a course as follows:

Column 1, sex: 1 for male and 2 for female
Column 2, class standing: 1 for freshman, 2 for sophomore, 3 for junior, 4 for senior.

Write a program which reads the information and prints the number of:

a. male freshmen, MALEFR
b. female seniors, FEMSEN
c. male juniors or sophomores, MJS

Program:

```
      PROGRAM LOGIF
      INTEGER SEX,CLASS,MALEFR, FEMSEN, MJS
      DATA MALEFR, FEMSEN, MJS/3*0/
      DO 100 I = 1,100
          READ(5,10)SEX,CLASS
              IF(SEX .EQ. 1 .AND. CLASS .EQ. 1)
     *          MALEFR = MALEFR + 1
              IF(SEX .EQ. 2 .AND. CLASS .EQ. 4)
     *          FEMSEN = FEMSEN + 1
              IF(SEX .EQ. 1 .AND. (CLASS .EQ. 1
     *          .OR. CLASS .EQ. 2))MJS = MJS + 1
          WRITE(6,20)MALEFR, FEMSEN, MJS
100   CONTINUE
10    FORMAT(2I1)
20    FORMAT(1X,'NO. OF MALE FRESHMEN =',I2,
     *'NO. OF FEMALE SENIORS =',I2,
     *'NO. OF MALE JUNIORS OR SOPHOMORES =',I2)
      END
```

Input and Output of Logical Data

The L descriptor is used to input or output logical data. The general form of the L specification is

Ln

where n is the length of the field occupied by the value. For example, the statements

```
      LOGICAL X,Y
      READ(5,10)X,Y
10    FORMAT(L4,1X,L5)
```

with data

```
TRUE FALSE
```

stores the values .TRUE. for X and .FALSE. for Y.

478 *Additional Features of FORTRAN*

The variables must be defined as logical type. However, the entries in the data line do not have to be the word .TRUE. or .FALSE. If the first nonblank character in the field is T or the first two are .T, the value .TRUE. will be stored for the variable. If the first nonblank character is F or the first two are .F, the value .FALSE. will be stored. If the first non-blank character is other than T, .T, F, or .F, the value is unpredictable and depends on the compiler being used. (Some of the compilers print a diagnostic; others will consider certain characters as true and others as false.) However, an all-blank field is considered as .FALSE..

When printing logical data, T or F will be printed right justified with blanks on the left. For example, the following program reads and prints two logical data:

```
        PROGRAM LOGIC
        LOGICAL I,J
        READ(5,10)I,J
        WRITE(6,20)I,J
10      FORMAT(2X,L3,3X,L5)
20      FORMAT(3X,2L3)
        END
```

Data:

```
TTTTTbbbbbFORTRAN
```

The program will print

```
bbbbTbbF
```

The following problem demonstrates using logical type data in a program.

EXAMPLE 11.4:

■ **Problem:** Forty students have taken a test with 50 true-false questions. Each student's name and responses are placed on a line as follows:

Name   Columns 1–12
50 responses, T or F   Columns 13–62

The correct responses (the key) are also recorded on a line and placed before the student data. Each question has a score of 2. Write a program which reads the information and prints the names of the students as well as their scores in report form.

Program:

```
        PROGRAM LOGIC
        CHARACTER*12 NAME
        LOGICAL ANSWER(50), KEY(50)
C READING THE KEY INTO ARRAY KEY
        READ(5,10)(KEY(I),I = 1,50)
```

```
      C READING, COMPARING, AND SCORING 40 STUDENTS
      C RESPONSES, ONE BY ONE.
            DO 100 I = 1,40
                READ(5,20)NAME,(ANSWER(J),J = 1,50)
                ISCORE = 0
                DO 200 K = 1,50
                    IF(ANSWER(K) .EQV. KEY(K))
     *              ISCORE = ISCORE + 2
200             CONTINUE
                WRITE(6,30)NAME,(ANSWER(J),J=1,50)
     *          ISCORE
100     CONTINUE
10      FORMAT(50L1)
20      FORMAT(A12,50L1)
30      FORMAT(1X,A12,5X,50L2,5X,'SCORE=',I3)
        END
```

Hierarchy of Operations

When an expression contains arithmetic operators, relational operators, and logical operators, the priority of execution is as follows:

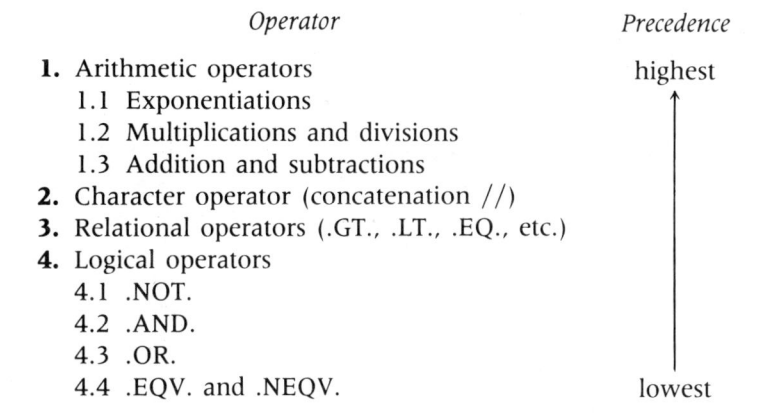

Operator	Precedence
1. Arithmetic operators	highest
1.1 Exponentiations	↑
1.2 Multiplications and divisions	
1.3 Addition and subtractions	
2. Character operator (concatenation //)	
3. Relational operators (.GT., .LT., .EQ., etc.)	
4. Logical operators	
4.1 .NOT.	
4.2 .AND.	
4.3 .OR.	
4.4 .EQV. and .NEQV.	lowest

Furthermore, the arithmetic operators as well as the relational operators are carried out from left to right.

Of course, a pair of parentheses will change these priorities. The expressions in parentheses have the first priority. If there are more than one pair of parentheses, the innermost one will be evaluated first. Often, parentheses must be used to force a desired sequence of operations. More importantly, extra spaces and parentheses should be used in order to make expressions more readable and reduce the chance of error. For example, the expression

    A+B.GT.B*C.OR.X**2.LT.5*X+2

can be written in a more readable form as follows

    (A + B) .GT. (B*C) .OR. (X**2) .LT. (5*X + 2)

This expression is evaluated in the following order:

(A + B) .GT. (B*C) .OR. (X**2) .LT. (5*X + 2)

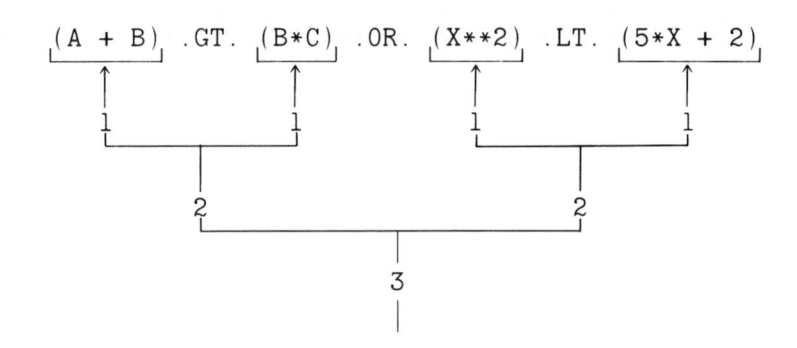

SOLVED PROBLEMS

11.5 Assume the variables A, B, C are defined as logical variables, identify errors in the following statements:

a. A = YES
b. B = TRUE
c. C = .ONE.
d. A = 5.0 * X + 11 * Y + 99.9
e. B = (6 * P .NE. Q + 10.0)
f. C = A .OR. .NOT. B
g. A = (Q .AND. P)
h. C = A NOT B
i. A = A .AND. .NOT. B
j. B = A .NOT. .AND. C
k. C = A .AND. .OR. B
l. A = .TRUE. .OR. .FALSE.
m. A = .NOT. B
n. B = .OR. A
o. IF(A .EQ. B)C = .TRUE.
p. A = A .OR. (.NOT. B .AND. .NOT. C)
q. IF(.NOT. (A .OR. B))GO TO 100
r. IF(A)GO TO 100
s. IF(A .EQV. .TRUE.)B = .FALSE.

☐ **Answers**

a. A logical constant cannot be YES.
b. The periods before and after TRUE are missing.
c. ONE cannot be a logical constant.
d. An arithmetic expression cannot be assigned to a logical variable.
e. This is correct.
f. This is also correct.
g. The logical operator .AND. cannot be used with the variables P and Q, since they are not logical variables.
h. Periods before and after NOT are missing.
i. This is correct.
j. Two logical operators cannot appear in sequence unless the second one is .NOT..

- **k.** The same type of error as **j**.
- **l.** This is correct.
- **m.** This is also correct.
- **n.** The first part of the expression is missing.
- **o.** Logical variables cannot be compared by using the relational operator .EQ. (EQV must be used).
- **p.** This is correct.
- **q.** This is also correct.
- **r.** This is also correct.
- **s.** This is also correct.

■ **11.6** Assume that the variables X, Y, and A are defined as logical variables, and the initial values of X, Y are set by

```
DATA X,Y /.TRUE.  ,   .FALSE./
```

What is the value of A in each of the following statements?

- **a.** A = X .AND. Y .OR. .TRUE.
- **b.** A = X .AND. .NOT. Y .AND.Y
- **c.** A = ((5 .GT. 6) .AND. X)
- **d.** A = (Y .OR. .NOT. (50 .GT. 60)) .AND. X
- **e.** A = ((50 .LT. 6) .OR. (4 .NE. 3).AND. X).AND. .NOT.Y
- **f.** A = (.NOT.(5 * 6 .LT. 40 .OR. X).OR. Y).AND. (X.AND.Y)

☐ **Answers**

- **a.** True
- **b.** False
- **c.** False
- **d.** True
- **e.** True
- **f.** False

■ **11.7** Identify errors in the following IF statements:

a.
```
IF(P .LT. .10 .AND. P .GT. .01)STOP
```

b.
```
IF(AMT .LT. 1000.0 .AND. .GT. 100.0)
*DIS = .05
```

c.
```
IF(IQ .LT. 70 .OR. .GT. 190)GO TO 300
```

d.
```
IF(PAY .EQ. 500.0 .AND. KODE .EQ. 2)
*GO TO 30
```

e.
```
IF  (NET .GT. 600 .AND. KODE .EQ. 5
*      .OR. NET .LT. 6000 .AND. KODE
*      .EQ. 4)GO TO 50
```

f.
```
IF(CLASS .EQ. 1.0 .AND. .OR. SEX
*.EQ. 'F')N = N + 1
```

482 Additional Features of FORTRAN

□ **Answers**

a. This is correct.
b. The second logical expression is not complete. It can be corrected to

```
IF(AMT .LT. 1000.0 .AND. AMT .GT.
*100.0)DIS = .05
```

c. Again, the second logical expression is not complete. It can be corrected to

```
IF(IQ .LT. 70 .OR. IQ .GT. 190)GO TO
*300
```

d. This is correct.
e. This is also correct, but by using parentheses for each expression, it can be made more readable.
f. The operators .AND., .OR. cannot be used in sequence.

■ **11.8** What is the value of the variables in the following statements?

a.
```
      LOGICAL AYE,BEE,JAY
      READ(5,10)AYE,BEE,JAY
10    FORMAT(L3,6X,2L5)
```

Data:

```
THIS IS A TEST FOR YOU
```

b.
```
      LOGICAL Q,R,X
      READ(5,20)Q,R,X,JA
20    FORMAT(L3,L5,L3,I3)
```

Data:

```
THE FOURTH 1000.0
```

■ **Answers**

a. AYE = .TRUE.
   BEE = .TRUE.
   JAY = .FALSE.

b. Q = .TRUE.
   R = .FALSE.
   X = .TRUE.
   JA = 100

□ **Complex Data**

Complex numbers have special applications in science and engineering. If you are not planning to use complex data, skip this section.

A complex number is written as

$a + bi$,

where a is called the real part, b is called the imaginary part, and i is defined as

$i = \sqrt{-1}$.

The standard FORTRAN compilers allow the processing of complex data. The rules concerning complex data are as follows:

1. *Complex constants:* A complex constant is written as a pair of real constants separated by a comma in parentheses. For example,

    (6.0 , 8.95)

    (-6,2E2 , 6.9)

    (0.0 , -1.5)

    (0. , -1.0)

    are complex constants, where the first number is the real part and the second number is the imaginary part. The *parentheses are part of the constants* and must appear. Each complex value is represented internally by two memory locations, one for the real part and the other for the imaginary part.

2. *Complex variables:* Complex variables must be defined by the COMPLEX statement. The general form of this statement is

    COMPLEX var1, var2, ...,

    where var1, var2, ... can be a variable, array, or function name. The following are examples:

    COMPLEX A,B,QUE,BETA,ZETA

    COMPLEX X,Y,NAY(10),RHO(100,2)

    COMPLEX FUNCTION CALC(A,B,C)

3. *Complex assignment:* Complex, real, and integer values can be assigned to a complex variable. The following is an example:

    COMPLEX X,Y,Z
    X = (62.5,3.0)
    Y = 38.0
    Z = 5*X + Y - A

    When a real or an integer value is assigned to a complex variable, the imaginary part is considered to be zero. For example, Y = 38.0 in the above example is equivalent to

    Y = (38.0, 0.0)

4. *Complex arithmetic:* Complex arithmetic can be performed with complex variables (or constants), or with a mixture of complex and real data. For example, if C, D, and E are complex, then, the following statements are valid:

    C = D + E

    C = D + 56.0

    D = 5*C + E*A - 52.8

    C = E*D

If an arithmetic expression contains complex and real values, the real value will be converted to its complex equivalent first and then complex arithmetic will be performed.

5. *Complex library functions:* Most of the complex operations are carried out by complex library functions. A list of commonly used functions are presented in Table 11.6. (It is assumed that the complex variable c is defined as $c = a + bi$.)

6. *Input and output of complex data:* Each complex value is read or written as two independent quantities. The following is an example:

```
PROGRAM CMPLX
COMPLEX RHO , KAA , PA
KAA = (69.3 , 5.9)
A = 39.5
READ(5,10)RHO
```

Table 11.6 Common Complex Library Functions

FUNCTION	PURPOSE	EXAMPLES	VALUE
CABS(c)	Returns the magnitude or the absolute value, of the complex variable c	E = CABS(C) F = CABS((3.0,4.0))	$\sqrt{a^2 + b^2}$ 5.0
REAL(c)	Returns the real part of the complex variable c	P=REAL(C) Q = REAL((2.5,3.8))	a 2.5
AIMAG(c)	Returns the imaginary part of the complex variable c	Q = AIMAG(C) P = AIMAG((9.8,13.2))	b 13.2
CONJG(c)	Returns a conjugate of a complex argument; that is, the result is that the imaginary part will be negated $(a, -bi)$	CONJG((8.5,5.9))	(8.5,−5.9)
CSQRT(c)	Returns the square root of a complex argument	CSQRT((5.0,6.0))	(2.53,1.18)
CMPLX(a,b)	Converts values a and b to a complex value. The result is complex; a and b must be of the same mode, and they can be reals, integers, or double precision. Also, only one argument can be used; in this case, the other is considered zero, or if the one is complex, the result is the same as the argument.	C = CMPLX(D,F) C = CMPLX(5.1,91.3)	(5.1,91.3)

```
10      FORMAT(F5.1,5X,F8.2)
        PA = RHO + KAA + A
        WRITE(6,20)KAA,RHO,PA
20      FORMAT(6X,3(F6.1,5X,F10.2,5X))
        END
```

For further examples illustrating these rules, refer to the solved problems below.

SOLVED PROBLEMS

11.9 Assuming the variables X, Y, Z are defined as complex variables, identify all the errors in the following statements:

    a. X = 5.2, 3.0
    b. Y = (62.0)
    c. Y = (13.0, 0.0)
    d. Z = (5,60)
    e. X = (56.0,9)
    f. Y = (6.9,2.0) + (23.5,10.0)
    g. Z = X + Y − 62.8
    h. X = (Y * Z/5.0 * (92.8,13.5))
    i. Y = (A,B)
    j. Z = X + (A + D,E)
    k. X = REAL((5.8,2.0))
    l. Y = AIMAG((90.3,3.0)
    m. Z = CABS((5.0,2.0))
    n. A = CMPLX(F,G)

☐ **Answers**

    a. Incorrect; parentheses are part of the constant and must always appear.
    b. Incorrect; both real and imaginary part must be present in a complex constant.
    c. This is correct.
    d. This is also correct.
    e. Incorrect; both quantities must be in the same mode.
    f. This is correct.
    g. This is also correct.
    h. This is also correct.
    i. Incorrect; A and B are not constants. This can be corrected, using a library function, to Y = CMPLX(A,B)
    j. Incorrect; the same error as in **i**. It can be corrected to Z = X + CMPLX(A + D,E)
    k. Incorrect; the result of REAL((5.8,2.0)) is real; thus it must be assigned to a real variable.
    l. Incorrect; the same type of error as in **k**.
    m. Incorrect; the same type of error as in **k**.
    n. Incorrect; the result of CMPLX(F,G) is complex. Thus it must be assigned to a complex variable.

MORE ABOUT FORMAT

G Descriptor

The *general* form, or *G descriptor*, can be used to input or output real and complex data. The general form of the G descriptor is

G*n.m*

where *n* indicates the width of the field which the number occupies, and *m* indicates the number of decimal places.

The following rules explain how a G field works:

1. G*n.m* will work like either F*n.m* or E*n.m* in the input, depending upon whether or not the data are recorded with an E. For example, in the statements

   ```
         READ(5,85)A,B
   85    FORMAT(G5.2,1X,G9.3)
   ```

   with data

   ```
   63.58⌿489.32E+5
   ```

   the format works like

   ```
   85    FORMAT(F5.2,1X,E9.3)
   ```

   and the value of A will be read as 63.58, while B is read as 489.32E+5.

2. When G*n.m* is used for output, *m* indicates the number of significant digits to be printed rather than the number of decimal places. Thus, the process of printing goes through two stages:

   a. the data will be rounded off to *m* significant digits, and then

   b. the data will be printed "as is" if the value is greater than .1 but small enough to fit into *n* − 4 columns (four spaces are reserved to print the exponent part, E+ee); otherwise, the data will be printed in E form. Table 11.7 presents some examples.

Notice that the numbers in the table have been rounded off to three significant digits. Then, if the number fits in 5 columns (9 − 4), it is printed as it is;

Table 11.7 Using the G Descriptor for Output

		VALUE AFTER ROUNDOFF		
INTERNAL VALUE	FORMAT USED	EXPONENT FORM	NUMERIC FORM	PRINTED DATA
65.35298	G9.3	.653E2	65.3	65.3
5.38962	G9.3	.539E1	5.39	5.39
2.5	G9.3	.250E1	2.50	2.50
2330000.0	G9.3	.233E7	2330000	.233E+07
9325867.23	G9.3	.933E7	9330000	.933E+07
285960000000.0	G9.3	.286E12	286000000000.	.286E+12
.00008	G9.3	.008E−2	.00008	.008E−02
.15	G9.3	.15E0	.15	.15

otherwise it is printed with E9.3 format. If the field is larger than required, the number is printed right justified with blanks on the left. But when the data are to be printed without exponents, each value will be printed with exactly four blanks at the right end of the field (spaces reserved for E + ee).

SOLVED PROBLEMS

11.10 What is the value of each variable read by the following READ statements?

a.

```
         REAL I
         READ(5,20)X,Y,I,F
20       FORMAT(4G5.2)
```

Data:

```
32.985321465892.59E7
```

b.

```
         REAL LONG, KING
         READ(5,30)LONG,PONY,KING,BOND
30       FORMAT(G3.1,3X,G5.2,T1,G2.1,T11,G7.2)
```

Data:

```
652398532162.31E5
```

c.

```
         READ(5,40)A,B,C
40       FORMAT(3G7.2)
```

Data:

```
3.6986789765218.953E9
```

d.

```
         REAL IRS
         READ(5,10)IRS,CORE
10       FORMAT(2G5.2,3X)
```

Data:

```
5632893593855
```

☐ **Answers**

a.

X = 32.98
Y = 532.14
I = 658.92
F = .59E7

b.

LONG = 65.2
PONY = 532.16
KING = 6.5
BOND = 62.31E5

c.

A = 3.69867
B = 89765.21
C = 8.953E9

d.

IRS = 563.28
CORE = 935.93

11.11 What value will be printed by using the following WRITE statements?

a.

```
        A = 32.52
        B = 563298.56
        WRITE(6,10)A,B
10      FORMAT(1X,G9.3,10X,G9.3)
```

b.

```
        X = 56200000
        Y = 0.005
        WRITE(6,30)X,Y
30      FORMAT(1X,'X=',G9.4,5X,'Y=',G9.4)
```

c.

```
        Z = .0008953167
        P = .5
        WRITE(6,40)Z,P
40      FORMAT(1X,'Z=',G11.5,5X,'P=',G10.4)
```

d.

```
        Q = 23.5
        R = 69.53693
        WRITE(6,50)Q,R
50      FORMAT(1X,'Q=',G9.4,5X,'R=',G9.4)
```

☐ **Answers**

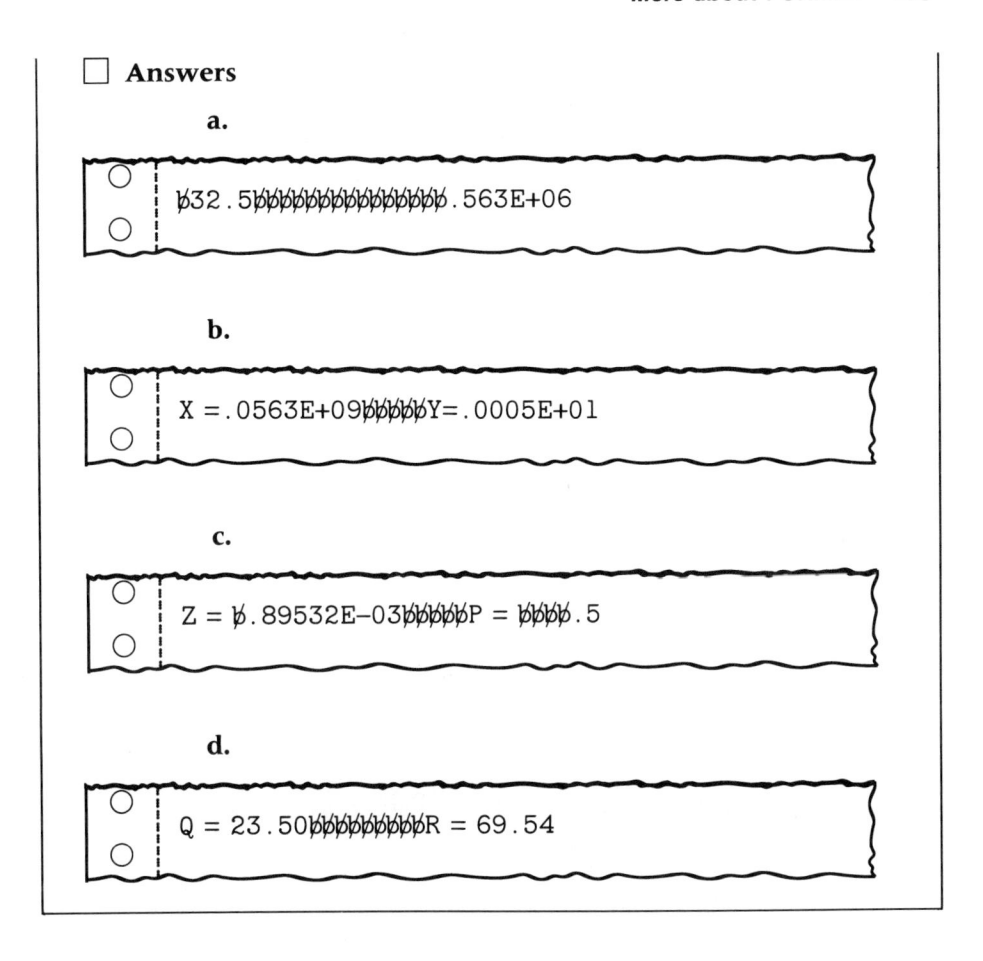

a.
```
 32.5bbbbbbbbbbbbbbbb.563E+06
```

b.
```
X =.0563E+09bbbbbY=.0005E+01
```

c.
```
Z = b.89532E-03bbbbbbP = bbbb.5
```

d.
```
Q = 23.50bbbbbbbbbbR = 69.54
```

☐ P Descriptor—Scale Factor

For some applications we need to shift the decimal point of a real number to the left or right when reading or writing the data. For example, we may wish to print an interest rate with a value of .125 as

    12.5 PERCENT

FORTRAN allows us to do this with the P descriptor. The P descriptor can be used both in input or output. Let's look at an example before explaining this topic further:

EXAMPLE 11.5

■ **Problem:** Suppose the value of RATE is .2575; write the program segment which prints the value as

    RATE = 25.75 PERCENT

Program Segment:

```
        WRITE(6,10)RATE
10      FORMAT(11X,'RATE=', 2PF6.2,1X,' PERCENT')
```

490 Additional Features of FORTRAN

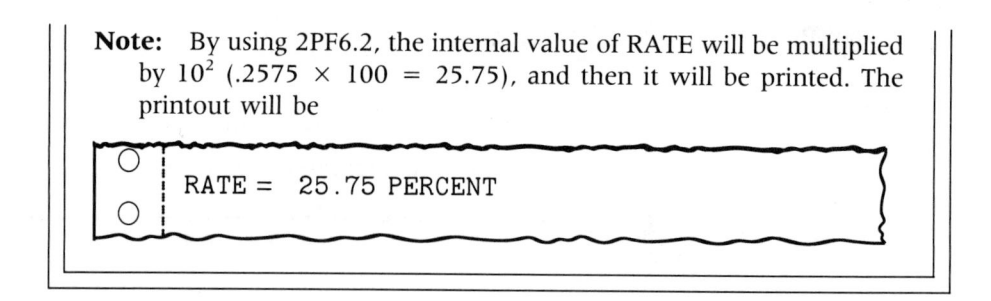

> **Note:** By using 2PF6.2, the internal value of RATE will be multiplied by 10^2 (.2575 × 100 = 25.75), and then it will be printed. The printout will be
>
> ```
> RATE =  25.75 PERCENT
> ```

The general form of the P descriptor is

*k*P

where *k* is an integer constant, positive or negative, and is called the *scale factor*. The scale factor may precede the field descriptors F, E, D, and G; or it can appear independently. The forms are *k*PF*n.m*, *k*PE*n.m*, *k*PD*n.m*, *k*PG*n.m*, and *k*P. The following are examples:

```
10      FORMAT(2PF6.2,-2PF6.2,4PF10.2,0PF3.1,I6)
20      FORMAT(-3PF5.1,3PE15.4,-5PE14.4,0P,F6.1)
30      FORMAT(4PF20.5,-3PG14.4,5P,F6.1)
40      FORMAT(0PF5.2,-3PF5.1,0P,F8.2)
```

The effect of the scale factor can be summarized as follows:

In input:

1. If P is used with F, D, G, and E field descriptors, and the data are not entered with the exponent, then the value will be divided by 10^k and then stored. For example, if an input item is recorded as

   ```
   28.5
   ```

   and it is read under the specification 2PF4.1, the internal value of the number is: 28.5 / 100 = .285.

2. If P is used with a G, E, or D specification and the exponent part is present, the scale factor will be ignored.

In output:

1. If P is used with an F descriptor, the internal value will be multiplied by 10^k and then will be printed. For example, if the internal value of a number is 5.325892, then, by using the specification 3PF8.3 to print it, the number will first be multiplied by 1000:

   5.325892 × 1000 = 5325.892,

   and then will be printed as

   5325.892

   Table 11.8 provides additional examples.
       Therefore, the effect of the scale factor is simply to shift the decimal point *k* places to the right (*k* positive) or to the left (*k* negative).

2. If the scale factor is used to print a number with an exponent part (with an E, D, or G specifications, such as *k*PE*n.m*), the decimal point will be shifted to the right (or to the left if *k* is negative), and the exponent part will be decreased (or increased) by *k*. Furthermore, the scale factor

Table 11.8 Example of Using P Specification for Output

INTERNAL VALUE	FIELD DESCRIPTOR	PRINTED
5.2	(3PF8.1)	5200.0
-3.1415926	(0PF10.4)	-3.1415
3.1415926	(-1PF10.4)	.3141
-3.1415926	(3PF10.4)	-3141.5926
56.28	(0PF6.2)	56.28

Table 11.9 Example of Using P Specification for Output

INTERNAL VALUE	FIELD DESCRIPTOR	PRINTED
.87238E-21	0PE12.3	.872E-21
.87238E-21	2PE12.3	87.24E-23
5.253456	-3PE20.4	.0005E+04
-.943281E-3	-1PE20.6	-.094328E-2

controls the position of the decimal point. For further details about this feature, refer to the FORTRAN manual for the system that you are using. The process works as follows: if $k \leq 0$, it prints k leading zeros and $m + k$ digits after the decimal point. If $k > 0$, it prints k digits to the left and $m - k + 1$ digits to the right of the decimal point. Table 11.9 provides some examples.

Once a scale factor is used in a FORMAT, it holds for all subsequent F, E, G, and D specifications in that FORMAT until another scale is encountered. For example, in

```
      WRITE(6,50)A,B,C,D,X
50    FORMAT(11X,2PF6.2,F4.1,F5.1,0PF6.1,F5.1)
```

the scale factor 2P applies to the field descriptors of A, B, and C, and 0PF6.1 restores normal processing for D and X. Thus, *to restore the normal scaling in a FORMAT, 0P must be used*. The following is another example:

```
      WRITE(6,60)X,Y,Z,U,P,Q,V
60    FORMAT(2P,E10.1,F8.2,G16.2,0P,4F13.2)
```

the scale factor 2P applies to the variables X, Y, Z. The scale factor 0P restores normal processing for U, P, Q, and V.

SOLVED PROBLEMS

11.12 What is the internal value of each variable when using the following READ statements:

a.
```
      READ(5,10)A,B,C,D,P
10    FORMAT(2PE10.4,1X,F6.2,1X,F4.1,1X,
     *-3PF6.4,1X,F7.1)
```

Data:

```
85.6243E+2 562.82 38.3 9.3284 5678.28
```

b.

```
        READ(6,20)X,Y,Z,U
20      FORMAT(-1PF6.1,1X,2PG5.2,1X,F6.1,1X,
       *G9.2)
```

Data:

```
982.35 39.45 238.31 325.68E-1
```

☐ **Answers**

a.

A = 85.6243E+2
B = 5.6282
C = .383
D = 9328.4
P = 5678280.

b.

X = 9823.5
Y = .3945
Z = 2.3831
U = 325.68E-1

■ **11.13** Assume the internal values of variables X, Y, and Z are as follows:

X = 693.289   Y = 56.23   Z = −.328

What values will be printed by using each of the following statements?

a.

```
        WRITE(5,10)X,Y,Z
10      FORMAT(F10.3,2PF8.2,F7.2)
```

b.

```
        WRITE(6,20)Y,Z
20      FORMAT(-2P,F7.4,0P,F6.3)
```

c.

```
        WRITE(6,30)X,Y,Z
30      FORMAT( 1PG14.6 , 2G12.4 )
```

☐ **Answers**

The printed values will be as follows:

> **a.** 693.289 for X, 5623.00 for Y, and −32.80 for Z
> **b.** .5623 for Y, and −.328 for Z
> **c.** 6932.8900 for X, 562.300 for Y, and −3.280 for Z

☐ BN and BZ Descriptors—Blank Interpretation

The BN and BZ can be used with the field descriptors I, F, E, G, and D in an input FORMAT to specify the blank spaces in an input field. In the absence of BN or BZ, blank spaces are ignored or can be interpreted as zero when specified by the OPEN statement. (The OPEN statement is discussed in the following section.) But if BN or BZ is used in a FORMAT, the effect is as follows:

1. If BN is encountered in an input format, all the blanks in succeeding numeric fields are ignored.
2. If BZ is encountered in an input format, all the blanks in succeeding numeric fields are interpreted as zeros.

For example,

```
        READ(5,10)K,L,M
10      FORMAT(I4,BZ,I4,BN,I4)
```

with data

```
56bb32bb4
```

will read the values as

K = 56   L = 3200   M = 4

■

ADDITIONAL FEATURES AND STATEMENTS

☐ EQUIVALENCE Statement

If two or more variables are to share the same storage unit, the EQUIVALENCE statement can be used. The general form of an EQUIVALENCE statement is

EQUIVALENCE(list 1), (list 2), ...,

where each list in parentheses is a series of variables which share the same memory location in the program unit. For example,

EQUIVALENCE(A,B,C,D), (LARGE,BIG)

at the beginning of a program unit, causes the variables A, B, C, D to refer to the same memory location, and LARGE to refer to the same memory location as BIG.

An entire array (or part of one) can be made equivalent by specifying the first element of the array. For example:

```
DIMENSION X(10), Y(10)
EQUIVALENCE(X(1),Y(1))
```

causes all the elements of X, starting from the first element, to be equivalent to all the elements of Y. Specifying the array name in the list indicates the first element of the array. For example, in

```
DIMENSION A(5), B(5)
EQUIVALENCE (A,B)
```

the EQUIVALENCE statement has the same effect as

```
EQUIVALENCE (A(1),B(1))
```

The process of matching pairs of elements starts with the first specified element in the equivalence statement, applies to all possible elements of the arrays, and terminates as soon as one of the arrays has reached its DIMENSION size. For example:

```
DIMENSION A(10),B(5)
EQUIVALENCE (A(1),B(1))
```

would cause the following associations:

```
A(1)   A(2)   A(3)   A(4)   A(5)
 ↕      ↕      ↕      ↕      ↕
B(1)   B(2)   B(3)   B(4)   B(5)
```

The following are further examples.

EXAMPLE A

```
DIMENSION A(50),B(7)
EQUIVALENCE (A(25), B(2))
```

would cause the following associations:

```
A(24)  A(25)  A(26)  A(27)  A(28)  A(29)  A(30)
 ↕      ↕      ↕      ↕      ↕      ↕      ↕
B(1)   B(2)   B(3)   B(4)   B(5)   B(6)   B(7)
```

Note that the entire array B [including B(1)] becomes equivalent to a part of the array A.

EXAMPLE B

```
DIMENSION A(8),AA(25)
EQUIVALENCE (A(3),AA(1))
EQUIVALENCE (A(4),P),(AA(6),Q), (A(10),R)
```

would cause the following associations:

```
     A(3)   A(4)   A(5)   A(6)   A(7)   A(8)   A(9)   A(10)
      ↕      ↕      ↕      ↕      ↕      ↕              ↑
     AA(1)  AA(2)  AA(3)  AA(4)  AA(5)  AA(6)           │
             ↕                           ↕              │
             P                           Q              ↓
                                                        R
```

EXAMPLE C

```
DIMENSION X(7),Y(10)
EQUIVALENCE (X(2),Y(4),B), (Y(10),A)
```

would cause the following associations:

```
X(1)   X(2)   X(3)   X(4)   X(5)   X(6)   X(7)
 ↕      ↕      ↕      ↕      ↕      ↕      ↕
Y(3)   Y(4)   Y(5)   Y(6)   Y(7)   Y(8)   Y(9)   Y(10)
        ↕                                          ↕
        B                                          A
```

EXAMPLE D

```
DIMENSION A(10), B(8), C(11)
EQUIVALENCE (A(2), B(3), C(1), X), (A(10), Y)
```

would cause the following associations:

A(2)	A(3)	A(4)	A(5)	A(6)	A(7)	A(8)	A(9)	A(10)
↕	↕	↕	↕	↕	↕			
B(3)	B(4)	B(5)	B(6)	B(7)	B(8)			
↕	↕	↕	↕	↕	↕			
C(1)	C(2)	C(3)	C(4)	C(5)	C(6)	C(7)	C(8)	C(9)
↕								↕
X								Y

EXAMPLE E

```
DIMENSION A(9), B(3,3)
EQUIVALENCE (A(1), B(1,1))
```

would cause the following associations:

A(1)	A(2)	A(3)	A(4)	A(5)
↕	↕	↕	↕	↕
B(1,1)	B(2,1)	B(3,1)	B(1,2)	B(2,2)

A(6)	A(7)	A(8)	A(9)
↕	↕	↕	↕
B(3,2)	B(1,3)	B(2,3)	B(3,3)

EXAMPLE F

```
DIMENSION X(4), Y(3,2)
EQUIVALENCE (X(1), Y(2,2))
```

would cause the following associations:

X(1)	X(2)	X(3)	X(4)
↕	↕	↕	↕
Y(2,2)	Y(3,2)	Y(1,3)	Y(2,3)

Note that the elements of arrays are associated according to the order in which they are stored.

The following cases illustrate some of the applications of the EQUIVALENCE statement:

1. When a variable name is spelled differently in several parts of a program, the EQUIVALENCE statement enables the programmer to make all the variables equivalent rather than correcting each individual one. For example if AVRAGE and AVERAG are used in different parts of the same program to refer to the same variable, then placing the statement

    ```
    EQUIVALENCE (AVRAGE, AVERAG)
    ```

    at the beginning of the program takes care of the problem.
2. When more than one programmer works independently on segments of a program, and they are using different names to refer to the same variable, the EQUIVALENCE statement can be used to make the variables equivalent.
3. When several arrays are used in a program but never more than one array at a time, making arrays equivalent saves memory space. The following is an example:

```
DIMENSION HIGHT(100),WEIGHT(100)
EQUIVALENCE (HIGHT(1), WEIGHT(1))
READ*, HIGHT
    ⋮
PRINT*,HIGHT
READ*, WEIGHT
    ⋮
PRINT*,WEIGHT
```

The above program uses only 100 memory locations for HIGHT and WEIGHT rather than 200. Of course, one must be careful not to use a single element for different purposes at the same time.

4. If a multidimensional array is to be converted into a linear array, the EQUIVALENCE statement is a convenient tool. For example, suppose A(50) is to be used instead of B(5,10); then the statement

```
EQUIVALENCE (A(1),B(1,1))
```

at the beginning of the program accomplishes the objective.

The difference between a COMMON statement and an EQUIVALENCE statement is that a COMMON statement causes the listed variables in different program units to share the same memory units. An EQUIVALENCE statement causes listed variables within a single program unit to share the same memory units.

Some other important rules about the EQUIVALENCE statement are as follows:

1. The EQUIVALENCE statement is nonexecutable and must appear at the beginning of a program in the type-specification statement group.
2. Only variables of the same type should be equivalenced. For example, a character-type variable can be equivalenced only to another character-type variable.
3. Each list in an EQUIVALENCE statement must contain at least two names.
4. Dummy arguments cannot be used in an EQUIVALENCE list.
5. Function names cannot appear in an EQUIVALENCE list.
6. The same storage unit must not occur more than once in an array seqeuence. For example,

```
DIMENSION X(10)
EQUIVALENCE (X(1),Y),(X(2),Y)
```

is not valid.

☐ IMPLICIT Statement

The IMPLICIT statement can be used to confirm or to change the "first letter" rule of the variable types. For example,

```
IMPLICIT INTEGER(P-W)
```

at the beginning of a program declares that all variable names which begin with the letters P through W (P, Q, R, S, T, U, V, W) are integers. The general form of an implicit statement is:

IMPLICIT type1(a_1), type2(a_2), . . . ,

where type$i$ can be INTEGER, REAL, DOUBLE PRECISION, COMPLEX, LOGICAL, or CHARACTER, and a_i is a single letter, or two letters separated by a hyphen representing a range of letters. The following are some examples:

```
IMPLICIT REAL(I),INTEGER(A-H), CHARACTER*10(X,Z)
IMPLICIT INTEGER(F-X),REAL(A), CHARACTER*12(C-E)
IMPLICIT REAL(I-J),LOGICAL(P-S), INTEGER(A,C)
IMPLICIT CHARACTER*16(M-Z), LOGICAL(A-C)
IMPLICIT CHARACTER*20(A,X-Z)
IMPLICIT REAL(A,C,U-Z),REAL(H-K),LOGICAL(B,D-G)
```

The IMPLICIT statement must be placed at the beginning of a program unit before any other statement.

A type statement after the IMPLICIT statement overrides the IMPLICIT statement. For example, the statement

```
IMPLICIT REAL(I-M)
INTEGER KAY
```

causes all the names beginning with I through M to be real type except the variable KAY.

☐ BLOCK DATA Subprograms

A BLOCK DATA subprogram can be used to initialize the variables in a labeled COMMON statement. A BLOCK DATA subprogram is a subprogram which starts with the word BLOCK DATA, ends with an END statement, and consists only of the following nonexecutable statements:

> IMPLICIT
> a type statement
> DIMENSION
> COMMON
> EQUIVALENCE
> DATA

The following examples illustrate the use of the BLOCK DATA subprogram.

EXAMPLE A

Main program:

```
COMMON /FIRST/ P,Q,R,S,T(10)
         ⋮
END
BLOCK DATA
COMMON /FIRST/A,B,C,D,E(10)
DATA A,B,C/3*0.0/ , D/1.0/
END
```

EXAMPLE B

Main program:

```
COMMON/A/X,Y,Z(100)/B/C,D
         ⋮
END
BLOCK DATA
COMMON/A/X,Y,Z(100)/B/C,D
DATA Z/100*.0/,C,D/2*0.0/
END
```

General Form of READ and WRITE

The general form of a formatted READ or WRITE statement is

READ(UNIT = u, FMT = fn, REC = rn, IOSTAT = ios, ERR = n1, END = n2) list

or

WRITE(UNIT = u, FMT = fn, REC = rn, IOSTAT = ios, ERR = n1, END = n2) list

where

> UNIT = u indicates the FORTRAN unit number which the data are transferred from. UNIT = may be omitted, but in that case u must be the first parameter in the list. u can be:
>
> 1. an integer number (or expression) having a value in the range of 1 to 999,
> 2. an asterisk,
> 3. the name of a character variable, array, or array element identifying an internal file or unit.
>
> FMT = fn indicates the FORMAT to be used. FMT = can be omitted if fn is the second parameter in the list. fn can be a variable as well as a FORMAT statement label. This will be explained further in the next section.
>
> REC = rn indicates the record length for the direct-access files. rn must be a positive integer.
>
> IOSTAT = ios indicates the status after the I/O operation is completed. ios is an integer variable into which one of the following values is placed after execution:
>
> > 0 if execution is completed normally,
> > >0 if an error condition is encountered,
> > <0 if an end-of-record is encountered and no error condition has occurred.
>
> ERR = n1 indicates an executable statement label to which control will be transferred if an error in data is encountered during I/O.
>
> END = n2 indicates an executable statement label to which control will be transferred at the end of data (when end-of-file is encountered during reading the data). This should be specified to avoid termination of execution when end-of-file is encountered.

The following are some examples:

```
READ(UNIT = 5, FMT = 100, ERR = 99, END = 200)A,B,C
READ(5,10,ERR = 900, END = 150)X,Y,Z
WRITE(UNITS = 6, FMT = 200, ERR = 125)P,Q,R
WRITE(*, FMT = 20)X,Y
READ(IDEVIC, FMT = 100, ERRS = 90)E,F
WRITE(6,30, ERR = 35)X,Y
READ(ICARD, 100)P,Q
READ(5,N)A,B,C
WRITE(ILINE, M)U,P,Q,R
```

☐ Variable FORMAT

The format identifier in a READ or WRITE statement can be any of the following forms:

1. An integer constant representing the FORMAT statement label, as it has been used throughout the text.
2. An asterisk indicating list-oriented READ or WRITE.
3. An integer-variable name (or an array-element name) representing the FORMAT-statement label. Of course, the integer variable must be defined prior to its use. The value must be assigned to the variable by an ASSIGN statement. The following are some examples:

```
          ASSIGN 10 TO N
          READ(5,N)A,B,C,...

          ASSIGN 500 TO M
          WRITE(6,M)X,Y,Z

          READ*,A
          ASSIGN 10 TO N
          IF(A .LT. 0.0)ASSIGN 20 TO N
          WRITE(6,N)
10        FORMAT(1X,'A IS POSITIVE')
20        FORMAT(1X,'A IS NEGATIVE')
          END
```

4. The name of a *character array* containing the format specifications. For example,

```
CHARACTER*1 X(4)
DATA X/'(','I','5',')'/     [X will be equal to  (│I│5│) ]
READ(5,X)K
    ⋮
```

is equivalent to

```
          READ(5,10)K
10        FORMAT(I5)
```

Note that the parentheses are included in the format specification. The following are further examples:

EXAMPLE A

```
          CHARACTER*1 FORM(12)
          READ(5,10)FORM
10        FORMAT(12A1)
          READ(5,FORM)A,B
              ⋮
```

First data line:

```
(2(F6.1,5X))
```

The second READ statement is equivalent to

```
          READ(5,20) A,B
20        FORMAT(2(F6.2,5X))
```

EXAMPLE B

```
          CHARACTER*1 A(14)
          READ(5,10) A
    10    FORMAT(14A1)
          READ(5,A) IX,Y,Z
             ⋮
```

First data line:

```
(I5,F6.3,E9.5)
```

The second READ statement in this example is equivalent to

```
          READ(5,100) IX,Y,Z
    100   FORMAT(I5,F6.3,E9.2)
```

EXAMPLE C

```
          CHARACTER*4 FORM(6)
          READ(5,20) FORM
    20    FORMAT(6A4)
          READ(5,FORM) X,Y,Z,U
             ⋮
```

First data line:

```
(2(F5.3,3X),2(F3.0,5X))
```

The second READ statement is equivalent to

```
          READ(5,100) X,Y,Z,U
    100   FORMAT(2(F5.3,3X),2(F3.0,5X))
```

The FORMAT specification can be assigned to a variable name or an array element as well as an array name. The following are some examples:

EXAMPLE D

```
          CHARACTER*16 FM
          DATA FM/'(I9,G8.2,2PF6.2)'/
          WRITE(6,FM) ID,FOUND,RATE
             ⋮
```

is equivalent to

```
          WRITE(6,100) ID,FOUND,RATE
    100   FORMAT(I9,G8.2,2PF6.2)
             ⋮
```

EXAMPLE E

```
   READ(5,'(2E14.4)') X,Y
```

is equivalent to

```
          READ(5,200) X,Y
    200   FORMAT(2E14.4)
```

OPEN and CLOSE Statements

An OPEN statement can be used to declare the files used in a program and associate them to the unit numbers. The general form of an OPEN statement is

OPEN(UNIT = u, IOSTAT = ios, ERR = ie, FILE = f,
STATUS = s, ACCESS = a, FORM = fm, RECL = r, BLANK = b)

where

u indicates the unit number; this number must then be used in the READ or WRITE statement which uses the file f.

ios indicates an integer variable to which a positive value is assigned if an error occurs during the OPEN, or zero if no errors occur.

ie indicates an executable statement label to which the control will be transferred if any error occurs.

f identifies the file name to be opened; the file name must be placed in apostrophes. The file then becomes associated with unit u.

s identifies the file status; it can take any of the following values:

'OLD' indicates f currently exists,
'NEW' indicates f is a new file,
'SCRATCH' indicates a temporary file which will be deleted on program termination (this must not appear if FILE = f is specified),
'UNKNOWN' indicates file status is unknown.

a identifies the method of access to the file; it can be either 'SEQUENTIAL' or 'DIRECT'; if omitted, 'SEQUENTIAL' is assumed.

fm identifies the formatting of the INPUT or OUTPUT and can be either 'FORMATTED' or 'UNFORMATTED'; if omitted, the value will be 'FORMATTED' for sequential-access files, and 'UNFORMATTED' for direct-access files.

r identifies the record length for a direct-access file; it can be omitted for a sequential-access file.

b tells whether or not blank values in numeric fields must be ignored. It can take either of the following two values:

'NULL' if it is desired to ignore the blanks in numeric fields (except that a field of all blanks will be treated as zeros).
'ZERO' if it is desired that blanks be treated as zeros. If BLANK = b is omitted, 'NULL' will be assumed.

For example, assume the data are placed in a file named XDATA, and it is to be assigned to unit 4. Then

```
OPEN(UNIT = 4,FILE = 'XDATA')
READ(4,100)A,B,C,D
```

defines the file and the unit, and the READ statement reads the data from XDATA. The following are additional examples:

```
OPEN(UNIT = 5,ERR = 90,FILE = 'DATA1',STATUS
*= 'OLD',ACCESS = 'SEQUENTIAL',FORM =
*'FORMATTED',BLANK = 'NULL')

OPEN(UNIT = 8,FILE = 'UNEMPL')

OPEN(UNIT = 9,FILE = 'PARTS')

OPEN(UNIT = 2,FILE = 'OUTPUT')

OPEN(5,FILE = 'INPUT')
```

Note that the word UNIT can be omitted if *n* is the first parameter in the list. A program may have more than one OPEN statement. The following program reads a series of data from a file named DATA1 and places the results in another file named RESULT:

```
      PROGRAM EXAMPL
      OPEN(8,FILE = 'DATA1')
      OPEN(7,FILE = 'RESULT')
      DO 100 I = 1,100
          READ(8,10,ERR = 90,END = 99)A,B
          S = A + B
          WRITE(7,20)A,B,S
100   CONTINUE
99    STOP
90    WRITE(7,30)
10    FORMAT(2F7.2)
20    FORMAT(3(F8.2,1X))
30    FORMAT(1X,'ERROR IN INPUT DATA')
      END
```

Once a file is opened, a CLOSE statement should be used to disconnect it. The general form of a CLOSE statement is

CLOSE(list)

where the list is similar to the list in the OPEN statement, but only UNIT, IOSTAT, ERR, STATUS are allowed. The STATUS can be 'KEEP' or 'DELETE'. The following are some examples:

CLOSE(UNIT = 8,ERR = 99,STATUS = 'DELETE')

CLOSE(UNIT = 2)

CLOSE(5,ERR = 85)

In most of the systems, any file which is not explicitly closed in the program will be automatically closed when the program terminates.

☐ EXTERNAL and INTRINSIC Statements

If an actual argument is itself the name of a subprogram, it should be identified as such by the EXTERNAL statement. For example, suppose FUN1 in the call statement

CALL RUSH(A,B,FUN1)

is a function's name. Then

EXTERNAL FUN1

at the beginning of the program identifies FUN1 as a function, rather than an ordinary variable. If the actual argument is the name of a library subprogram, the INTRINSIC statement must be used to identify it as such. For example:

```
INTRINSIC SQRT
   ⋮
CALL SUB1(A,B,SQRT)
   ⋮
SUBROUTINE SUB1(X,Y,ZZZ)
X = ZZZ(Y)
   ⋮
```

PAUSE Statement

This causes the computer to halt the execution of a program and wait. If the user types some predetermined symbol or word such as GO, the execution resumes where it was left off. The user can type another word such as DROP to terminate. This offers the programmer an opportunity to think during execution.

SOLVED PROBLEMS

11.14 Specify the effect of each of the following statements. Also identify any errors.

    a.   `EQUIVALENCE(TERM,TRM,TERM1)`

    b.   `DIMENSION AAA(100),BBB(50),CCC(50)`
            `EQUIVALENCE(AAA(1),BBB(1))`
            `EQUIVALENCE(AAA(51),CCC(1))`

    c.   `IMPLICIT CHARACTER*20(A-B)`

    d.   `BLOCK DATA`
            `REAL KAI`
            `COMMON/CONST/KAI,PIE`
            `DATA KAI,PIE/2.193268, 3.141593/`
            `END`

    e.   `M = 100`
            `READ(5,M,ERR = 210,END = 555)A,B,C`

    f.   `CHARACTER FMT(22)`
            `READ(5,10)FMT`
            `READ(5,FMT)X,Y,Z`
            ⋮
      10   `FORMAT(21A)`
            `END`

First data line:

`(F6.1,1X,F3.0,1X,F5.0)`

    g.   `CHARACTER*14 FM`
            `DATA FM/'(F5.1,1X,F6.2)'/`
            `READ(5,FM)A,B`

    h.   `CHARACTER*14 FM`
            `READ(5,'(F5.1,1X,F6.2)')A,B`

    i.   `OPEN(UNIT = 5,ERR = 99,FILE =`
          `*'DATA2',STATUS = 'OLD',ACCESS =`
          `*'SEQUENTIAL')`

Answers

    a. The statement causes the variables TERM, TRM, and TERM1 to refer to the same memory.

b. The first fifty elements of array AAA will be equivalent to elements of array BBB. The next fifty elements will be equivalent to CCC.
c. Any variable name which starts with either A or B is defined as character type (20 characters long). For example, all the variables ART, BON, BAY are character type.
d. The block data subprogram initializes the variables KAI and PIE.
e. Variables A, B, C will be read from unit 5. In case of an error in the data, control will be transferred to statement 210. After the data are finished, control will be transferred to statement 555. The data are read according to FORMAT statement 100. Nevertheless, the FORMAT statement label must be assigned with an ASSIGN statement such as

ASSIGN 100 TO N

f. Variables X, Y, and Z will be read with the field descriptors F6.1, F3.0, and F5.1, respectively.
g. Variables A and B will be read with the field descriptors F5.1 and F6.2 respectively.
h. It does the same function as **g.** above.
i. The sequential access file DATA2, an existent file, will be assigned to unit 5. In case of an error, control will be transferred to statement 99.

EXERCISES

11.1 Complete the following table. Use E notation and three decimal places to present the data. Assume that the computer being used will not accept more than seven significant digits.

Number	E notation	Format
a. 369820000	36.982E7	E8.3
b. 56238		
c. +6328000000000		
d. +.3282		
e. −58.38932		
f. 61.5623891		
g. 8932.85 × 10^{39}		
h. 28395628292		
i. −32		
j. +.000829		
k. +.932832579		
l. 5623985328		
m. 562.3289231		
n. −.0000005628		

11.2 What is the value of each variable read by the following READ statements?

a.

```
        READ(5,10)X,Y,Z
10      FORMAT(3E9.3)
```

Data:

```
63.293E+200923E-2263.0000E5
```

b.

```
     READ(5,20)A,B
20   FORMAT(E9.0 , 1X , E9.0 )
```

Data:

```
.00643E+5 58.9324E5
```

c.

```
     READ(5,30)A,B,I,J
30   FORMAT(2X,F6.2,E10.3,I2,I1)
```

Data:

```
560321546320953E10526
```

☐ **Double Precision**

11.3 Indicate whether each variable should be single or double precision in each of the following statements (assuming seven significant digits for real constants).

a.

```
C = 6.32896E+35
```

b.

```
Y = 9328.32D0
```

c.

```
Z = 5623.53E-17
```

d.

```
A = 1234567.8D+10
B = 63289D-5
C = A + B
```

e.

```
A = 37.0/15.0
```

f.

```
P = 37.0D0/15.0D0
```

g.

```
R = 20.0D0/3.0D0 - 19.0D0/3.0D0
```

h.

```
     READ(5,10)A,B,C
10   FORMAT(E10.2,D9.1,F7.2)
```

Logical Data

11.4 Assume the variables P, Q, and R are declared as

LOGICAL P,Q,R.

Identify errors in each of the following statements:

a. P = ON
b. Q = OFF
c. R = TRUE
d. P = F
e. P = FLAG
f. P = Q .OR. .AND. R
g. R = 5.0 * X ** 2 + 3.0 * X
h. R = A .OR. B
i. Q = .OR. R
j. IF(P .EQ. Q) GO TO 10
k. IF(Q)R = 5.0
l. R = P .EQ. Q

11.5 Assume the variables L1, L2, X, and Y are defined as

LOGICAL L1,L2
DATA L1,L2/.TRUE.,.FALSE./,X,Y/1.0,2.0/

Complete the following table:

Expression	First Part	Second Part	Outcome
a. L1 .AND. L2	True	False	False
b. X .GT. Y .OR. X .GT. 0.0	False	True	True
c. X .LT. Y .AND. Y .LT. 0.0			
d. X .GT. Y .OR. X .LT. 0.0			
e. X .GT. Y .AND. X .GT. 0.0			
f. X .GT. Y .AND. X .LT. 0.0			
g. X .GT. Y .OR. .NOT. X .GT. 0.0			
h. Y .GT. X .EQV. Y .GT. 0.0			
i. X .GE. Y .EQV. Y .GT. X			
j. X .GE. Y .EQV. .NOT. X .LT. 0.0			
k. X .GE. Y .NEQV. Y .GT. X			
l. X .GT. Y .NEQV. .NOT. Y .GT. X			
m. L1 .OR. L2			
n. L1 .AND. L2 .OR. L1 .AND. L2			
o. (L1 .AND. .NOT. L2) .AND. *(L1 .OR. L2)			
p. (L2 .OR. .NOT. L1) .OR. (L1 .OR. *.NOT. L2)			

IF and Logical Expressions

11.6 Are the following IF statements correct? (Also, find the errors in **b.**, **d.**, and **g.**)

a. IF(A .LT. 10.0 .AND. B .GT. 10.0) STOP
b. IF(HRS .GT. 40.0 .AND. .LT. 60.0) GO TO 10
c. IF(GROSS .LT. 10000.0 .OR. IDEPT .LT. 3) RATE = .12

d. IF(I .LT. 100 .AND. .NOT. .GT. 75) GO TO 100
e. IF(X .GT. Y .AND. (A .LE. 0.0 .OR. B .LE. 0.0)
f. IF(A .EQ. C .OR. A .LT. B .OR. A .LT. D)P = Q
g. IF(AMT .EQ. 1000.0 .OR. .LT. 10000.0 .OR. .GT. 50000.0)DISC = .25

☐ **Complex Data**

11.7 Assume the variables A, B, and C are defined as

COMPLEX A,B,C

Identify errors in each of the following statements:

a. A = 93.10,32.2
b. B = (89.1)
c. C = 89,32
d. A = (P,Q)
e. B = A + (R,S)
f. A = REAL((6.2,3.0))
g. B = AIMAG((37.32,9.82))
h. C = CABS(B)
i. L = CMPLX(P,Q)

☐ **G Descriptor**

11.8 What is the value of each variable read by the following READ statements?

a.
```
        READ(5,80) X,U,P,Q
80      FORMAT(F5.2,E8.2,2G5.2)
```

Data:

```
6.352678.50E92.35698215
```

b.
```
        READ(5,90) FOX,GEE,M
90      FORMAT(G7.1,1X,G12.4,2X,G2.1)
```

Data:

```
63218E9532835.93E+1267100
```

c.
```
        READ(5,100)RAY,PAY,KAY
        FORMAT(3G9.5)
```

Data:

```
563285E-3058846E20000063200
```

11.9 Show what values will be printed by the following WRITE statements:

a.

```
        X = 5.35
        Y = 89325.32
        WRITE(6,25)X,Y
25      FORMAT(11X,2G15.4)
```

b.

```
        A = .35
        Y = .0069
        WRITE(6,35)A,Y
35      FORMAT(11X,2G8.2)
```

c.

```
        AY = 52.8
        EX = 3.5500E-5
        WRITE(6,45)AY,EX
45      FORMAT(11X,2G11.4)
```

d.

```
        PI = .2500E-8
        RI = 95.63898
        WRITE(6,55)PI,RI
        FORMAT(11X,2G12.5)
```

☐ **Scale Factor P**

11.10 What is the internal value of each variable read by the following statements?

a.

```
        READ(5,100)X,Y,P,Q
100     FORMAT(2PF6.1,1X,E9.1,1X,E9.1,1X,F8.1,1X,
       *-2PF5.1)
```

Data:

> 678.92 9283.5E10 .56328E-5 6328.527 9328.

b.

```
        READ(5,200)E,F,G,H,I
200     FORMAT(2PG5.1,1X,G8.I,1X,G9.1,1X,-3PG11.2,
       *1X,G3.2)
```

Data:

> 93.28 328359.7 9328.0E22 5628.90E-10 632.89

11.11 Assume the internal value of the variables A, B, and C are as follows:

A = 6932.586   B = -.928   C = +328.0

Show what values will be printed by using the following statements:

a.

```
        WRITE(6,10)A,B,C
10      FORMAT(F10.2,5X,3PF8.2,5X,F8.2)
```

b.

```
        WRITE(6,20),A,B,C
        FORMAT(-2PF9.3,5X,F8.2,5X,F8.2)
```

c.

```
        WRITE(6,30)A,B,C
30      FORMAT(2PG10.7,5X,G9.3,5X,-2PG9.3)
```

☐ Additional Features

11.12 Explain the purpose of each of the following statements:

a.
```
        EQUIVALENCE(V,P,Q),(QUE,QUEUE)
```

b.
```
        DIMENSION ABC(25),CARE(65)
        EQUIVALENCE(ABC(1),CARE(41))
```

c.
```
        IMPLICIT REAL(I-N)
```

d.
```
        IMPLICIT REAL(A-Z)
```

e.
```
        IMPLICIT REAL(A-N),INTEGER(M-Q),CHARACTER*15
       *(R-Z)
```

f.
```
        BLOCK DATA
        COMMON/GRP1/X,Y,Z,A(100)
        DATA X,Y,Z,A/103*0.0/
        END
```

g.
```
        READ(UNIT = 5,FMT = 100,ERR = 99,END = 200)
       *A,B,C
```

h.
```
        WRITE(6,FMT = 300,ERR = 69)X,Y,Z
```

i.
```
        ASSIGN 300 TO IFMT
        READ(5,IFMT,ERR = 99)QUE,TRY
```

j.
```
        READ(IUNIT,IFORM,ERR = 999,END = 100)X,Y,Z
```

k.
```
        WRITE(IPRT,MAY,ERR = 90)C,D
```

l.

```
      CHARACTER*12 FM
      DATA FM/'(3(F5.1,1X))'/
      READ(5,FM)A,B,C
         ⋮
```

m.

```
      CHARACTER VAR(16)
      READ(5,10) VAR
      READ(5,VAR) NAY,MAY,POOR
10    FORMAT(16A1)
         ⋮
```

First data line:

```
(I6,3X,I3,E10.2)
```

n.

```
      READ(5,'(F8.2)') TEE
```

o.

```
      OPEN(6,ERR = 99,FILE = 'SALES',STATUS = 
     *'OLD')
```

p.

```
      OPEN(7,FILE = 'CLASS')
```

PROGRAMMING EXERCISES

11.13 Write a complete FORTRAN program which calculates a table of values of the function: $X/(1 - X^2)$ for $X = 2, 3, ..., N$. The values must be accurate to 13 decimal places. Print a table of values such as

X	X/(1 - X*X)
2	.6666666666666
3	.3750000000000
⋮	⋮

11.14 Write a program which calculates $N!$ for N from 1 to 20. You must use double-precision variables because the factorial of a number larger than 10 is very big.

11.15 A group of students (less than 100) are taking a true-false question test. There are 33 questions on the test, and each question is worth 3 points. Write a program which reads the names of the students (columns 1–15) and the responses (columns 16–48). Print the name, scores, letter grade, and group of each student. The letter grade is determined on a scale of 1–99 (A: 89–99, B: 79–89, C: 69–79, D: 59–69, F: below 59), and the group is determined as follows:

Score	Group
87–99	Excellent
70–87	Good
Below 70	Poor

Print also the grand average and the numbers of A's, B's, C's, D's, and E's at the end of the report. The first data record is a header showing the number of students taking the test, and the second record is the key.

11.16 Each student's information record contains:

Name	Columns 1–15
Age	Columns 17–18
Sex (M or F)	Column 20
Class:	Columns 22–23
FR: Freshman	
SO: Sophomore	
JR: Junior	
SR: Senior	
GR: Graduate	
GPA (on a scale of 0.0 to 4.0)	Column 25

There are less than 100 records. The first one is a header record. Write a program which reads the information and prints the following statistics:

1. The number of students who are over 21 years old and have a GPA more than 3.5.
2. The names and the number of students who are senior or graduate whose GPA is over 3.65.
3. The names and the number of female students who are over 21 and have a GPA over 3.1.
4. The names and the number of male students who are under 19 years and are seniors.
5. The names and the number of graduate students who are under 21 years old and have GPAs over 3.75.
6. The total number of male undergraduate students.
7. The total number of female undergraduate students.
8. The total number of male graduate students.
9. The total number of female graduate students.

11.17 A marketing survey is conducted to find out whether a certain brand of beer is more popular than other brands. After giving the samples to a group of about 300 tasters, the result was recorded as follows:

1. Age of the taster   Columns 1–2
2. Result of the test: Column 4

   T for "tasty"
   F for "flop" (dislike)

Write a program which reads the data and prints the following statistics:

1. The percentage of tasters who liked the beer.
2. The percentage of the tasters who liked the beer in each of the following age categories:

   a. under 21
   b. over 21 to 26
   c. over 26 to 30

d. over 30 to 35
e. over 35 to 40
f. over 40

3. The percentage of the tasters who are under 21 or over 40.

The last data item is a trailer record with 00 in the age field.

■
SUMMARY OF CHAPTER 11

You have learned:

1. The E notation is used to indicate the displacement of the decimal point of a real number. For example, 3562.3E23 indicates

    356230000000000000000000000

2. The digits of a number (not including the decimal point or leading and trailing zeros) are referred to as significant digits. The number of significant digits and the magnitudes of data which can be stored in a memory word of a computer are limited, depending upon the computer system being used.

3. The precision of a number can be increased by the double-precision feature of FORTRAN. When a variable is specified in a DOUBLE PRECISION statement, its maximum number of significant digits is approximately two times that of a single-precision variable.

4. Logical data are used for applications where the data are in the form of true/false, yes/no, or positive/negative. Logical variables must be specified by the LOGICAL statement. A logical variable can assume only two values:

    .TRUE.
    .FALSE.

    The features of logical data are:

    a. A logical expression can be built by using relational operators (.GT., .GE., .LT., etc.) and numerical data. For example, (X .LT. 2.6) and (Y .GE. Z) are valid logical expressions.
    b. A compound logical expression can be formed by using logical operators (.AND., .NOT., .OR., .EQV., and .NEQV.) and logical expressions. For example,

    (I .LT. 10) .AND. (I .GE. 20)

    and

    (AMOUNT .GT. 10000.0) .OR. (AMOUNT .LT. 50000.0)

    are valid compound logical expressions. The interpretation of the logical operators is as follows:
    The outcome of

    Logexp1 .AND. Logexp2

    is true only if both expressions are true.
    The outcome of

    Logexp1 .OR. Logexp2

    is true if either Logexp1 or Logexp2 is true.

The operator .NOT. in

.NOT. Logexp

forms the complement (or negation) of a logical expression.
  c. A compound logical expression can be used in an IF statement. The following are examples:

  IF((AMT .LT. 500.0) .AND. (AMT .GT. 100.0))RATE *=.05

  IF((KODE .EQ. 2) .OR. (KODE .EQ. 5)) DED = 100.0

  IF(LOG1 .AND. LOG2) GO TO 99

  (assuming LOG1 and LOG2 are logical variables)
  d. The order of execution of operators in an expression is as follows:
    1. Arithmetic operators (**,*,/,+,−)
    2. Character operator (//)
    3. Relational operators (.GT., .LT., .EQ., etc.)
    4. Logical operators (.NOT. first; then .AND., .OR.)
5. Complex data have special applications in science and engineering. The following is an example of using complex data:

```
      PROGRAM LOG
      COMPLEX RHO, KAA
      RHO = (3.5,2.1)
      KAA = RHO + 4.5
      WRITE(6,20) RHO,KAA
  20  FORMAT(1X,2(F6.1,5X,F4.1))
      END
```

6. Some additional features of FORMAT are as follows:
  a. The G descriptor can be used to input or output E-form, real, and complex data, depending upon the type of the variable in an I/O list. When G$n.m$ is used for output, m indicates the number of significant digits to be printed, rather than the number of decimal places.
  b. The P descriptor (scale factor) can be used to shift the decimal point of a real number to the left or right when reading or writing.
  c. A BN or BZ descriptor in the input format specifies whether blank spaces in the data fields are to be considered as zero or ignored.
7. Some useful additional features of FORTRAN are:
  a. The EQUIVALENCE statement, to make two or more variables equivalent in a program unit.
  b. The IMPLICIT statement, to set the first-letter rules for the type of the variables.
  c. The BLOCK DATA subprogram, to initialize the variables which are in a COMMON statement.
  d. The general form of the READ statement

  READ(UNIT = u, FMT = fn, REC = rn, IOSTAT = ios, ERR = $n1$, END = $n2$)list

  in which the format identifier fn can be a variable, a character array, or a FORMAT statement label.
  e. The OPEN statement, to declare a file used in a program and associate it to an appropriate unit number.
  f. The CLOSE statement, to disconnect the file.
  g. The EXTERNAL statement, to identify an actual argument as a subprogram's name rather than a regular variable.

514 Additional Features of FORTRAN

h. The INTRINSIC statement, to identify an actual argument as a library subprogram rather than a regular variable.

SELF-TEST REVIEW

11.1 Write a READ statement which reads the variables A, B, and C from the following data line:

```
63289.32E21 6.89328589D+24 -6.9823E+12
```

11.2 Write a program which calculates and prints 5/3 to 12 decimal places.

11.3 Which of the following statements is *incorrect*? Assume the variables L1 and L2 are logical variables.
  a. L1 = .TRUE. .OR. .FALSE.
  b. L2 = 5.0 .GT. 1.0
  c. L1 = (6.0 .GT. A + 5.0) .OR. (6.0 .LT. B − 3.0)
  d. L2 = A * B .OR. 3 * X
  e. L1 = L1 .OR. L2

11.4 Assume variables L1, L2, A, and B are defined as

```
LOGICAL L1,L2
DATA L1,L2/.TRUE., .FALSE./A,B/1.0,2.0/
```

What is the outcome of the following expressions?
  a. (L1 .AND. .NOT. L2)
  b. (L1 .OR. L2)
  c. (A .LT. B .AND. A .GT. 0.0)
  d. (A .GT. B .OR. .NOT. B .LT. 0.0)
  e. (L1 .AND. .NOT. L2) .AND. (A .LT. B)

11.5 Which of the following statements are incorrect?
  a. IF(NET .LT. 100 .AND. .OR. .LT. 500)GO TO 10
  b. IF(GROSS .GT. 10.0 .AND. GROSS .LT. 100.0) STOP
  c. IF(AMT .LT. 1000.0 .AND. KODE .EQ. 1 .AND. AGE .LT. 30.0) STOP
  d. IF(CALL .EQ. 1.0 .AND. BET .LT. 100.0 .OR. 200.0) GO TO 10

11.6 Complete the following FORMAT statement. Use the G descriptor.

```
      READ(5,10)A,B,C,CAY
10    FORMAT(                    )
```

Data:

```
6328.9 562.832E+32 62 56
```

11.7 What value will be printed by the following program?

```
PROGRAM SMALL
X = .63280
WRITE(6,10)X,X
```

```
      10    FORMAT(1X,3PF10.2,2X,F10.2)
            END
```

11.8 Assume you have used the variables GRSINC, GROSIN, GINCOM, GRSCOM to indicate gross income in a long program. How do you correct it?

11.9 Suppose you would like to read the values of several variables from a data file named SPEED; after processing you would like to place the results in the file named CHECK. Write the OPEN statements for this purpose.

11.10 Write a program segment to calculate the discount rate based on the following table (use compound logical IF):

	Rate for KODE = 1
Amount less than $500.00	.05
Amount over $500.00 but less than $1000.0	.06
Amount over $1000.00	.08

☐ ANSWERS

11.1
```
            DOUBLE PRECISION B
            READ(5,10)A,B,C
      10    FORMAT(E11.2,1X,D14.8,1X,E11.4)
```

11.2
```
            PROGRAM EX
            DOUBLE PRECISION X
            X = 5.0D0/3.0D0
            WRITE(6,10)X
      10    FORMAT(1X,D20.12)
            END
```

11.3 **d.** is incorrect.

11.4 **a.** True
 b. True
 c. True
 d. True
 e. True

11.5 **a.** and **d.**

11.6
```
      10    FORMAT(G6.1,1X,G11.3,1X,G2.0,1X,G2.0)
```

11.7
```
            632.80    632.80
```

11.8 By placing the statement

```
EQUIVALENCE(GRSINC,GROSIN,GINCOM,GRSCOM)
```

at the beginning of the program.

11.9
```
      OPEN(5,FILE = 'SPEED')
      OPEN(6,FILE = 'CHECK')
```
11.10
```
      IF((AMT .LT. 500.0) .AND. (KODE .EQ. 1)) THEN
                          RATE = .05
      ELSE IF((AMT .LT. 1000.0) .AND. (KODE .EQ. 1)) THEN
                          RATE = .06
      ELSE IF((AMT .GT. 1000.0) .AND. (KODE .EQ. 1)) THEN
                          RATE = .08
      ENDIF
```

Chapter 12
Structured Programming

WRITING BETTER PROGRAMS

   Introduction
   Structured Design
      Top-Down Design
      Modularity
   Readability
   Documentation
   Reliability

A SAMPLE PROGRAM

WRITING BETTER PROGRAMS

☐ **Introduction**

Structured programming is a method of designing and writing organized programs so that they are easier to code, easier to understand, and less prone to logical errors.

One goal of structured programming is to avoid bad program organization. Another goal is to produce programs that can be read, maintained, and modified easily by other programmers. There have been many reports that the costs of maintaining and modifying a program are greater than the costs of producing it—in fact, three to five times more. Therefore, more readable programs will ultimately make data processing more efficient and less costly. This can be achieved by forgetting about tricks to make programs run faster, and by following a set of simple rules.

Structured programming has a lot of theory behind it. However, in this chapter we discuss a practical approach to structured programming for your FORTRAN programs, without getting into a deep theoretical discussion. A sample program is presented also to demonstrate the concept.

A structured approach should exhibit the following characteristics:

1. Structured design
    a. Top-down design
    b. Modularity
2. Readability
3. Reliability
4. Full documentation

☐ **Structured Design**

Before getting into any technical definitions of structured design, let us run through a simple analogy. Assume a builder makes a contract to build a house. Building cannot actually start without a lot of planning. First, decisions must be made on how the house must be built: the size, style, number of the rooms, etc. Also, many constraints must be considered: the area of the building site, rules and regulations, cost, etc.

After all these decisions have been made and there is a general plan, as well as the blueprints, the builder can break the job of building the house down into many separate jobs. For example:

1. Foundation
2. Framing
3. Exterior trim
4. Utilities
5. Interior trim

Looking at this more closely we find that we can break some of these jobs down into even smaller jobs:

2. Framing
    a. Floors
    b. Exterior walls
    c. Interior walls
    d. Roof

and still smaller jobs:

b. Exterior walls
   - **i.** Studs and plates
   - **ii.** Doors
   - **iii.** Windows

By now, you can see what we are leading to. We took the large job of building a house, and broke it down into many smaller ones. This is the main idea behind structured design, and we will consider it the first rule in structured programming:

> **RULE 12.1**
> 1. *Plan before coding.*
> 2. *Look at the overall problem.*
> 3. *Break the problem down into a sequence of smaller, easier problems (modules).*

No matter how big or small a program is, it can be broken down into modules; even if it consists of such simple steps as input, processing, output. Such an approach is especially helpful today because there are many businesses which have very large, complex programs consisting of many pages. By breaking these large programs into modules, writing them becomes a relatively easy task.

There is another important aspect that we must consider. Referring back to the construction example, you cannot build the walls until the foundation is complete; nor can you build the roof until the walls are complete. Parts are interrelated and most parts are dependent upon previous ones.

Top-Down Design

The concept of top-down design is very simple. The programmer focuses first on the overall structure of the problem, and then designs the major modules of the program. Once that is done, he or she works on the next lower level. This process continues until the entire problem is divided into functional modules and submodules.

In top-down development, both the design and the actual programming are done from top to bottom. That is, the specifications and the program codes are verified as being correct at the top level before going to the lower level of specification. The top-down approach focuses on modules in partially completed programs. Throughout their development, continual integration and testing are performed as program modules are developed.

The levels of abstraction in top-down design become levels of modules in a program. For example, the abstract term "payroll" is useful for those who wish to communicate generally without having to deal with such specifics as employee information, check-writing procedures, etc. These specifics can be considered lower levels of abstraction.

A hierarchy chart is a useful tool in structured programming for overviewing the entire program (Figure 12.1). It shows the levels of abstraction in each stage. A hierarchy chart does not show decision making, logic, or flow of execution as a flowchart does. (See Figure 12.1 for a depiction of the differences.) Flowcharts show procedures; hierarchy charts show functions. A hierarchy chart allows the programmer to concentrate on *what* needs to be done, before determining *how* it is to be done. It also groups related functions together. The top module's function is to summarize the lower modules. Each subsequent level of defined modules is related to a level of abstraction of the problem.

520 *Structured Programming*

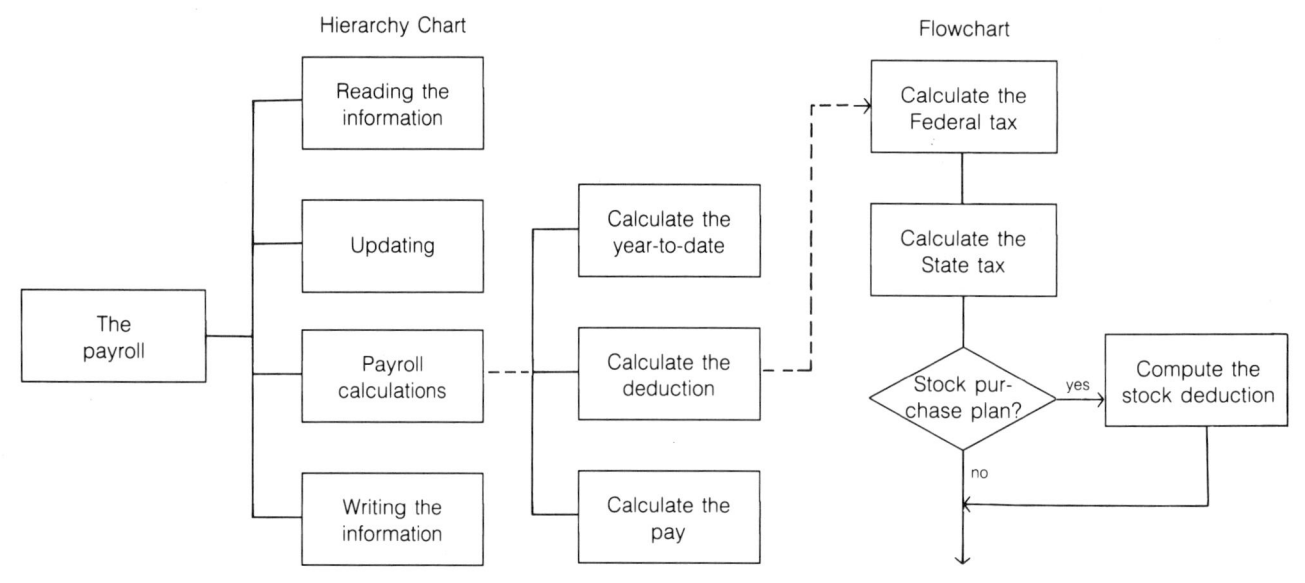

Figure 12.1 Comparing a hierarchy chart with a flowchart.

To design a program in top-down fashion, a clear understanding of the requirements of the program is imperative. Communication with the people who will use the program is essential. The communication should take place before, during, and after the program design.

The guidelines for top-down design are as follows:

1. Plan and design first; do not do any coding until the plan is completed.
2. Start at "the top"—an overall look at the entire problem. Determine what the program will do before developing the details of how the functions will be performed.
3. Refine the problem by working "down" to modules and submodules.
4. Refine and further develop each module or submodule in greater detail through successive levels.
5. Both the design and the coding should be done from top to bottom.

Modularity

A module is a self-contained function which performs a task (or tasks) in support of the major function of the program. In terms of programming code, a module is a collection of logically related instructions which can be grouped together. Limiting the size of a program unit and insuring that each unit performs a well-defined function can:

1. simplify the design,
2. increase readability,
3. make the debugging easier,
4. make maintenance and modification easier.

Some other advantages of the modular approach are:

1. Each module can be written once and used by many programs.
2. Debugging can be performed for each module rather than for the entire program; thus errors can be easily seen and corrected.
3. Teamwork can be facilitated. That is, a large program can be written by several people, which can result in earlier completion.
4. In companies where there are individual programmers who excel in certain areas of programming, they can be assigned parts of the program for which they are most qualified.

5. A library of commonly used routines can be created.
6. Checkpoints for measuring progress can be provided.

A module should not be chosen arbitrarily. It is not just "a chunk of a program." A module is a well-defined unit of the entire program. It should serve its own function. The combination of modules builds the organization of the entire program. A hierarchy diagram is a good tool to show how the modules fit together.

Some of the desiderata for a module are:

1. A module should perform a well-defined function.
2. A module must return to its caller, to the unit which was called from, rather than to, the other parts of the program.
3. A module should have a single entry and a single exit.
4. A module should be small—typically less than a page.

Sometimes modularity has to be weighed against computer resource costs. Subprograms are more expensive to run than regular branching. Typically, however, the program development, debugging, and maintenance contribute more to the overall cost than the computer resources required.

☐ Readability

Certainly by now you have noticed that programs can be written with different styles, methods, and algorithms. Furthermore, as discussed in Chapter 10, the efficiency of a program depends on several factors, the most important one being the human factor. There were times when writing clever and tricky codes which took less storage and run time was considered the best way to utilize computer resources. As a result, many inefficient programs were written. These programs were difficult not only to understand, debug, and maintain, but also to modify. Almost every program written will subsequently be modified to meet changing specifications or to eliminate bugs. Should a program be intricate to begin with, such future changes will only add to the complications in reading it.

It is currently accepted that the efficiency of a program can be increased by reducing human effort in the development, debugging, or modifying process; this can be accomplished by writing a readable program, avoiding tricks, and providing useful documents.

Good programming starts with good planning. Devote ample time to the planning process before rushing into coding. All the following steps (discussed in Chapter 3) should be carefully followed:

1. Understand the problem fully; define it precisely in writing.
2. List (a) the output items and (b) the input items.
3. Design the layout for the output and the input.
4. Design the solution procedure; use top-down design and a modular approach. An appropriate algorithm should be found at this time for each module. The best way to minimize costs is to choose an efficient algorithm.
5. Write the entire program in easy-to-understand pseudo-FORTRAN (or pseudocode), or a flowchart.
6. Code and test each module.

The value of modular programming cannot be overemphasized. Fortunately, FORTRAN provides a simple way to assign a name to a module or submodule: by means of subprograms. Consequently, subroutines and functions are extremely helpful in modular programming. In addition, a repetitive function makes an especially valuable subprogram, since its use ultimately adds to the program's efficiency.

As previously discussed, the modules should not be defined arbitrarily. They should serve well-defined purposes. Do not attempt to accomplish more than one task in a module. Be certain that the connection between modules is clear. Run and test each module sequentially and separately. In particular, *make sure that the input data are read correctly before doing anything else.* This can be done easily by mirror-printing techniques for the data-reading module.

Write the codes clearly, even if it makes the program less efficient when running. Remember that the computer's time is cheaper than yours. As mentioned earlier, the best way to minimize the computer running time is by choosing an efficient algorithm. A good way to see if your program is readable is to read it aloud to a friend (or even to yourself); if he or she understands the pattern of the logic, very likely the program is right. Do not hesitate to rewrite any part which is not clear. Never use a GO TO statement to fix a coding error; rewrite the entire part.

The important points discussed so far can be summarized as follows:

1. Plan ahead (i.e., plan and design the details before coding).
2. Use top-down design and a modular approach.
3. For a repetitive task, use a subprogram.
4. Run and test all modules, one by one.
5. Make sure the input data are read correctly before doing anything else.
6. Never use a GO TO for patching an error; instead, rewrite the entire part.

Also, remember that the physical appearance of a program plays an important role in the readability of its logical structure. It is not difficult to make a program easy to run and easy to debug if you remember a few simple rules:

1. Each variable's symbolic name should suggest its meaning. For example, A, S, L, D may be obscure; AREA, SPEED, LENGTH, and DSTNCE are much better, more meaningful.
2. Always initialize the variables.
3. Do not use a GOTO, IF-GOTO, or arithmetic IF statement unless it is absolutely necessary.
4. Take advantage of indentation; it is a cheap tool to make a program more readable. Using indentation in IF-THEN-ELSE and DO loops is a must.
5. FORTRAN allows several statements per line separated by the $ character. However, it is best to put only one statement on a line.
6. Always use structured IF for comparisons or decisions. Furthermore:
   a. avoid THEN-IF, or THEN-GOTO, or ELSE-GOTO;
   b. do not use mixed modes for checking equality (3 is hardly ever equal to 3.0);
   c. indent the appropriate statements under THEN or ELSE.
7. Use structured DO loops for looping. Furthermore:
   a. terminate all the DO loops with a CONTINUE statement;
   b. indent the statements in any DO loop by at least 3 spaces;
   c. inside nested DO loops, indent inner loops with respect to their outer loops;
   d. do not use more than one exit from a loop.
8. Make statement labels significant. For example, all 3-digit labels (over 100) could be for executable statements; all the labels for the CONTINUE statements could end with a zero (220, 320, 330, etc.), and 2-digit labels could be for FORMAT statements. If the labels of a given kind progress in increments, say of 10, they are flexible and easy to find.
9. FORMAT statements should be grouped together and placed at the end of the program unit.
10. If there is a need to continue a statement to the next line, always break

the statement after a comma. The character in the column six of the continuation line should be one that cannot be mistaken for other characters in that line. For example, you can always use an asterisk or dollar sign for this purpose; on the other hand, in statements of many lines, sequence numbers (1, 2, 3, ...) in each line help keep the lines in sequence. It is also a good idea to indent the continuation line to the column right after the key word. Example:

```
DATA A,B,D,D/4*0.0/
*    D,E,F,G,H/5*0.0/
```

11. If a slash is used in a FORMAT to start a new record, start the information on a new line as well. Example:

```
WRITE(6 , 10)
FORMAT (51X,'ABC COMPANY' //
*       46X , 'PRICE CALCULATION'/
*       36X , 'UNITS' , 9X , 'PRICE' , 9X , 'TOTAL')
```

12. Use extra parentheses in an arithmetic expression to group the variables and avoid ambiguity. Also use parentheses inside a compound logical expression to help make the order of logical operators clear.
13. Blanks should be used to separate the components of a statement. Use a blank between a key word and the rest of the statement. Also use a blank before and after the following characters:
    a. A comma (,).
    b. An equal sign (=).
    c. Addition and subtraction operators (+ , −).
    d. Relational and logical operators.
    Do not use blanks:
    a. before or inside the parentheses of an array subscript or a subprogram reference.
    b. before or after the multiplication, division, and exponentiation operators (*, /, **).
    For example,

    ```
    RESULT(I) = (A + B - C*D)/E**2
    ```

14. Use blank lines or a line of asterisks to separate a module from the rest of the program.
15. Follow the standard developed and adopted by your organization for the physical appearance of the programs. For example, most organizations use a COMMENT block at the beginning of a program including the following items:
    a. The purpose and a brief description of the program and the algorithms used.
    b. The name of the programmer and the date the program was developed.
    c. The variables dictionary.
    d. The input and output files—the file names, the medium, and the record format.
       Further comments should be used judiciously throughout the program to explain the major steps. Comments are particularly necessary
    a. before each branching;
    b. to explain the connection between each module;
    c. before a statement which does something obscure or confusing;
    d. before a statement to which control is transferred from far away (explain how it is reached);
    e. to show the layout of the data.
       Using too many comments (overcommenting) is not advisable either;

it is especially important that comments not be used simply to repeat a FORTRAN statement.

The following is an example of a program before and after readability considerations are implemented.

Before:

```
         PROGRAMAVERAG
         DIMENSIONX(100)
         S=0.0
         N=0
         DO10I = 1,100
         READ(5,20)X(I)
20       FORMAT(F5.2)
         IF(X(I).LT.0.0)GOTO30
         N=N+1
10       CONTINUE
30       DO40I=1,N
         S=S+X(I)
40       CONTINUE
         A=S/N
         DO50I=1,N
50       WRITE(6,60)X(I)
60       FORMAT(1X,F6.2)
         WRITE(6,70)A
70       FORMAT(1X,'THE AVERAGE IS:    ',F7.2)
         STOP
         END
```

After:

```
C AUTHOR: CHRISTINE CADY
C
         PROGRAM AVERAG
C
C THIS PROGRAM CALCULATES THE AVERAGE OF A GROUP OF
C DATA.
C ******************************************************
C VARIABLE DICTIONARY:
C NUM   THE ARRAY CONTAINING THE DATA
C SUM   THE SUM OF THE DATA
C N     THE NUMBER OF DATA
C AVG   THE AVERAGE OF THE DATA
C ******************************************************
C DEFINING ARRAY NUM WITH A MAXIMUM OF 100 ELEMENTS
C AND INITIALIZING VARIABLES
C
         REAL NUM(100)
         DATA SUM , N /0.0, 0/ , NUM/100*0.0/
C
C READING THE DATA, ONE PER LINE, A NEGATIVE NUMBER
C TERMINATES THE LOOP
C
         DO 100 I = 1,100
            READ (5,10) NUM(I)
               IF (NUM(I) .LT. 0.0) GOTO 150
            N = N + 1
```

```
100     CONTINUE
C
C SUMMING THE DATA AFTER READING ALL THE DATA
C
150     DO 200 I = 1,N
            SUM = SUM + NUM(I)
200     CONTINUE
C
C CALCULATING THE AVERAGE: DIVIDING THE SUM BY THE
C NUMBER OF DATA
C
        AVG = SUM/FLOAT(N)
C
C WRITING THE SCORES AND THE AVERAGE, SCORES ARE TO BE
C PRINTED TEN PER LINE
        WRITE(6,20)
C
            WRITE (6,30)(NUM(I),I = 1,N)
C
        WRITE (6,40) AVG
        STOP
C ***********************************************
C FORMAT STATEMENTS
C
10      FORMAT (F5.2)
20      FORMAT (11X, 'THE DATA ARE:'//)
30      FORMAT (10(6X,F6.2))
40      FORMAT (//1X, 'THE AVERAGE IS:   ', F7.2)
        END
```

☐ Documentation

Documentation is the process of reporting how the program was developed, designed, coded, and tested. It also includes a report explaining how the program can be used. Documentation of a program is important for several reasons. First, almost any program will subsequently be modified. If there is no documentation (or poor documentation), a tremendous amount of time will be wasted in attempts to determine how the program works. Secondly, most programs are to be used by someone other than the original programmer. Without documentation, it is almost impossible for them to do so. Thus documentation may be provided for:

1. The reviewer, who needs to maintain and modify the program.
2. The user, to show how to communicate the data to the program.
3. The operator, to show what files are to be used if necessary.

Documentation is a major part of the development of a program. It should begin with the project and progress with it. One of the advantages of such gradual documentation is that it can provide a basis for continuing the project should there be a change in programmers. Documentation should include, but not be limited to:

1. What the program does, how it does it, and what its limitations are.
2. The methods of computation (algorithms or processing).
3. A detailed, logical description of the program with appropriate flowcharts, pseudocodes, or hierarchy charts.

4. The planning documents, such as print layouts and record layouts.
5. A list of the codes, including the internal documents.
6. Samples of the input and output, and the results of the tests performed.

The user document should include detailed instructions for using the program and feeding in the data. It should also include limitations on the number of variables and an explanation of error messages, if any.

It is important that each organization develop and adopt standards for documentation. This will greatly enhance efficiency by providing a uniform tool for further program development and maintenance. Without standards, it is difficult to measure the quality of program work.

☐ Reliability

Reliability in a broad sense refers to minimizing the chance of errors in all aspects of program development, design, and coding. However, when speaking of reliability in programs we are usually referring to the requirement that they work correctly under a wide range of conditions.

Thus, any program must go through extensive testing before it can be considered reliable and ready for use. Test data should be created for all cases, especially the critical values. It is particularly important that a program be tested with:

1. "Good" data: the appropriate normal data.
2. "Bad" data: the data for which the program is sensitive.
3. Flexible number of data:
   a. No data at all.
   b. Only one datum.
   c. The maximum possible number of data.

A SAMPLE PROGRAM

The following problem has been developed to demonstrate the application of the structured design which has been covered in this chapter.

☐ Step 1: The Problem

An instructor would like to have some statistics about students' test scores, specifically:

1. the mean,
2. the variance,
3. the standard deviation,
4. the median,
5. the range.

The objective is therefore to write a program which can calculate this information. Analysis also shows that there will be a maximum of 200 students, each with one test score between 0.00 and 100.00.

☐ Step 2: Input-Output Formulation

It is determined that the variables listed in Table 12.1 are necessary for the output.

Table 12.1 Output Analysis Form for the Sample Program

ITEM	SYMBOLIC NAME OF VARIABLE	TYPE	FIELD LENGTH (COLUMNS)
1. Students' test scores (sorted, in an array)	SCORE (array)	Real	6
2. Mean of the scores	MEAN	Real	6
3. Variance of the scores	VAR	Real	7
4. Standard deviation of the scores	STD	Real	6
5. Median of the scores	MED	Real	6
6. Range of the scores	RANGE	Real	6

The mean, variance, standard deviation, median, and range can be calculated from the test scores. Therefore, we need only the students' test scores, as shown in Table 12.2, for input. A negative number is to be entered for the trailer record.

☐ Step 3: Layout of the Input and Output

The output is to be as shown in Figure 12.2.
    The placement of the input data is to be as follows:

The students' test scores:
5 columns without the decimal point   Columns 1–5

☐ Step 4: Process Design

The program is broken down into the following modules and submodules:

1. Housekeeping
   a. Defining arrays, characters, variables, types
   b. Initializing the variables
2. Reading the input data and a trailer record
3. Sorting the scores
4. Calculating the statistics
   a. Calculating the mean
   b. Calculating the variance, standard deviation
   c. Calculating the median
   d. Calculating the range
5. Writing the information
   a. Writing the headings
   b. Writing the sorted data
   c. Writing the statistics

The hierarchy chart is shown in Figure 12.3.

Table 12.2 Input Analysis Form for the Sample Program

ITEMS	SYMBOLIC NAME OF VARIABLE	TYPE	FIELD LENGTH (COLUMNS)
1. Students' test scores (without decimal point)	SCORE (array)	Real	5

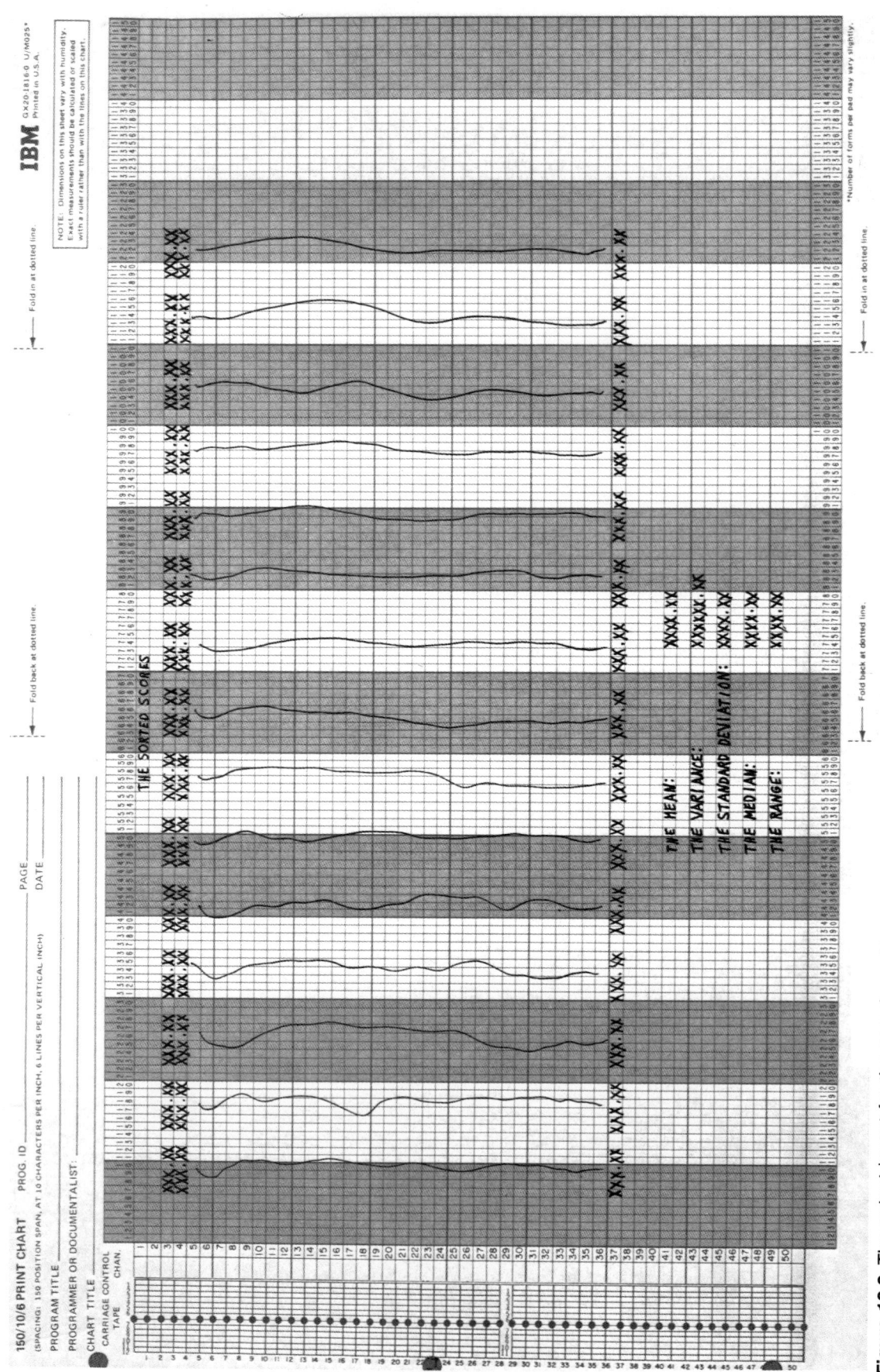

Figure 12.2 The output layout for the sample problem.

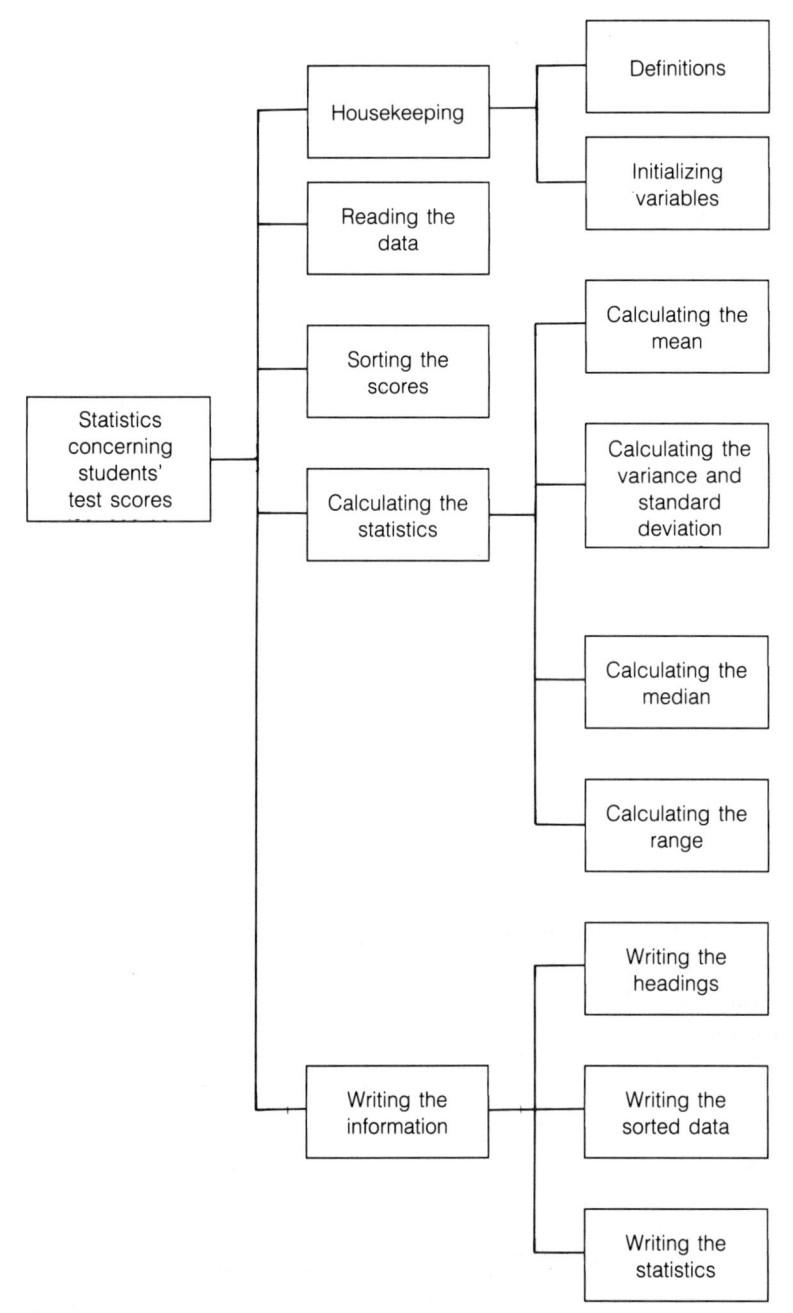

Figure 12.3 The hierarchy chart for the sample problem.

The following algorithms are decided on for the sorting and calculating modules:

1. The elements of the array SCORE can be sorted in ascending order using the bubble-sort algorithm.

    For I = 1 to I = N − 1, repeat the following steps:

    Set LAST equal to N − I.
    Starting with K = 1 to K = LAST, compare SCORE(K) and SCORE(K + 1); if SCORE(K) > SCORE(K + 1), then interchange the values of SCORE(K) and SCORE(K + 1).

2. The mean can be calculated by

MEAN = SUM/N

where

MEAN = the mean,
SUM = the sum of the scores,
N = the number of scores.

3. The variance can be calculated by using the following statement in a loop created for this purpose (VSUM will be set equal to zero before the loop):

VSUM = VSUM + (SCORE − MEAN)**2

and then using the following formula after the loop:

VAR = VSUM/N

The standard deviation can then be calculated by

STD = SQRT(VAR)

where

VAR = the variance,
SCORE = the array containing the scores,
MEAN = the mean,
VSUM = the sum of the squares of the differences between the individual scores and the mean,
N = the number of scores,
STD = the standard deviation,
SQRT = the library function that calculates the square root.

4. The median can be calculated as follows:
   a. If the remainder of N/2 is zero, then

   MED = (SCORE(K) + SCORE(K + 1))/2.0

   where K = N/2
   b. If the remainder is not zero, then

   MED = (SCORE(K) + SCORE (K + 1))/2.0

   where K = N/2 (assuming the data are already sorted). Here
   MED = the median,
   SCORE = the array containing the scores.

5. The range can be calculated as follows:

RANGE = DIM (SCORE(N),SCORE(1))

where

RANGE = the range,
SCORE(N) = the highest score in the sorted array,
SCORE(1) = the lowest score in the sorted array,
DIM = the library function that calculates the positive difference between two numbers

(assuming the data are already sorted in ascending sequence).

☐ Step 5: Pseudocode

1. Define MEAN and MED as real variables.
2. Define the dimension of SCORE as 200.
3. Initialize N to 0.
4. Open the files.
5. For K = 1 to I = 200, repeat the following steps:
   - 5.1 Read SCORE(I)
     - 5.1.1 If end-of-file encountered and no data, then print a message and stop.
       If end-of-file encountered and data have been read, print NO TRAILER CARD, and go on to step 6.
   - 5.2 If SCORE(I) < 0, go to step 6.
   - 5.3 Increase the counter block by 1.
6. Sort the data by using subroutine SORT; arguments: SCORE and N.
7. Calculate the mean by using subroutine MEASUB; arguments: SCORE, N, and MEAN.
8. Calculate the variance and standard deviation by using subroutine VARSUB; arguments: SCORE, N, MEAN, VAR, and STD.
9. Calculate the median by using subroutine MEDSUB; arguments: SCORE, N, and MED.
10. RANGE = DIM (SCORE(N),SCORE(1)).
11. Write headings.
12. Write SCORE(I) for I = 1 to N, 15 scores per line.
13. Write MEAN,VAR,STD,MED, and RANGE.
14. Stop.
15. FORMAT statements.
16. End of the main program.
17. Subroutine for sorting the scores, called SORT, using dummy arguments: SCORE and N.
    - 17.1 Define the dimension of SCORE as N.
    - 17.2 M = N − 1.
    - 17.3 For I = 1 to I = M, repeat the following steps:
      - 17.3.1 LAST = N − I.
      - 17.3.2 Repeat the following step for K = 1 to LAST:
        If SCORE(K) > SCORE(K + 1), then interchange SCORE (K + 1) and SCORE(K) using subroutine CHANGE.
    - 17.4 Return the arguments.
    - 17.5 End of subroutine SORT.
18. Subroutine for calculating the mean, called MEASUB, with dummy arguments SCORE, N, and MEAN.
    - 18.1 Define MEAN as a real variable.
    - 18.2 Define the dimension of SCORE as N.
    - 18.3 Initialize SUM to 0.0.
    - 18.4 For I = 1 to I = N, repeat the following step:

      SUM = SUM + SCORE(I).

    - 18.5 MEAN = SUM/FLOAT(N).
    - 18.6 Return the arguments.
    - 18.7 End of subroutine MEASUB.
19. Subroutine for calculating the variance and standard deviation, called VARSUB, with dummy arguments SCORE, N, MEAN, VAR, and STD.
    - 19.1 Define MEAN as a real variable.
    - 19.2 Define the dimension of SCORE as N.

19.3 Initialize VSUM to 0.0.
19.4 For I = 1 to I = N, repeat the following step:

VSUM = VSUM + (SCORE(I) − MEAN)**2

19.5 VAR = VSUM/FLOAT(N).
19.6 STD = SQRT(VAR).
19.7 Return the arguments.
19.8 End of subroutine VARSUB.

20. Subroutine for calculating the median, called MEDSUB, with dummy arguments SCORE, N, and MED.
20.1 Define MED as a real variable.
20.2 Define the dimension of SCORE as N.
20.3 If MOD(N,2) = 0 then do the following:

K = N/2
MED = (SCORE(K) + SCORE(K + 1))/2.0

Otherwise do the following:

K = (N + 1)/2
MED = SCORE(K)

20.4 Return the arguments.
20.5 End of subroutine MEDSUB.

21. Subroutine for interchanging two numbers called CHANGE with dummy arguments A and B:
21.1 TEMP = A
21.2 A = B
21.3 B = TEMP
21.4 Return the arguments.
21.5 End of subroutine CHANGE.

☐ Step 6: The Program

```
C AUTHOR: ANNETTE COOK
C DATE: JUNE 19--
C
        PROGRAM STATS
C **********************************************************
C THIS PROGRAM IS DESIGNED TO READ A GROUP OF SCORES,
C SORT THEM, AND CALCULATE THE MEAN, VARIANCE, STANDARD
C DEVIATION, MEDIAN, AND RANGE OF THE SCORES. THE
C SORTED SCORES AND THE STATISTICS WILL BE PRINTED.
C **********************************************************
C VARIABLE DICTIONARY:
C
C SCORE    THE ARRAY CONTAINING THE SCORES, FILE =
C          'TESTS', ONE SCORE PER LINE
C N        THE NUMBER OF DATA, MAXIMUM = 200.
C MEASUB   THE SUBROUTINE WHICH WILL CALCULATE THE MEAN
C MEAN     THE MEAN OF THE SCORES
C VARSUB   THE SUBROUTINE WHICH WILL CALCULATE THE
C          VARIANCE AND THE STANDARD DEVIATION
C VAR      THE VARIANCE OF THE SCORES
C STD      THE STANDARD DEVIATION OF THE SCORES
C
```

```fortran
C SQRT     THE LIBRARY FUNCTION THAT CALCULATES THE
C          SQUARE ROOT
C
C MEDSUB   THE SUBROUTINE WHICH WILL CALCULATE THE
C          MEDIAN
C MED      THE MEDIAN OF THE SCORES
C RANGE    THE RANGE OF THE SCORES
C DIM      THE LIBRARY FUNCTION THAT CALCULATES THE
C          POSITIVE DIFFERENCE BETWEEN TWO NUMBERS
C **********************************************
C
C DEFINING VARIABLE TYPES AND ARRAY DIMENSIONS
C
      REAL MEAN, MED
      DIMENSION SCORE(200)
C
C INITIALIZING VARIABLES
C
      DATA N/0/
C
C OPENING THE INPUT FILE AND OUTPUT FILE, THE DATA ARE
C TO BE PLACED ON THE 'TESTS' FILE, THE OUTPUT ON THE
C 'OUTPUT' FILE.
C
      OPEN (5,FILE = 'TESTS')
      OPEN (6, FILE = 'OUTPUT')
C
C READING THE SCORES, EACH SCORE IS TO BE PLACED ON A
C DIFFERENT LINE IN COLUMNS 1-5; THE LAST RECORD IS TO
C BE A NEGATIVE NUMBER. THE NUMBER OF SCORES SHOULD
C NOT BE MORE THAN 200.
C
      DO 10 I = 1,200
         READ(5,100,END = 99) SCORE(I)
         IF (SCORE(I) .LT. 0.0) GOTO 20
         N = N + 1
10    CONTINUE
99       IF (N .EQ. 0) THEN
             WRITE (6,200)
             STOP
         ELSE
             WRITE (6,300)
         END IF
C CALLING THE SUBROUTINE TO SORT THE SCORES
C
20    CALL SORT (SCORE,N)
C
C CALLING THE SUBROUTINE TO CALCULATE THE MEAN
C
      CALL MEASUB(SCORE,N,MEAN)
C
C CALLING THE SUBROUTINE TO CALCULATE THE VARIANCE AND
C STANDARD DEVIATION
C
      CALL VARSUB(SCORE,N,MEAN,VAR,STD)
```

534   Structured Programming

```
C
C CALLING THE SUBROUTINE TO CALCULATE THE MEDIAN
C
      CALL MEDSUB(SCORE,N,MED)
C
C CALCULATING THE RANGE
C
      RANGE = DIM(SCORE(N), SCORE(1))
C
C WRITING THE HEADINGS, THE SORTED SCORES, AND THE
C STATISTICS
C
      WRITE (6,400)
      WRITE (6,500) (SCORE(I), I = 1,N)
      WRITE (6,600) MEAN,VAR,STD,MED,RANGE
C
      STOP
C FORMAT STATEMENTS
C
100   FORMAT (F5.2)
200   FORMAT (11X, 'THERE IS NO DATA')
300   FORMAT (11X, 'THERE IS NO TRAILER RECORD, BUT THE
     *PROGRAM CONTINUES' )
400   FORMAT ('1',  /, 56X, 'THE SORTED SCORES:',/)
500   FORMAT(1X,200(6X,15(F6.2,2X)/1X))
C
600   FORMAT (///, 50X, 'THE MEAN:', 16X, F6.2 /
     *          '0', 49X, 'THE VARIANCE:', 12X, F8.2 /
     *          '0', 49X, 'THE STANDARD DEVIATION:', 2X,
     *          F6.2 /
     *          '0', 49X, 'THE MEDIAN:', 14X, F6.2 /
     *          '0', 49X, 'THE RANGE:', 15X, F6.2)
C
      END
C ***********************************************************
C SUBROUTINE FOR SORTING THE SCORES
C ***********************************************************
      SUBROUTINE SORT(SCORE,N)
C
C DEFINING ARRAY DIMENSIONS
C
      DIMENSION SCORE(N)
C
C SORTING THE SCORES BY BUBBLE SORT PROCEDURE
C
      M = N - 1
      DO 10 I = 1, M
         LAST = N - I
         DO 10 K = 1, LAST
            IF (SCORE(K) .GT. SCORE(K + 1)) THEN
               CALL CHANGE(SCORE(K), SCORE(K + 1))
            ENDIF
10    CONTINUE
C
      RETURN
      END
```

```
C **********************************************************
C SUBROUTINE FOR CALCULATING THE MEAN
C **********************************************************
      SUBROUTINE MEASUB(SCORE,N,MEAN)
C
C DEFINING VARIABLE TYPES AND ARRAY DIMENSIONS
C
      REAL MEAN
      DIMENSION SCORE(N)
C
C INITIALIZING VARIABLES
C
      SUM = 0.0
C
C CALCULATING THE MEAN
C
      DO 10 I = 1, N
          SUM = SUM + SCORE(I)
10    CONTINUE
      MEAN = SUM/FLOAT(N)
C
      RETURN
      END
C **********************************************************
C SUBROUTINE FOR CALCULATING THE VARIANCE AND STANDARD
C DEVIATION
C **********************************************************
      SUBROUTINE VARSUB (SCORE,N,MEAN,VAR,STD)
C
C DEFINING VARIABLE TYPES AND ARRAY DIMENSIONS
C
      REAL MEAN
      DIMENSION SCORE(N)
C
C INITIALIZING VARIABLES
C
      VSUM = 0.0
C
C CALCULATING THE VARIANCE
C
      DO 10 I = 1, N
          VSUM = VSUM + (SCORE(I) - MEAN)**2
10    CONTINUE
      VAR = VSUM/FLOAT(N)
C
C CALCULATING THE STANDARD DEVIATION
C
      STD = SQRT(VAR)
C
      RETURN
      END
C **********************************************************
C SUBROUTINE FOR CALCULATING THE MEDIAN
C **********************************************************
      SUBROUTINE MEDSUB(SCORE,N,MED)
C
```

```
C DEFINING VARIABLE TYPES AND ARRAY DIMENSIONS
C
      REAL MED
      DIMENSION SCORE(N)
C
C CALCULATING THE MEDIAN
C
      IF (MOD(N,2) .EQ. 0)   THEN
     *                       K = N/2
     *                       MED = (SCORE(K) +
     *                              SCORE(K + 1))/2.0
      ELSE
     *                       K = (N + 1)/2
     *                       MED = SCORE(K)
      ENDIF
C
      RETURN
      END
C ***********************************************************
C SUBROUTINE FOR INTERCHANGING TWO NUMBERS
C ***********************************************************
      SUBROUTINE CHANGE(A,B)
C
      TEMP = A
           A = B
      B = TEMP
C
      RETURN
      END
```

Appendix A
An Overview of Computer Organization

Knowledge of the internal construction of a computer system is not a requirement for using FORTRAN, but some understanding of its basic functions and components is helpful to the programmer. This section gives a brief overview of a computer system.

COMPUTER SYSTEMS

A computer is an electronic machine and can be compared to a calculator: both can process information—accept, store, and manipulate data. Like a calculator, a computer is a "dumb machine"—it has to be told what to do through explicit instructions or programs. The difference between a computer and a calculator is that a computer can do the processing extremely fast, and has a much greater capacity for storing instructions and data.

Programs which cause automatic processing, and/or application programs, are generally referred to as *software*, whereas the equipment itself is referred to as *hardware*. The hardware can accept, process, and output the information automatically by means of software. Four basic components in a computer system carry out these functions (see Figures A.1 and A.2):

1. Input devices
2. Output devices
3. Central processing unit (CPU)
4. Auxiliary storage devices

First, instructions (the program) and data are entered into the computer through an input device. Next, the information is processed by the CPU, and finally the information is presented through an output device. The information can also be placed in auxiliary storage for later processing.

☐ Input Devices

Input devices are the medium of communication between human beings and the computer. Typical input devices include the following:

1. *Cathode-ray tube (CRT) terminals* look like TV monitors which have been attached to a typewriter keyboard. The information is displayed on the screen as it is typed.
2. *Card readers,* the most common input devices in the 1970s, read the information from a deck of punch cards. These cards are punched on a *keypunch* machine, which also has a keyboard similar to a typewriter.
3. *Hard-copy terminals* (printing terminals) look and operate like a typewriter. When this type of terminal is used, the results are printed on paper, rather than on the display screen as with a CRT.
4. *Special input devices* have various purposes. The following are some examples:

   · Optical character reader (*OCR*), for reading test score sheets or other specially marked papers.

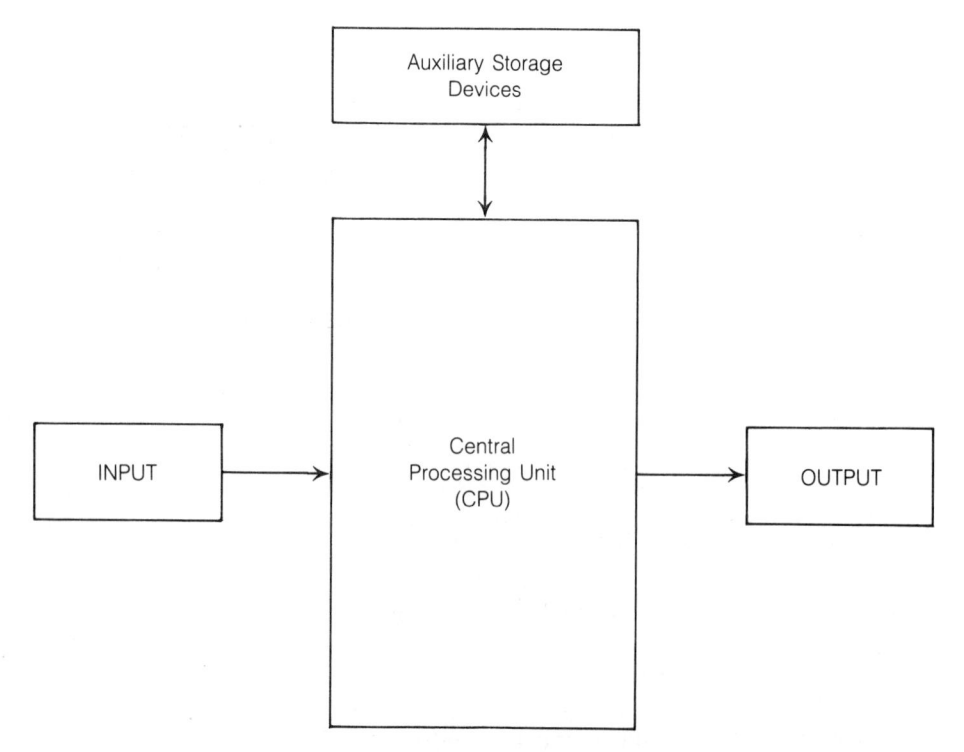

Figure A.1 Basic components of a computer system.

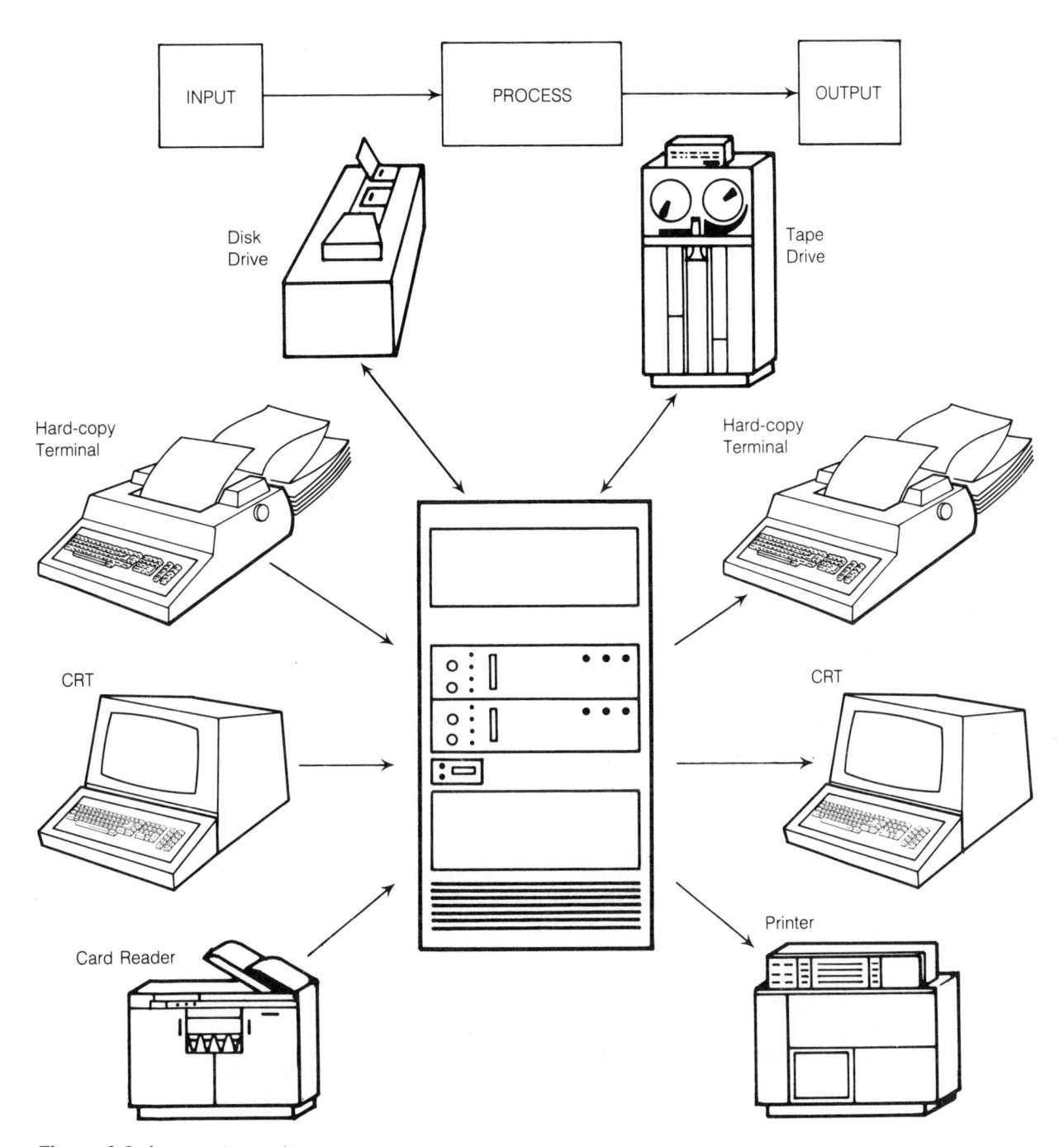

Figure A.2 A computer system.

- Magnetic character reader, for reading data (such as account numbers of checks) which are printed in magnetic ink.
- Point-of-sale (POS) terminals, which are electronic cash registers.

☐ Output Devices

Output devices provide the user with the desired information. Some of the commonly used output devices are the following:

1. *Printers* (line printers) are commonly used for rapid printing of information, especially reports.

540 *Appendix A*

2. *CRT terminals* are used for output as well as input. They are quickly becoming the most widely used input-output devices.
3. *Hard-copy terminals* also serve as output devices. They provide the user with a hard copy (printed copy) of the processed information.

☐ Central Processing Unit

After the information is fed into the computer, it is processed through the Central Processing Unit. The CPU is composed of three basic parts (see Figure A.3):

1. The main memory
2. The arithmetic-logic unit
3. The control unit

The Main Memory
A/L Unit
Control

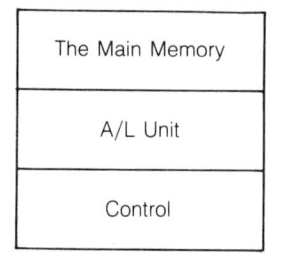

Figure A.3 The CPU.

The main memory stores the programs and data. Generally, the information will be stored in the main memory first. Each piece of information being stored is then referred to by an address.

The arithmetic-logic unit performs all arithmetic operations, such as addition and subtraction. It also performs logical operations, such as comparing two quantities (in an IF statement).

The control unit has control ovver the entire computer system. It controls the input-output devices, the operation of the arithmetic-logic unit, and data transformation to and from the main storage section.

☐ Auxiliary Devices

The auxiliary devices store information for later processing. These devices are then both input and output devices. Two typical auxiliary devices are:

1. Tape drives, which store or retrieve the information to and from a tape.
2. Disk drives, which store or retrieve the information to and from a magnetic disk.

You may be wondering how data are stored in a computer system. The following section will provide a background.

INTERNAL DATA STORAGE

The storage of a computer is composed of very many small electronic components which can sense the presence or absence of some electrical phenomenon. Thus each tiny component can be turned "on" or "off." This ability of the component allows the computer to store and process information.

The unit of information which can be stored in those tiny components is called a bit (contraction of binary digit). A bit can be either one or zero, translated easily into "on" or "off" by the computer:

○   ●
On   Off
Bit: one   Bit: zero

A series of bits can be used to store numbers, letters, and other characters. A combination of several bits forms a *byte*. Usually, a combination of four bytes forms a unit of storage called a *word* (see Figure A.4). Table 11.3 (page 466) presents the number of bits per word for several computer systems.

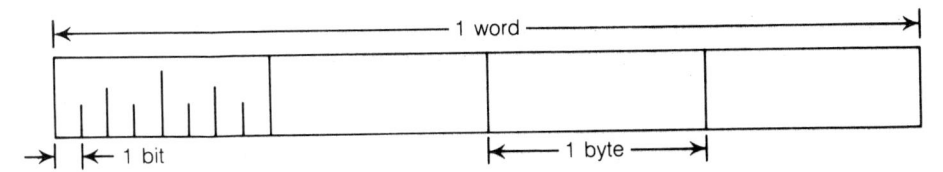

Figure A.4 Bits, bytes, and words.

☐ Binary Numbers

Binary numbers are base-two numbers and have only two values for each digit: 0 and 1. These can easily be translated into "on" or "off" by the computer. The rightmost digit of a binary number represents the 1's place, the next digit to the left represents the 2's place, the next represents the 4's place, and so on (Figure A.5).

Figure A.5 Binary numbers.

Therefore, the binary number 11101 is equal to

$$(1 \times 1) + (0 \times 2) + (1 \times 4) + (1 \times 8) + (1 \times 16)$$

which equals 29 in the decimal system. Further examples of binary numbers and their equivalents in decimal are listed in Table A.1.

Table A.1 Example of Binary numbers and their equivalents in decimal

DECIMAL NUMBER	BINARY EQUIVALENT
1	1
2	10
3	11
4	100
5	101
6	110
7	111
8	1000
9	1001
10	1010
40	101000
41	101001

542   Appendix A

Table A.2   Example of EBCDIC

CHARACTER	EBCDIC REPRESENTATION	CHARACTER	EBCDIC REPRESENTATION	CHARACTER	EBCDIC REPRESENTATION
0	1111 0000	A	1100 0001	N	1101 0101
1	1111 0001	B	1100 0010	O	1101 0110
2	1111 0010	C	1100 0011	P	1101 0111
3	1111 0011	D	1100 0100	Q	1101 1000
4	1111 0100	E	1100 0101	R	1101 1001
5	1111 0101	F	1100 0110	S	1110 0010
6	1111 0110	G	1100 0111	T	1110 0011
7	1111 0111	H	1100 1000	U	1110 0100
8	1111 1000	I	1100 1001	V	1110 0101
9	1111 1001	J	1101 0001	W	1110 0110
		K	1101 0010	X	1110 0111
		L	1101 0011	Y	1110 1000
		M	1101 0100	Z	1110 1001

☐ Storing Character Data

Internally, the computer represents character data in terms of certain codes. One of the most common coding forms is called the *Extended Binary Coded Decimal Interchange Code* (EBCDIC). This system of coding uses an 8-bit byte to represent each letter, number, or special character. Table A.2 represents some of the binary codes in this system.

☐ Storing Integer Data

Integer data are internally represented in the binary system. Normally, each number is stored in a fixed-length word. For example, in an IBM computer, each integer may be stored in a 32-bit word. Thus the integer number 13, which is 1101 in binary, will be represented as

00000000000000000000000000001101

There is an important difference between numeric data and character data. In the binary system, the number 13 is represented as shown above; but as character data, 13 will be stored in EBCDIC in 2 bytes—one for storing the first digit and another for storing the second digit. Therefore, the number 13 will be stored as

1111 0001	1111 0011
2nd byte	1st byte

☐ Storing Real Data

Real data (also called floating-point data) are changed to an exponential form (as explained in Chapter 11), with the decimal point placed to the left of the significant digits. For example, 273.06 will be changed to 0.27306E3, and .00061 will be changed to .61E−3. The number is then stored in binary in one word so that the fractional digits are stored in the extreme right bits of the word, the exponent part is stored next, and the sign is stored in the extreme left bit of the word (Figure A.6).

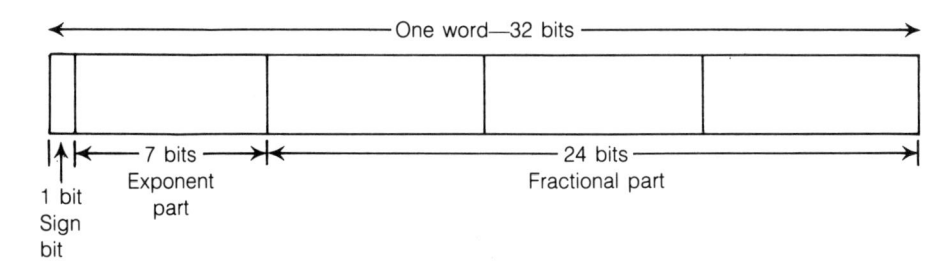

Figure A.6 Storing a number in binary.

Because word size is limited, there is a maximum number of digits that can be stored for a number (the *number of significant digits*). There is also a limit on the exponent part of the number (the *magnitude*). For example, for a 32-bit-word IBM machine, the maximum number of significant digits is approximately seven decimal digits and the range of the magnitude is between 10^{-78} and 10^{+75}. (Table 11.3 gives these limits for some other computer systems.)

Appendix B
Interactive FORTRAN

Some computer systems support running an interactive FORTRAN program on line with some simple system commands.

On a typical terminal, after the user logs in, the system responds with the word

READY

The user then types the command which accesses the interactive FORTRAN such as

FORTRAN

After the system accepts the command, the programmer types the program line by line. The commands, the rules, and the process of accesssing the compiler depends on the system being used. For example, when using the CDC system's interactive FORTRAN, each statement starts with a line number. The line numbers must be in sequence with any desired increment. They occupy columns 1 to 5. FORTRAN statements, including statement labels, start on or after column 7. The letter "C" for comment and the continuation character "+" must be placed in column 6.

When an interactive FORTRAN program is run, the computer automatically prints a question mark "?" for each READ statement which it encounters, and waits for the data to be typed in.

The following is an example of an interactive FORTRAN program:

```
00010   CHARACTER*3 ANSWER
00020   PRINT*, "IS TODAY YOUR BIRTHDAY"
00030   READ *, ANSWER
00040   IF (ANSWER .NE. "YES") GO TO 99
00050   PRINT*, "HAPPY BIRTHDAY. HOW OLD ARE YOU"
00060   READ*, AGE
00070C  THIS STATEMENT CALCULATES THE AGE IN DAYS
00080   DAY = 365 * AGE
00090   PRINT*, "CONGRATULATIONS. YOU ARE ", DAY, " DAYS OLD"
00100   STOP
00110   99 PRINT*, "COME BACK WHEN YOU HAVE A BIRTHDAY"
00120   END
```

When this program is run, the computer prints the messages, and whenever the READ is encountered, it automaticaly prints a question mark and pauses until the response is typed in. The computer then executes the rest of the program.

Index

Adjustable array size, 360
Algorithm, 105, 414, 416
Alphanumeric data, 213
   A-field, 213
Apostrophe, 59, 233
Arguments
   actual, 352
   dummy, 352
Arithmetic logic limits, 540
Arithmetic operation, 6
   priorities, 10
Array, 266
   initializing, 283
   input and output, 276
   lower bound, 268, 313–314
   one-dimensional, 266–310
   in subroutines, 356–361
   two-dimensional, 312–346
   type, 267
   upper bound, 268, 313–314
Arrays and functions, 378–381
Auxiliary storage devices, 540

Bar chart, 435

Batch processing, 82
Binary numbers, 428, 541–543
Binary search, 417, 422
Bisection method, 429
Bit, 541–543
Blank spaces
   in data, 54, 493
   in a line, 84
Block DATA subprogram, 497
BN, BZ, 493
Branching, 23
Bubble sort, 419
Byte, 541, 542

CALL statement, 350
Carriage control, 48, 88
Cathode-ray tube (CRT), 83
   terminal, 538
Central Processing Units, 538, 540
Character
   A-field, 214
   collating sequence, 369
   compared, 218
   constants, 216

   expression, 216
   input and output, 214
   left justified, 217
   omitting the length, 218
   sorting, 421
   statement, 213
   substring reference, 216
   type data, 213–224
   variables, 213
CLOSE statement, 502
Coding, 85
Coding form, 85
Column, 84
   in arrays, 312
   by column, 319
Comments, 91
Common block, 386
   labeled, 390
COMMON statement, 386
Compilation, 96
Compiler, 96
Complex data
   assignment, 482
   constants, 482

547

(Complex data *continued*)
   input and output, 484
   library function, 484
   variables, 482
Compound logical expression, 473
Computer
   language, 3
   organization, 537–543
   program, 3
   system, 537–540
Concatenation, 216
Constants, 13
Continuation of a line, 84
CONTINUE statement, 179, 185
Control variable, 179
Counter block, 25, 278
CRT. See *Cathode-ray tube*

Data card, 19, 84
Data line, 19, 84
Data section, 18
DATA statement, 225, 321, 284, 473
Data storage, 540–543
Data type, 212, 464
Debugging, 9
Decimal point, omitting, 53
Decision support system, 445
Deck, 83
Dimension. See *Arrays*
DIMENSION statement, 267, 313
Disk drive, 540
DO loop, 178–200
   abnormal termination, 187
   components, 179
   control variable, 181
   form, 180
   nested, 195
   normal termination, 187
Documentation, 91, 109, 525–526
Double precision, 469–471
   constants, 469
   input and output, 470
   statement, 469
Dummy argument, 352

EBCDIC, 542
Echo printing, 282
END statement, 65
Elementary row operations, 431
ELSE-IF block, 141
EQUIVALENCE statement, 493
E-specification, 465
Executable statement, 98
Execution error, 101
Exponent-form, 464
EXTERNAL statement, 502

Field, 40, 87
Field-description, 40
Floating point data, 542

Flowchart, 102
FORMAT
   A-descriptor, 213
   D-descriptor, 470
   E-descriptor, 465
   F-descriptor, 40
   G-descriptor, 486
   H-descriptor, 233
   I-descriptor, 40
   L-descriptor, 477
   P-descriptor, 489
   statement, 41
   T-descriptor, 230
Format-free input and output, 240
Format, repeating, 226
Format scanning, 227–229
Formatted PRINT statement, 64, 242
Formatted READ statement, 41–42
Formatted WRITE statement, 45–46
FORTRAN 77, 97
FORTRAN IV, 97
Functions, 366–386
   built-in, 367
   library, 367
   statement, 367, 371
   subprogram, 367, 375

Gauss Elimination Method, 431
GO TO statement, 23
   computed GO TO, 148
Grammatical error, 98
Graphing, 434

Hard-copy terminals, 83, 538
Hardware, 537
Header-record, 133, 278
Hierarchy chart, 519, 527
Hierarchy of operations, 479
Higher-level language, 96
Histogram, 435
Horner's Method, 415

IF statement
   arithmetic, 146
   logical, 124
   with logical expressions, 475
   structured, 139
IF-THEN-ELSE, 140
IMPLICIT statement, 496
Implied DO loop, 278–281
Indentation, 84, 125, 139, 140, 178, 522, 523
Index for arrays, 266
Index variable, 179
Infinite loop, 65
Input. See *READ statement*
Input analysis form, 108, 110, 153, 527
Input data, 19
Input devices, 538

Integer data, 34, 542
INTEGER statement, 212
Interactive program, 84, 545–546
INTRINSIC functions, 367
INTRINSIC statement, 502

Job, 83
JCL, 83

Language, 3
Least squares, 427
Library functions, 367
Linear array, 266
List-directed input and output, 240
Logic-charts, 105
Logical constants, 472
Logical data, 472–482
   constants, 472
   input and output, 477
Logical errors, 101
Logical expression, 472
Logical operators, 473
   priorities, 475
LOGICAL statement, 472
Looping, 22, 177–209

Machine language, 95
Magnetic character reader, 539
Magnitude of data, 466, 543
Main memory, 540
Management information system, 444
Matrix, 312
Median, 426
Merging, 424
Mirror printing, 277, 282
Mixed modes of operation, 35
Modularity, 520
Monte Carlo simulation, 439
Multidimensional array, 312–347

Nassi-Schneiderman diagram, 105
Nested DO loop, 195
Nested IF block, 141
Nested implied DO loop, 319
Nonexecutable statement, 98

Object program, 96
On-line processing, 83
One-dimensional arrays, 266–310
OPEN statement, 501
Optical character reader (OCR), 538
Order of statements, 396
Output analysis form, 108, 110, 153, 527
Output devices, 539–540

Parentheses, use of, 6, 9, 10, 479
PAUSE statement, 503
Planning, 108, 518–526

Planning tools, 102
Point-of-sale terminals, 539
Position of instructions, 84
Precision of data, 466
Print-chart, 85
Printer, 539
PRINT statement, 5, 45, 64, 241–242
Program, definition, 3
PROGRAM statement, 5
Programming
   cycle, 107
   plan, 418
   planning, 107
Pseudocode, 104

Random numbers, 437–439
READ statement, 18, 498
Readability, 521–524
Real data, 34, 212
REAL statement, 212
Record, 86
Record length, 63
Regression analysis, 427
Relational operators, 126
Reliability, 526
RETURN statement, 356
ROW, in arrays, 312
Row by row, 319

Scale factor, 489

Searching, 422
Sentinel value, 133
Sequence numbers, 25
Short-list method, 281
Significant digits, 466
Simulation, 437
Simultaneous linear equation, 431
Sequential search, 422
Slash
   in arithmetic expressions, 6
   in FORMAT, 89–91, 235–239
   in format-free input and output, 242
Software, 537
Sorting, 419
Source program, 83
Standards, 526
Statement
   executable, 98
   label, number, 22, 84
   nonexecutable, 98
   orders, 396
STOP statement, 65
Straight search, 417
String, 213
Structured design, 518
Structured flowchart. See *Logic-chart*
Structured IF, 139
Structured programming, 137, 518–526
Style, programming, 136

Subprograms, 350–412
Subroutines, 350–366
Subscript, 266–274, 312
Substring reference, 216
Summation notation, 426
Symbolic name, 12
Syntax errors, 98

Tape drive, 540
Terminal, 83, 539–540
Termination, DO loop, 187
Top-down design, 519
Trailer card, 133
Trailer record, 133, 278
Two-dimensional arrays, 312
Type statement, 268, 313

Unconditional jump, 23
Unlabeled common block, 386

Variables, 12, 13
   symbolic names of, 12, 91, 522
Vector, 312

WATFIV, 97
WATFOR, 97
WRITE statement, 41
Word, 541–543
Word, memory, 466

Summary of FORTRAN Statements (*continued from page i*)

KEYWORD	EXAMPLE	EXPLANATION OF THE EXAMPLES
IF	IF(X) 10,20,30	Jump to statement 10, 20, or 30 if X is less than, equal to, or greater than zero, respectively.

IF–THEN–ELSE	IF(B .GT. A)THEN
	    BIG=B
	ELSE
	    BIG=A
	ENDIF
		If the condition is satisfied, perform BIG=B; otherwise perform BIG=A.

IMPLICIT	IMPLICIT REAL(L–N)	Defines all variables beginning with the letters L to N to be real.
INTEGER	INTEGER XXX	Defines XXX to be an integer variable.
INTRINSIC	INTRINSIC SQRT	Allows the library function name SQRT to be an argument of a function or subroutine.
LOGICAL	LOGICAL A	Defines A to be a logical variable.
OPEN	OPEN(6,FILE='OUTPUT')	Opens file OUTPUT and associates it with unit 6.
PAUSE	PAUSE	Temporary stop of execution of the program.
PRINT	PRINT*,A	Outputs the value of the variable A.
PROGRAM	PROGRAM MYPROG	The first statement in a program; it informs the computer that the name of the program is MYPROG.
READ	READ*,A	Inputs a value for the variable A.
	READ(5,20)A	Inputs a value for the variable A using device 5 according to FORMAT statement 20.
	READ(5,20,END=99)A	Same as above, but if an end-of-file is encountered, jump to statement 99.
REAL	REAL M	Defines M to be a real variable.
RETURN	RETURN	Transfers control from a subroutine or function back to the statement immediately following the statement which invoked it.
STOP	STOP	Indicates where the execution of the program is to be terminated.
SUBROUTINE	SUBROUTINE SUB(X,Y)	The first statement of the subroutine SUB.
WRITE	WRITE(6,30)A	Outputs the value of the variable A using device 6 according to FORMAT statement 30.